Welding Handbook

Seventh Edition, Volume 2

Welding Processes –
Arc and Gas Welding and Cutting,
Brazing, and Soldering

W.H. Kearns, Editor

AMERICAN WELDING SOCIETY
2501 Northwest 7th Street
Miami, Florida 33125

Library of Congress Number: 75-28707
International Standard Book Number: 0-87171-148-6

American Welding Society, 2501 N.W. 7th Street, Miami, FL 33125

Note: By publication of this handbook, the American Welding Society does not insure anyone utilizing this handbook against liability arising from the use of such handbook. A publication of a handbook by the American Welding Society does not carry with it the right to make, use, or sell patented items. Each prospective user should make an independent investigation.

Printed in the United States of America

Welding Handbook

Seventh Edition, Volume 2

*Welding Processes–
Arc and Gas Welding and Cutting,
Brazing, and Soldering*

The Five Volumes of the
Welding Handbook, Seventh Edition

Contents

CONTENTS

Welding Handbook Committee

A.W. Pense, Chairman   Lehigh University
W. L. Wilcox, Vice-Chairman   Gilbert-Commonwealth
W.H. Kearns, Secretary   American Welding Society
I.G. Betz   Department of the Army
J.R. Condra   E.I. DuPont de Nemours & Co.
R.L. Frohlich   Westinghouse Electric Corp.
A.F. Manz   Union Carbide Corp.
L.J. Privoznik   Westinghouse Electric Corp.
R.E. Somers   Bethlehem Steel Corp.
I.L. Stern   American Bureau of Shipping
R.T. Telford*   Union Carbide Corp.
S. Weiss*   University of Wisconsin—Milwaukee

---

*Membership expired May 31, 1977

Preface

This is the second of five volumes planned for the Seventh Edition of the *Welding Handbook*. Volume 1, *Fundamentals of Welding,* was published in 1976. It contains much of the basic information on welding, such as the physics of welding, heat flow, welding metallurgy, testing, and residual stresses and distortion.

Volume 2 is the first of two volumes on welding and allied processes. It covers arc and gas welding and cutting, brazing, and soldering. Volume 3, when published, will include resistance welding, solid state welding, electron beam welding, laser beam welding and cutting, adhesive bonding, thermal spraying, and other processes.

In developing the plans for the Seventh Edition, the *Welding Handbook* Committee decided to cover the processes in two volumes instead of three (Sections 2, 2A, 3A, and 3B, Sixth Edition). Consequently, coverage of the processes in this edition is more concise but adequate for most industrial applications. Also, the Committee decided to present subjects related to materials, design, manufacturing, and quality control in Volumes 4 and 5 of the Seventh Edition. This rearrangement will delay the revision of several chapters in Sections 1, 2, 3A, and 3B of the Sixth Edition. Therefore, the reader should retain each Section of the Sixth Edition until all of its contents are revised.

One very important subject in Section 1, Sixth Edition, is safe practices for welding, cutting, brazing, and soldering. In many cases, safety with respect to a specific process is mentioned in the appropriate process chapter of this volume. However, these discussions are not intended to be complete and all-inclusive. The reader should investigate the subject thoroughly for his particular application.

A Handbook index of major subjects, located by edition, volume or section number, and chapter number follows the volume index. Its purpose is to give the reader the location of current information on a subject in either the Sixth or Seventh Edition of the *Handbook*.

As in Volume 1, this volume contains dual dimensioning with SI (metric units) and U.S. customary units. The SI units were obtained by conversions of the U.S. customary units, which are shown in parentheses. Thus, the measurements were originally made or specified in U.S. customary units. Conversion to SI units was done on the basis of judgement by members of the Chapter Committees or the Editor. If any of the measurements are to be used in metric applications, the reader should convert the U.S. customary units to the accuracy that he deems necessary for the intended application.

The preparation of this volume was a voluntary effort by the Chapter Committees and The *Welding Handbook* Committee. In addition, other individuals received various chapters and submitted comments and suggestions. All of the participants generously contributed their own valuable time and abilities to the *Handbook* preparation. The American Welding Society extends the appreciation of all its members to the contributors and their employers for support of this task.

The Editor acknowledges the assistance of Hallock Campbell and Carl Willer in editing several of the chapters, and Richard French, Leslie Harrington, and Sandra Wagenheim for the copy editing and production of this volume.

We welcome your comments on the *Handbook*. Several minor changes in style were made to better identify figure titles and improve page appearance. Please address your comments to: The Editor, *Welding Handbook,* American Welding Society, 2501 N.W. 7th Street, Miami, FL 33125.

Dr. A.W. Pense, *Chairman*
Welding Handbook Committee

W.H. Kearns, *Editor*
Welding Handbook

1

Arc Welding Power Sources

Prepared by

D.J. CORRIGAL
Miller Electric Manufacturing Company

Chapter Reviewers

M. CLISSA
  Chemetron Corporation

E.H. DAGGETT
  Babcock and Wilcox Company

W.S. DORSEY
  *Union Carbide Corporation
  Linde Division*

A.F. MANZ
  *Union Carbide Corporation
  Linde Division*

D.C. MARTINEZ
  Teledyne-Walterboro

N.J. NORMANDO
  *Mid-States Welder
  Manufacturing Company*

R.S. SABO
  Lincoln Electric Company

G.E. SCOTT
  Hobart Brothers Company

1

Arc Welding Power Sources

INTRODUCTION

There are many types of power sources available to meet the requirements of the various arc welding processes. The arc welding power sources described in this chapter include those for shielded metal arc, gas metal arc, flux cored arc, gas tungsten arc, submerged arc, electroslag, electrogas, plasma arc, and arc stud welding. This chapter is intended to be a guide to further understanding and selection of the proper power source. The information given here can provide the framework for such study and knowledge. The applications outlined are typical and they serve only to illustrate and explain the relationship of power source to process.

The selection of a power source must be related to the job requirements. The first step in the selection procedure is to detail the requirements of the selected welding process. Other factors then enter. Such things as available power, future requirements, maintenance, economic considerations, portability, environment, available skills, safety, manufacturer's performance, service availability, and standardization will have some influence on the ultimate choice of a particular power source. Many of these considerations are beyond the scope of this chapter. This chapter will be restricted to those areas of recognized classification to assist in understanding and further insight.

CLASSIFICATION

GENERAL

Certain features of arc welding power sources allow broad classification according to one or more of the following characteristics. With respect to typical output, a power source may deliver alternating current (ac), direct current (dc), or both. It may also have the characteristic of providing either constant current or constant voltage. With respect to the input, a power source may derive its power from utility lines or a prime mover such as an internal com-

bustion engine. Power sources deriving power from utility lines are transformer or motor-generator types. Direct connection to primary utility lines for welding is not an acceptable safe practice because they do not provide current and voltage conditions suitable for arc welding requirements. Also, there is extreme danger of electrical shock. Certain configurations of power sources lend themselves more favorably to deliver certain types of current. Transformer type power sources will deliver only alternating current. Transformer-rectifier power sources

Table 1.1 – Types of arc welding power sources

Type of power source	Volt-ampere characteristic		Welding current output		
	Constant current	Constant voltage	AC only	DC only	AC or dc
Transformer	X	X	X		
Transformer-rectifier	X	X		X	X
Motor-generator	X	X	X	X	X
Motor-alternator	X	X	X	X	X

may deliver either ac or dc. Electric motor-generator power sources usually deliver direct current output. A motor-alternator will deliver ac or dc when equipped with rectifiers. Table 1.1 provides a summary of available power sources for arc welding. Other characteristics are also available.

A description of a power source should include identification under each of these sub-categories. For example, an identification of a gas tungsten arc welding power source might be "transformer-rectifier, constant current, ac/dc." A more complete description should include welding current rating, duty cycle rating, service classification, and power line requirements. Special features, such as remote control, high-frequency stabilization, current pulsing capability, starting and finishing current vs time programing, wave balancing capabilities, and line load compensation, could also be included. Current or voltage control method might also be incorporated. Typical controls are saturable reactor, magnetic amplifiers, solid state, or magnetic-solid state hybrid types. A power source identification might also list many other desirable and compatible options.

Figure 1.1 shows the basic elements of a welding power source supplied from utility lines. The arc welding power source itself does not customarily include the fused disconnect switch, although this is a necessary protective element. An engine-driven power supply would require elements different from those shown in Fig. 1.1. It would require an engine, an engine speed regulator, an alternator with or without rectifier, or a generator, and an output control means.

The voltage supplied by utility power companies for industrial purposes is too high to use directly in arc welding. Therefore, means are incorporated in an arc welding power source to reduce the high input voltage down to a suitable output voltage range (usually 20 to 80 volts). Either a transformer, or a motor connected to a generator, provides the facility of reducing 208, 240, 480, or 600 volt utility power, for example, to the rated terminal or open-circuit voltage used for arc welding. The same device (transformer or motor generator) also provides the high welding current (50 to 1500 amperes) from relatively low-current power lines.

CONSTANT CURRENT AND VOLTAGE CLASSIFICATIONS

Arc welding power sources are commonly classified as constant current or constant voltage. The classifications are based on the static characteristics of the power supply, not the dynamic characteristics. Generally, the word constant is true only to a degree. "Constant voltage" power sources are generally much closer to constant voltage output than "constant current" sources are to constant current output. In either case, power sources are available that will hold output voltage or current relatively constant. Constant current power sources are also referred to as variable voltage sources, and constant voltage power sources are often called constant potential supplies.

The National Electrical Manufacturers Association (NEMA) Publication EW-1, "Electric Arc-Welding Apparatus", defines a constant-current arc-welding machine as follows: "A constant-current arc-welding machine is one which has means for adjusting the arc current and which has a static volt-ampere curve that tends to produce a relatively constant output current. The arc voltage, at a given welding current, is responsive to the rate at which a con-

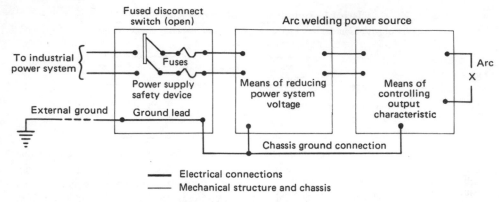

Fig. 1.1—Elements of an arc welding power source

sumable electrode is fed into the arc except that, when a nonconsumable electrode is used, the arc voltage is responsive to the electrode-to-work distance. A constant-current arc-welding machine is usually used with welding processes which employ manually held electrodes, continuously fed consumable electrodes, or nonconsumable electrodes."

The characteristics of this type of supply are such that if the arc length varies because of external influences and slight changes in arc voltage result, the welding current remains substantially constant. Generally, manual welding with a covered electrode or a tungsten electrode, where variations in arc length take place due to the human element, employs this type of power source.

Although each current setting yields a separate and individual volt-ampere curve when tested under steady conditions, the curves are not usually parallel. In the vicinity of the operating point, the percent change in current is less than the percent change in voltage.

The no-load or open-circuit voltage of constant-current arc welding power sources is considerably higher than the arc voltage.

Constant current power sources are not used exclusively for manual shielded metal arc welding processes. They are also used with semiautomatic or automatic processes where a constant arc length is maintained either by automatic changes in the feed rate of a consumable electrode or by automatic electrode holder adjustment using a nonconsumable electrode. This

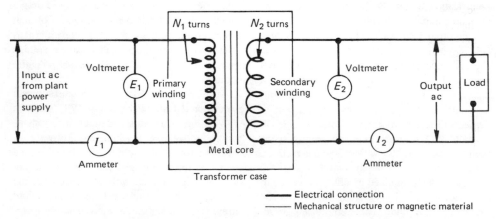

Fig. 1.2—Principal electrical elements of a welding transformer

type of power source is suitable for automatically controlled processes for which a reasonably constant current is required.

In the NEMA standard, a constant voltage source is defined as follows: "A constant-voltage arc-welding machine is one which has means for adjusting the arc voltage and which has a static volt-ampere curve that tends to produce a relatively constant output voltage. The arc current, at a given welding voltage, is responsive to the rate at which a consumable electrode is fed into the arc. A constant-voltage arc-welding machine is usually used with welding processes which employ a continuously fed consumable electrode."

A welding arc powered by a constant voltage source, utilizing a consumable electrode and a constant speed wire feed, is essentially a self-regulating system. It tends to stabilize itself despite momentary changes, such as arc length variations and fluctuations in the power supply. The arc length and welding current are inter-related to correct for rapid changes in length. For example, arc length variation is fundamentally determined by the differences between melting rate and feed rate, and arc voltage is directly related to arc length. If arc length (voltage) varies for any reason, the current will change quickly to a higher or lower value. The current change will alter the melting rate, thereby returning the arc length to its initial value. Thus, a good constant voltage source is capable of providing large current variations while still maintaining nearly constant arc voltage, and a constant speed system for wire feed can be used to good advantage. The arc current will be approximately proportional to wire feed rate for all wire sizes.

A power source that provides both constant current and constant voltage is defined by NEMA as: "A constant-current/constant-voltage arc-welding machine has the selectable characteristics of a constant-current arc-welding machine and a constant-voltage arc-welding machine."

PRINCIPLES OF OPERATION

The means of reducing power system voltage in Fig. 1.1 may be an electric generator driven by an electric motor, or it may be a transformer. When an electric generator is used, the arc welding power supply is usually designed for dc welding only. In this case, the electromagnetic means of controlling the volt-ampere characteristic of the arc welding power source is usually an integral part of the generator and not a separate element, as shown in Fig. 1.1.

Various dc generator configurations are employed. The dc generator may use a separate exciter and field strength control to establish the desired volt-ampere characteristics. Another type uses a separate exciter and differential or cumulative compounding for selection of volt-ampere characteristics. This equipment is discussed in detail later in this chapter.

WELDING TRANSFORMER

Figure 1.2 shows the basic elements of a welding transformer. For a transformer, the sig-nificant relationships between winding turns and input and output voltages and currents are as follows:

$$\frac{N_1}{N_2} = \frac{E_1}{E_2} = \frac{I_2}{I_1}$$

where N_1 is the number of turns on the primary winding of the transformer, N_2 is the number of turns on the secondary winding, E_1 is the input voltage, E_2 is the output voltage, I_1 is the input current, and I_2 is the output (load) current. The element that determines the volt-ampere characteristics is not shown in Fig. 1.2.

Taps in the transformer secondary winding may be furnished, as shown in Fig. 1.3, to control the open-circuit (no-load) output voltage. In this case, the tapped transformer permits the adjustment or control of the number of turns, N_2, in the secondary winding of the transformer. When the number of turns is decreased on the secondary, output voltage is lowered because a smaller proportion of the transformer secondary windings is in use. The tap selection, therefore,

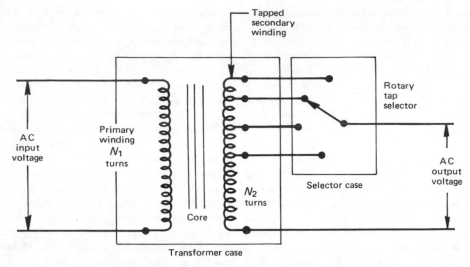

Fig. 1.3—Welding transformer with tapped secondary winding

controls the open-circuit voltage. As shown by the equation, the primary-secondary current ratio is inversely proportional to the primary-secondary voltage ratio. Thus, large secondary (welding) currents are available from relatively low line currents.

The transformer may be designed so that the tap selection will directly adjust the output volt-ampere characteristics for a proper welding condition. More often, however, impedance is inserted in series with the transformer to provide this characteristic, as shown in Fig. 1.4. Some types of power sources use a combination of these arrangements with the taps adjusting the open-circuit (or no-load) voltage of the welding machine and the impedance providing the desired volt-ampere characteristics.

Figure 1.4 is typical of a large variety of power sources. It is shown as a single-phase transformer for simplicity but could equally well be an alternator instead of the transformer. In constant current power supplies, the voltage

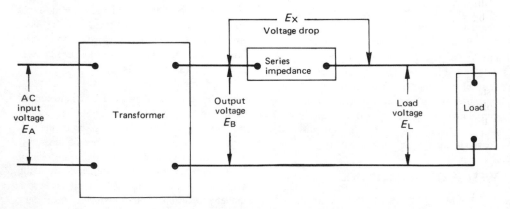

Fig. 1.4—Typical series impedance control of output current

drop, E_X, across the impedance increases greatly as the load current is increased. The increase in voltage drop, E_X, causes a large reduction in the load voltage, E_L. Adjustment of the value of the series impedance controls its voltage drop and the relation of load current to load voltage. This is called "current control" or, in some cases, "slope control." Voltage E_B essentially equals the no-load (open-circuit) voltage of the power supply.

In constant voltage power sources, the voltage drop, E_X, across the impedance (reactor) increases only slightly as the load current increases. The drop in the load voltage is small. Adjustment in the value of reactance gives slight control of the relation of load current to load voltage. This method of slope control with simple reactors is also a method of voltage control obtainable with saturable reactors or magnetic amplifiers.

Figure 1.5 shows an ideal vector relationship of the alternating voltages for the circuit of Fig. 1.4 when a reactor is used as an impedance device. The voltage drop across the impedance plus the load voltage equals the no-load voltage only when added vectorially. In the example pictured, open-circuit voltage of the transformer is 80 volts; the voltage drop across the reactor is about 69 volts when the load (equivalent to a resistor) voltage is 40 volts. The vectorial addition is necessary because the alternating load and impedance voltages are not in time phase.

The voltage drop across a series impedance in an ac circuit is added vectorially to the load voltage to equal the transformer secondary voltage. By varying the voltage drop across the impedance, the load voltage may be changed. This peculiar characteristic (vectorial addition) of impedance voltage in ac circuits is related directly to the reason that both reactors and resistances may be used to produce a drooping voltage characteristic. An advantage of a reactor is that it consumes little or no power, despite the fact that a current flows through it and a voltage can be measured across it.

When resistors are used, there is a power loss as well as a temperature rise. Theoretically, in a purely resistive circuit (no reactance), the voltage drop across the resistor could be added

arithmetically to the load voltage to equal the output voltage of the transformer. For example: A welding machine with approximately a constant current characteristic, 80 V open circuit, and powering a 25 V, 200 A arc would need to dissipate 55 V × 200 A or 11 000 watts (W) in the resistor to supply 5000 W to the arc. This is because, in the resistive circuit, the voltage and current are said to be in phase. In the reactor circuit, phase shift accounts for the greatly reduced power loss. In the reactor circuit, there are only the iron and copper losses, and these are very small by comparison.

Variable inductive reactance or variable mutual inductance may be used to control the volt-ampere characteristics in typical transformer or transformer-rectifier arc welding power sources. The equivalent impedance of a variable inductive reactance or mutual inductance is, of course, located in the ac electrical circuit of the power source in series with the secondary circuit of the transformer, as shown in Fig. 1.4. The basic transformer and one of its many electrical versions, a tapped transformer, are shown in Figs. 1.2 and 1.3, respectively. Another major advantage of inductive reactance over

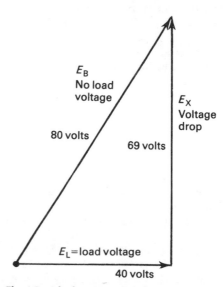

Fig. 1.5—Ideal vector relationship of the alternating voltage output using reactor control

resistance is that the phase shift produced in the alternating current by the reactor improves ac arc stability for a given open-circuit voltage. This is an advantage with gas tungsten arc and shielded metal arc welding processes.

The reactance of a reactor can be varied by several means. It can be varied by changing taps on a coil or by other electrical and mechanical schemes, some of which are described later. Varying the reactance will vary the voltage drop across the reactor with respect to the current through it. Thus, for any single value of inductive reactance, a single output volt-ampere curve can be plotted. Each value of reactor adjustment will generate a different output curve. This creates the control feature of the power supply.

In addition to the adjustment of reactance, it is also possible to adjust the mutual inductance between the primary and secondary coils. This may be done by moving the coils in relation to one another or by using a movable shunt that can be inserted or withdrawn from the transformer. These methods change the magnetic coupling of the coils to produce an adjustable mutual inductance characteristic.

In those welding power sources having a transformer and a rectifier for ac or dc welding, the rectifiers are located between the adjustable impedance or transformer taps and the output terminal. The transformer-rectifier type of arc welding power supply usually incorporates a stabilizing inductance or choke, located in the dc welding circuit, to improve arc stability.

WELDING GENERATOR

The no-load voltage of a dc generator used for arc welding may be controlled by adjusting a relatively small current in the main field winding. The current-range control adjusts the dc generator series or bucking field winding that carries the welding current. Polarity may be reversed by changing the interconnection between the exciter and the main field. An inductor or filter reactor is not usually needed in the welding circuit to improve arc stability in this type of welding equipment. The several turns of series winding on the field poles of the rotating

generator provide more than enough inductance to ensure satisfactory arc stability. This unit is described in more detail in later sections of this chapter.

An alternator power source, a rotating type power source in which ac is produced and either used directly or rectified into dc, can use a combination of the above adjustment means. It is possible to use a tapped reactor for gross adjustment of the welding output and then control the field strength for fine adjustment.

There are many schemes available to the welding power source designer to control power source output. A number of power sources utilizing solid state technology are available. They are discussed later.

VOLT-AMPERE CHARACTERISTICS

All welding power sources have two kinds of operating characteristics, each of which affects their welding performance in different ways. These are their dynamic characteristics and their static characteristics.

The static output characteristics can be readily measured by conventional test procedures. A set of output voltage versus output current characteristic curves (volt-ampere curves) are usually used to describe the static characteristics.

The dynamic characteristic of an arc welding power source is determined by the transient variations in output current and voltage that appear in the arc. Dynamic characteristics describe instantaneous variations or those that occur during very short intervals of time, such as 0.001 second. Static characteristics are measured over longer periods of time under steady state conditions. Arc stability is determined by the combination of the static and dynamic characteristics of the arc welding power supply.

The inherent transient characteristic of a welding arc is the principal reason for the great importance of the dynamic feature of an arc welding power source. Most welding arcs are continuously changing conditions. In particular, transients occur during the striking of the arc, during the transfer of metal across the arc, and during arc extinction and reignition during each

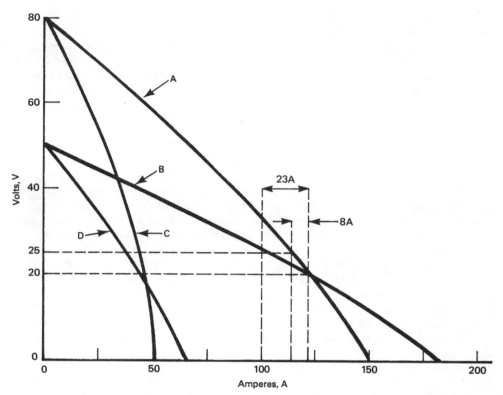

Fig. 1.6—Typical volt-ampere characteristics of a "drooping" power source with adjustable open-circuit voltage

half cycle of ac welding. Other causes of transients include variations in arc length, arc temperatures, and emission characteristics.

These arc transients can appear and disappear in a time interval comparable to the interval during which a significant change in ionization of the arc column can occur (0.001 second). This time interval is too short for the static characteristics of the welding power supply to react. Therefore, steady state or static volt-ampere characteristics have little significance in determining dynamic characteristics of an arc welding system.

Among the arc welding power supply characteristics that do have an effect on arc stability are those that provide

(1) Local transient energy storage, such as parallel capacitance circuits or a dc series inductance

(2) Feedback controls in automatically regulated systems

(3) Modifications of wave form or circuit frequencies

(4) Response time

(5) Open-circuit voltage

An improvement in arc stability is typically the goal of modifications or control of these characteristics. Beneficial results include

(1) Better control of the amount and stability of metal transfer

(2) Reduction in metal spatter

(3) Improvement in the uniformity of metal transfer

(4) Reduced weld pool turbulence

(5) Improved dynamic response to arc changes

The static volt-ampere characteristic is the one generally published by the power supply manufacturer. There is no universally recognized method by which dynamic characteristics are specified. The user should obtain assurance from the manufacturer that both the static and dynamic characteristics of the power supply will meet the intended application.

Constant Current

Figure 1.6 shows typical volt-ampere output curves for a conventional constant current power source. It is sometimes called a "drooper" because of the substantial downward (negative) slope of the curves. The power source might have open-circuit voltage adjustment in addition to output current control. A change in either control will change the slope of the volt-ampere curve.

The effect of the slope of the V-A curve on power output is shown in Fig. 1.6. Set on 80 V open circuit, curve A, an increase in arc voltage from 20 to 25 V (25 percent) would result in a decrease in current from 123 to 115 A (6.5 percent). The change in current is small. Therefore, with a consumable electrode welding process, electrode melting rate would remain fairly constant with a change in arc length.

Setting the power source for 50 V open circuit and a lower slope will give volt-ampere curve B. The same increase in arc voltage from 20 to 25 V would decrease the current from 123 to 100 A (19 percent), a significantly greater change. In manual welding, the flatter V-A curve would give the welder the opportunity to substantially vary the current by changing the arc length. This may be useful in out-of-position welding to control electrode melting rate and molten pool size.

The current setting may be changed to a lower value that would result in volt-ampere curves with greater slope, as illustrated by curves C and D. The amperage change with voltage would be less at the lower current setting (more nearly "constant current").

Constant Voltage

A typical volt-ampere curve for a constant voltage power source is shown in Fig. 1.7. This type of power source does not have true constant voltage output. It has a slightly downward (negative) slope. There is usually sufficient internal electrical impedance in the welding circuit to cause a minor voltage droop in the output volt-ampere characteristics. Changing the impedance in the welding circuit will also alter the slope of the volt-ampere curve.

As illustrated at point A in Fig. 1.7, an increase or decrease in voltage to B or C (5 V or 25 percent) produces a large change in amperage (100 A or 50 percent). This V-A characteristic is suitable for constant feed electrode processes, such as gas metal arc and flux cored arc welding, to maintain a constant arc length. A slight change in arc length (voltage) will cause a fairly large change in welding current. This will automatically increase or decrease electrode melting rate to regain the desired arc length (voltage). Adjustments are sometimes provided with constant voltage power sources to change or modify the slope or shape of the V-A curve.

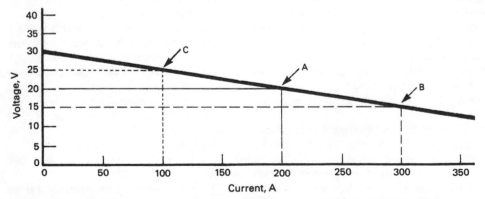

Fig. 1.7—Volt-ampere output relationship for a constant voltage power source

These may also be designed to change the dynamic characteristics simultaneously.

The curve shown in Fig. 1.7 can also be used to explain the difference between static and dynamic power supply characteristics. For example, it has been established that, in GMAW short circuiting transfer, the welding electrode actually touches the weld pool, causing a short circuit. At this point, the arc voltage is zero with only the circuit resistance or reactance to limit current. If the power supply responded instantly, very high current would immediately begin to flow through the welding circuit. The instantaneous high current would melt the short-circuited electrode free with explosive force, dispelling the weld metal. Of course, this is not desirable in practice. The dynamic characteristics designed into this type power source compensate for this action by limiting the rate of current rise and decreasing the explosive force.

DUTY CYCLE

Since many welding jobs do not require the arc to be operating 100 percent of the time, a power source that supplies a given current for a short interval of time need not be as large and rugged as one required to supply the same current continuously. Duty cycle is one of the most important rating points of a welding power supply. It takes into consideration this difference in load requirements. A welding power supply is usually required to deliver power during limited periods of time only, unless it is used for automatic welding processes. Because a welder might stop welding to change electrodes, load and unload his work, or change his position, a welding power supply is often idle during a considerable part of the time.

Duty cycle expresses, as a percentage, the portion of the time that the power supply must deliver its rated output in each of a number of successive ten minute intervals without exceeding a predetermined temperature limit. Thus, a 60 percent duty cycle (a standard industrial rating) means that the power supply can deliver its rated load output for six out of every ten minutes. (Operation at rated load steadily for 36 minutes out of one hour is not a 60 percent duty cycle.) A 100 percent duty cycle power supply is designed to produce its rated output continuously without exceeding the prescribed temperature limits of its components.

Duty cycle is a major factor that determines the type of service for which a power supply is designed. Heavy industrial units designed for manual welding are normally rated at 60 percent duty cycle. For automatic and semiautomatic processes, the rating is usually 100 percent duty cycle. Light duty power supplies are usually 20 percent duty cycle. Ratings at other duty cycle values are also available. If the output rating of the power supply at the stated duty cycle matches the job requirements, there is reason to believe it will perform adequately in that service.

Fundamentally, the duty cycle is a ratio of the load-on time capability to the no-load time required for cooling in order that the internal windings and components and their electrical insulation system will be within their designed temperature ratings. In any duty-cycle rating, the maximum allowable temperature of the components in the unit is the determining factor. These maximum temperatures are specified by various organizations and agencies whose interest lies in the field of insulation standards. The maximum temperature criteria do not change with the duty cycle or current rating of the power supply. Thus, it can be expected that the internal temperatures of similarly insulated components in a welding machine rated at 200 A, 28 V, 40 percent duty cycle will be very close to the internal temperatures of a second machine with the same current and voltage rating and 100 percent duty cycle when both are operated at their rated load and duty cycles.

An important point is that the duty cycle of a power supply is based on the output current and not on the kilovolt-ampere or kilowatt rating. The following two formulas are given for estimating the new duty cycle at other than rated output or for estimating other than rated output current at other than the specified duty cycle:

$$T_a = \left[\frac{I}{I_a} \right]^2 T \quad (1)$$

$$I_a = \left[\frac{T}{T_a} \right]^{\frac{1}{2}} I \quad (2)$$

T is rated duty cycle in percent; T_a is the required duty cycle in percent; I is rated current at rated duty cycle; and I_a is the maximum current at required duty cycle. The power supply should never be operated above its rated current or duty cycle unless approved by the manufacturer.

Example: At what duty cycle can a 200 A power supply rated at 60 percent duty cycle be operated at 250 A output? Using equation (1)

$$T_a = \left[\frac{200}{250} \right]^2 \times 60$$
$$= (0.80)^2 \times 60\% = (0.64) \times 60\%$$
$$T_a = 38\%$$

Therefore, this unit must not be operated more than 3.8 minutes out of each ten minute period at 250 A. The 250 A should not exceed the current rating of any power supply component.

Example: The aforesaid power supply is to be operated continuously (100 percent duty cycle). What output current must not be exceeded? Using equation (2)

$$I_a = 200 \left[\frac{60}{100} \right]^{\frac{1}{2}}$$
$$I_a = 200 \times 0.775$$
$$I_a = 155 \text{ A}$$

If this power supply is operated continuously, no more than 155 A output should be used.

In the past, for very high current power supplies (750 A and higher), another duty-cycle rating was sometimes used. This was identified as the one-hour duty rating. To determine the rated output of these machines, they were loaded for one hour at rated output and then tested. Then the output was reduced immediately to 75 percent of the rated current value, and operation was continued for an additional three hours; at this time, the test period was terminated. Component temperatures were measured at the end of the first one-hour period and at the conclusion of the test. These temperatures had to be within established allowable limits. Some manufacturers may still provide power supplies with a one-hour duty rating.

OPEN-CIRCUIT VOLTAGE

Open-circuit voltage is the voltage at the output terminals of a welding power source when it is energized but has no current output. Open-circuit voltage is one of the design factors influencing the performance of all welding power sources. In a transformer, open-circuit voltage is a function of the primary input voltage and the ratio of primary-to-secondary coils. This fact naturally affects the physical size of the welding power source. Although a high open-circuit voltage may be desirable from the standpoint of arc initiation and stability, the factors of safety and cost preclude values higher than needed to meet the requirements of the welding process.

Open-circuit voltage of a generator or alternator type welding machine is a design condition relating to the strength of the magnetic field, the speed of rotation, the number of turns in the load coils, etc. These types of power sources generally have controls that vary or adjust the open-circuit voltage.

NEMA Standard EW-1 contains specific requirements for maximum open-circuit voltage. When rated voltage is applied to the primary winding of a transformer or when a generator-type arc welding machine is operating at maximum rated no-load speed, the open-circuit voltage is limited to the values shown in Table 1.2. Three-phase transformer-rectifier type, three-phase alternator-rectifier type, and generator type power sources normally have less than ten percent ripple voltage. Single-phase alter-

Table 1.2—Open-circuit voltages for various types of arc welding power sources

Type of power source	Maximum open-circuit voltage
For manual and semiautomatic applications	
Alternating current	80 rms
Direct current—over 10% ripple voltage [a]	80 rms
Direct current—10% or less ripple voltage [a]	100 avg.
For automatic applications	
Alternating current	100 rms
Direct current	100 avg.

a. Ripple voltage, $\% = \dfrac{\text{ripple voltage, rms}}{\text{avg. total voltage}}$

nator-rectifier and single-phase transformer-rectifier type power sources normally have greater than ten percent ripple voltage.

NEMA Class I and Class II power sources normally have open-circuit voltage at, or close to, the maximum specified. Class III ac power sources frequently have two or more open-circuit voltages. One arrangement is to have a high and low range of amperage output from the power source. The low range normally has approximately 80 V open circuit, with the high range somewhat lower. Another arrangement is the tapped secondary coil method, described earlier, in which, at each current setting, the open-circuit voltage changes about 2 to 4 volts.

In ac power sources of the transformer or alternator type, the alternating current (60 cycle) reverses direction of flow each 1/120 second. Plotting the direction and strength of the current through a complete cycle of alternations produces a sine wave form. Figure 1.8 shows typical sine wave forms of a dual range machine with open-circuit voltages of 80 and 55 V rms. A key fact is that each sine wave form shown goes through a complete cycle in 1/60 second.

Since the current must change direction after each half cycle, it is apparent that, for an instant, at the point at which the current wave form crosses the zero line, the current must be zero. At that time, the current flow in the arc ceases. An instant later, the current should reverse direction of flow. However, during the period in which current is decreasing and reaches zero, the arc plasma cools, reducing the thermal energy

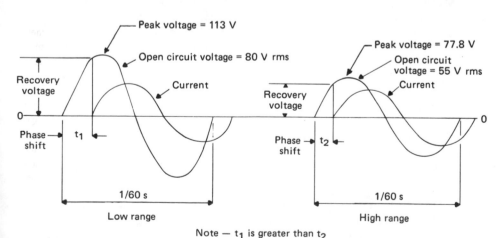

Note — t_1 is greater than t_2

Fig. 1.8—Typical voltage and current waves of a dual range ac power source

level. This retards ionization of the arc gas stream. For the welding current to begin flowing in the opposite direction, ionization of the material in the arc gap must either be maintained or reinitiated by the voltage across the arc gap. This voltage, shown in Fig. 1.8, is the recovery voltage. The greater this recovery voltage, the shorter the period during which the arc is extinguished.

Figure 1.8 also shows the phase relations between voltage and equal currents for two different open-circuit voltages, assuming the same arc voltage (not shown) in each case. As can be seen in this figure, recovery voltage is greater with 80 V open circuit, than with 55 V open circuit because of the higher source voltage. Greater phase shift causes zero current at a time when recovery voltage is near the peak of the open-circuit voltage wave form. If resistance, rather than reactance, were used to regulate welding current, the power source voltage and current would be in phase. Then both the recovery voltage and current would be zero, causing an unstable arc.

Factors that permit use of low open-circuit voltages include the ingredients in some electrode coverings, which help to maintain ionization, and also the favorable metal transfer characteristics of some electrodes, which prevent sudden gross increases in arc length.

Direct current power sources operate much the same way with respect to the immediate return to open-circuit voltage for arc recovery. Normally, reactance or inductance is built into these power sources to enhance this effect. In a dc system, once the arc is established, the welding current does not pass through zero. Thus, rapid voltage increase is not critical and resistors are suitable current controls for dc welding.

NEMA POWER SOURCE REQUIREMENTS

The National Electrical Manufacturers Association (NEMA) publication EW-1 (latest revision) covers the requirements for electric arc welding apparatus including power sources.

NEMA CLASSIFICATIONS

NEMA classifies arc welding power sources primarily on the basis of duty cycle. There are three classes:

(1) "A NEMA Class I arc-welding machine is characterized by its ability to deliver rated output at duty cycles of 60, 80, or 100 percent. If a machine is manufactured in accordance with the applicable standards for Class I machines, it shall be marked 'NEMA Class I (60),' 'NEMA Class I (80),' or 'NEMA Class I (100).'

(2) "A NEMA Class II arc-welding machine is characterized by its ability to deliver rated output at duty cycles of 30, 40, or 50 percent. If a machine is manufactured in accordance with the applicable standards for Class II machines, it shall be marked 'NEMA Class II (30),' 'NEMA Class II (40),' or 'NEMA Class II (50).'

(3) "A NEMA Class III arc-welding ma-chine is characterized by its ability to deliver rated output at a duty cycle of 20 percent. If a machine is manufactured in accordance with the applicable standards for Class III machines, it shall be marked 'NEMA Class III (20).' "

NEMA Class I and II power supplies are further defined as completely assembled arc welding power supplies which are comprised of the characteristics of the following:

(1) A constant current, a constant voltage, or a constant current/constant voltage machine and

(2) A single-operator machine and

(3) One of the following:

    (a) DC generator arc welding machine

    (b) AC generator arc welding machine

    (c) DC generator-rectifier arc welding machine

    (d) AC/DC generator-rectifier arc welding machine

    (e) AC transformer arc welding machine

    (f) DC transformer-rectifier arc welding machine

    (g) AC/DC transformer-rectifier arc welding machine

Table 1.3—NEMA-rated output current (size) for arc welding machines

Rated output current, A		
Class I	Class II	Class III
200	150	180-230
250	175	235-295
300	200	
400	225	
500	250	
600	300	
800	350	
1000		
1200		
1500		

A NEMA Class III power supply is further defined as a completely assembled arc welding power supply which is comprised of the characteristics of the following:

(1) A constant current machine and

(2) A single-operator machine and

(3) One of the following:

    (a) AC transformer arc welding machine

    (b) DC transformer-rectifier arc welding machine

    (c) AC/DC transformer-rectifier arc welding machine

OUTPUT AND INPUT REQUIREMENTS

In addition to duty cycle, NEMA specifies the output ratings and performance capabilities of power supplies of each class. Table 1.3 gives the output current ratings (size) for Class I, II, and III arc welding machines. The rated load volts (E) for Class I and II machines under 500 A can be calculated using $E = 20 + 0.04 I$, where I is the rated load current. For sizes of 600 A and above ratings, the rated load voltage is 44. The output ratings in amperes and load volts and also the minimum and maximum output currents and load volts for power supplies are given in NEMA publication EW-1 (latest edition).

The electrical input requirements of NEMA Class I and II transformer arc welding machines for 50 and 60 Hz are

60 Hz—200, 230, 460, and 575 V

50 Hz—220, 380, and 440 V

For NEMA Class III transformer arc welding machines, the electrical input requirement is 60 Hz—230 volts. The transformer primary windings are usually tapped to permit selection of two or three alternate voltage supplies, such as 200, 230, and 460 V.

The voltage and frequency standards for welding generator drive motors are the same as for NEMA Class I and II transformer primaries.

NAMEPLATE DATA

The minimum nameplate data for an arc welding power supply specified in NEMA publication EW-1 are

(1) Manufacturer's type designation and/or identification number

(2) NEMA class designation

(3) Maximum open-circuit voltage (OCV)

(4) Rated load volts

(5) Rated load amperes

(6) Duty cycle at rated load

(7) Maximum speed in rpm at no load (generator or alternator)

(8) Frequency of power supply

(9) Number of phases of power supply

(10) Voltage(s) of power supply

(11) Amperes input at rated load output

The instruction book supplied with each power source is the prime source of data concerning electrical input requirements. General data are also given on the power source nameplate. These data are most often found in tabular

Table 1.4—Typical nameplate specifications for an ac-dc arc welding power source

	Welding current, A				Open-circuit voltage ac & dc	Amperes input at rated load output—60 Hz single phase				
	AC		DC			Amperes				
Model	Gas tungsten arc	Shielded metal arc	Gas tungsten arc	Shielded metal arc		200 V	230 V	460 V	kVA	kW
300 Ampere	5-48	5-48	5-60	5-45	80	115	104	52	23.9	21.8
	20-230	20-245	20-250	16-200						
	190-435	200-460	230-460	150-350						

Table 1.5 — Typical primary conductor and fuse size recommendations

Model	Input wire size, AWG[a]				Fuse size in amperes			
	200 V	230 V	460 V	575 V	200 V	230 V	460 V	575 V
300	No. 2	No. 2	No. 6	No. 8				
A	(No. 6)[b]	(No. 6)[b]	(No. 8)[b]	(No. 8)[b]	200	175	90	70

a. American wire gage
b. Indicates ground conductor size

form along with other pertinent data that might apply to the particular unit. Table 1.4 shows typical information for a NEMA-rated 300 A GTAW power supply. The welding current ranges are given with respect to welding process. The power source may use one of three line input voltages with the corresponding current input listed for each voltage when the machine is producing its rated load. The kilovolt-ampere (kVA) and kilowatt (kW) input data are also listed. The power factor, F_p, can be calculated.

$$F_p = \frac{kW}{kVA}$$

The manufacturer will also provide other useful data concerning input requirements, such as primary conductor size and recommended fuse size. Power sources cannot be protected with fuses of equal value to their primary current demand. If this is done, nuisance blowing of the fuses or tripping of circuit breakers will result. Table 1.5 shows typical fuse and input wire sizes for the 300 A power source of Table 1.4. However, all pertinent codes should be consulted in addition to these recommendations.

ALTERNATING CURRENT POWER SOURCES, CONSTANT CURRENT

Alternating current power sources are normally single-phase transformers that take ac power from the building power line and transform the voltage and amperage to values suitable for arc welding. The transformer also serves to isolate the welding circuit from the plant power lines.

Another source of ac welding power is an alternator (often called an ac generator), which converts mechanical energy into electrical power suitable for arc welding. The mechanical power may be obtained from various sources, such as an internal combustion engine or an electric motor.

A typical alternator design normally places the magnetic field coils on the rotor and the armature coils in the stator. This configuration precludes the necessity of the commutator and the brushes used with dc output generators. The frequency of the output welding current is controlled by the speed of rotation of the rotor assembly and by the number of poles in the alternator design. A two-pole alternator must operate at 3600 rpm to produce 60 Hz current, whereas a four-pole alternator design must operate at 1800 rpm to produce 60 Hz current.

TRANSFORMER POWER SOURCES

The purpose of any welding transformer is to change high voltage, low amperage power to low voltage, high amperage welding power. Because various welding applications have different welding power requirements, means for control of welding current or arc voltage, or both, must be provided in a welding transformer power source. The methods commonly used to control the welding circuit output are described in the following sections.

Movable-Coil Control

A movable-coil transformer consists essentially of an elongated core on which are located primary and secondary coils. Either the primary coil or the secondary coil may be movable, while the other one is fixed in position. Most ac transformers of this design have a fixed position secondary coil. The primary coil is normally

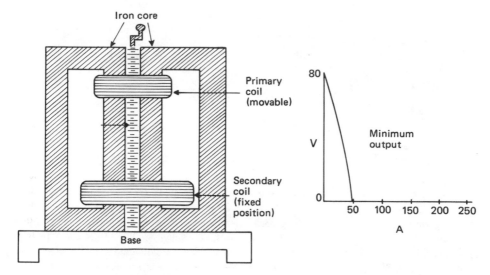

Fig. 1.9—Movable-coil ac power source with coils set for minimum output

attached to a lead screw and, as the screw is turned, the coil moves closer to or farther from the secondary coil.

The varying distance between the two coils regulates the inductive coupling of the magnetic lines of force between them. The farther the two coils are apart, the more vertical the volt-ampere output curve and the lower the maximum short circuit current value. Conversely, when the two coils are closer together, the maximum short circuit current is higher and the slope of the volt-ampere output curve is lower.

Figure 1.9 shows one form of a movable-coil transformer with the coils far apart for minimum output and the steep slope of the volt-ampere curve. Figure 1.10 shows the coils as close together as possible. The volt-ampere curve is indicated at maximum output with less slope than the curve of Fig. 1.9.

Another form employs a pivot motion.

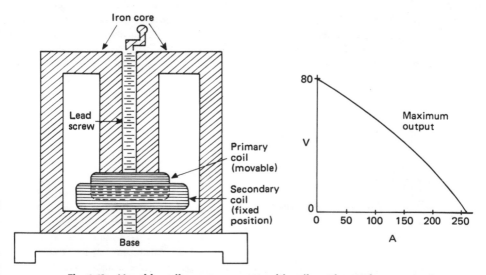

Fig. 1.10—Movable-coil ac power source with coils set for maximum output

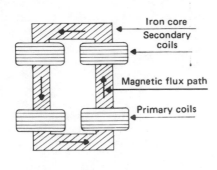

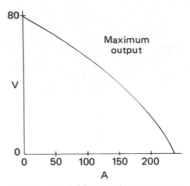

Iron core
Secondary coils
Magnetic flux path
Primary coils

Maximum output

Fig. 1.11—Movable-shunt ac power source with shunt removed for maximum power

When the two coils are at a right angle to each other, output is at a minimum. When the coils are aligned with one coil nested inside the other, output is at maximum.

Movable-Shunt Control

The movable-shunt method of control is often used with ac transformers. It also may be used with ac-dc power sources. In this design, both the primary coils and secondary coils are fixed in position. A laminated iron core shunt is moved between the primary and secondary coils. It is made of the same material as that used for the transformer core. The shunt acts to divert magnetic flux around the coils. (The term flux means the same as magnetic lines of force in this usage.)

As illustrated in Fig. 1.11, the arrangement of the magnetic lines of force, or magnetic flux, is unobstructed when the iron shunt is not between the primary and secondary coils. As the shunt

is moved in between the primary and secondary coils, as shown in Fig. 1.12, some magnetic lines of force are diverted through the iron shunt rather than to the secondary coil. The output volt-ampere curve is adjusted from minimum to maximum within the amperage range of the welding power source by moving the iron shunt. When the shunt is withdrawn from between the primary-secondary coils, the output current is at maximum. As the iron shunt moves in between the primary and secondary coils, the slope of the volt-ampere curve increases and the available welding current is decreased.

Movable-Core Reactor

The movable-core reactor type of ac welding machine consists of a constant voltage transformer and a reactor. The inductance of the reactor is varied by mechanically moving a section of its iron core. The machine is shown in Fig. 1.13. When the movable section of the core is in a

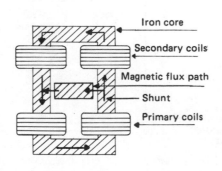

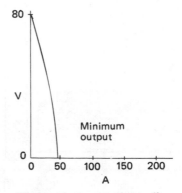

Iron core
Secondary coils
Magnetic flux path
Shunt
Primary coils

Minimum output

Fig. 1.12—Movable-shunt ac power source with shunt between power coils

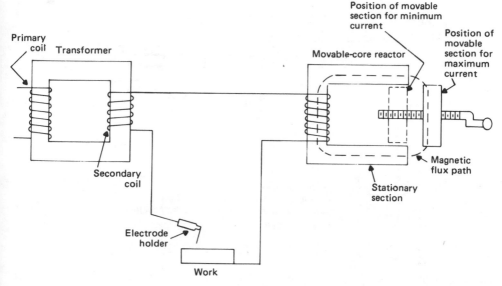

Fig. 1.13—Movable-core reactor type ac power source

withdrawn position, the permeability of the magnetic path is very low due to the air gap. The result is a low inductive reactance that permits a large welding current to flow. When the movable-core section is advanced into the stationary core, as shown by broken lines in Fig. 1.13, the increase in permeability causes an increase in inductive reactance and, thus, welding current is reduced.

Tapped Secondary Coil Control

A tapped secondary coil may be used for control of the volt-ampere output of a transformer (Fig. 1.3). This method of adjustment is often used with NEMA Class III power sources. Basic construction is somewhat similar to the movable-shunt type, except that the shunt is permanently located inside the main core and the secondary coils are tapped to permit adjustment of the number of turns. Decreasing secondary turns reduces open-circuit voltage and, also, the inductance of the transformer, causing welding current to increase.

Saturable Reactor Control

A saturable reactor control is called an electrical control because it employs an isolated low voltage, low amperage dc circuit to change the effective magnetic characteristics of the reactor cores. A self-saturating saturable reactor is referred to as a magnetic amplifier because a relatively small control power change will produce a sizeable output power change. The principal features of this type of control circuit are that (1) it makes remote control of output from the power source relatively easy and (2) it normally requires less maintenance.

In this design, the main transformer has no moving parts. The volt-ampere characteristic of this type power source is determined by the designs of the transformer and the saturable reactor. By adding a dc control circuit to the reactor system, it is possible to adjust the output volt-ampere curve from minimum to maximum. A large amount of alternating current is controlled using a relatively small amount of direct current.

A simple saturable reactor power source is shown in Fig. 1.14. The reactor coils are connected opposing with respect to the dc control coils. If this were not done, high circulating currents and high voltages would be present in the control circuit because of transformer action. With the opposing connection, the instantaneous voltages and currents tend to cancel out. Saturable reactor type controls are often used for gas tungsten arc welding power supplies. With ac,

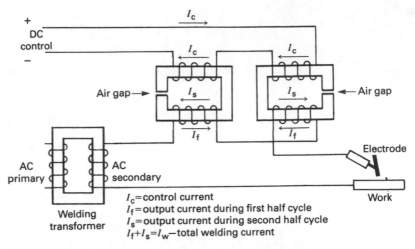

I_c=control current
I_f=output current during first half cycle
I_s=output current during second half cycle
$I_f+I_s=I_w$—total welding current

Fig. 1.14—Saturable reactor type ac welding power source

the wave form for GTAW is quite important. Saturable reactors tend to cause severe distortion of the sine wave supplied from the transformer. Placing an air gap in the reactor core is one method of reducing this distortion. Another method is to insert a large choke in the dc control circuit. Either method, or a combination of both, will produce desirable results.

The amount of current adjustment in a saturable reactor is based on the ampere-turns of the various coils. Ampere-turns is defined as the number of turns in the coil multiplied by the current in amperes flowing through the coil. In the basic saturable reactor, the law of equal ampere-turns applies. To increase output in the welding circuit, a current must be made to flow in the control circuit. The amount of change will be proportional as follows:

$$\Delta I_w = \frac{I_c N_c}{N_w} \text{ (approximately)}$$

where

I_w = welding current, A
I_c = current, A, in the control circuit
N_c = number of turns in the control circuit
N_w = number of turns in the welding current circuit

The minimum current of the power source is established by the number of turns in the welding current reactor coils and the amount of iron in the reactor core. For a low minimum current, either a lot of iron or a relatively large number

of turns, or both, are required. If a large number of turns are used, then either a large number of control turns or a high control current, or both, are necessary. To reduce the requirement for large control coils, large amounts of iron, or high control currents, the saturable reactors often employ taps on the welding current coils, creating multirange machines. The higher ranges would have fewer turns in these windings and, thus, correspondingly higher minimum currents.

Magnetic Amplifier Control

Technically, the magnetic amplifier is a self-saturating saturable reactor. It is called a magnetic amplifier because it appears to amplify and, in fact, it does, when only the power or ampere-turns of the control are considered. It reduces the requirements for high control currents and large control coils. While a magnetic amplifier machine is often multirange, the ranges of control will be much broader than those possible with an ordinary saturable reactor control. Referring to Fig. 1.15, it can be seen that by using a different connection for the welding current coils and rectifying diodes in series with the coils, the load ampere-turns are used to assist the control ampere-turns in magnetizing the cores. A smaller amount of control ampere-turns will cause a correspondingly larger welding current to flow because the welding current will essentially "turn itself on." The control windings are polarity sensitive.

Power Factor

Constant ac power sources are characterized by low power factor due to their large inductive reactance. This is often objectionable because of high line currents under heavy loads and also the electrical rate penalty with low power factor to most industrial users. Power factor may be improved by the addition of capacitors to the primary circuit of the welding power source. The addition of capacitors to an inductive circuit, such as a transformer type power source, improves power factor, thereby demanding less primary current from the plant power lines while welding is being performed. It will, however, draw a high current, under light or no load conditions.

Large alternating current transformer power sources may be equipped with capacitors for power factor correction to approximately 75 percent at rated load. At lower than rated load current settings, the power factor may have a leading characteristic. When the transformer is operating at no load or very light loads, the capacitors are drawing their full corrective kVA, thus contributing power factor correction to the remainder of the load on the total electrical system.

When a number of transformer type welding power sources are operating at light loads, care should be taken that the combined power factor correction capacitance will not upset the voltage stability of the line. If three-phase primary power is used, the load on each phase of the primary system should be balanced for best performance. Power factor correction, under normal conditions, has no bearing on welding performance.

Auxiliary Features

Constant-current ac power sources are available in many configurations with respect to their auxiliary features. Generally, these features are incorporated to better adapt the unit to a specific process or application or to make it more convenient to operate. The manufacturer

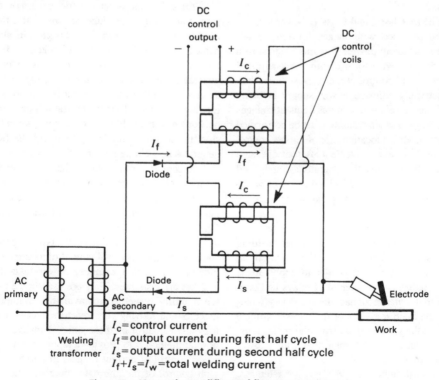

I_c = control current
I_f = output current during first half cycle
I_s = output current during second half cycle
$I_f + I_s = I_w$ = total welding current

Fig. 1.15—Magnetic amplifier welding current control

should be consulted for the available features when considering a power source.

Primary contactors or manually operated power switches are usually included in ac power sources to turn the unit on and off. Most NEMA Class I and Class II units are furnished with a terminal board or other means for connection of various rated primary line voltages. Input supply cords are not normally supplied with NEMA Class I and Class II welding power sources. The smaller NEMA Class III power sources are generally equipped with a manually operated primary switch and an input supply cord.

Some ac power sources incorporate a system for supplying a higher than normal current to the arc for a fraction of a second at the start of a weld. This "hot start" feature provides starting surge characteristics, similar to those of motor-generator sets, to assist in initiating the arc, particularly at current levels under 100 A. Other power sources may be equipped with a start control to provide adjustable "soft" starts to minimize transfer of tungsten from the electrode with the GTAW process.

NEMA Class I and Class II power sources may be provided with means for remote adjustment of output power. This may consist of a motor-driven device for use with crank-adjusted units or a hand control at the work station when an electrically adjusted power supply is being used. When a weldment requires frequent changes of amperage or when welding must be performed in an inconvenient location, use of remote control adjustment is very convenient. A foot-operated remote control frees the operator's hands and permits gradual increase and reduction of welding current. This is of great assistance in crater filling.

Safety voltage controls are available to reduce the open-circuit voltage of ac arc welding power sources. They reduce the open-circuit voltage at the electrode holder to about 30 V. Voltage reducers may consist of relays and contactors that either reconnect the secondary winding of the main transformer for a lower voltage or switch the welding load from the main transformer to an auxiliary transformer with a lower no-load voltage.

Units designed for the gas tungsten arc welding process usually incorporate electrically operated valves and timers to control the shielding gas and coolant flows to the electrode holder. Also, high frequency units may be added to assist in starting and stabilizing the ac arc.

Applications

Constant ac power sources have wide application in industry. In addition to gas tungsten arc welding, they are useful for shielded metal arc welding, especially in deep grooves, because alternating current minimizes arc blow. This type of power source is also used for submerged arc welding, carbon arc cutting and gouging, and electroslag welding. These units find wide usage in homes, garages, farms, repair shops, and sign shops. Some of the lowest priced power sources available today are constant ac.

Some ac power sources are also the most expensive and sophisticated machines in the market place. When designed for GTAW, the package often becomes very elaborate.

ALTERNATOR AC POWER SOURCES

An alternator is an electric generator designed and built to produce ac power. It differs from the standard ac generator design. The alternator rotor assembly contains the magnetic field coils instead of the armature coils as in other generators. Slip rings are used to conduct low dc power into the rotating member to produce a rotating magnetic field. The stator (stationary portion) has the welding current coils wound in slots in the iron core. The rotation of the field generates ac welding power in these coils.

Except for information relating to control means and transformers, the information presented on transformer power sources is equally applicable to alternator type ac power sources. One other exception is the data concerning electrical power input. Alternator type ac power sources are normally available as units driven by internal combustion engines, power take-offs, or other drive sources. They are used in the field when utility power is not available. Saturable reactors and moving core reactors may be used for output control of these units. However, the normal method is to provide a tapped reactor for broad control of current ranges, in combination with control of the alternator magnetic field to produce fine control within these ranges.

DIRECT CURRENT POWER SOURCES, CONSTANT CURRENT

Welding power sources that are called constant direct current machines can be divided into two categories: transformer-rectifiers and generators. The transformer-rectifier machines are static in nature, transforming ac to dc power. Generators convert mechanical energy of rotation to electrical power.

TRANSFORMER-RECTIFIER TYPES

The principal feature of these types of power sources is the rectification of the transformer output to dc welding power. Two types of transformers are normally used. These are single-phase and three-phase transformers. Single-phase units are designed to produce ac, dc, or both. Three-phase units usually produce dc only, although there are a few designs that produce both ac and dc.

Single-Phase Machines

A single-phase transformer is used to change the alternating voltage of the building power line to a suitable lower voltage. An example would be a transformer wound with a 460 V primary and a lower voltage secondary to give about 80 V dc after rectification and filtering. In ac-dc welding machines, a switch is provided so that the rectifiers can be connected to or removed from the welding circuit to furnish either ac, dcsp (electrode negative), or dcrp (electrode positive) at the output terminals. A large filter choke may be used in the welding circuit to help stabilize the arc. These machines involve some engineering compromises, but they are very useful and have wide acceptance because of their versatility.

Three-Phase Machines

A three-phase transformer is used to change the high primary voltage to a lower secondary voltage. An example would be a 460 volt primary and a 55 volt secondary. Rectifying the 55 volt three-phase ac will provide dc with an open-circuit voltage of approximately 80 volts. The rectification circuitry used to convert three-phase ac to dc can take several forms, depending on the type of welding current control system used. Some of these circuits are described later.

Electrical Characteristics

An important electrical characteristic is the relation of the output current to the output voltage. Both a static (steady-state) relationship and a dynamic (transient) relationship are of interest. The static relationship is usually shown by volt-ampere curves, such as those in Fig. 1.16. The curves usually represent the maximum and minimum curves for each current range setting. As discussed in a previous section, the dynamic relationship is difficult to define and measure for all load conditions. The dynamic characteristics determine the stability of the arc under actual welding conditions. They are influenced by circuit design and control.

General Design

The usual voltages of the ac supply mains in the United States are 208, 240, 480, and 600 with a frequency of 60 Hz. Transformers are seldom designed to work on all the above voltages, although connections for two or three of those voltages are often provided in a single machine. This is done by arranging the primary coils in sections with taps so that the leads from each section can be connected in series or parallel with other sections to suitably match the incoming line voltage. On three-phase machines,

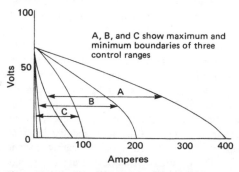

Fig. 1.16—Static volt-ampere characteristics of a typical constant current rectifier power source

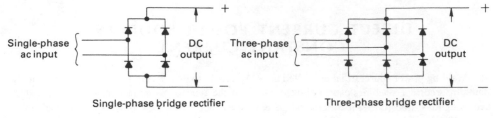

Fig. 1.17—Typical rectifier bridge circuits

the primary can be connected in delta or wye. The secondary is frequently connected in delta because a delta connection is very satisfactory for low voltage and high current from the standpoint of wire size.

The method of controlling the current is usually in the ac section of the machine between the transformer and the rectifiers. The control of the current uses the principle of variable inductance or impedance. The methods of varying the impedance for current controls are

(1) Moving coil

(2) Moving shunt

(3) Saturable reactor or magnetic amplifiers

(4) Tapped reactor

(5) Moving reactor core

(6) Solid state

In addition to these six systems, there is the type that employs resistors in series with the dc portion of the welding circuit. Methods (1), (2), and (5) are classed as mechanical controls; methods (3) and (6), as electric control; method (4) and the resistor type, as tap controls. These are the same methods used for control of transformer-type constant current power sources.

Rectifiers are used to convert alternating current to direct current. The arrangements of rectifying elements, called bridge rectifiers, are shown in Fig. 1.17. The section on Solid State Circuitry contains information on the design and application of rectifiers in power sources.

An inductance is usually used in the dc welding circuit to control excessive surges in load current.[1] These current surges may occur due to dynamic changes in arc load and the inherent ripple with rectification of alternating current into direct current. A three-phase rectifier has

relatively little ripple so that the size of the inductor is determined by the need to control arc load surges. With single-phase rectification, the ripple is quite high because the voltage goes to zero twice each cycle or 120 times per second with 60 Hz power. Therefore, the inductors of single-phase machines will be larger than those of three-phase machines of the same rating. Power sources of this type usually have a switch on the dc output so that the polarity of the voltage at the machine terminals can be reversed without reversing the welding cables.

Frequently, transformer-rectifier power sources have the total current range broken into segments or steps. This is a coarse adjustment. The range selector is usually a switching device. In general, the wider the machine output current range, the more switch positions necessary. As many as three to five ranges are common. The current vernier control is used by the operator to vary the current output within each range.

Mechanically controlled machines sometimes use a hand wheel that rotates an adjusting screw which transmits motion to the controlling element. For electric control, a rheostat, potentiometer, variable transformer, or other device is used to vary the control current to a saturable reactor. Electrically controlled machines are easily adjusted from a remote location. Mechanical controls require the addition of a reversible motor drive for remote current control.

Open-circuit voltage for constant current, rectifier type power sources depends on the intended welding application. It can range from 50 to 100 V. Most NEMA Class I and Class II machines are usually fixed in the 70 and 80 V range.

Electrical power inputs for transformer-rectifier power sources are similar to transformer type ac power sources. Three-phase transformer-

1. A welding circuit inductor is called by other terms, such as filter choke, stabilizer, and reactor.

rectifier power sources are seldom power factor corrected, unless specified as an option.

Auxiliary Features

The auxiliary features available are normally similar to those available for constant ac power sources. Not all features are available on all power sources. The manufacturer should be consulted for complete information.

In addition to previously listed features, many dc power sources have available, as standard or optional equipment, current pulsing capabilities. Pulse power sources are capable of alternately switching from high to low welding current repetitively. Normally, high and low current values, pulse duration, and pulse repetition rate are independently adjustable. This feature is very useful for out-of-position welding and critical gas tungsten arc welding applications.

GENERATORS AND ALTERNATORS

Generator type power sources convert mechanical energy into electrical power suitable for arc welding. The mechanical power can be obtained from an internal combustion engine, an electric motor, or from a power take-off from other equipment. For welding, there are two basic types of rotating power sources, the generator and the alternator. Both have a rotating member, called a rotor or an armature, and a stationary member, called a stator. A system of excitation is needed for both types.

The basic principle of any rotating power source is that when electrical conductors are moved mechanically through a magnetic field, electrical power is induced in the conductors. Physically, it makes no difference whether the magnetic field moves or the conductor moves, just so the coil experiences a changing magnetic intensity. In actual practice, most generators have a stationary field and moving conductors, and an alternator has a moving field and stationary conductors.

A welding generator normally has a stator containing the coils that produce the magnetic field. These are connected so that a like number of poles of the opposite polarity are produced adjacent to each other. A two-pole arrangement would have one north pole on one side of the

stator and a south pole on the other side, 180 degrees apart. A four-pole arrangement would have a north pole, a south pole, a north pole, and a south pole, each positioned 90 degrees apart. The field coils are normally many turns of fine wire. It is common practice to wind a second set of field coils in juxtaposition with the first set. The second set can be connected to the generator output in some manner, either aiding or bucking the regular field. This produces a field that will vary with generator output and can be used as a feedback system of output current control.

The armature conductors of a welding generator are relatively heavy because they carry the welding current. The commutator is located at one end of the armature. It is a group of conducting bars arranged parallel to the rotating shaft to make switching contact with a set of stationary carbon brushes. These bars are connected to the armature conductors. The whole arrangement is constructed in proper synchronization with the magnetic field so that, as the armature rotates, the commutator performs the function of mechanical rectification.

An alternator power source is very similar, except that generally the magnetic field coils are wound on the rotor, and the heavy welding current winding is wound into the stator. These machines are also called revolving or rotating field machines. Placing the heavy conductors in the stator eliminates the need for carbon brushes and a commutator to carry high current. The output, however, is ac, which requires external rectification for dc application. Rectification is usually done with a bridge, using silicon diodes. An alternator usually has brushes and slip rings to provide the low dc power to the field coils. It is not usual practice in alternators to feed back a portion of the welding current to the field circuit. Both single- and three-phase alternators are available to supply ac to the necessary rectifier system. The dc characteristics are similar to those of single- and three-phase transformer-rectifier units.

An alternator or generator may be either self-excited or separately excited, depending on the source of the field power. Either may utilize a small auxiliary alternator or generator, with the rotor on the same shaft as the main rotor, to provide exciting power. On many engine-driven units, a portion of exciter field power is available to operate tools or lights necessary to the welding operation. In the case of a generator, this auxiliary power is usually 115 V dc. With an alternator type power source, 120 or 120/240 V ac is usually available. Voltage frequency depends on the engine speed.

Output Characteristics

Both generator and alternator type power sources generally provide welding current adjustment in broad steps called ranges. A rheostat or other control is usually placed in the field circuit to adjust the internal magnetic field strength for fine adjustment of power output. The fine adjustment, because it regulates the strength of the magnetic field, will vary the open-circuit voltage. When adjusted near the bottom of the range, the open-circuit voltage will normally be substantially lower than at the high end of the range, as shown in Fig. 1.18.

Figure 1.18 shows a family of volt-ampere curve characteristics of either an alternator or generator type power supply. With many alternator power supplies, the broad ranges are ob-

tained from taps on a reactor in the ac portion of the circuit. As such, the basic machine does not often have the dynamic response required for shielded metal arc welding. Thus, a suitable inductor is generally inserted in series connection in one leg of the dc output from the rectifier. Welding generators do not normally require an inductor.

There is a limited range of overlap normally associated with rotating equipment where the desired welding current can be obtained over a range of open-circuit voltages. If welding is done in this area, the welder has the opportunity to better tailor his power supply to the job. With lower open-circuit voltage, the slope of the curve is lower. Then the welder can regulate the welding current to some degree by varying the arc length. This can assist him in weld pool control, particularly for out-of-position work.

Some welding generators carry this feature beyond the limited steps illustrated above. Generators that are compound wound with separate and continuous current and voltage controls can provide the operator with a selection of volt-ampere curves at nearly any amperage capability within the total range of the machine.

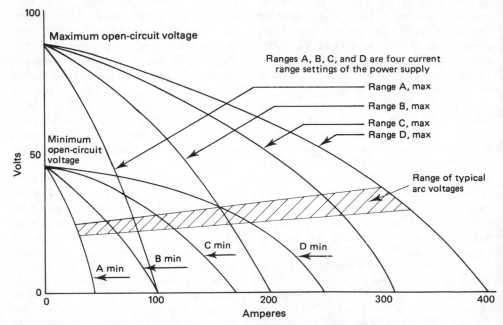

Fig. 1.18—Volt-ampere relationship for a typical constant-current rotating type power source

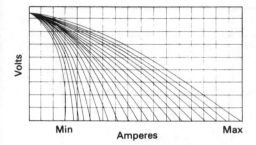

Fig. 1.19—Effect of current control variations on generator output

Thus, the welder can set the desired arc voltage with one control and the arc current with another. This adjusts the generator power source to provide a static volt-ampere characteristic that can be "tailored" to the job throughout most of its range. The volt-ampere curves that result when each control is changed independently are shown in Figs. 1.19 and 1.20.

Welding power sources are available that produce both constant current and constant voltage. These units are used for field applications where both are needed at the job site and utility power is not available.

Sources of Mechanical Power

Generators are available with ac drive motors of several voltage and frequency ratings and also with dc motors. Welding generators are usually single units with the drive motor and generator assembled on the same shaft.

Induction motor-driven welding generators are normally available for 200, 230, 460, and 575 V three-phase, 60 Hz input. Other standard input requirements are 220, 380, and 440, 50 Hz. Few are made with single-phase motors, since transformer type welding power supplies usually fill the need for single-phase operation. The most commonly used driving motor is the 230/460 V, three-phase, 60 Hz induction motor.

Typical curves for overall efficiency, power factor, and current input of a 230/460 volt, three-phase, 60 Hz induction motor-generator set are shown in Fig. 1.21. The motors of dc welding generators usually have a good power factor (80 to 90 percent) when under load and from 30 to 40 percent lagging power factor at no load. No-load power input ranges between 2 and 5 kW,

depending upon the rating of the motor-generator set. The power factor of induction motor-driven welding generators may be improved by the use of static capacitors similar to those used on welding transformers. Welding generators have been built with synchronous motor drives in order to correct the low power factor.

Rotating type power supplies are valuable for field erection work when no electric power is available. For this use, a wide variety of internal combustion engines is available. Welding generators are produced without an electric motor or engine drive. They are equipped with a shaft extension for a belt or a direct connection to any suitable source of mechanical power.

Parallel Operation

Although increased current capacity can be obtained by connecting welding generators in parallel, parallel connection is not advised unless the manufacturer's specific instructions are followed. Such caution is necessary because successful paralleling depends upon matching the output voltage, output setting, and polarity of each machine. In the case of self-excited generators, the problem is further complicated by the necessity to equalize the excitation between the generators.

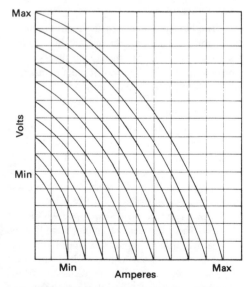

Fig. 1.20—Effect of voltage control variations on generator output

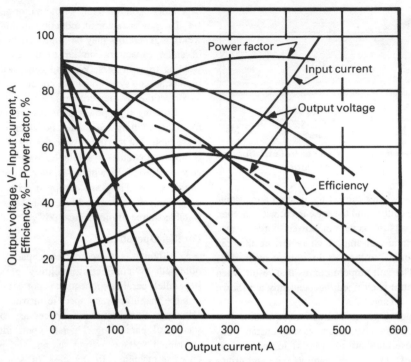

Fig. 1.21—Typical characteristic curves of a 300 ampere dc motor-generator power supply

The blocking nature of the rectifiers make dc alternator units easy to operate in parallel. Care should be taken to make sure that connections are the same polarity. All units parallelled must be set to deliver equal outputs.

Auxiliary Features

Rotating type power sources are available with many auxiliary features. Units may be equipped with a remote control attachment. It may be either a hand control or a foot control that the operator takes to the work station to make power source adjustments while welding.

Gas engines are often equipped with idling devices. These devices are automatic in that the engine will run at some idle speed until the electrode is contacted to the work. Under idle, the open-circuit voltage of the generator is low. Touching the electrode to the work energizes a sensing circuit that automatically accelerates the engine to operating speed. When the arc is broken, the engine will return to idle after a set time.

Engine-driven generators are often equipped with a provision for auxiliary electric power. This power is available at all times on some units and only during idle time on other units. Other auxiliary features that can often be obtained on engine-driven welding machines are polarity switches, to easily change from straight to reverse polarity, high frequency arc starters, tachometers, and output meters.

Applications

Units are available that can be applied to any of the arc welding processes requiring a constant dc power source. In addition to shielded metal arc welding, these power sources are normally used for gas tungsten arc, gas metal arc, flux cored arc, and submerged arc welding; arc surfacing; and air carbon arc gouging.

DIRECT CURRENT POWER SUPPLIES, CONSTANT VOLTAGE

Constant voltage (potential) power sources are either rotating or transformer-rectifier types. Most rotating (generator) power sources are combination units that will produce both constant current and constant voltage power. Transformer-rectifier types are designed to produce constant voltage only.

Transformer-rectifier type constant voltage power sources are normally three-phase units. However, small single-phase units, usually rated 200 A or below, are available for light duty applications.

Generators that can supply constant voltage welding power are normally the separately excited, modified compound wound type. The compounding of constant voltage units, which is different than that of constant current units, produces flat volt-ampere output characteristics. These machines may have solid state devices in the excitation circuit to optimize performance and to provide convenient facilities for remote control.

Alternator units are also available that have characteristics of both the transformer and the generator types. Often these are three-phase rectifier units that are combination constant voltage/constant current power sources.

ELECTRICAL CHARACTERISTICS

Constant potential power sources are characterized by their typically flat volt-ampere curve. A negative slope of 1 or 2 V per 100 A is not uncommon. This means that the maximum short circuiting current is usually very high, often in the range of thousands of amperes. Units with volt-ampere curves having slopes of up to 8 V per 100 A are still referred to as constant voltage power supplies (Fig. 1.7).

There are many varieties and combinations of constant voltage power sources. Often these power sources have an adjustment to tailor the slope of the volt-ampere curve to the welding process. A fixed slope is sometimes built into a power supply. Of prime importance are the dynamic characteristics of these power supplies.

Adjustment of the slope not only changes the static but also the dynamic characteristics of the machine. Adjustable inductors may be placed in the dc portion of the circuit to obtain separate control of the static and dynamic features. Theoretically, the presence of the dc inductor will not alter the static characteristics but will affect the dynamic characteristics. The dynamic characteristics are very important for short-circuiting transfer in gas metal arc welding.

GENERAL DESIGN

There are many designs of constant voltage type machines available. The advantage of any particular type is related to the application and to the expectations of the user.

The open-circuit voltage of some transformer-rectifier machines is adjusted by changing taps on the transformer. Another type of machine has its secondary coils wound in such fashion that carbon brushes, driven by a lead screw, slide along the secondary coil conductors. A second control is often provided to adjust the volt-ampere characteristics to meet the requirements of the welding process. Because of the effect on the volt-ampere output curve, this is called "slope control."

Slope control is generally obtained by changing taps on reactors in series with the ac portion of the circuit. Slope control may be provided by carbon brushes, attached to a lead screw, contacting the reactor turns. This variable reactor provides continuous adjustment of slope. Another method of control uses magnetic amplifiers or solid state devices to electrically regulate output voltage. These machines may have either voltage taps or slope taps in addition to electrical control. Some features of electrical control are the ease of adjustment, the ability to use remote control, and the absence of moving parts. Also, some electrically controlled machines permit adjustment of output during welding. This is helpful when "crater filling" or changing welding conditions. The combination of taps with electrical control to give fine output adjustment between taps is a suitable arrangement in a service application where the machine requires little attention during welding. Fully electrically controlled machines are easier to set up and readjust when welding requirements change rapidly.

Electrically controlled machines often do not have separate slope controls. A fixed, all-purpose slope is designed into them. However, machines designed for gas metal arc short circuiting transfer generally have some dc inductance to improve performance and to provide the dynamic characteristics required. Such inductance can be variable or fixed.

Single-phase power sources generally require some type of ripple filter arrangement in the welding circuit. Usually this filter is a bank of electrolytic capacitors across the rectifier output. The purpose is to provide a smooth dc output, capable of clearing a short circuit. An inductor is used to control the output of the capacitors. Without some inductance, the discharge of the capacitors through a short circuit would be much too violent for good welding operation.

Constant voltage power sources have a wide range of open-circuit voltages. Electrically controlled machines may have as high as 75 V open circuit. Tapped or adjustable transformer types have open-circuit voltage which may be varied from 30 to 50 V maximum to 10 V minimum. While there is no fixed rule for volt-ampere slope in the welding range, most machines have slopes of from 1 to 3 V per 100 A.

On constant voltage generators, slope control is usually provided by a tapped resistor in the welding circuit. This is desirable because of the inherent slow dynamic response of the generator to changing arc conditions. Resistance type slope control limits maximum short circuit current and increases the dynamic response rate. Reactor slope control will also limit maximum short circuiting current. However, it will slow the rate of response of the power source to changing arc conditions even more.

CONTROL DEVICES

Constant voltage power sources are usually equipped with voltmeters and ammeters as standard equipment. Most also have primary contactors with the control coil connected to a terminal strip or panel receptacle to interface with wire feed equipment. Electrically controlled models usually have remote control capabilities. Other features available on certain machines are line voltage compensation and also accessories to interface with electrode wire feed equipment to change both feed rate and welding current.

ELECTRICAL RATING

Primary ratings are similar to those discussed earlier. Constant voltage machines usually have a favorable power factor and do not require power factor correction. Open-circuit voltage, while subject to NEMA specification, is usually well below the established maximum. Current ratings of NEMA Class I machines range from 200 to 1500 A.

Constant voltage power sources are normally classified as NEMA Class I or Class II. It is the usual practice to rate them at 100 percent duty cycle, except for some of the light duty units of 200 A and under.

APPLICATIONS

These machines are normally manufactured for all welding processes employing a continuously fed electrode such as gas metal arc welding, flux cored arc welding, submerged arc welding with small diameter electrodes, and electroslag welding. They can be used for welding most metals in a wide range of thicknesses.

Units are available with self-contained electrode reel storage and electrode feed mechanisms. These units are designed for ease of operation. They have proved useful for short circuiting transfer on light gage metals in small shops, auto repairing, light manufacturing, and ornamental iron working.

SPECIAL POWER SOURCES

MULTIPLE-OPERATOR WELDING

Multiple-operator welding equipment is economical where there are a number of welding stations in a small area. Multiple-operator welding equipment is used to advantage in shipbuilding and construction, for example, with the shielded metal arc welding process.

Multiple-operator installations (MO systems) are supplied from either motor-generator sets, transformer-rectifier type power sources, or from transformers. Commercially available units vary from 500 to 2500 A for motor-generator units, from 500 to 1500 A for transformer-rectifier type installations, and from 500 to 2000 A for transformers. Overload devices and circuit breakers protect the equipment from damage. The usual practice is to provide a constant potential power source voltage of 70 to 80 V with provisions for paralleling two or more units for combined current output. The manufacturer's instructions should be followed to assure proper parallel operation.

Large copper bus bars are run from the power source to welding centers. Lines are then connected to welding outlet panels. Sometimes, individual panels are installed for each welding operator. As many as ten circuits may be grouped in one panel. Each circuit is basically a variable resistor for direct current or a reactor for alternating current that is connected in series with the electrode holder.

Individual Modules

One type of multiple-operator power source consists of a bank of individual power modules that, housed in a common cabinet, provide remotely controlled dc welding power to individual stations at distances up to 60 m (200 ft) from the main unit. Such a welding machine consists of separate modules powered by one three-phase power transformer. Where the output of an individual module is not sufficient for a particular welding job, two or more modules may be paralleled. Common work connection of all modules is provided. Each module consists of a dc control coil, ac control coil, rectifier stack, control rectifier, current control rheostat, inductor, and thermal protection thermostat. Each individual welder can use either polarity that he requires. Since each module is isolated, it can be individually controlled.

Adjustable Resistor Banks

A second type of power source is a constant voltage transformer-rectifier unit or motor-generator set providing 75 to 80 V to a group of adjustable resistor banks often called grids. Three-phase full-wave bridge rectifiers are used to change the output of transformer types from ac to dc welding power.

Design Concept

The basic idea behind multiple-operator equipment is the fact that the actual operating duty cycle of a manual welding machine may be quite low (20 to 25 percent). A person doing manual shielded metal arc welding must change electrodes frequently, check fit-up, change position, and chip slag. Thus, his actual arc time will be relatively short in relation to his total work period. Multiple-operator equipment attempts to utilize this low duty cycle to optimize equipment loading. For example, in construction work when it is desirable to have many welding operations, minimum welding equipment cost, and maximum flexibility and portability, multiple-operator systems are often used.

It is unlikely that all welding will be going on simultaneously. Welders perform at different operating speeds. The average covered electrode will be consumed in a few minutes. Consequently, only a relatively few welders will be welding at any given time. For example, if the average arc current required is 160 A and the average ten minute duty cycle of a welding machine is 25 percent, the average current per welding machine is 160 × 0.25 or 40 A. Thus, in any ten minute period an 800 A power source could supply an average 40 A to about 20 welding machines on the job site.

Features

In large installations, the use of multiple-operator power supplies usually results in a reduction in capital equipment and installation costs. Maintenance costs are also reduced since only one power supply must be maintained in place of many. The individual resistor bank or reactor panel can usually be located close to the welder to enable him to make current adjustments conveniently.

SUBMERGED ARC WELDING

The power source used for submerged arc welding (SAW) may be constant current dc, dc constant voltage, or ac. Direct current power supplies may be either motor-generators or transformer-rectifiers. Constant current power sources are used with arc voltage control of electrode feed rate. If electrode feed speed is used to regulate welding current, a constant voltage power source is used to provide the required arc length. Submerged arc welding generally is done at high currents (350 to 1200 A) so the power source must have a high current rating at high duty cycle.

A standard NEMA-rated transformer-rectifier type or motor-generator type dc power source can be used for submerged arc welding if the unit is rated adequately for the application. Power supplies may be paralleled according to the manufacturer's instructions to obtain the necessary welding current capacity. Duplex units are available, consisting of two single-operator units assembled and connected for single or parallel operation. The use of duplex machines or single units with high current rating is preferred over the use of standard power sources connected in parallel.

Constant-voltage dc power sources used for submerged arc welding should have an open-circuit voltage in the 50 volt range and an adequate current rating for the application. The welding current is automatically controlled by the feed rate of the electrode wire. One of the advantages of this method is the simple control system used. This system provides a uniformly stable arc voltage, which is of particular advantage for high-speed, light-gage welding, and also more consistant arc starting because of the high initial current surge.

With motor-generator power sources, the high power requirements of some submerged arc applications cause high loading of the drive motor. At a given current, the input to the generator is roughly proportional to its load voltage. Care should be taken to select units having adequate motor power rating. This precaution also applies to some transformer-rectifier type machines when actual arc voltage exceeds its rated output voltage.

The flow of welding current from a generator may be started and stopped by a magnetic contactor in the welding circuit or by a relay in the generator field circuit, depending upon its design and characteristics. Transformer type power sources control the flow of current by means of a contactor in the primary line of the machine.

Magnetic deflection of the arc (arc blow), a characteristic of direct current welding, usually limits the magnitude of dc which can be used in submerged arc welding. Alternating current minimizes arc blow. Reference should be made to Chapter 6, "Submerged Arc Welding," (p. 189), for additional details.

Transformers with ratings up to and including 2000 A are available with special features for adapting them to submerged arc welding application. Open-circuit voltage should be at least 80 V, but preferably 85 to 100.

Welding current for multiple arc welding systems may be supplied in a number of different ways. For parallel arc welding, a power supply may be connected in the conventional single-electrode manner. The two or more welding electrodes may be fed by a single-drive head through a common contact nozzle or jaw. A single transformer or dc power source may be used for supplying two independent electrodes feeding into the same weld pool. Each output terminal of the power supply is connected to one of the welding heads. This is called a series arc system. It requires a power supply with a high open-circuit voltage.

High-speed tandem welding generally utilizes two independent welding heads. They are supplied by multiple transformer units connected to a three-phase line using either a closed delta connection or a Scott connection. Since heavy currents are frequently used in tandem

Three-phase ac

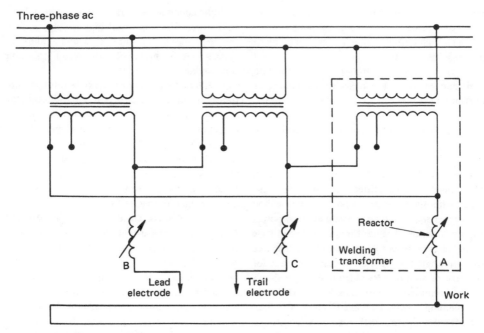

Fig. 1.22—Arc welding transformer connections for three-phase delta system

welding, these systems distribute the power load on the three phases.

The closed delta system requires the use of three transformers with separate current control reactors. The transformer secondaries are connected in closed delta ahead of the re-actors, as shown in Fig. 1.22. This system pro-vides for the adjustment of welding currents in the two arcs, the ground current, and the phase-angle displacement between the three currents. Adjustment of these conditions is important to obtain desired arc deflection (magnetically), weld penetration, and weld contour. Arc cur-rents cannot be independently adjusted. A change in one will change the other two be-cause of the phase angles.

The Scott-connected system uses two trans-formers with the primaries and secondaries connected, as shown in Fig. 1.23. These should be specifically designed. Standard units can be used if at least one of them has a center tap connection in the primary. The units should have 85 to 100 V open circuit. This system overcomes the interrelated current adjustment inherent in the closed delta system. It gives independent control of arc currents.

STUD ARC WELDING

Stud arc welding (SW) must be done with a dc power source. The process requires higher capacity, better consistency of operation, and better dynamic current control than is normally available with conventional power sources. Therefore, special power sources are usually

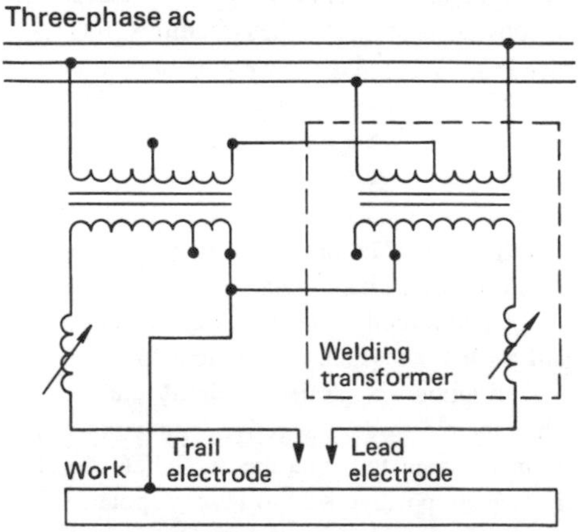

Three-phase ac

Fig. 1.23—Arc welding transformer connections for Scott-connected system

designed for SW. The general characteristics desired in a stud arc welding power source are

(1) High terminal voltage in the range of 70 to 100 V open circuit

(2) A drooping volt-ampere characteristic such that 25 to 35 V appears across the arc at maximum load

(3) A rapid rate of current rise

(4) High current output for a relatively low duty cycle

Each of the various types of special power sources available has its own characteristics. Consequently, it is difficult to compare these types with other sources of power. One method of comparison would be to evaluate each power supply in terms of current output and stud base diameter. The power source should be capable of delivering the required welding current for the size of stud to be welded. Table 1.6 shows the maximum steel stud sizes for NEMA-rated arc welding power sources. Chapter 8, "Stud Arc Welding," (p. 261), contains additional information on power supplies for the process.

GAS TUNGSTEN ARC WELDING

Almost any constant-current type power source, either ac or dc, may be used with the gas tungsten arc welding (GTAW) process. However, power sources that are designed for this particular process are preferred. They may be as simple as a mechanically controlled ac unit with a built-in high frequency unit for arc stabilization or as sophisticated as a three-phase, dc power source that has facilities for completely programming the welding operation.

The choice of welding current, either ac or dc, depends on the type of metal to be welded, the type of shielding gas used, the welding techniques, etc. Chapter 3, "Gas Tungsten Arc Welding," (p. 77), describes the type of current used for various base metals.

Open-circuit voltage is usually between 70 and 80 V rms, which is sufficient for consistent arc initiation on straight polarity half cycles when the electrode is negative. This voltage may be insufficient to ignite the arc when the electrode goes positive on the reverse polarity half cycles. The result is a very unstable, erratic arc unless a suitable high voltage is imposed on the

welding circuit at the start of each reverse polarity half cycle. The impressed high voltage ionizes the inert gas in the arc region and ignites the reverse polarity half cycle. The result is a stable arc condition.

Alternating current must pass through zero before it can reverse direction. At the instant the arc is struck, the current begins flowing in one direction. As the current reverses direction, no current flows for an instant, causing the arc to go out. Then, depending upon the electrical characteristics of the system and the particular arc conditions, the arc will either reignite or remain extinguished for that half cycle.

Normally, the current is unbalanced in the GTAW process. Current flows more readily in one direction than in the other. This is because the hot tungsten electrode can emit electrons better than the molten weld pool. This difference in current flow is significant with some metals, namely aluminum and magnesium. During current reversals involving change from electrode negative to electrode positive, voltages as high as 150 V rms may be required for reliable arc reignition. The exact voltage differs for different metals and varies with surface condition, welding current, type of shielding gas, and electrode type.

Figure 1.24 shows the following electrical conditions:

(A) Complete rectification of several half cycles

(B) Normal unbalanced current

(C) Balanced current

Rectification occurs when the arc does not ignite on the reverse polarity half cycle (Fig. 1.24A). Normal unbalance occurs when the arc ignites but the power source does not have ade-

Table 1.6 — Maximum steel stud size for typical stud arc welding power supply

Power supply	Stud weld base diameter, max	
	mm	in.
400 A dc NEMA-rated unit	11	7/16
600 A dc NEMA-rated unit	13	1/2
Two 400 A units in parallel	16	5/8
Two 600 A units in parallel	19	3/4

quate voltage to overcome the additional arc resistance to current flow in the reverse polarity half cycles (Fig. 1.24B). Balanced current occurs when the power source has been modified to overcome the additional resistance to current flow (Fig. 1.24C). Complete rectification is not usable because the arc is unstable and cleaning action is lost. Often the current will exhibit intermittent complete rectification. This can be used for noncritical welding. Stable current unbalance (Fig. 1.24B) is very useful and good quality welds can be made with it. Complete current balance provides the best cleaning action. It will produce excellent welds.

Several methods have been developed to provide the voltage necessary to ignite the reverse polarity half cycles and stabilize the arc. The most common methods are

(1) A superimposed high-frequency voltage of 3000 to 5000 V (very low current flow).

(2) A welding transformer having a relatively high open-circuit voltage of 150 to 200 V rms (requires protection from operator contact).

(3) Surge voltage where a 200 to 400 V capacitor charge is injected into the welding circuit at the start of each reverse polarity half cycle.

(4) A "ringing" circuit connected across the power supply terminals. It is a resonant circuit, consisting of a capacitor and an inductance, that oscillates when the arc extinguishes. It creates a very steep high-voltage wave front to reignite the arc.

These systems tend to produce stable arcs. One more factor is the time of passage of arc current through zero. A power source with a more rapid reversal of current will inherently be more stable than a slower one with a normal sine wave. The rapid current reversal assists arc reignition during cycle reversal. More rapid reversal may be achieved by "wave shaping" or a frequency increase.

Balanced current flow is usually achieved by the use of series-connected condensers in the welding circuit. An alternate method is to place batteries in the circuit in such a way that their voltage will be additive to the reverse polarity half cycle. Balanced wave current flow will produce the following:

(1) Best oxide cleaning action

(2) A more stable arc and uniform heating

(3) Minimum heating within the welding transformer

The use of balanced current will require

(1) Larger tungsten electrodes (rms current is higher than when the ac wave is unbalanced)

(2) Higher open-circuit voltages generally associated with some wave-balancing methods

(3) Additional cost of a wave-balancing system

Although desirable for some applications, balanced wave is not essential for most manual welding operations. It is, however, desirable for many mechanized welding uses.

Special machines are available for ac GTAW that provide a current balance control. With this control, a stable welding condition can be produced at a selectable condition of balance or unbalance.

Direct current power sources designed for GTAW have an open-circuit voltage between 70 and 80. The current rating of the power source depends on the application. Current ratings up to 600 A are available. These machines are usually equipped with a built-in system for arc initiation, gas and water control valves, and a welding current control circuit. They may also have remote foot or hand controls.

The welding of thin structural members and thin wall tubing led to the development of electronically controlled power supplies for auto-

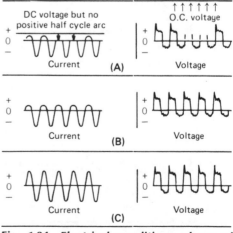

Fig. 1.24—*Electrical conditions observed with ac gas tungsten arc welding*

matic gas tungsten arc welding. These machines have nearly vertical volt-ampere curves in the usual welding region. This insures that changes in arc length (voltage) will not cause changes in the arc current.

Many dc GTAW power sources have current vs time programming facilities. They can start with a low welding current value, increasing to a higher current value for producing the weld, and then decreasing to a final current value. They are especially useful in pipe and tube welding where the weld overlaps at start and finish. The change in current with time is referred to as upslope and downslope, as shown in Fig. 1.25.

Some dc GTAW power sources have the ability to pulse the welding current from one controllable current level to another controllable current level, as shown in Fig. 1.25. Pulsed GTAW is characterized by a repetitive variation in arc current from a peak (high) value to a background (low) value. Both peak and background current levels are adjustable in a wide range. Peak current time plus background current time represent the duration of one pulse cycle. The peak current duration and background current duration are adjustable. Typical pulse cycle frequency will range from ten pulses per second to one pulse in two seconds, depending on the application of the power supply.

The purpose of pulsing is to alternately heat and cool the molten weld metal. The heating cycle (peak current) is based on achieving a suitable molten weld pool size during the peak pulse without excessive side wall fusion or melt-thru, depending on the joint being welded. Background current and duration are determined by the desired cooling of the weld pool. The purpose of the cooling (background current) portion of the cycle is to speed up the solidification rate and reduce the size of the molten pool. Thus, pulsing is alternately increasing and decreasing the size of the molten pool.

The fluctuation in molten pool size and penetration is related to the pulsing variables, such as arc travel speed; the type, thickness, and mass of the base metal; filler metal size; position of welding; and method of welding (manual, automatic). Because the size of the molten pool is partially controlled by the current pulsing action, the need for arc manipulation to control the molten pool is reduced or eliminated. Thus, pulsed current is a useful tool for manual out-of-position GTAW, such as in-place pipe joints, and also for automatic butt welding of thin wall tubing.

ELECTROSLAG AND ELECTROGAS WELDING

The equipment used for electroslag and electrogas welding is very similar to that required for submerged arc or flux cored arc welding. The same power sources can be used for either process, with one exception: ac power supplies are not used with the electrogas process. Both ac and dc power supplies are used with the electroslag process. Standard power sources used for either process should have an open-circuit voltage up to 80 V and be capable of delivering 600 A continuously (100 percent duty cycle). The power supplies should be equipped with remote controls. The number of power supplies required depends on the number of

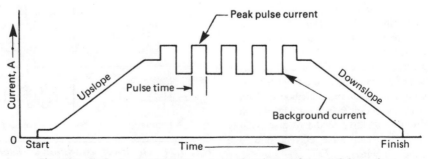

Fig. 1.25—Typical pulsed GTAW program showing upslope and downslope

welding electrodes being used to fill the joint. One power supply is required for each welding electrode.

Special constant-voltage dc power supplies designed for electroslag and electrogas welding are available. Typical power supplies are transformer-rectifiers having 74 V open circuit and rated at 750 A at 50 V output, 100 percent duty cycle. The primary input is 60 Hz, three-phase, 230/460 V.

PLASMA ARC WELDING

Conventional direct current power supplies with drooping volt-ampere characteristics and 70 V open circuit are suitable for most plasma arc welding (PAW) applications where argon or mixtures of argon and hydrogen are used. However, if helium or an argon-hydrogen mixture containing more than five percent hydrogen is used, higher open-circuit voltage is required for reliable arc ignition. Specially designed units are available with adequate open-circuit voltage for good arc stability.

Higher open-circuit voltage may be obtained by connecting two power supplies in series. An alternate approach, requiring the use of only one power supply, is to inititate the plasma arc in pure argon and then switch to the desired argon-hydrogen or helium mixture for the welding operation. Plasma arcs can be started by any one of several means, including high frequency units, impulse systems, pilot arc, etc.

The power supplies may be either dc trans-former-rectifiers or dc motor-generators. Transformer-rectifiers are preferred.

Welding currents used for plasma arc welding range from less than one to over 500 A. Power sources are normally rated no more than 500 A. Low current power sources (0.1 to 10 A) are usually combined with the plasma arc controls in a single cabinet.

Because of the relative insensitivity of the plasma arc process to arc length variations, arc voltage control equipment is not normally used. However, arc voltage control can be used with the transferred PAW process for applications such as welding contoured joints. The voltage control should be inoperative when current or gas upslope and downslope are used.

Specially designed power sources are available that meet the requirements of plasma arc surfacing and cutting. They may be used for plasma arc welding, but they are not generally suitable for other welding processes. The machines are usually rated between 500 to 1000 A at 100 percent duty cycle with open-circuit voltages of up to 400 V. Many have reconnection capability to reduce the voltage and increase the current by a factor of two.

PULSED GAS METAL ARC WELDING

The pulsed gas metal arc welding process is used with electrodes and shielding gases that normally operate only in the high current density or spray-transfer region. The process uses a lower

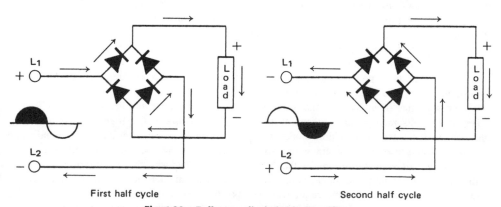

First half cycle   Second half cycle

Fig. 1.26—Full wave diode bridge rectifier

average welding current than conventional spray transfer. The pulsed power controls the size and frequency of the metal droplets and the rate at which they are transferred across the arc. This permits the use of lower average welding currents for a given electrode size.

A typical pulsed arc welding machine normally consists of a three-phase dc transformer-rectifier in parallel with a single-phase half-wave rectifier. The three-phase unit provides background current and the single-phase unit provides peak current. Both transformer-rectifiers are mounted in a single configuration with appro-priate controls for individual adjustment of background and peak currents.

The peak current is selected to accommodate the electrode size and feed rate. The current is set just above the spray transfer threshold amper-age for the specific type and diameter of electrode used. Background current is set in the globular transfer range. Spray type transfer occurs during the peak current time. Globular transfer does not have time to occur at background current level. Thus, filler metal deposition rates are between those for continuous spray transfer and globu-lar transfer.

SOLID STATE CIRCUITRY

Solid state derives its name from solid state physics—the science of the crystalline solid. Technology has developed methods of treating certain crystals, such as selenium, germanium, and silicon, to enhance their electrical prop-erties. The most important of these materials is silicon.

Transformer-rectifier or alternator-rectifier power sources rely on rectifiers to convert ac to dc. The early models of these welding ma-chines used selenium rectifiers. Because of econ-omy, reliability, and efficiency, most rectifiers are now made of silicon.

A single rectifying element is called a diode. A diode is an electrical one-way valve. When placed in an electric circuit, it will allow current to pass in one direction only. The action is such that current flows only when the anode is posi-tive with respect to the cathode. Using an arrangement of diodes, it is possible to convert ac to dc. Fig. 1.26 shows a four-diode rectifier converting single-phase ac to dc.

In rectifier type power sources, current adjustment should be done by some means as-sociated with the input side of the rectifier. There is always some resistance to current flow through the device. In most welding diodes, this resistance results in a drop of several volts across each device. This voltage drop means that heat will be generated within the diode. Unless the heat is dissipated, the diode temperature will in-crease until failure occurs. Diodes are normally mounted on heat sinks (aluminum plates) to remove heat.

Diodes have limits on how much voltage they can block in the reverse direction (anode negative and cathode positive). This determines the voltage rating of the device. Welding power source diodes are usually selected with a blocking rating at least twice the open-circuit voltage to provide a safe operating margin. A diode can accommodate current peaks well beyond their normal steady state rating, but a high reverse voltage of even a short duration can cause failure of the device. Most rectifier power sources have a resistor, capacitor, or other electronic device to suppress voltage transients that could damage the rectifiers.

Solid state devices with special character-istics are used to control welding power directly. The welding current or voltage wave form is controlled by the solid state devices rather than by saturable reactors, moving shunts, moving coils, etc. One of the most important of these devices is the silicon controlled rectifier, called a thyristor.

The silicon controlled rectifier (SCR) is a diode variation with a trigger called a gate, as shown in Fig. 1.27. An SCR is nonconducting until an electrical signal is applied to the gate. When this happens, the device turns into a diode, and it will conduct current as long as the anode is positive with respect to the cathode. The device cannot control the current once it conducts. It

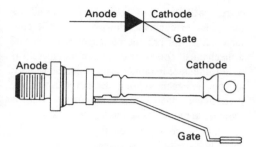

Fig. 1.27—Silicon controlled rectifier

cannot be turned off by a signal to the gate. If the current stops or the anode becomes negative with respect to the cathode, conduction will stop. Conduction will not take place again until another gate signal is received. The gate signal must be positive with respect to the cathode.

Using this feature of the SCR, a welding power source can be designed that derives its control from the ability of a gate signal to selectively turn on the SCR. A single-phase circuit is shown in Fig. 1.28.

In Fig. 1.28, if point B is positive with respect to point E, no current will flow until such time that both SCR 1 and SCR 3 receive a gate signal to turn on. At that instant, current will flow through the load. At the end of the half cycle, when the polarity of B and E reverse, a reverse voltage will be impressed across SCR 1 and SCR 3 and they will turn off. A gate signal now applied to SCR 2 and SCR 4 by the control will cause these two to conduct, again applying power to the

load circuit. To adjust the amount of power in the load, it is necessary to precisely time where, in any given half cycle, conduction is to initiate. If high power is required, conduction must start early in the half cycle. If low power is required, conduction is delayed until late in a half cycle. This is known as phase control. The effect is shown in Fig. 1.29. The power supplied in pulses to the load is proportional to the shaded area under the wave form envelope.

Figure 1.29 shows that significant intervals may exist when there is no power supplied to the load. This will cause arc outages, especially at low power levels. Therefore, wave filtering is required. Figure 1.28 shows a large inductance, Z, in the load circuit. For a single-phase circuit to operate over a significant range of control, Z must be very large. If, however, SCRs are used in a three-phase circuit, the nonconducting intervals would be reduced significantly. The Inductance (Z) would be sized accordingly. This is, in fact, characteristic of SCR controlled power supplies. A three-phase SCR system is a practical control means for welding power sources.

The timing must be accurately controlled. This is another function of the control shown in Fig. 1.28. The control must precisely signal each SCR when to conduct. To adapt the system satisfactorily to welding service, another feature is necessary. That feature is feedback. Depending upon the parameter to be controlled and the degree of control required, a corresponding feedback is necessary. If the machine is to have

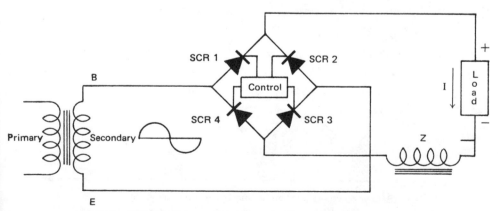

Fig. 1.28—Single-phase dc power source using an SCR bridge for control

constant voltage characteristics, the feedback must consist of some signal proportional to the arc voltage. This would tell the control the precise arc voltage at any instant so that the control can properly time and sequence the initiation of the SCR to hold the preset voltage. The same is true of constant current.

There are a number of SCR controlled welding machines commercially available. These are largely three-phase machines of either the constant voltage or constant current type.

There are some distinct features of SCR phase-controlled power supplies. Because the welding characteristics are applied largely by electronics, many capabilities can be added to the power supply. Automatic line voltage compensation is very easily accomplished. If the power line voltage varies, welding power can be held precisely as set. Volt-ampere curves can be shaped and tailored for a particular welding process and its application. These machines can have the ability to adapt their static characteristic to any welding process from one approaching a truly constant current to one approaching truly constant voltage. Other capabilities are pulsing, controlled starting, controlled current with respect to arc voltage, controlled arc voltage with respect to current, and a high initial current or voltage pulse. However, all of these features may cause some sacrifice of dynamic performance.

Another feature of SCR units is the ability to use an SCR as a secondary contactor. Welding current will flow only when the control allows the SCRs to conduct. However, an SCR contactor does not provide electrical isolation that a mechanical contactor or switch provides. It is a useful feature in rapid cycling operations, such as spot and tack welding. However. a primary mechanical contactor, qr some other device, is required to provide isolation for electrical safety.

There are several configurations of SCRs that can be used for arc welding. Figure 1.30 shows a three-phase bridge with six SCR devices. This arrangement produces a 360 Hz ripple frequency under load. This yields precise control and quick response because each half cycle of the three-phase input has separate control. Dynamic response is enhanced because it reduces the size of the inductor which is used to smooth out the welding current. At full output, the dynamic performance approaches that of mechanically adjusted power supply systems.

Figure 1.31 shows a three-phase bridge rectifier with three diodes and three SCRs. This configuration has a slower dynamic response. It requires a larger inductor than the six SCR unit because of greater current ripple. A fourth diode, called a "freewheeling diode," is used to recirculate the inductive currents from the inductor so that the SCRs will turn off, i.e. commutate. It has the advantage of economy over the six SCR unit because it uses fewer SCRs. It also uses a lower cost control unit.

The transistor is another solid state device

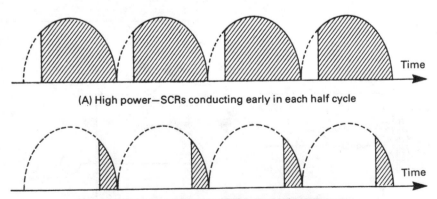

(A) High power—SCRs conducting early in each half cycle

(B) Low power—SCRs conducting late in each half cycle

Fig. 1.29—Phase control using an SCR bridge

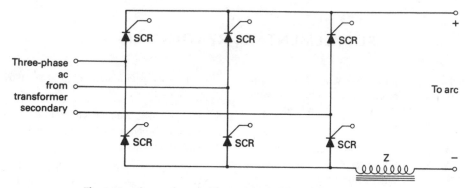

Fig. 1.30—Three-phase bridge using six SCRs (full wave control)

that is used in welding power supplies. These are limited to power supplies having precise control of a number of variables. The transistor has several basic differences from the SCR. One is that conduction through the device is proportional to the control signal applied. With no signal, there is no conduction. When a small signal is applied, there is a corresponding small conduction; with a large signal, there is a correspondingly large conduction. Unlike the SCR, the control can turn off the device without waiting for polarity reversal or an "off" time.

There are several ways to apply transistors to welding machines. The most common way is to use transistors in their active region. This means that the transistors are operated at a level between off and full on (saturation). In this mode, the transistor functions as an electronically controlled series resistance. Transistors have relatively low operating temperature restrictions. If allowed to go above these temperature restrictions, they will either malfunction or fail altogether.

Transistor operated machines are normally used for welding applications that require precise and exact controls. The speed of operation of a transistor is very rapid, making it ideal for pulsed GTAW. Pulse repetition rates in the ultrasonic region are quite possible. However, conducting such current pulses to the arc at normal welding currents is very difficult, at best.

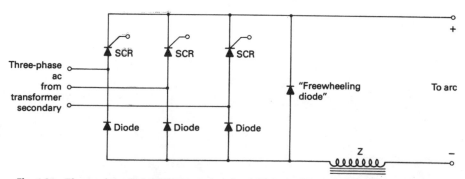

Fig. 1.31—Three-phase hybrid bridge using three SCRs and four diodes (half-wave control)

SUPPLEMENTARY READING LIST

Daggett, E.H., and Zircher, W.E. New developments in pulsed spray welding. *Welding Journal* 49 (10), Oct. 1970, pp. 780-787.

Hackman, R.L., and Manz, A.F. D.C. Welding Power Sources for Gas Shielded Metal Arc Welding. Bulletin 97. New York: Welding Research Council, July 1964.

Kuhr, J.N., and Aldenhoff, B.J., Development of a solid state constant potential arc welding power source using silicon controlled rectifiers. *Welding Journal* 44 (10), Oct. 1965, pp. 471-s—475-s.

Lawrence, B.D., and Jackson, C.E., Variable frequency gas-shielded pulsed-current arc welding. *Welding Journal* 48 (3), Mar. 1969, pp. 97-s—104-s.

Lesnewich, A. MIG Welding with Pulsed Power. Bulletin 170. New York: Welding Research Council, Feb. 1972.

Manz, A.F. *Welding Power Handbook*. New York: Union Carbide Corporation, 1973.

Pierre, E.R. *Welding Processes and Power Sources*. Appleton, Wis.: Power Publications Company, 1974.

Reinks, F., and Ashauer, R.C. Development and evaluation of a modulated power control for fusion welding. *Welding Journal* 50 (5), May 1971, pp. 222-s—230-s.

2
Shielded Metal Arc Welding

Prepared by a committee consisting of

J.C. FALLICK, *Chairman*
Sun Shipbuilding & Dry Dock Company

E.R. SZUMACHOWSKI
Teledyne McKay

R. SNYDER
Babcock & Wilcox Company

Shielded Metal Arc Welding

FUNDAMENTALS OF THE PROCESS

DEFINITION AND GENERAL DESCRIPTION

Shielded metal arc welding (SMAW) is an arc welding process in which coalescence of metals is produced by heat from an electric arc that is maintained between the tip of a covered electrode and the surface of the base metal in the joint being welded.

The core of the covered electrode consists of either a solid metal rod of drawn or cast material or one fabricated by encasing metal powders in a metallic sheath. The core rod conducts the electric current to the arc and provides filler metal for the joint. The covering shields the molten metal from the atmosphere as it is transferred across the arc and improves the smoothness or stability of the arc. The covering also has other functions, depending on the type of electrode, but these two are the primary functions of the covering on every electrode.

Arc shielding is obtained from the gases which form as a result of the decomposition of certain of the ingredients in the covering. The ingredients which provide the shielding vary according to the type of electrode. The shielding employed, along with other ingredients in the covering and the core wire, largely controls the mechanical properties, chemical composition, and metallurgical structure of the weld metal, as well as the arc characteristics of the electrode.

PRINCIPLES OF OPERATION

Shielded metal arc welding is by far the most widely used of the various arc welding processes. It employs the heat of the arc to melt the base metal and the tip of a consumable covered electrode. The electrode and the work are part of an electric circuit known as the welding circuit, as shown in Fig. 2.1. This circuit begins with the electric power source and includes the welding cables, an electrode holder, a ground clamp, the work, and an arc welding electrode. One of the two cables from the power source is attached to the work. The other is attached to the electrode holder.

Welding commences when an electric arc is struck between the tip of the electrode and the work. The intense heat of the arc melts the tip of the electrode and the surface of the work beneath the arc. Tiny globules of molten metal rapidly form on the tip of the electrode, then transfer through the arc stream into the molten weld pool.[1] In this manner, filler metal is deposited as the electrode is progressively consumed. The arc is moved over the work at an appropriate arc length and travel speed, melting and fusing a portion of the base metal and adding filler metal as the arc progresses. Since the arc is one of the

1. Metal transfer across the welding arc is described in Chapter 2, "Physics of Welding," *Welding Handbook*, Vol. 1, 7th ed., pp. 59-66.

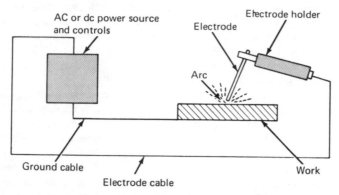

Fig. 2.1—Elements of a typical welding circuit for shielded metal arc welding

hottest of the commercial sources of heat (temperatures above 5000° C [9000° F] have been measured at its center), melting takes place almost instantaneously, as the arc contacts the metal. If welds are made in either the flat or the horizontal position, metal transfer is induced by the force of gravity, gas expansion, electric and electromagnetic forces, and surface tension. For welds in other positions, gravity works against the other forces.

The process requires sufficient electric current to melt both the electrode and a proper amount of base metal. It also requires an appropriate gap between the tip of the electrode and the base metal or the molten weld pool. These requirements are necessary to set the stage for coalescence. The sizes and types of electrodes for shielded metal arc welding define the arc voltage requirements (within the overall range of 16 to 40 V) and the amperage requirements (within the overall range of 20 to 550 A). The current may be either alternating or direct, but the power source must be able to control the level of current within a reasonable range in order to respond to the complex variables of the welding process itself.

Covered Electrodes

In addition to establishing the arc and supplying filler metal for the weld deposit, the electrode introduces other materials into or around the arc, or both. Depending upon the type of electrode being used, the covering performs one or more of the following functions:

(1) Provides a gas to shield the arc and prevent excessive atmospheric contamination of the molten filler metal as it travels across the arc.

(2) Provides scavengers, deoxidizers, and fluxing agents to cleanse the weld and prevent excessive grain growth in the weld metal.

(3) Establishes the electrical characteristics of the electrode.

(4) Provides a slag blanket to protect the hot weld metal from the air and enhance the mechanical properties, bead shape, and surface cleanliness of the weld metal.

(5) Provides a means of adding alloying elements to change the mechanical properties of the weld metal.

Functions 1 and 4 prevent the pickup of oxygen and nitrogen from the air by the molten filler metal in the arc stream and by the weld metal as it solidifies and cools.

The covering on shielded metal arc electrodes is applied by either the extrusion or the dipping process. Extrusion is much more widely used. The dipping process is used primarily for cast and some fabricated core rods. In either case, the covering contains most of the shielding, scavenging, and deoxidizing materials. Most SMAW electrodes have a solid metal core. Some are made with a fabricated or composite core consisting of metal powders encased in a metallic sheath. In this latter case, the purpose of some or even all of the metal powders is to produce an alloy weld deposit.

In addition to improving the mechanical properties of the weld metal, the covering on the electrode can be designed for welding with alternating current. With ac, the welding arc goes out and is reestablished each time the current reverses its direction. For good arc stability, it is necessary to have a gas in the arc stream that will remain ionized during each reversal of the current. This ionized gas makes possible the reignition of the arc. Gases that readily ionize are available from a variety of compounds, including those that contain potassium. It is the incorporation of these compounds in the electrode covering that enables the electrode to operate on ac.

To increase the deposition rate, the coverings of some carbon and low alloy steel electrodes contain iron powder. The iron powder is another source of metal available for deposition, in addition to that obtained from the core of the electrode. The presence of iron powder in the covering also makes more efficient use of the arc energy. Metal powders other than iron are frequently used to alter the mechanical properties of the weld metal.

The thick coverings on electrodes with relatively large amounts of iron powder increase the depth of the crucible at the tip of the electrode. This deep crucible helps to contain the heat of the arc and to maintain a constant arc length by using the "drag" technique. When iron or other metal powders are added in relatively large amounts, the deposition rate and welding speed usually increase.

Iron powder electrodes with thick coverings reduce the level of skill needed to weld. The tip of the electrode can be dragged along the surface of the work while maintaining a welding arc. For this reason, heavy iron powder electrodes frequently are called "drag electrodes." Deposition rates are high; but, because slag solidification is slow, these electrodes are not suitable for out-of-position use.

Arc Shielding

The arc shielding action, illustrated in Fig. 2.2, is essentially the same for the different types of electrodes, but the specific method of shielding and the volume of slag produced vary from type to type. The bulk of the covering materials in some electrodes is converted to gas by the heat of the arc, and only a small amount of slag is produced. This type of electrode depends largely upon a gaseous shield to prevent atmospheric contamination. Weld metal from such electrodes can be identified by the incomplete or light layer of slag which covers the bead.

For electrodes at the other extreme, the bulk of the covering is converted to slag by the heat of the arc, and only a small volume of shielding gas is produced. The tiny globules of metal being transferred across the arc are entirely coated with a thin film of molten slag. This molten slag floats to the surface of the weld puddle because it is lighter than the metal. The slag solidifies after the weld metal has solidified. Welds made with these electrodes are identified by the heavy slag deposits that completely cover the weld beads. Between these extremes is a wide variety of electrode types, each with a different combination of gas and slag shielding.

The variations in the amount of slag and gas shielding also influence the welding characteristics of the different types of covered electrodes. Electrodes that have a heavy slag carry high am-

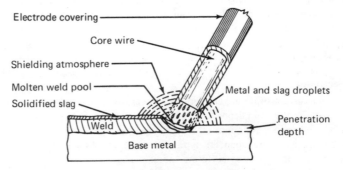

Electrode covering

Core wire

Shielding atmosphere

Molten weld pool

Solidified slag

Metal and slag droplets

Penetration depth

Weld

Base metal

Fig. 2.2—Shielded metal arc welding

perage and have high deposition rates. These electrodes are ideal for making large beads in the flat position. Electrodes that develop a gaseous arc shield and have a light layer of slag carry lower amperage and have lower deposition rates. These electrodes produce a smaller weld pool and, accordingly, are better suited to make welds in the vertical and overhead positions. Because of the differences in their welding characteristics, one type of covered electrode usually will be best suited for a given application.

PROCESS CAPABILITIES AND LIMITATIONS

Shielded metal arc welding is one of the most widely used processes, particularly for short welds in production, maintenance and repair, and field construction. The equipment is relatively simple, inexpensive, and portable. The filler metal, and the means of protecting it and the weld metal from harmful oxidation during welding, are provided by the covered electrode. Auxiliary gas shielding or granular flux is not required. The shielding characteristics of the electrodes make the process less sensitive to wind and draft than a gas shielded arc welding process.

The position of welding is not limited by the process but by the type and size of the electrode. The process is suitable for most of the commonly used metals and alloys. It can be used in areas of limited access. Shielded metal arc welding electrodes are available to weld carbon and low alloy steels; stainless steels; cast irons; aluminum, copper, and nickel and their alloys.

Low melting metals, such as lead, tin, and zinc, and their alloys, are not welded with SMAW because the intense heat of the arc is too high for them. Also, the "reactive" metals, such as titanium, zirconium, tantalum, and columbium, are not welded with covered electrodes. These metals are very sensitive to oxygen contamination and the shielding obtained with covered electrodes is not adequate for them.

Covered electrodes are produced in lengths of 230 to 460 mm (9 to 18 in.). As the arc is first struck, the current flows the entire length of the electrode. The amount of current that can be used, therefore, is limited by the electrical resistance of the core wire. Excessive amperage overheats the electrode and breaks down the covering. This, in turn, changes the arc characteristics and the shielding that is obtained. Because of this limitation, the deposition rates of covered electrodes are lower than those of a continuous electrode welding process where the electrode is not covered.

Because the electrodes are produced in straight lengths, they can be consumed only to some certain minimum length. When that length has been reached, the welder must interrupt his work to change electrodes. This causes the arc time and, hence, the overall deposition rate to be lower with covered electrodes than with a continuous electrode process. Also, the slag usually must be removed from the end of the bead before the welder can continue that bead with a new electrode. For many applications, the completed weld must be cleaned for subsequent processing or to prevent corrosion.

EQUIPMENT

POWER SOURCES

Type of Output Current

Either alternating current (ac) or direct current (dc) may be employed for shielded metal arc welding, if appropriately designed electrodes and power sources are used. The specific type of current employed will influence the performance of every covered electrode. Each of the two types has its features and its limitations, and these must be considered when the type of current to be used for any specific application is being selected. Factors which need to be considered are as follows:

(1) *Voltage Drop*. Voltage drop in the welding cables is lower with ac. This makes ac more suitable if the welding is to be done at long distances from the power supply. However, long cables which carry ac should not be coiled because the inductive losses encountered in such cases can be substantial.

(2) *Low Current*. With small diameter electrodes and low welding currents, dc provides better operating characteristics and a more stable arc.

(3) *Arc Starting*. Striking the arc is generally easier with dc, particularly if small diameter electrodes are used. With ac, the welding current passes through zero each half cycle, and this presents problems for arc starting and arc stability.

(4) *Arc Length*. Welding with a short arc length (low arc voltage) is easier with dc than with ac. This is an important consideration, except for the heavy iron powder electrodes. With those electrodes, the deep crucible formed by the heavy covering automatically maintains proper arc length when the electrode tip is dragged on the surface of the joint.

(5) *Arc Blow*. Alternating current rarely presents a problem with arc blow because the magnetic field is constantly reversing (120 times per second). Arc blow can be a significant problem with dc welding of steel because of unbalanced magnetic fields around the arc.[2]

2. The influence of magnetic fields on arcs and arc blow is discussed on pp. 73-74 and also in Chapter 2, "Physics of Welding," *Welding Handbook*, Vol. 1, 7th ed., pp. 56-59.

(6) *Welding Position*. Direct current is somewhat better than ac for vertical and overhead welds because lower amperage can be used. With suitable electrodes, however, satisfactory welds can be made in all positions with ac.

(7) *Metal Thickness*. Both sheet metal and heavy sections can be welded with a steady dc arc. The welding of sheet metal with ac is less desirable because of the problem of starting and maintaining the arc at the low currents that must be used.

A study of the type of welding that is to be done will generally tell whether alternating or direct current should be used. This, in turn, will result in the selection of either an ac or a dc power source or, if applications justify it, a combination ac-dc power source.[3] In any case, the power source must be a constant current type rather than a constant voltage type.

Significance of the Volt-Ampere Curve

Figure 2.3 shows typical volt-ampere output characteristics for both ac and dc power sources. It is difficult for a welder to hold an absolutely constant arc length with SMAW. For this reason, constant voltage power sources are not suitable for this process. Because of their flat volt-ampere curve, even a small change in arc length (voltage) produces a relatively large change in amperage. A constant current power source, on the other hand, produces only a small change in amperage with a change in arc length. Because of this, a constant current power source is preferred for manual welding. The steeper the slope of the volt-ampere curve (within the welding range), the smaller the change in current for a given change in arc voltage.

Constant current power permits maximum welding speed and highest quality welds for applications that involve large diameter electrodes and high welding currents. In such cases, a steep volt-ampere curve is desirable.

Where more precise control of the size of the molten pool is required (out-of-position

3. See Chapter 1.

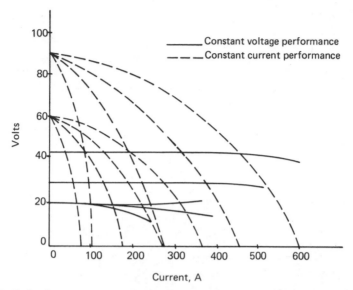

Fig. 2.3—Typical volt-ampere curves for constant current and constant voltage power sources

welds and root passes of joints with varying fit-up, for example), a flatter volt-ampere curve is desirable. This enables the welder to change the welding current within a specific range simply by changing arc length. In this manner, he has some control over the amount of filler metal that he deposits. Figure 2.4 portrays these different volt-ampere curves for a typical welding power source. Even though the difference in the slope of the various curves is substantial, the power source is still considered a constant current power source. The changes shown in the volt-ampere curve are accomplished by adjusting both the open-circuit voltage (OCV) and the current settings on the power source.

Open-Circuit Voltage

Open-circuit voltage, which is the voltage set on the power source, does not refer to arc voltage. Arc voltage is determined by arc length for any given electrode. Open-circuit voltage, on the other hand, is the voltage generated by the welding machine when no welding is being done. The arc voltage is the voltage between the electrode and the work during welding. Open-circuit voltages generally run between 50 and 100 V, whereas arc voltages are between 17 and 40 V. The open-circuit voltage drops to the arc voltage

when the arc is struck and the welding load comes on the machine. The arc length and the type of electrode being used determine just what this arc voltage will be. If the arc is lengthened, the arc voltage will increase and the welding current will decrease. The change in amperage which a change in arc length produces is determined by the slope of the volt-ampere curve within the welding range.

Some power sources do not provide for control of the open-circuit voltage because this control is not needed for all welding processes. It is a useful feature for SMAW, yet it is not necessary for all applications of the process.

Power Source Selection

Several factors need to be considered when a power source for SMAW is selected. The primary factors are

(1) The type of welding current required

(2) The amperage range required

(3) The positions in which welding will be done

(4) The primary power available at the work station

Selection of the type of current, ac, dc, or both, will be based largely on the types of electrodes to be used and the kind of welds to be made.

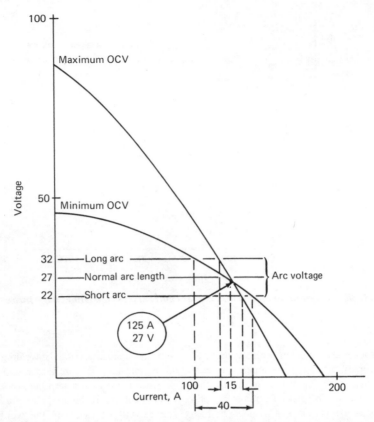

**Fig. 2.4—The effect of volt-ampere curve slope on current output with a change in arc voltage
(Lower slope gives a greater change in welding current for a given change in arc voltage.)**

For ac, a transformer or an alternator type of power source may be used. For dc, transformer-rectifier or motor-generator power sources are available. When both ac and dc will be needed, a single-phase transformer-rectifier or an alternator-rectifier power source may be used. Otherwise, two welding machines will be required, one for ac and one for dc.

The amperage requirements will be determined by the sizes and types of electrodes to be used. When a variety will be encountered, the power supply must be capable of providing the amperage range needed. The duty cycle must be adequate.[4]

The positions in which welding will be done should also be considered. If vertical and over-

head welding are planned, adjustment of the slope of the V-A curve probably will be desirable (see Fig. 2.4). If so, the power supply must provide this feature. This usually requires controls for both the output voltage and the current.

A supply of primary power is needed. If line power is available, it should be determined whether the power is single-phase or three-phase. The welding power source must be designed for either single- or three-phase power, and it must be used with the one it was designed for. If line power is not available, an engine-driven generator or alternator must be used.

ACCESSORY EQUIPMENT

Electrode Holder

An electrode holder is simply a clamping device to allow the welder to hold and control

---

4. See Chapter 1, pp. 11-12, for an explanation of duty cycle.

the electrode, as shown in Fig. 2.5. It also serves as a device for conducting the welding current from the welding cable to the electrode. An insulated handle on the holder separates the welder's hand from the welding circuit. The current is transferred to the electrode through the jaws of the holder. To assure minimum contact resistance and to avoid overheating of the holder, the jaws must be kept in good condition. Overheating of the holder not only makes it uncomfortable for the welder, but also it can cause excessive voltage drop in the welding circuit. Both can impair the welder's performance and reduce the quality of the weld.

The holder must grip the electrode securely and hold it in position with good electrical contact. Installation of the electrode must be quick and easy. The holder needs to be light in weight and easy to handle, yet it must be sturdy enough to withstand rough use. Most holders have insulating material around the jaws to prevent grounding of the jaws to the work.

Electrode holders are produced in sizes to accommodate a range of standard electrode diameters. Each size of holder is designed to carry the current required for the largest diameter electrode that it will hold. The smallest size holder that can be used without overheating is the best one for the job. It will be the lightest, and it will provide the best operator comfort.

Ground Clamp

A ground clamp is a device for connecting the work lead or ground cable to the work. It should produce a strong connection, yet be able to be attached quickly and easily to the work. For light duty, a spring-loaded clamp may be suitable. For high currents, however, a screw clamp may be needed to provide a good connection without overheating the clamp.

Welding Cables

Welding cables are used to connect the electrode holder and the ground clamp to the power source. They are part of the welding circuit (see Fig. 2.1). The cable is constructed for maximum flexibility to permit easy manipulation, particularly of the electrode holder. It also must be wear and abrasion resistant.

Fig. 2.5—Welding a structure with the shielded metal arc welding process

Welding cable consists of many fine copper or aluminum wires stranded together and enclosed in a flexible, insulating jacket. The jacket is made of synthetic rubber or of a plastic that has good toughness, high electrical resistance, and good heat resistance. A protective wrapping is placed between the stranded conductor wires and the insulating jacket to permit some movement between them and provide maximum flexibility.

Welding cable is produced in a range of sizes (from about Awg. 6 to 4/0[5]). The size of the cable required for a particular application depends on the maximum amperage to be used for welding, the length of the welding circuit (welding and ground cables combined), and the duty cycle of the welding machine. Table 2.1 shows the recommended size of copper welding cable for various power sources and circuit lengths. When aluminum cable is used, it should be two Awg sizes larger than copper cable for the application. Cable

---

5. American wire gage sizes.

sizes are increased as the length of the welding circuit increases to keep the voltage drop and the attendant power loss in the cable at acceptable levels.

If long cables are necessary, short sections can be joined by suitable cable connectors. The connectors must provide good electrical contact with low resistance, and their insulation must be equivalent to that of the cable. Lugs, at the end of each cable, are used to connect the cables to the power source. The connection between the cable and a connector or lug must be strong with low electrical resistance. Soldered joints and mechanical connections are used. Aluminum cable requires a good mechanical connection to avoid overheating. Oxidation of the aluminum significantly increases the electrical resistance of the connection. This, of course, can lead to overheating, excessive power loss, and cable failure.

Care must be taken to avoid damage to the jacket of the cable, particularly for the electrode cable. Contact with hot metal or sharp edges may penetrate the jacket and ground the cable.

Helmet

The purpose of the helmet is to protect the welder's eyes, face, forehead, neck, and ears from the direct rays of the arc and from flying sparks and spatter. Some helmets have an optional "flip lid" which permits the dark filter plate over the opening in the shield to be flipped up so the welder can see while he chips the slag from the weld. This protects the welder's face and eyes from flying slag. Slag can cause serious injury if it strikes a person, particularly while it is hot. It can be harmful to the eyes whether it is hot or cold, but it is especially harmful when it is hot.

Helmets are generally constructed of a dark-colored, pressed fiber or fiberglass insulating material. A helmet should be light in weight and should be designed to give the welder the greatest possible comfort. The welder in Fig. 2.5 has a helmet on. The observer is using a hand shield.

Welding helmets are provided with filter plate windows, the standard size being 51 by 130 mm (2 by 4-1/8 in.). Larger openings are available. The filter plate should be capable of absorbing infrared rays, ultraviolet rays, and most of the visible rays emanating from the arc. Filter plates that are now available absorb 99 percent or more of the infrared and ultraviolet rays from the arc.

The shade of the filter plate suggested for use with electrodes up to 4 mm (5/32 in.) diameter is No. 10. For 4.8 to 6.4 mm (3/16 to 1/4 in.) electrodes, Shade No. 12 should be used. Shade No. 14 should be used for electrodes over 6.4 mm (1/4 in.).

The filter plate needs to be protected from molten spatter and from breakage. This is done by placing a plate of clear glass, or other suitable material, on each side of the filter plate. Those who are not welders but work near the arc also

Table 2.1 — Recommended copper welding cable sizes

Power source		Awg cable size for combined length of electrode and ground cables				
Size in amperes	Duty cycle, %	0 to 15 m (0 to 50 ft)	15 to 30 m (50 to 100 ft)	30 to 46 m (100 to 150 ft)	46 to 61 m (150 to 200 ft)	61 to 76 m (200 to 250 ft)
100	20	6	4	3	2	1
180	20-30	4	4	3	2	1
200	60	2	2	2	1	1/0
200	50	3	3	2	1	1/0
250	30					
300	60	1/0	1/0	1/0	2/0	3/0
400	60	2/0	2/0	2/0	3/0	4/0
500	60	2/0	2/0	3/0	3/0	4/0
600	60	2/0	2/0	3/0	4/0	a

a. Use two 3/0 cables in parallel.

need to be protected. This protection usually is provided by either permanent or portable screens. Failure to use adequate protection can result in eye burn for the welder or for those working around the arc. Eye burn, which is similar to sunburn, is extremely painful for a period of 24 to 48 hours. It generally will not permanently injure the eyes, but it does cause intense discomfort. Unprotected skin, exposed to the arc, may also be burned. A physician should be consulted in the case of severe arc burn, regardless of whether it is of the skin or the eyes.

If welding is being performed in tight places with poor ventilation, auxiliary air should be supplied to the welder. This should be done through an attachment to the helmet.

The method used must not restrict the welder's manipulation of the helmet, interfere with his field of vision, or make welding difficult for him. Additional information on eye protection and ventilation is given in ANSI Z49.1, Safety in Welding and Cutting, published by the American Welding Society.

Protective Clothing

From time to time during welding, sparks or globules of molten metal are thrown out from the arc. This is always a point of concern, but it becomes more serious when welding is performed out of position or when extremely high welding currents are used. To ensure protection from burns under these conditions, the welder should wear flame-resistant gloves, a protective apron, and a jacket (see Fig. 2.5). It may also be desirable to protect the welder's ankles and feet from slag and spatter. Cuffless pants and high work shoes or boots are recommended. For additional information, see ANSI Z49.1.

Miscellaneous Equipment

Cleanliness is important in welding. The surfaces of the workpieces and the previously deposited weld metal must be cleaned of dirt, slag, and any other foreign matter that would interfere with welding. To accomplish this, the welder should have a steel wire brush, a hammer, a chisel, and a chipping hammer. These tools are used to remove dirt and rust from the base metal, cut tack welds, and chip slag from the weld bead.

The joint to be welded may require backing to support the molten weld pool during deposition of the first layer of weld metal. Backing strips or nonmetallic backing materials are sometimes used, particularly for joints which are accessible from only one side.

MATERIALS

BASE METALS

The SMAW process is used in joining and surfacing applications on a variety of base metals. The suitability of the process for any specific base metal depends on the availability of a covered electrode whose weld metal has the required composition and properties. Electrodes are available for the following base metals:

(1) Carbon steels
(2) Low alloy steels
(3) Corrosion resisting steels
(4) Cast irons (ductile and gray)
(5) Aluminum and aluminum alloys
(6) Copper and copper alloys

(7) Nickel and nickel alloys

Electrodes are available for application of wear, impact, or corrosion resistant surfaces to these same base metals.

COVERED ELECTRODES

Covered electrodes are classified according to the requirements of specifications issued by the American Welding Society. Certain agencies of the Department of Defense also issue specifications for covered electrodes. AWS specifications are intended for broad commercial use. The AWS specification numbers and their electrode classifications are given in Table 2.2. The elec-

Table 2.2—AWS specifications for covered electrodes

Type of electrode	AWS specification
Carbon steel	A5.1
Low alloy steel	A5.5
Corrosion resistant steel	A5.4
Cast iron	A5.15
Aluminum and aluminum alloys	A5.3
Copper and copper alloys	A5.6
Nickel and nickel alloys	A5.11
Surfacing	A5.13 and A5.21

trodes are classified on the basis of the chemical composition or mechanical properties, or both, of their undiluted weld metal. Carbon steel, low alloy steel, and stainless steel electrodes are also classified according to the type of welding current they are suited for and sometimes according to the positions of welding that they can be used in. Specific information on this topic can be obtained from the filler metal specifications and the *Handbook*.[6]

Carbon Steel Electrodes

In AWS A5.1, a simple numbering system is used for electrode classification. In E6010, for example, the *E* designates an electrode. The first two digits (*60*) signify the minimum tensile strength of the undiluted weld metal in ksi, in the as-welded condition. The third digit represents the welding position (the *1*, in this case, refers to all positions). The last digit refers to the covering type and the type of current with which the electrode can be used.

There are two strength levels of carbon steel electrodes: the 60 series and the 70 series. The minimum allowable tensile strength of the weld metal for the 60 series is 427 MPa (62 ksi), although additional elongation may allow some of these to go as low as 414 MPa (60 ksi). For the 70 series, it is 496 MPa (72 ksi) and, again, some of these may go as low as 483 MPa (70 ksi), with additional elongation. Chemical composition is not specified, except for maximum limits on certain elements in the 70 series. Charpy V-notch impact requirements are given for some electrodes in both series.

6. See Chapter 94, "Filler Metals," in Section 5 and also *Metals and Their Weldability, Welding Handbook*, Section 4, 6th ed. (The material will be revised in Vol. 4, 7th ed.)

Certain of the carbon steel electrodes are designed to operate only on dc. Others are for either ac or dc. Polarity on dc usually is reverse (electrode positive), although a few of the electrodes are intended for straight polarity. Some of these may be used with either straight or reverse polarity.

Most of the electrodes are designed for welding in all positions. However, those which contain large amounts of iron powder or iron oxide in the coatings are generally restricted to groove welds in the flat position and horizontal fillet welds. The coverings on these electrodes are very heavy, which precludes their operation in the vertical and overhead positions.

Several electrodes of the 70 series are low hydrogen type. Their coatings are formulated with ingredients that are low in moisture and cellulose and, hence, in hydrogen content. Hydrogen is responsible for low ductility and for underbead cracking sometimes encountered in highly restrained welds. For this reason, low hydrogen electrodes are used to weld hardenable steels. They are also used for high sulfur steels and to provide weld metal having good low temperature notch toughness.

The specification does not set a limit on the moisture content of these electrodes, but less than 0.6 percent is recommended. To control moisture, proper storage and handling is required. Typical storage and baking conditions are given in AWS A5.1.

Low Alloy Steel Electrodes

AWS A5.5 classifies low alloy steel covered electrodes according to a numbering system which is similar to that just described for carbon steel electrodes. It uses, in addition, a suffix such as A1 to designate the chemical composition (alloy system) of the weld metal. Thus, a complete electrode classification is E7010-A1. Another is E8016-C2. Alloy systems into which the electrodes fall are carbon-molybdenum steel, chromium-molybdenum steel, nickel steel, and manganese-molybdenum steel. Weld metal strength levels range from 480 to 830 MPa (70 to 120 ksi) minimum tensile strength, in 70 MPa (10 ksi) increments. In this specification, weld metals that are commonly used in the as-welded condition are classified on the basis of their properties in that condition. Similarly, those that are commonly used in the stress-relieved condition are classified on the basis of their properties after a stress relief heat treatment.

In this connection, it should be noted that the stress relief called for in AWS A5.5 consists of holding the test assembly at temperature for one hour. Fabricators using holding times that are significantly different from one hour at temperature may have to be more selective in the electrodes they use. They may also have to run special tests to make certain that the mechanical properties of the weld metal they select are adequate after their specified heat treatment. They may also have to do this when a different temperature is used. Notch toughness requirements are included for many of the weld metals and all classifications must meet radiographic standards.

The military specifications for low alloy steel electrodes sometimes use designations that are similar to those in the AWS specification. Also, some electrodes are produced that are not classified in AWS specifications, but which are designed for specific materials or which broadly match standard AISI low alloy steel base metal compositions, such as 4130. The A5.5 specification sets limits on the moisture content of low hydrogen electrodes packaged in hermetically sealed containers. These limits range from 0.2 percent to 0.6 percent by weight, depending on the classification of the electrode. The higher the strength level, the lower the limit on the moisture content. The reason for this is that moisture is a primary source of hydrogen, and hydrogen can produce cracking in most low alloy steels, unless high preheats and long, slow cooling cycles are employed. The higher the strength of the weld and the base metal, the greater the need for low moisture levels to avoid cracking. Exposure to high humidity (in the range of 70 percent relative humidity or higher) may increase the moisture content of the electrode in only a few hours.

Corrosion Resisting Steel Electrodes

Covered electrodes for welding corrosion resisting steels are classified in AWS A5.4. Classification in this specification is based on the chemical composition of the undiluted weld metal, the positions of welding, and the type of welding current for which the electrodes are suit-

able. The classification system is similar to the one used for carbon and low alloy steel electrodes. Taking E310-15 and E310-16 as examples, the prefix E indicates an electrode. The first three digits refer to the alloy type (with respect to chemical composition). They may be followed by a letter or letters to indicate a modification, such as E310Mo-15. The last two digits refer to the position of welding and the type of current for which the electrodes are suitable. The -1 indicates that the electrodes are usable in all positions through 4 mm (5/32 in.) diameter. The number 5 indicates that the electrodes are suitable for use with dcrp (electrode positive). The number 6 means that electrodes are suitable for either ac or dcrp (electrode positive). Electrodes over 4 mm (5/32 in.) diameter are for use in the flat and horizontal positions.

The specification does not describe the covering ingredients, but -15 coverings usually contain a large proportion of limestone (calcium carbonate). This ingredient provides the CO and CO_2 that are used to shield the arc. The binder which holds the ingredients together in this case is sodium silicate. The -16 covering also contains limestone for arc shielding. In addition, it usually contains considerable titania (titanium dioxide) for arc stability. The binder in this case is likely to be potassium silicate.

Differences in the proportion of these ingredients result in differences in arc characteristics. The -15 electrodes (lime type coverings) tend to provide a more penetrating arc and to produce a more convex and coarsely rippled bead. The slag solidifies relatively rapidly so that these electrodes often are preferred for out of position work, such as pipe welding. On the other hand, the -16 coverings (titania type) produce a smoother arc, less spatter, and a more uniform, finely rippled bead. The slag, however, is more fluid, and the electrode usually is more difficult to handle in out-of-position work.

As a whole, the stainless steels can be separated into three basic types: austenitic, martensitic, and ferritic. Some of these steels are precipitation hardenable. The austenitic group (2XX and 3XX) is, by far, the largest one. Most of the wrought and cast materials in this group are readily weldable. Of the covered electrodes, types E308, E308L, E309, E316, and E316L are the most frequently used. Normally, the composition of the weld metal from a stainless steel electrode is similar to the base metal composition that the electrode is designed to weld. At the same time, the composition is always significantly different from that of the base metal. The primary purpose of this difference is to avoid weld metal cracking.

The composition of many of the austenitic stainless steel weld metals is balanced to produce some small amount of ferrite. This is done to prevent fissuring or hot cracking of the weld metal. Minimum ferrite content in the range of 3 to 5 Ferrite Number (FN), previously referred to as 3 to 5 percent, normally is adequate. Ferrite content as high as 20 FN may be acceptable for some welds when no postweld heat treatment is employed. The Schaeffler diagram, or the DeLong modification of a portion of that diagram, can be used to predict the ferrite content of stainless steel weld metals. Direct measurement of the ferrite content with a magnetic instrument is another method. (See AWS A4.2, Standard Procedures for Calibrating Magnetic Instruments to Measure Delta Ferrite Content of Austenitic Stainless Steel Weld Metal.)

Certain austenitic stainless steel weld metals (Types 310, 320, and 330, for example) do not form ferrite because their nickel content is too high. For these materials, the phosphorus, sulfur, and silicon content of the weld metal is limited or the carbon content increased as a means of minimizing fissuring and cracking.

Appropriate welding procedures can also be used to reduce fissuring and cracking. Low amperage, for example, is beneficial. Some small amount of weaving, as a means of promoting cellular grain growth, may be helpful. Proper procedures in terminating the arc should be used to avoid crater cracks.

AWS A5.4 contains two covered electrode classifications for the straight chromium stainless steels (4XX series). One contains 11 to 13.5 percent chromium, the other, 15 to 18 percent. The carbon content of both is 0.1 percent maximum. Both weld metals are air hardening and their weldments require preheat and postheat treatment to provide the ductility which is necessary in most engineering applications.

The specification also contains three elec-

trode classifications that are used to weld the four to ten percent chromium-molybdenum steels. These materials, too, are air hardening and preheat and postheat treatment are required for sound, serviceable joints.

Nickel and Nickel Alloy Electrodes

Covered electrodes for shielded metal arc welding of nickel and its alloys have compositions which are generally similar to those of the base metals they are used to join. Here again, there are some differences, however. The electrodes normally have additions of elements such as titanium, manganese, and columbium to deoxidize the weld metal and thereby prevent cracking.

AWS A5.11 classifies the electrodes in groups according to their principal alloying elements. The letter E at the beginning indicates an electrode and the chemical symbol Ni identifies the weld metals as nickel base alloys. Other chemical symbols are added to show the principal alloying elements. Then, successive numbers are added to identify each classification within its group. ENiCrFe-1, for example, contains significant additions of chromium and iron.

Most of the electrodes are intended for use with dcrp (electrode positive). Some are also capable of operating with ac, to overcome problems that may be encountered with arc blow (when nine percent nickel steel is welded, for example). Most of the electrodes are suitable for the horizontal and flat positions only. Electrodes of some groups, 3.2 mm (1/8 in.) and smaller sizes, can be used in all positions.

The electrical resistivity of the core wire in these electrodes is exceptionally high. For this reason, excessive amperage will overheat the electrode and damage the covering, causing arc instability and unacceptable amounts of spatter. Each classification and size of electrode has an optimum amperage range.

Aluminum and Aluminum Alloy Electrodes

AWS A5.3 contains two classifications of covered electrodes for the welding of aluminum base metals. These classifications are based on the mechanical properties of the weld metal in the as-welded condition and the chemical composition of the core wire. One core wire is commercially pure aluminum (1100), and the other is an aluminum-five percent silicon alloy (4043). Both electrodes are used with dcrp (electrode positive).

The covering on these electrodes has three functions. It provides a gas to shield the arc, a flux to dissolve the aluminum oxide, and a protective slag to cover the weld bead. Because the slag can be very corrosive to aluminum, it is important that all of it be removed upon completion of the weld.

The presence of moisture in the covering of these electrodes is a major source of porosity in the weld metal. To avoid this porosity, the electrodes should be stored in a heated cabinet until they are to be used. Those electrodes that have been exposed to moisture should be reconditioned (baked) before they are used.

One difficulty which may occur in welding is the fusing of slag over the end of the electrode if the arc is broken. In order to restrike the arc, this fused slag must be removed.

Covered aluminum electrodes are used primarily for noncritical welding applications and for repair work. They should be used only on base metals for which either the 1100 or 4043 filler metals are recommended. These weld metals do not respond to precipitation hardening heat treatments. If they are used for such material, each application should be carefully evaluated.

Copper and Copper Alloy Electrodes

AWS A5.6 classifies copper and copper alloy electrodes on the basis of their all-weld-metal properties and the chemical composition of their undiluted weld metal. The designation system is similar to that used for nickel alloy electrodes. The major difference is that each individual classification within a group is identified by a letter. This letter is sometimes followed by a number, as in ECuAl-A2, for example. The groups are: CuSi for silicon bronze, CuSn for phosphor bronze, CuNi for copper-nickel, and CuAl for aluminum bronze. These electrodes, generally, are used with dcrp (electrode positive).

Copper electrodes are used to weld unalloyed copper and to repair copper cladding on steel or cast iron. Silicon bronze electrodes are used to weld copper-zinc alloys, copper, and some iron

base materials. They are also used for surfacing to provide corrosion resistance.

Phosphor bronze and brass base metals are welded with phosphor bronze electrodes. These electrodes are also used to braze weld copper alloys to steel and cast iron. The phosphor bronzes are rather viscous when molten, but their fluidity is improved by preheating to about 200° C (400° F). The electrodes and the work must be dry.

Copper-nickel electrodes are used to weld a wide range of copper-nickel alloys and also copper-nickel cladding on steel. In general, no preheat is necessary for these materials.

Aluminum bronze electrodes have broad use for welding copper base alloys and some dissimilar metal combinations. They are used to braze weld many ferrous metals and to apply wear and corrosion resistant bearing surfaces. Welding is usually done in the flat position with some preheat.

Electrodes for Cast Iron

AWS A5.15 classifies covered electrodes for welding cast iron. Gray iron castings may be welded with special covered electrodes. These electrodes are steel and the casting should be preheated to a temperature ranging from 150° to 750° C (300° to 1400° F). The specific temperature depends on the size and complexity of the casting and the machinability requirements. Small pits and cracks can be welded without preheat, but the weld will not be machineable. Welding is done with low amperage dcrp (electrode positive) to minimize dilution with the base metal. Preheating is not used in this case, except to minimize the residual stresses in other parts of the casting.

Nickel and nickel alloy electrodes made specifically for welding cast iron are widely used to repair castings and to join the various types of cast iron to themselves and to other metals. The hardness of the weld metal depends on the amount of base metal dilution.

Phosphor bronze and aluminum bronze electrodes are used to braze weld cast iron. The melting temperature of their weld metals is below that of cast iron. The casting should be preheated to about 200° C (400° F), and welding should be done with dcrp (electrode positive), using the lowest amperage that will produce good bonding between the weld metal and groove faces. The cast iron surfaces should not be melted.

Surfacing Electrodes

Most hard surfacing electrodes are designed to meet AWS A5.13 or AWS A5.21 specifications. A wide range of SMAW electrodes is available (under these and other AWS filler metal specifications) to provide wear, impact, heat, or corrosion resistant layers on a variety of base metals. All of the covered electrodes specified in A5.13 have a solid core wire. Those specified in A5.21 have a composite core. The electrode designation system in both specifications is similar to that used for copper alloy electrodes with the exception of tungsten carbide electrodes. The E in the designation for these electrodes is followed by WC. The mesh size limits for the tungsten carbide granules in the core follow these to complete the designation. The core, in this case, consists of a steel tube filled with the tungsten carbide granules.

Surfacing with covered electrodes is used for cladding, buttering, buildup, and hard surfacing. The weld deposit in these applications is intended to provide one or more of the following for the surfaces to which they are applied:

(1) Corrosion resistance
(2) Metallurgical control
(3) Dimensional control
(4) Wear resistance
(5) Impact resistance

Covered electrodes for a particular surfacing application should be selected after a careful review of the required properties of the weld metal when it is applied to a specific base metal. Further information on this and on surfacing in general is given in Chapter 14.

Electrode Conditioning

SMAW electrode coverings are hygroscopic.

Some coverings are more hygroscopic than others. The moisture they pick up on exposure to a humid atmosphere dissociates to form hydrogen and oxygen during welding. The atoms of hydrogen dissolve in the weld and the heat-affected zone and may cause cold cracking. This type of crack is more prevalent in hardenable steel base metals and high strength steel weld metals. Moisture in the electrode covering may also cause porosity in the weld metal with most materials.

To minimize moisture problems, covered electrodes need to be properly packaged, stored, and handled. This is particularly important for low hydrogen electrodes. Moisture control of the mineral coverings on these electrodes is needed for ferrous and nonferrous electrodes alike. It is of critical importance for low hydrogen electrodes which are to be used to weld hardenable steel base metals. The importance of moisture control for these electrodes increases as the strength of either the weld metal or the base metal increases. Holding ovens are used for low hydrogen electrodes once those electrodes have been removed from their sealed container and have not been used within a certain period of time. This period varies from as little as half an hour to as much as eight hours depending on the strength level of the electrode, the humidity during exposure, and even the specific covering on the electrode. The time for any electrode must be reduced as the humidity increases. The temperature of the holding oven should be within the range of 65° to 150° C (150° to 300° F). Electrodes that have been exposed too long may have to be baked at a sustantially higher temperature to drive off the absorbed moisture. The specific recommendations of the manufacturer of the electrode need to be followed because the time and temperature limitations can vary from manufacturer to manufacturer, even for electrodes within a given classification. Excessive heating can damage the covering on an electrode.

APPLICATIONS

Materials

The SMAW process can be used to join most of the common metals and alloys for which covered electrodes are available. The list includes the carbon steels, the low alloy steels, the stainless steels, and cast iron, as well as copper, nickel, and aluminum and their alloys. Shielded metal arc welding is also used to join a wide range of chemically dissimilar but metallurgically compatible materials.

The process is not used for materials for which shielding of the arc by the gaseous products of an electrode covering is unsatisfactory. The reactive (Ti, Zr) and refractory (Cb, Ta, Mo) metals fall into this group.

Thicknesses

The shielded metal arc process is adaptable to any material thickness within certain practical and economic limitations. For material thicknesses less than about 2 mm (1/16 in.), the base metal will melt through and the molten metal will fall away before a common weld pool can be established, unless special fixturing and welding procedures are employed. There is no upper limit on thickness, but because of the limited deposition rate of this process, the continuous electrode processes weld at lower cost on sections thicker than approximately 38 mm (1-1/2 in.). Most of the SMAW applications are on thicknesses between 3 and 38 mm (1/8 and 1-1/2 in.), except where irregular configurations are encountered. Such configurations put an automated welding process at an economic disadvantage. In such instances, the shielded metal arc process is commonly used to weld materials at thick as 250 mm (10 in.).

Position of Welding

One of the major advantages of SMAW is that welding can be done in any position on most of the materials for which the process is suitable. This makes the process useful on joints that cannot be placed in the flat position. Despite this advantage, welding should be done in the flat position whenever possible because it is easier to do in that position. Less skill is required, and larger electrodes with their higher amperages and correspondingly higher deposition rates can be used. Greater skill is needed to control the molten weld pool in the vertical and overhead positions. Moreover, smaller electrodes with their lower amperages are necessary to provide the operating characteristics required in those positions. Joint design is also a significant factor. In the vertical and overhead positions, the design may have to be different from that which would be suitable in the flat position.

Location of Welding

The simplicity of the equipment makes SMAW an extremely versatile process with respect to the location and environment of the operation. Welding can be done indoors or outdoors, on a production line, a ship, a bridge, a building framework, an oil refinery, a cross-country pipeline, or any such types of work. No gas or water hoses are needed and the welding cables can be extended quite some distance from the power source. In remote areas, gasoline or diesel powered units can be used. Despite this versatility, the process should always be used in an environment which shelters it from the wind, the rain, and the snow.

JOINT DESIGN AND PREPARATION

TYPES OF WELDS

Welded joints are designed primarily on the basis of the strength and safety required of the weldment under the service conditions imposed on it. The manner in which the service stresses will be applied and the temperature of the weldment in service must always be considered. A joint required for dynamic loading may be quite different from one permitted in static loading. Dynamic loading requires consideration of fatigue strength and resistance to brittle fracture. These, among other things, require that the joints be designed to reduce or eliminate points of stress concentration. The design should also balance the residual stresses and obtain as low a residual stress level as possible. The weld must produce adequate joint strength. Joints subject to dynamic loading or to corrosion or erosion must be free of such irregularities as undercut and overlap. Regardless of their service requirements, the joints need to be designed for economy and accessibility during fabrication and erection. Accessibility not only reduces the welding cost but also provides a better opportunity for good workmanship and good control of distortion and residual stresses. The effect of joint design on some of these considerations is discussed below.

Groove Welds

Groove weld joint designs of different types are used. Selection of the most appropriate design for a specific application is influenced by

(1) Suitability for the structure under consideration

(2) Accessibility to the joint for welding

(3) Cost of welding

(4) Position in which welding is to be done

A square groove is the most economical to prepare. It only requires squaring-off of the edge of each member. This type of joint is limited to those thicknesses with which satisfactory strength and soundness can be obtained. For SMAW, that thickness is usually not greater than about 6 mm (1/4 in.) and then only when the joint is to be welded in the flat position from both sides. The type of material to be welded is also a consideration.

When thicker members are to be welded, the edge of each member must be prepared to a contour that will permit the arc to be directed to the point where the weld metal must be deposited. This is necessary to provide fusion to whatever depth is required.

For economy as well as to reduce distortion and residual stresses, the joint design should have a root opening and a groove angle that will provide adequate strength and soundness with the deposition of the least amount of filler metal. The key to soundness is accessibility to the root and sidewalls of the joint. J-groove and U-groove joints are desirable for thick sections. In very thick sections, the savings in filler metal and welding time alone are sufficient to more than offset the added cost of this joint preparation. The angle of the sidewalls must be large enough to prevent slag entrapment.

Fillet Welds

Where the service requirements of the weldment permit, fillet welds frequently are used in preference to groove welds. Fillet welds require little or no joint preparation, although groove welds sometimes require less welding. Intermittent fillet welding may be used when a continuous weld would provide more strength than is required to carry the load.

A fillet weld is often combined with a groove weld to provide the required strength and reduce the stress concentration at the joint. Minimum stress concentration at the toes of the weld is obtained with concave fillets.

WELD BACKING

When full penetration welds are required and welding is done from one side of the joint, weld backing may be required. Its purpose is to provide something on which to deposit the first layer of metal and thereby prevent the molten metal in that layer from escaping through the root of the joint.

Four types of backing are commonly used. They are

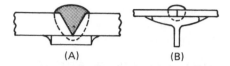

Fig. 2.6—Fusible metal backing for a weld; (A) backing strip; (B) structure backing

(1) Backing strip
(2) Backing weld
(3) Copper backing bar
(4) Nonmetallic backing

Backing Strip

A backing strip is a strip of metal placed on the back of the joint, as shown in Fig. 2.6(A).The first weld pass ties both members of the joint together and to the backing strip. The strip may be left in place if it will not interfere with the serviceability of the joint. Otherwise, it should be removed, in which case the back side of the joint must be accessible. If the back side is not accessible, some other means of obtaining a proper root pass must be used.

The backing strip must always be made of a material that is metallurgically compatible with the base metal and the welding electrode to be used. Where design permits, another member of the structure may serve as backing for the weld. Figure 2.6(B) provides an example of this. In all cases, it is important that the backing strip as well as the surfaces of the joint be clean to avoid porosity and slag inclusions in the weld. It is also important that the backing strip fit properly. Otherwise, the molten weld metal can run out through any gap between the strip and the base metal at the root of the joint.

Copper Backing Bar

A copper bar is sometimes used as a means of supporting the molten weld pool at the root of the joint. Copper is used because of its high thermal conductivity. This high conductivity helps prevent the weld metal from fusing to the backing bar. Despite this, the copper bar must have sufficient mass to avoid melting during deposition of the first weld pass. In high production use, water can be passed through holes in the bar to remove the heat that accumulates during continuous welding. Regardless of the method of cool-

ing, the arc should not be allowed to impinge on the copper bar, for if any copper melts, the weld metal can become contaminated with copper. The copper bar may be grooved to provide the desired root surface contour and reinforcement.

Nonmetallic Backing

Nonmetallic backing of either granular flux or refractory material is also a method that is used to produce a sound first pass. The flux is used primarily to support the weld metal and to shape the root surface. A granular flux layer is supported against the back side of the weld by some method such as a pressurized fire hose. A system of this type is generally used for production line work, although it is not widely used for SMAW.

Refractory type backing consists of a flexible, shaped form that is held on the back side of the joint by clamps or by pressure sensitive tape. This type of backing is sometimes used with the SMAW process, although special welding techniques are required to consistently produce good results. The recommendations of the manufacturer of the backing should be followed.

Backing Weld

A backing weld is one or more backing passes in a single groove weld joint. This weld is deposited on the back side of the joint before the first pass is deposited on the face side. The concept is illustrated in Fig. 2.7. After the backing weld, all subsequent passes are made in the groove from the face side. The root of the joint may be ground or gouged after the backing weld is made to produce sound, clean metal on which to deposit the first pass on the face side of the joint. The backing weld can be made with the

Groove weld made
after welding other side

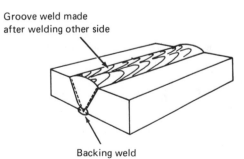

Backing weld

Fig. 2.7—A typical backing weld

same process or with a different process from that to be used for welding the groove. If the same process is used, the electrodes should be of the same classification as those to be used for welding the groove. If a different process such as gas tungsten arc welding is used, the welding rods should deposit weld metal having composition and properties similar to those of the SMAW weld metal. The backing weld must be large enough to support any load that is placed on it. This is especially important when the weldment must be repositioned after the backing weld has been deposited and before the groove weld is made.

FIT-UP

Joint fit-up involves the positioning of the members of the joint to provide the specified groove dimensions and alignment. The points of concern are the root opening and the alignment of the members along the root of the weld. Both of these have an important influence on the quality of the weld and the economics of the process. After the joint has been properly aligned throughout its length, the position of the members should be maintained by clamps or tack welds. Finger bars or U-shaped bridges can be placed across the joint and tack welded to each of the members.

If the root opening is not uniform, the amount of weld metal will vary from location to location along the joint. As a result, shrinkage, and hence distortion, will not be uniform. This can cause problems when the finished dimensions have been predicated on the basis of uniform, controlled shrinkage.

Misalignment along the root of the weld may cause lack of penetration in some areas or poor root surface contour, or both. Inadequate root opening can cause lack of complete joint penetration. Too wide a root opening makes welding difficult and requires more weld metal to fill the joint. This, of course, also involves additional cost. In thin members, an excessive

root opening may cause excessive melt-thru on the back side. It may even cause the edge of one or both members to melt away.

RECOMMENDED PROPORTIONS OF WELD GROOVES

The weld grooves shown in Figs. 2.8 to 2.10 illustrate typical designs and dimensions of joints for shielded metal arc welding of steel. The proportions and limitations shown are those that generally have been found suitable for producing sound welds with the least amount of weld metal. Some modification of the dimensions may be required for different materials or for special applications in order to obtain the quality of joint that may be necessary in any particular case.

RUNOFF TABS

In some applications, it is necessary to completely fill out the groove right to the very ends of the joint. In such cases, runoff tabs are used. They, in effect, extend the groove beyond the ends of the members to be welded. The weld is carried over into the tabs. This assures that the entire length of the joint is filled to the necessary depth with sound weld metal. A typical runoff tab is shown in Fig. 2.11. Runoff tabs are excellent appendages on which to start and stop welding. Any defects in these starts and stops are located in areas that later will be discarded.

Selection of the material for runoff tabs is important. The composition of the tabs should not be allowed to adversely affect the properties of the weld metal. For example, for stainless steel which is intended for corrosion service, the runoff tabs should be of a compatible grade of stainless steel. Carbon steel tabs would be less costly, but fusion with the stainless steel filler metal would change the composition of the weld metal at the junction of the carbon steel tab and the stainless steel members of the joint. The weld metal at this location probably would not have adequate corrosion resistance.

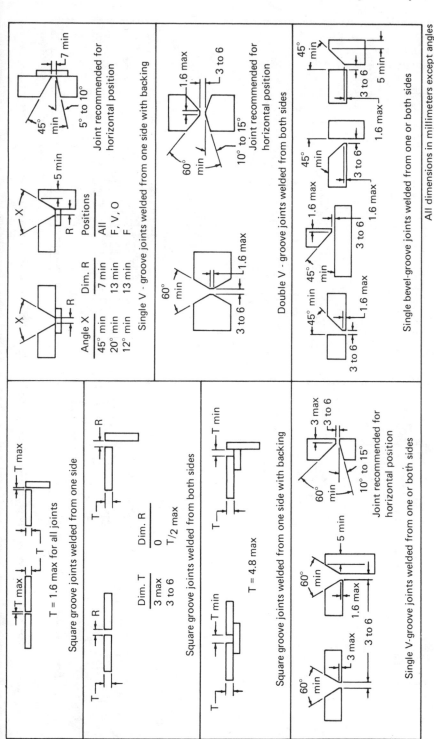

Fig. 2.8—Recommended proportions of grooves for shielded metal arc welding of steel (See chart under Fig. 2.10 for dimensional conversions.)

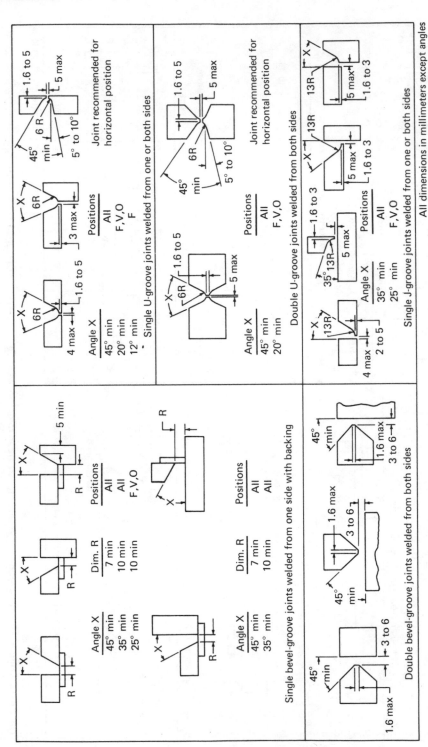

Fig. 2.9—Recommended proportions of groove welds for shielded metal arc welding of steel (See chart under Fig. 2.10 for conversions.)

PREHEATING

Preheating the area to be welded before and during welding can be an important factor in making sound welds in some materials. Preheat should not be used when it is not necessary, since that takes time and energy. Excessive preheat temperatures should be avoided because they are wasteful and costly and because they might degrade the properties and the quality of the joint. Excessive preheat increases welder discomfort, which tends to reduce the quality of his work. The use of preheat and the specific temperature employed should be based on welding code requirements, competent technical information, or the results of tests. In general, the temperature will depend on the material to be welded, the electrodes to be used, and the degree of restraint in the joint.

Hardenable steels, high strength steels, and electrodes other than low hydrogen electrodes will require higher preheat. The moisture content of low hydrogen electrodes must be kept low when minimal preheat temperatures are to be used. This becomes even more important as the strength levels of the base metal and the weld metal increase.

Preheat is sometimes used for shielded metal arc welding of materials which have very high thermal conductivity, such as copper and aluminum alloys. Preheat, in this case, reduces the amperage required for welding, improves penetration, and aids the fusion of the weld metal and the base metal.

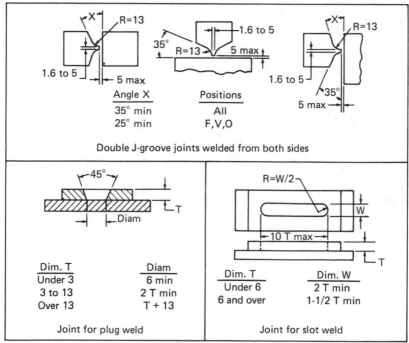

All dimensions in millimeters except angles

Dimensional conversions for Figs. 2.8, 2.9, and 2.10

mm	in.	mm	in.
1.6	1/16	6.4	1/4
3.2	1/8	9.5	3/8
4.8	3/16	13	1/2

Fig. 2.10—Recommended proportions of groove welds for shielded metal arc welding of steel

WELDING PROCEDURES

ELECTRODE DIAMETER

The diameter of the electrodes to be used for a given job depends largely on the thickness of the material to be welded, the position in which welding is to be done, and the type of joint to be welded. Large electrodes, with their correspondingly high currents, are used on thick materials. The high currents used with these electrodes are helpful in obtaining complete fusion and proper penetration of the weld metal in the joint. The large electrodes also have higher deposition rates than the smaller sizes.

When welding in the horizontal, vertical, and overhead positions, the molten metal tends to flow out of the joint because of gravitational forces. This tendency varies with weld pool size. The tendency can be reduced by keeping the weld pool small which generally requires the use of small electrodes. Arc forces and electrode manipulation also aid in controlling the molten weld metal in these positions.

Weld groove design must also be considered when electrode size is selected. The electrode used in the first few passes must be small enough for easy manipulation in the root of the joint. In V-grooves, small diameter electrodes are frequently used for the initial pass to control melt-thru and bead shape. Larger electrodes can be used to complete the weld, taking advantage of their deeper penetration and higher deposition rates.

Finally, the experience of the welder often has a bearing on the electrode size. This is particularly true for out-of-position welding, since the welder's skill governs the size of the molten puddle that he can control.

The largest possible electrode that does not violate any pertinent heat input limitations or deposit too large a weld should be used. Welds that are larger than necessary are more costly and, in some instances, actually are harmful. Any sudden change in section size or in the contour of a weld, such as that caused by overwelding, creates stress concentrations. It is obvious that the correct electrode size is the one that, when used with the proper amperage and travel speed, produces a weld of the required size in the least amount of time.

WELDING CURRENT

Shielded metal arc welding can be accomplished with either alternating or direct current, when an appropriate electrode is used. The type of welding current, the polarity, and the constituents in the electrode covering influence the melting rate of all covered electrodes. For any given electrode, the melting rate is directly related to the electrical energy supplied to the arc. Part of this energy is used to melt a portion of the base metal and part is used to melt the electrode.

Direct Current

Direct current always provides a steadier arc and smoother metal transfer than ac does. This is because the polarity of dc is not always changing as it is with ac. Most covered electrodes operate better on reverse polarity (electrode positive), although some are suitable for (and even are intended for) straight polarity (electrode negative). Reverse polarity produces deeper penetration, but straight polarity produces a higher electrode melting rate.

The dc arc produces good wetting action by the molten weld metal and uniform weld bead size even at low amperage. For this reason, dc is particularly suited to welding thin sections. Most combination ac-dc electrodes operate better on dc than on ac, even though they are designed to operate with either type of current.

Direct current is preferred for vertical and overhead welding and for welding with a short arc. The dc arc has less tendency to short out as globules of molten metal are transferred across it.

Arc blow may be a problem when magnetic metals (iron and nickel) are welded with dc. One way to overcome this problem is to change to ac.

Alternating Current

For SMAW, alternating current offers two advantages over dc. One is the absence of arc blow and the other is the cost of the power source.

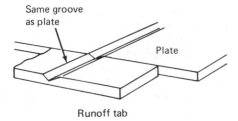

Fig. 2.11—*Runoff tab at end of a weld joint*

Without arc blow, larger electrodes and higher welding currents can be used. Certain electrodes (specifically, those with iron powder in their coverings) are designed for operation at higher amperages with ac. The highest welding speeds for SMAW can be obtained in the flat position with these electrodes on ac. Fixturing materials, fixture design, and grounding location may not be as critical with ac.

An ac transformer costs less than an equivalent dc power source. The cost of the equipment alone should not be the sole criterion in the selection of the power source, however. All operating factors need to be considered.

Amperage

Covered electrodes of a specific size and classification will operate satisfactorily at various amperages within some certain range. This range will vary somewhat with the thickness and formulation of the covering.

Deposition rates increase as the amperage increases. For a given size of electrode, the amperage ranges and the resulting deposition rates will vary from one electrode classification to another. This variation for several classifications of carbon steel electrodes of one size is shown in Fig. 2.12.

With a specific type and size of electrode, the optimum amperage depends on several factors such as the position of welding and the type of joint. The amperage must be sufficient to obtain good fusion and penetration yet permit

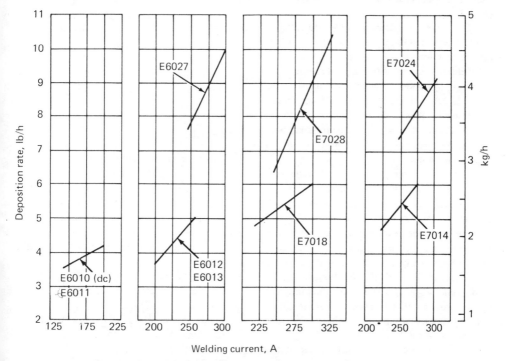

Fig. 2.12—*The relationship between deposition rate and welding current for various types of 4.8 mm (3/16 in.) carbon steel electrodes*

proper control of the molten weld pool. For vertical and overhead welding, the optimum amperages would be likely to be on the low end of the allowable range.

Amperage beyond the recommended range should not be used. It can overheat the electrode and cause excessive spatter, arc blow, undercut, and weld metal cracking. Figures 2.13(B) and (C) show the effect of amperage on bead shape.

ARC LENGTH

The arc length is the distance from the molten tip of the electrode core wire to the surface of the molten weld pool. Proper arc length is important in obtaining a sound welded joint. Metal transfer from the tip of the electrode to the weld pool is not a smooth, uniform action. Instantaneous arc voltage varies as droplets of molten metal are transferred across the arc, even with constant arc length. However, any variation in voltage will be minimal when welding is done with the proper amperage and arc length. The latter requires constant and consistent electrode feed.

The correct arc length varies according to the electrode classification, diameter, and covering composition; it also varies with amperage and welding position. Arc length increases with increasing electrode diameter and amperage. As a general rule, the arc length should not exceed the diameter of the core wire of the electrode. The arc usually is shorter than this for electrodes with thick coverings, such as iron powder or "drag" electrodes.

Too short an arc will be erratic and may short circuit during metal transfer. Too long an arc will lack direction and intensity, which will tend to scatter the molten metal as it moves from the electrode to the weld. The spatter may be heavy and the deposition efficiency low. Also, the gas and flux generated by the covering are not as effective in shielding the arc and the weld metal from air. The poor shielding can cause porosity and contamination of the weld metal by oxygen or nitrogen, or both. The quality of the weld will be poor.

Control of arc length is largely a matter of welder skill, involving the welder's knowledge, experience, visual perception, and manual dexterity. Although the arc length does change to some extent with changing conditions, certain fundamental principles can be given as a guide to the proper arc length for a given set of conditions.

For downhand welding, particularly with heavy electrode coverings, the tip of the electrode can be dragged lightly along the joint. The arc length, in this case, is automatically determined by the coating thickness and the melting rate of the electrode. Moreover, the arc length is uniform. For vertical or overhead welding, the arc length is gaged by the welder. The proper arc length, in such cases, is the one that permits the welder to control the size and motion of the molten weld pool.

For fillet welds, the arc is crowded into the joint for highest deposition rate and best penetration. The same is true of the root passes in groove welds in pipe.

When arc blow is encountered, the arc length should be shortened as much as possible. The various classifications of electrodes have widely different operating characteristics, including arc length. It is important, therefore, for the welder to be familiar with the operating characteristics of the types of electrodes he uses in order to recognize the proper arc length and to know the effect of different arc lengths. The effect of a long and a short arc on bead appearance with a mild steel electrode is illustrated in Figs. 2.13(D) and (E).

TRAVEL SPEED

Travel speed is the rate at which the electrode moves along the joint. The proper travel speed is the one which produces a weld bead of proper contour and appearance, as shown in Fig. 2.13(A). Travel speed is influenced by several factors. Some of these are

(1) Type of welding current, amperage, and polarity
(2) Position of welding
(3) Melting rate of the electrode
(4) Thickness of material
(5) Surface condition of the base metal
(6) Type of joint
(7) Joint fit-up
(8) Electrode manipulation

When welding, the travel speed should be

adjusted so that the arc slightly leads the molten weld pool. Up to a point, increasing the travel speed will narrow the weld bead and increase penetration. Beyond this point, higher travel speeds can decrease penetration; cause the surface of the bead to deteriorate and produce undercutting at the edges of the weld; make slag removal difficult; and entrap gas (porosity) in the weld metal. The effect of high travel speed on bead appearance is shown in Fig. 2.13(G). With low travel speed, the weld bead will be wide and convex with shallow penetration, as illustrated in Fig. 2.13(F). The shallow penetration is caused by the arc dwelling on the molten weld pool instead of leading it and concentrating on the base metal. This, in turn, affects dilution. When dilution must be kept low (as in cladding), the travel speed, too, must be kept low.

Travel speed also influences heat input, and this affects the metallurgical structures of the weld metal and the heat-affected zone. Low travel speed increases heat input and this, in turn, in-creases the size of the heat-affected zone and reduces the cooling rate of the weld. Forward travel speed is necessarily reduced·with a weave bead as opposed to the higher travel speed that can be attained with a stringer bead. Higher travel speed reduces the size of the heated-affected zone and increases the cooling rate of the weld. The increase in the cooling rate can increase the strength and hardness of a weld in a hardenable steel, unless preheat of a level sufficient to prevent hardening is used.

ELECTRODE ORIENTATION

Electrode orientation, with respect to the work and the weld groove, is important to the quality of a weld. Improper orientation can result in slag entrapment, porosity, and undercutting. Proper orientation depends on the type and size of electrode, the position of welding, and the geometry of the joint. A skilled welder automatically takes these into account when he determines the orientation to be used for a specific

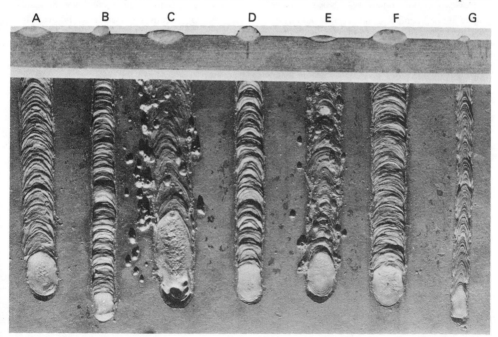

Fig. 2.13—The effect of welding amperage, arc length, and travel speed on covered electrode weld beads: (A) proper amperage, arc length, and travel speed; (B) amperage too low; (C) amperage too high; (D) arc length too short; (E) arc length too long; (F) travel speed too slow; (G) travel speed too fast

joint. Travel angle and work angle are used to specify electrode orientation.

Travel angle is the angle which the electrode makes with a reference line that both lies in a plane through the axis of the weld and is perpendicular to that axis. *Work angle* is the angle which the electrode makes with a surface of the base metal in a plane perpendicular to the axis of the weld. When the electrode is pointed in the direction of welding, the *forehand* technique is being used. The travel angle, then, is known as the *push angle*. The *backhand* technique involves pointing the electrode in the direction opposite that of welding. The travel angle, then, is called the *drag angle*. These angles are shown in Fig. 2.14.

Typical electrode orientation and welding technique for groove and fillet welds, with carbon steel electrodes, are listed in Table 2.3. These may be different for other materials. Correct orientation provides good control of the molten weld pool, the desired penetration, and complete fusion with the steel base.

A large travel angle may cause a convex, poorly shaped bead with inadequate penetration, whereas a small travel angle may cause slag entrapment. A large work angle can cause undercutting, while a small work angle can result in lack of fusion.

WELDING TECHNIQUE

The first step in SMAW is to assemble the proper equipment, materials, and tools for the job. Next, the type of welding current and the polarity, if dc, need to be determined and the power source set accordingly. The power source must also be set to give the proper volt-ampere characteristic (open-circuit voltage) for the size and type of electrode to be used. After this, the work is positioned for welding and, if necessary, clamped in place.

The arc is struck by tapping the end of the electrode on the work near the point where welding is to begin, then quickly withdrawing it a small amount to produce an arc of proper length. Another technique for striking the arc is to use a scratching motion similar to that used in striking a match. When the electrode touches the work, there is a tendency for them to stick together.

The purpose of the tapping and scratching motion is to prevent this. When the electrode does stick, it needs to be quickly broken free. Otherwise, it will overheat, and attempts to remove it from the workpiece will only bend the hot electrode. Freeing it then will require a hammer and chisel.

The technique of restriking the arc once it has been broken varies somewhat with the type of electrode. Generally, the covering at the tip of the electrode becomes conductive when it is heated during welding. This assists in restriking the arc if it is restruck before the electrode cools. Arc striking and restriking are much easier for electrodes with large amounts of metal powders in their coverings. Such coverings are conductive when cold. When using heavily covered electrodes which do not have conductive coatings, such as E6020, low hydrogen, and stainless steel electrodes, it may be necessary to break off the projecting covering to expose the core wire at the tip for easy restriking.

Striking the arc with low hydrogen electrodes requires a special technique to avoid porosity in the weld at the point where the arc is started. This technique consists of striking the arc a few electrode diameters ahead of the place where welding is to begin. The arc is then quickly moved back, and welding is begun in the normal manner. Welding continues over the area where the arc originally was struck, re-fusing any small globules of weld metal that may have remained from striking the arc.

During welding, the welder maintains a normal arc length by uniformly moving the electrode toward the work as the electrode melts. At this same time, the electrode is moved uniformly along the joint in the direction of welding, to form the bead.

Any of a variety of techniques may be employed to break the arc. One of these is to rapidly shorten the arc, then quickly move the electrode sideways out of the crater. This technique is used when replacing a spent electrode, in which case welding will continue from the crater. Another technique is to stop the forward motion of the electrode and allow the crater to fill, then gradually withdraw the electrode to break the arc. When continuing a weld from a crater, the arc should be struck at the forward end of the crater. It should then quickly be moved to the back of

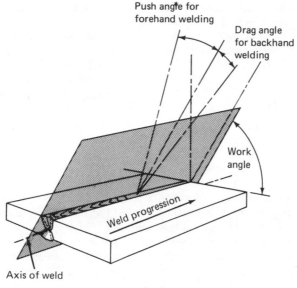

(A) Groove weld

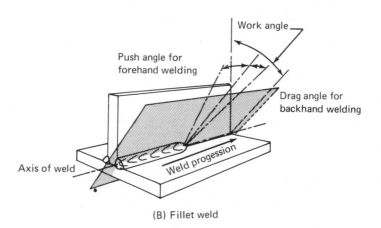

(B) Fillet weld

Fig. 2.14—Orientation of the electrode

the crater and slowly brought forward to continue the weld. In this manner, the crater is filled, and porosity and entrapped slag are avoided. This technique is particularly important for low hydrogen electrodes.

SLAG REMOVAL

The extent to which slag is removed from each weld bead before welding over the bead has a direct bearing on the quality of a multiple pass weld. Failure to thoroughly clean each bead increases the probability of trapping slag and, thus, producing a defective weld. Complete and efficient slag removal requires that each bead be properly contoured and that it blend smoothly into the adjacent bead or base metal.

Small beads cool more rapidly than large ones. This tends to make slag removal from small beads easier. Concave or flat beads that wash smoothly into the base metal or any adjoining beads minimize undercutting and avoid a sharp notch along the edge 'of the bead where slag could stick. Finally, it is most important that welders be able to recognize areas where slag entrapment is likely to occur. Skilled welders understand that complete removal of slag is necessary before continuing a weld.

WELDING GROUND

Proper grounding of the workpiece is a necessary consideration in shielded metal arc welding. The location of the ground is especially important with dc welding. Improper location may promote arc blow, making it difficult to control the arc. Moreover, the method of attaching the ground is important. A poorly attached ground will not provide consistent electrical contact, and the connection will heat up. This can lead to an interruption of the circuit and a breaking of the arc. A copper contact shoe secured with a C-clamp is best. If copper pickup by this attachment to the base metal is detrimental, the copper shoe should be attached to a plate that is compatible with the work. The plate, in turn, is then secured to the work. For rotating work, contact should be made by shoes sliding on the work or through roller bearings on the spindle on which the work is mounted. If sliding shoes are used, at least two shoes should be employed. If loss of contact occurred with only a single shoe, the arc would be extinguished.

ARC STABILITY

A stable arc is required if high quality welds are to be produced. Such defects as inconsistent fusion, entrapped slag, blowholes, and porosity can be the result of an unstable arc.

The important factors influencing arc stability are

(1) The open-circuit voltage of the power source

(2) Transient voltage recovery characteristics of the power source

(3) Size of the molten drops of filler metal and slag in the arc

(4) Ionization of the arc path from the electrode to the work

(5) Manipulation of the electrode

The first two factors are related to the design and operating characteristics of the power source. The next two are functions of the welding electrode. The last one represents the skill of the welder.

The arc of a covered electrode is a transient arc, even when the welder maintains a fairly constant arc length. The welding machine must be able to respond rapidly when the arc tends to go out, or it is short circuited by large droplets of metal bridging the arc gap. In that case, a surge of current is needed to clear a short circuit. With ac, it is important that the voltage lead the current in going through zero. If the two were in phase, the arc would be very unstable. This phase shift must be designed into the welding machine.

Some electrode covering ingredients tend to stabilize the arc. These are necessary ingredients for an electrode to operate well on ac. A few of these ingredients are titanium dioxide, feldspar, and various potassium compounds (including the binder, potassium silicate). The inclusion of one or more of these arc stabilizing compounds in the covering provides a large number of readily ionized particles and thereby contributes to ionization of the arc stream. Thus, the electrode,

Table 2.3—Typical shielded metal arc electrode orientation and welding technique for carbon steel electrodes

Type of joint	Position of welding	Work angle, deg	Travel angle, deg	Technique of welding
Groove	Flat	90	5-10[a]	Backhand
Groove	Horizontal	80-100	5-10	Backhand
Groove	Vertical-up	90	5-10	Forehand
Groove	Overhead	90	5-10	Backhand
Fillet	Horizontal	45	5-10[a]	Backhand
Fillet	Vertical-up	35-55	5-10	Forehand
Fillet	Overhead	30-45	5-10	Backhand

a. Travel angle may be 10° to 30° for electrodes with heavy iron powder coatings.

the power source, and the welder, all contribute to arc stability.

ARC BLOW

Arc blow, when it occurs, is encountered principally with dc welding of magnetic materials (iron and nickel). It may be encountered with ac, under some conditions, but those cases are rare and the intensity of the blow is always much less severe. Direct current, flowing through the electrode and the base metal, sets up magnetic fields around the electrode which tend to deflect the arc from its intended path. The arc may be deflected to the side at times, but usually it is deflected either forward or backward along the joint. Back blow is encountered when welding toward the ground near the end of a joint or into a corner. Forward blow is encountered when welding away from the ground at the start of the joint, as shown in Fig. 2.15.

Arc blow may become so severe at times that a satisfactory weld cannot be made. Incomplete fusion and excessive weld spatter result. To those who are welding with iron powder electrodes or to those with other electrodes that produce a large amount of slag, foward blow is especially troublesome. It permits the molten slag, which normally is confined to the edge of the crater, to run forward under the arc.

The bending of the arc under these conditions is caused by the effects of an unbalanced magnetic field. When there is a greater concentration of magnetic flux on one side of the arc than on the other, the arc always bends away from the greater concentration. The source of the magnetic flux is indicated by the electrical rule which states that a conductor carrying an electric current produces a magnetic flux in circles around the conductor. These circles are in planes perpendicular to the conductor and are centered on the conductor.

In welding, this magnetic flux is superimposed on the steel and across the gap to be welded. The flux in the plate does not cause difficulty, but unequal concentration of flux across the gap or around the arc causes the arc to bend away from the heavier concentration. Since the flux passes through steel many times more readily than it does through air, the path of the flux tends to remain within the steel plates. For this reason, the flux around the electrode, when the electrode is near either end of the joint, is concentrated between the electrode and the end of the plate. This high concentration of flux on one side of the arc, at the start or the finish of the weld, deflects the arc away from the ends of the plates.

Forward blow exists for a short time at the start of a weld, then it diminishes. This is because the flux soon finds an easy path through the weld metal. Once the magnetic flux behind the arc is concentrated in the plate and the weld, the arc is influenced mainly by the flux in front of it as this flux crosses the root opening. At this point, back blow may be encountered. Back blow can occur right up to the end of the joint. As the weld approaches the end, the flux ahead of the arc becomes more crowded, increasing the back blow. Back blow can become extremely severe right at the very end of the joint.

The welding current passing through the work creates a magnetic field around it. The field is perpendicular to the path of current between the arc and the ground clamp. The flux field around the arc is perpendicular to the one in the work. This concentrates the magnetic flux on the ground side of the arc and tends to push the arc away. The two flux fields mentioned above are, in reality, one field. That field is perpendicular to the path of the current through the cable, the work, the arc, and the electrode.

Unless the arc blow is unusually severe, certain corrective steps may be taken to eliminate it or, at least, to reduce its severity. All or only some of the following steps may be necessary:

(1) Place ground connections as far as possible from the joints to be welded.

(2) If back blow is the problem, place the ground connection at the start of welding, and weld toward a heavy tack weld.

(3) If forward blow causes trouble, place

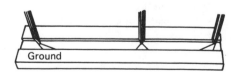

Fig. 2.15—The effect of ground location on magnetic arc blow

the ground connection at the end of the joint to be welded.

(4) Position the electrode so that the arc force counteracts the arc blow.

(5) Use the shortest possible arc consistent with good welding practice. This helps the arc force to counteract the arc blow.

(6) Reduce the welding current.

(7) Weld toward a heavy tack or runoff tab.

(8) Use the backstep sequence of welding.

(9) Change to ac, which may require a change in electrode classification.

(10) Wrap the ground cable around the workpiece in a direction such that the magnetic field it sets up will counteract the magnetic field causing the arc blow.

QUALITY OF THE WELD

A welded joint must possess those qualities which are necessary to enable it to perform its expected function in service. To accomplish this, the joint needs to have the required physical and mechanical properties. It may have to have a certain microstructure and chemical composition to meet these properties. The size and shape of the weld also are involved as is the soundness of the joint. Corrosion resistance also may be required. All of these are influenced by the base materials, the welding materials, and the manner in which the weld is made.

Shielded metal arc welding is a manual welding process, and the quality of the joint depends also on the skill of the welder who makes it. For this reason, the materials to be used must be selected with care, the welder must be proficient, and the procedure he uses must be correct.

Welded joints, by their nature, contain discontinuities of various types and sizes. Below some acceptable level, these are not considered harmful. Above that level they are considered defects. The acceptance level varies with the severity of the service to be encountered.

Discontinuities that are sometimes encountered in welds made by the SMAW process are

(1) Porosity
(2) Slag inclusions
(3) Incomplete fusion
(4) Undercut
(5) Cracks

Porosity

This term is used to describe gas pockets or voids in the weld metal. These voids result from gas that forms from certain chemical reactions that take place during welding. They contain gas rather than solids, and, in this respect, they differ from slag inclusions.

Porosity usually can be prevented by using proper amperage and holding a proper arc length. Dry electrodes are also helpful in many cases. The deoxidizers which a covered electrode needs are easily lost during deposition when high amperage or a long arc is used. This leaves a supply which is insufficient for proper deoxidation of the molten metal.

Slag Inclusions

This term is used to describe the oxides and nonmetallic solids that sometimes are entrapped in weld metal, between adjacent beads, or between the weld metal and the base metal. During deposition and subsequent solidification of the weld metal, many chemical reactions occur. Some of the products of these reactions are solid nonmetallic compounds which are insoluble in the molten metal. Because of their lower specific gravity, these compounds will rise to the surface of the molten metal unless they become entrapped within the weld metal.

Slag formed from the covering on shielded metal arc electrodes may be forced below the surface of the molten metal by the stirring action of the arc. Slag may also flow ahead of the arc if the welder is not careful. This can easily happen when welding over the crevasse between two parallel but convex beads or between one convex bead and a side wall of the groove. It can also happen when the welding is done downhill. In such cases, the molten metal may flow over the slag, entrapping the slag beneath the bead. Factors such as highly viscous or rapidly solidifying slag or insufficient welding current set the stage for this.

Most slag inclusions can be prevented by good welding practice and, in problem areas, by proper preparation of the groove before depositing the next bead of weld metal. In these cases, care must be taken to correct contours that are difficult to adequately penetrate with the arc.

Incomplete Fusion

This term, as it is used here, refers to the failure to fuse together adjacent beads of weld metal or weld metal and base metal. This condition may be localized or it may be extensive, and it can occur at any point in the welding groove. It may even occur at the root of the joint.

Incomplete fusion may be caused by failure to raise the base metal (or the previously deposited bead of weld metal) to the melting temperature. It may also be caused by failure to dissolve, because of improper fluxing, any oxides or other foreign material that might be present on the surfaces which must fuse with the weld metal.

Incomplete fusion can be avoided by making certain that the surfaces to be welded are properly prepared and fitted and are smooth and clean. In the case of incomplete root fusion, the corrections are to make certain that the root face is not too large; the root opening is not too small; the electrode is not too large; the welding current is not too low; and the travel speed is not too high.

Undercut

This term is used to describe either of two situations. One is the melting away of the sidewall of a welding groove at the edge of the bead, thus forming a sharp recess in the sidewall in the area in which the next bead is to be deposited. The other is the reduction in thickness of the base metal at the line where the beads in the final layer of weld metal tie into the surface of the base metal (e.g., at the toe of the weld).

Both types of undercut usually are due to the specific welding technique used by the welder. High amperage and a long arc increase the tendency to undercut. Incorrect electrode position and travel speed also are causes, as is improper dwell time in a weave bead. Even the type of elec-

trode used has an influence. The various classifications of electrodes show widely different characteristics in this respect. With some electrodes, even the most skilled welder may be unable to avoid undercutting completely in certain welding positions, particularly on joints with restricted access.

Undercut of the sidewalls of a welding groove will in no way affect the completed weld if the undercut is removed before the next bead is deposited at that location. A well-rounded chipping tool or grinding wheel will be required to remove the undercut. If the undercut is slight, however, an experienced welder who knows just how deep his arc will penetrate may not need to remove the undercut.

Undercut at the surface should not be permitted except when it is very shallow because it may materially reduce the strength of the joint. This is particularly true in applications subject to fatigue. Fortunately, this type of undercut can be detected by visual examination of the completed weld, and it can be corrected by blend grinding or depositing an additional bead.

Cracks

Cracking of a welded joint, when it occurs, can be classified as either hot cracking or cold cracking. Cracking of either type, or of both types, can occur in the weld metal or in the base metal, or in both. When cracking occurs in one bead of weld metal, it can follow each subsequent pass through the weld. To prevent this, such cracks must be removed before subsequent beads are deposited.

Hot cracking is a function of chemical composition. The main cause of hot cracking is constituents in the weld metal which have a relatively low melting temperature and which accumulate at the grain boundaries during solidification. A typical example is iron sulfide in steel. The cracks are intergranular or interdendritic. They form as the weld metal cools. As solidification progresses in the cooling weld metal, the shrinkage stresses increase and eventually draw apart those grains which still have some liquid at their boundaries. Coarse-grained, single-phase structures have a marked propensity to this type of cracking. The cures are to improve the purity of the material, add a small amount of a second structure (ferrite in austenitic stainless steel, for example), and use the proper welding procedure. This includes low preheat and interpass temperatures. Welding current should be reduced within the operating range. Grain refinement and better deoxidation are helpful. Manganese additions are beneficial in some materials, such as steels.

Cold cracking is the result of inadequate ductility or the presence of hydrogen in hardenable steels. Inadequate toughness in the presence of a mechanical or metallurgical notch and stresses of sufficient magnitude cause this condition. These stresses do not have to be very high in some materials—large-grained ferritic stainless steel, for instance.

To prevent cold cracking in hardenable steels, the use of dry low hydrogen electrodes and proper preheat is required. Preheat is also required for those materials which are naturally low in ductility or toughness. Materials which are subject to extreme grain growth (28 percent chromium steel, for instance) must be welded with low heat input and low interpass temperatures. Notches need to be avoided.

More information on the weld quality of welded joints can be found in the *Welding Handbook,* Chapter 5, Volume 1, 7th edition and Chapter 6, Section 1, 6th edition. *Welding Inspection,* published by AWS, also is a good reference.

SUPPLEMENTARY READING LIST

Barbin, L.M. The new moisture resistant electrodes. *Welding Journal* 56 (7), July 1977, pp. 15-18.

Chew, B. Moisture loss and gain by some basic flux covered electrodes. *Welding Journal* 55 (5), May 1976, pp. 127-s'—134-s.

Gregory, E.N. Shielded metal arc welding of galvanized steel. *Welding Journal* 48 (8), Aug. 1969, pp. 631-638.

Jackson, C.E. Fluxes and Slags in Welding. Bulletin 190. New York: Welding Research Council, Dec. 1973.

Metals Handbook, 8th ed., Vol. 6, *Welding and Brazing.* Metals Park, Ohio: American Society for Metals, 1971, pp. 46-77.

Silva, E.A. and Hazlett, T.H. Shielded metal arc welding underwater with iron powder electrodes. *Welding Journal* 50 (6), June 1971, pp. 406-s—415-s.

Stout, R.D., and Doty, W.D. *Weldability of Steels,* 2nd ed., ed. Epstein, S., and Somers, R.E. New York: Welding Research Council, 1971.

The Procedure Handbook of Arc Welding, 12th ed. Cleveland: Lincoln Electric Company, 1973.

3

Gas Tungsten Arc Welding

Prepared by a committee consisting of

L.E. STARK, *Chairman*
    The Babcock & Wilcox Co.
E.F. GORMAN
    Airco, Inc.

J.A. HOGAN
    Hypertherm, Inc.
E.P. VILKAS
    Astro-Arc Co.

3

Gas Tungsten Arc Welding

FUNDAMENTALS OF THE PROCESS

DEFINITIONS AND GENERAL DESCRIPTION

Gas tungsten arc welding (GTAW) is a process wherein coalescence of metals is produced by heating them with an arc between a tungsten (nonconsumable) electrode and the work. Gas tungsten arc welding is generally done with a single electrode, but multiple electrodes are sometimes used. Shielding of the electrode and weld zone is obtained from a gas or gas mixture. Filler metal may or may not be added. Figure 3.1 shows the relative positions of the GTAW torch, the arc, the tungsten electrode, the gas shield, and the welding rod (wire) as it is being fed into the arc and weld pool. The welding rod guide is used only for mechanized or automatic welding. For manual welding the rod is hand held. A backing bar (as shown) may or may not be used depending on the joint design.

The gas tungsten arc welding process is sometimes called "TIG" (tungsten inert gas) welding but the preferred letter designation is GTAW.

The basic features of the equipment used for the process are shown in Fig. 3.2. The major equipment components required for GTAW are (1) the welding machine (power source), (2) the welding electrode holder and the tungsten electrode, and (3) the shielding gas supply and controls. Several optional accessories are available. These include a foot rheostat which permits the welder to control current while welding, water

circulating systems to cool the electrode holder, arc timers, and other accessories which are described later.

The hot tungsten electrode and the weld metal will oxidize rapidly during welding if exposed to air. Therefore, the shielding gas must be chiefly inert consisting of helium, argon, or a mixture which will protect both the electrode and the weld pool from oxidation.

PRINCIPLES OF OPERATION

The heat for gas tungsten arc welding is produced by an electric arc between the nonconsumable electrode and the part to be welded. The electrode used to carry the current is a pure tungsten or tungsten alloy rod. The heated weld zone, the molten metal, and the tungsten electrode are shielded from the atmosphere by a blanket of inert gas fed through the electrode holder. A weld is made by applying the arc so that the abutting workpieces and filler metal are melted and joined together as the weld metal solidifies.

The electric arc is produced by the passage of current through the ionized inert shielding gas. The ionized atoms lose electrons and are left with a positive charge. The positive gas ions flow from the positive to the negative pole of the arc. The electrons travel from the negative to the positive pole. The power expended in an arc, expressed in electrical units, is the product of the current passing through the arc and the voltage drop across the arc. The electrical features of a

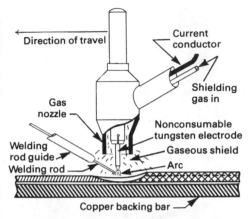

Fig. 3.1 — Gas tungsten arc welding

welding arc are discussed in Chapter 2, "Physics of Welding," *Welding Handbook,* Volume 1, 7th edition.

Before welding is started, all oil, grease, paint, rust, dirt, or other contaminants should be removed from the areas to be welded. This may be accomplished by mechanical means or by the use of vapor or liquid cleaners.

Striking the arc may be done in the following ways: (1) by touching the electrode to the work momentarily and quickly withdrawing it a short distance, (2) by means of an apparatus that will cause a spark to jump from the electrode to the work, or (3) by means of an apparatus that initiates and maintains a small pilot arc, which provides an ionized path for the main arc. High frequency arc stabilizers, which are required when alternating current is used, will provide for the second type of arc starting. High frequency arc initiation occurs when a high frequency, high voltage signal is superimposed on the welding circuit. High voltage (low current) ionizes the shielding gas between the electrode and work, thereby making the gas conductive and initiating the arc. Inert gases are not conductive until ionized. For dc welding, the high frequency voltage is cut off after arc initiation. However, with ac welding it usually remains on during welding, particularly for aluminum welding.

For manual welding, once the arc is started, the electrode holder (torch) is held at a travel angle of about 15 degrees. For mechanized welding, the electrode holder is generally positioned vertical to the surface. To start manual welding, the arc is usually moved in a small circle until a pool of molten metal of suitable size is obtained. Once adequate fusion is achieved at any one point, a weld is made by gradually moving the electrode along the parts to be welded so as to progressively melt the adjoining surfaces. Solidification of the molten metal follows progression of the arc along the joint and completes the welding cycle.

Welding is usually stopped by shutting off the current with foot- or hand-controlled switches. Welding may also be stopped by quickly withdrawing the electrode from the workpiece, but this method disturbs the gas shielding and exposes the tungsten and the weld pool to oxidation.

Foot and hand controls that permit the welder to start, adjust, and stop welding current are available. They provide the welder with a means of controlling welding current as required to obtain good fusion and penetration.

The base metal thickness and joint design determine whether or not filler metal need be added to the joints. When filler metal is added during manual welding, it is applied by manually feeding the welding rod into the pool of molten metal ahead of the arc. The technique for manual welding is illustrated in Fig. 3.3.

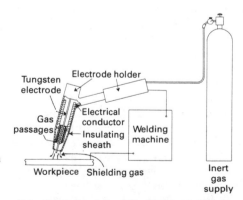

Note: A water-cooled welding torch is used when cooling from the inert gas shield is inadequate.

Fig. 3.2 — Gas tungsten arc welding equipment arrangement

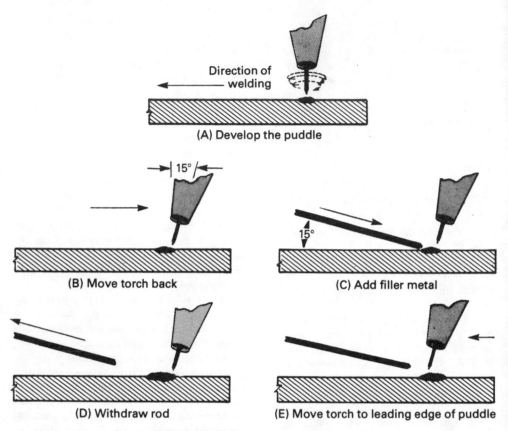

(A) Develop the puddle

(B) Move torch back

(C) Add filler metal

(D) Withdraw rod

(E) Move torch to leading edge of puddle

Fig. 3.3—Technique for manual gas tungsten arc welding

The welding rod and torch must be moved progressively and smoothly so the weld pool, the hot welding rod end, and the hot solidified weld are not exposed to air that will contaminate the weld metal area or heat-affected zone. Generally, a large shielding gas envelope will prevent exposure to air.

The welding rod is usually held at an angle of about 15 degrees to the surface of the work and slowly fed into the molten pool. During welding, the hot end of the welding rod must not be removed from the protection of the inert gas shield. A second method is to press the welding rod against the work, in line with the weld, and melt the rod along with the joint edges. This method is used often in multiple pass welding of V-groove joints. A third method, used frequently in weld surfacing and in the making of large

welds, is to feed filler metal continuously into the molten weld pool by oscillating the welding rod and arc from side to side. The welding rod moves in one direction while the arc moves in the opposite direction, but the welding rod is at all times near the arc and feeding into the molten pool. When filler metal is required in automatic welding, the welding rod (wire) is fed mechanically through a guide into the molten weld pool in a uniform manner.

The selection of welding position is determined by the mobility of the weldment, the availability of tooling and fixtures, and the welding cost. The minimum time, and therefore cost, for producing a weld is usually achieved in the flat position. Maximum joint penetration and deposition rate are obtained in this position, because a large volume of molten metal can be supported. Also, an acceptably shaped rein-forcement is easily obtained in this position.

Good penetration can be achieved in the vertical-up position, but the rate of welding is slower because of the effect of gravity on the molten weld metal. Penetration in vertical-down welding is poor. The molten weld metal droops, and lack of fusion occurs unless high welding speeds are used to deposit thin layers of weld metal. The welding torch is usually pointed forward at an angle of about 75 degrees from the weld surface in the vertical-up and flat positions. Too great an angle causes aspiration of air into the shielding gas and consequent oxidation of the molten weld metal.

The joints that may be welded by this process include all the standard types, such as square-groove and V-groove joints, T-joints, and lap joints. As a rule, it is not necessary to bevel the edges of base metal that is 3.2 mm (1/8 in.) or less in thickness. Thicker base metal is usually beveled and filler metal is always added.

The gas tungsten arc welding process can be used for continuous welds, intermittent welds, or for spot welds. It can be done manually or automatically by machine.

The major operating variables summarized briefly are

(1) Welding current, voltage, and power source characteristics

(2) Electrode composition, current carry-ing capacity, and shape

(3) Shielding gas—welding grade argon, helium, or mixtures of both

(4) Filler metals that are generally similar to the metal being joined and suitable for the intended service

ADVANTAGES AND LIMITATIONS

The gas tungsten arc welding process is very good for joining thin base metals because of the excellent control of heat input. As in oxy-acetylene welding, the heat source and the addition of filler metal can be separately controlled. Because the electrode is nonconsumable, the process can be used to weld by fusion alone without the addition of filler metal. It can be used on almost all metals, but it is generally not used for the very low melting metals such as solders, or lead, tin, or zinc alloys. It is especially useful for joining aluminum and magnesium which form refractory oxides, and also for the reactive metals like titanium and zirconium, which dissolve oxygen and nitrogen and become embrittled if exposed to air while melting.

The process lends itself to high quality welding. In very critical service applications or for very expensive metals or parts, the materials should be carefully cleaned of surface dirt, grease, and oxides in preparation for welding.

Some limitations of the gas tungsten arc process are

(1) The process is slower than consumable electrode arc welding processes.

(2) Transfer of molten tungsten from the electrode to the weld causes contamination. The resulting tungsten inclusion is hard and brittle.

(3) Exposure of the hot filler rod to air using improper welding techniques causes weld metal contamination.

(4) Inert gases for shielding and tungsten electrode costs add to the total cost of welding compared to other processes. Argon and helium used for shielding the arc are relatively expensive.

(5) Equipment costs are greater than that for other processes, such as shielded metal arc welding, which require less precise controls.

For these reasons, the gas tungsten arc welding process is generally not commercially competitive with other processes for welding the heavier gages of metal if they can be readily welded by the shielded metal arc, submerged arc, or gas metal arc welding processes with adequate quality.

EQUIPMENT

ELECTRICAL

Of the many variables that affect gas tungsten arc welding, those with a particularly strong effect on the process are the electrical variables of current, voltage, and power source characteristics. Besides their obvious relation to the amount, distribution, and control of arc-produced heat, they also play a role in arc stability and in the removal of refractory oxide from the surfaces of certain metals. They interact with each other and are affected by other process variables that are discussed in this chapter. Their interaction is in large measure determined by the nature of the arc (see Chapter 2, "Physics of Welding," *Welding Handbook,* Volume 1, 7th edition).

Steady Direct Current

Two electrical connections can be used for steady direct current. The tungsten electrode may be connected to either the negative or the positive terminal of the power supply. The electrically charged particles in the arc flow in the direction shown in Fig. 3.4. As indicated in this figure, a positive electrode (reverse polarity) must have a larger diameter than a negative electrode for the same current. The heat of the dc arc is concentrated at the anode or positive terminal. The larger electrode is required to dissipate the heat developed at the anode.

A negatively connected electrode (straight polarity) can be used with argon, helium, or a mixture of the two to weld metals. The refractory oxide on the surfaces of aluminum, magnesium, and their alloys hinders fusion. To weld them with direct current, a positive electrode (reverse polarity) can be used. The oxide film is removed from the metal surface using dcrp (electrode positive). The current carrying capacity of a positive electrode is about 1/10 that of a negative electrode. Therefore, reverse polarity current is generally limited to welding sheet metal. Using dcrp (electrode positive) permits welding of thin gages with sufficient amperage for a stable arc. However, weld penetration is shallower and the weld bead is wider than with dcsp (electrode

negative). Adequate fusion of aluminum and its alloys can be obtained with a negative electrode (straight polarity) and a very short arc length when pure helium is used and the surface oxides are removed prior to welding. This technique is used almost exclusively in machine welding.

Pulsed Direct Current

Pulsed direct current with a negative electrode has been used to weld pipe automatically. When the current is pulsed, a circumferential weld in pipe can be made in the horizontal fixed pipe position with the same average current and voltage at all points around the pipe except at the beginning and end. Without pulsing, the welding conditions would have to be adjusted as the arc moved from the flat position through the vertical to the overhead position. This type of current reduces both distortion and the difficulty of bridging gaps. The pulsing usually takes place at an approximate rate of several pulses per second.

Alternating Current

With alternating current, the advantage of a positive electrode can be gained without the current limitation that is encountered with a positive electrode in dc welding. This advantage is offset to some extent by the need to provide means for the maintenance of a stable arc. Figure 3.4 compares ac welding with dc welding.

When the tungsten electrode and weld metal have different abilities to emit electrons, arc currents during each half cycle will be unequal. The tungsten emits electrons more readily because it attains a much higher temperature than the metal being welded. With a conventional transformer of the type used for shielded metal arc welding, the difference in resistance to current flow tends to produce unbalanced alternating current. The current amplitude during the half cycle when the electrode is negative is greater than the amplitude during the half cycle when the electrode is positive. Rectification of the current, which is a complete absence of the half cycle during which the tungsten electrode is

Current type	DC	DC	AC (balanced)
Electrode polarity	Negative	Positive	
Electron and ion flow			
Penetration characteristics			
Oxide cleaning action	No	Yes	Yes—once every half cycle
Heat balance in the arc (approx.)	70% at work end 30% at electrode end	30% at work end 70% at electrode end	50% at work end 50% at electrode end
Penetration	Deep; narrow	Shallow; wide	Medium
Electrode capacity	Excellent e.g., 3.18 mm (1/8 in.)—400 A	Poor e.g., 6.35 mm (1/4 in.)—120 A	Good e.g., 3.18 mm (1/8 in.)—225 A

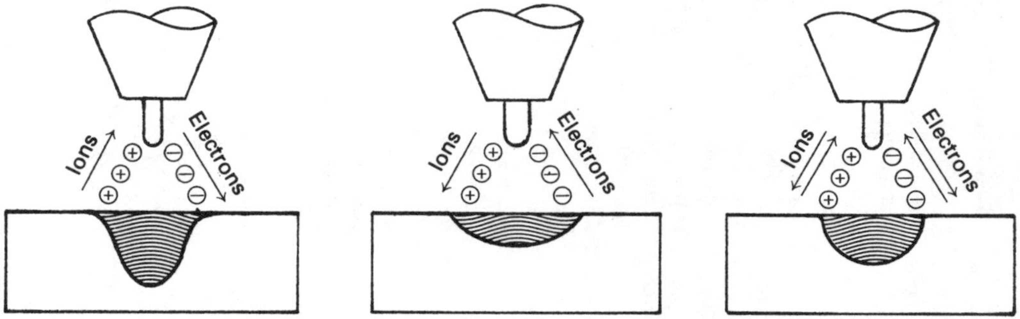

Fig. 3.4—Characteristics of current types for gas tungsten arc welding

positive, also may occur. Both of these conditions are described in Fig. 3.5.

To obtain balanced current flow, either series-connected capacitors or a dc voltage source can be inserted in the welding circuit. Although desirable for some applications, balanced current flow is not essential for most manual welding operations. It is, however, desirable for high speed mechanized welding. The advantages of balanced current flow are

(1) Better oxide-cleaning action is obtained.

(2) Smoother, better welding action results.

(3) The reduction in output rating of a given size of conventional welding transformer is not required. (The unbalanced core magnetization that is produced by the dc component of an unbalanced current flow is minimized.)

The disadvantages of balanced current flow are

(1) A larger tungsten electrode is required.

(2) The higher open-circuit voltages that are generally associated with some wave balancing means may constitute a safety hazard.

(3) Wave-balanced welding power sources cost more than conventional ones.

To avoid rectification, the open-circuit voltage of the transformer can be increased. An open-circuit voltage of about 100 V (rms) is needed with helium shielding. The necessary voltage can also be obtained by adding, in series with the transformer, a large high frequency voltage supply. The high frequency voltage is generally on the order of several thousand volts, and its frequency can be as high as several megahertz. The current is very low. The high frequency voltage may be applied continuously or periodically during welding. In the latter case, a burst of high frequency voltage is set to occur during the time when the welding current passes through zero. The voltage is high so that it may be used for arc initiation without touching the electrode to the work. Instead of a short burst of high frequency voltage at zero current, an accurately placed single pulse of voltage may be used. High frequency arc initiation is described under "Arc Starting Means," p. 95.

Square wave ac power sources are available. These power sources allow adjustment of the wave form so that penetration and cleaning can be adjusted to suit the application. The lower "peak" current of the square wave form tends to increase the usable current range of the electrode.

Power Source Characteristics

The power source used for ac and dc gas tungsten arc welding is usually a drooping-voltage type. In this type power source, the slope of the volt-ampere curve is relatively steep so that a change in arc voltage (arc length) will not create a major change in arc current.

The relation of the power source volt-ampere characteristic to the operating conditions of the arc is shown in Fig. 3.6. Superimposed on power source volt-ampere curves for two different settings are two volt-ampere curves that represent steady state load conditions for two different arc lengths, L1 and L2. The shapes of the curves are typical for gas tungsten arc welding. The exact position of such a curve is a function of arc length and composition of the shielding gas. For an arc length L1 and a power source setting 2, the welding voltage and current are shown by the intersection A of the power source and the arc length curves. If the arc length is increased slowly from L1 to L2 without changing the power source setting, the

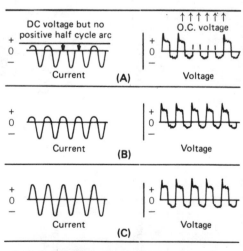

Fig. 3.5—Voltage and current wave forms for ac welding: (A) partial and complete rectification; (B) with arc stabilization; (C) with current balancing

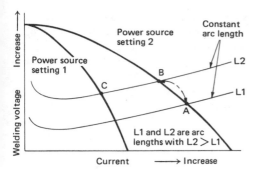

Fig. 3.6—Relation of power source character-
istics to operating conditions of the arc
the arc

operating conditions during the change will follow the power source curve to point B, which
shows the new welding voltage and current for
the increased arc length. If the increase is rapid,
the power source reactance will cause the current
to remain constant during the first instant of
change. If, on the other hand, the power source
setting is then changed from 2 to 1 (to achieve a
lower current) while the arc length is held constant, the operating conditions will change from
B to C.

True constant current type power supplies,
i.e., vertical volt-ampere characteristic, are
also available. The constant current feature is
usually obtained with SCR control of the output
transformer or reactor. In this type of power supply, the welding current is unaffected by variations in arc length (voltage). This feature is
especially desirable for applications where
maintaining a fixed arc length is less critical.

ELECTRODES

Types

Electrodes for gas tungsten arc welding are
classified as pure tungsten, tungsten with one or
two percent thoria, tungsten with 0.15 to 0.4
percent zirconia, and a tungsten electrode with an
integral lateral segment of tungsten throughout
its length which contains one to two percent
thoria. These tungsten electrodes are commonly
available in diameters ranging from 0.25 to
6.35 mm (0.010 to 0.250 in.) and lengths ranging from 76 to 610 mm (3 to 24 in.) with either
clean or ground finish. Clean finish refers to
chemical cleaning to remove surface impurities
after the drawing or swaging operation. Ground
finish refers to removal of surface imperfections
by grinding. Requirements for tungsten arc
welding electrodes are given in AWS A5.12,
Specification for Tungsten Arc-Welding Electrodes. Table 3.1 gives the chemical requirements for the electrodes.

Tungsten electrodes of 99.5 percent purity
(EWP) are less expensive and are generally used
on less critical operations than tungsten electrodes that contain thoria or zirconia. A pure
tungsten electrode has a relatively low current
carrying capacity with ac power and has a low
resistance to contamination.

Tungsten electrodes with one or two percent
thoria are superior to pure tungsten electrodes
in several respects. They have higher electron
emissivity, better current carrying capacity,
longer life, and greater resistance to contamination. With these electrodes, arc starting is easier,
and the arc is more stable.

Tungsten electrodes that contain some

Table 3.1—Chemical requirements of gas tungsten arc welding electrodes

AWS classification	Tungsten, min, percent	Thoria, percent	Zirconia, percent	Total other elements, max, percent
EWP	99.5	—	—	0.5
EWTh-1	98.5	0.8 to 1.2	—	0.5
EWTh-2	97.5	1.7 to 2.2	—	0.5
EWTh-3[a]	98.95	0.35 to 0.55	—	0.5
EWZr	99.2	—	0.15 to 0.40	0.5

a. A tungsten electrode with an integral lateral segment throughout its length which contains
1.0 to 2.0 percent thoria. The average thoria content of the electrode shall be as specified in this
table.

zirconia (EWZr) have properties that generally fall between those of pure and thoria-containing tungsten. However, there is some indication of better performance in certain types of welding with alternating current. EWZr electrodes, when used for ac welding, combine the desirable stability characteristics of pure tungsten with the capacity and starting characteristics of thoriated tungsten.

Current Carrying Capacity

The current carrying capacity of all types of tungsten electrodes is affected by the type of electrode holder, the extension of the electrode from the holder, the position of welding, the shielding gas, and the type of welding current. Approximate current ranges for electrodes are given in Table 3.2. Since the maximum current capacity of an electrode depends on a large number of factors, the typical current ranges should be used only as guides.

About two-thirds of the heat is generated at the anode and one-third at the cathode. Therefore, an electrode can carry much higher current without overheating when it is negative (straight polarity) than when it is positive (reverse polarity). Similarly, the dc capacity of an electrode when it is negative is greater than its ac capacity, and the ac capacity with a balanced wave is less than it is with an unbalanced wave.

The dc capacity of a negative electrode is not increased as much as might be expected by the addition of thoria or zirconia, even though the addition of these elements increases electron emission and reduces the temperature of the electrode tip. The reason is that the capacity is limited by electrical resistance heating. At excessive currents, the electrode overheats and melts. Tungsten electrodes with thoria or zirconia are selected over pure tungsten electrodes for good arc-starting characteristics and contamination resistance.

Shape

With ac welding, a molten hemisphere forms at the tip of a pure tungsten electrode at its minimum usable current, and it does not become perceptibly larger as the current is increased to full usable capacity. At still higher currents, it increases its size as it forms an unwanted droplet. The molten hemisphere tip is most desirable for welding.

Thoriated tungsten electrodes do not ball so readily and, therefore, cannot be used for low currents without a tapered point. With their greater current carrying capacity, they can operate readily with a point or taper for more reliable arc starting with high frequency ignition and for a more stable arc. The degree of taper also affects weld penetration; decreasing the included angle of the electrode taper tends to reduce the width of the weld bead and increase weld penetration.

ELECTRODE HOLDER AND NOZZLE

The tungsten electrode holder must have sufficient welding current capacity to prevent overheating. Collets accommodate the correct sizes of tungsten electrodes. The direction and amount of inert gas covering the weld is controlled by gas nozzles threaded into the head of the electrode holder. The gas nozzles are made of various heat resistant materials in different diameters, shapes, and lengths. Length and shape are selected on the basis of joint accessibility and the required clearance between the nozzle and the work. The nozzle should be large enough to provide adequate inert gas coverage of the molten weld pool and adjacent hot base metal.

Specially designed gas nozzles are available that reduce gas turbulence at the end of the nozzle. With them, welding can be done with the nozzle as far as 25 mm (1 in.) from the work. This improves the welder's ability to see the weld puddle and allows him to reach difficult places, such as inside corners.

In some applications, two or more tungsten electrode holders are used either in tandem on the same side of the weld, or in line with one another on opposite sides of the weld. In the tandem technique, the first electrode provides preheat to the base metal, which helps to provide deeper penetration and higher weld speeds. The opposed torches may be used in a similar manner with the preheat torch approximately 2.4 mm (3/32 in.) ahead of the other. Individual power sources for each arc are required for the best control and quality of weld.

Table 3.2—Typical current ranges for tungsten electrodes[a]

| Electrode diameter | | Direct current, A | | Alternating current, A | | | | | |
| | | Straight polarity EWP EWTh-1 EWTh-2 EWTh-3 | Reverse polarity EWP EWTh-1 EWTh-2 EWTh-3 | Unbalanced wave | | | Balanced wave | | |
mm	in.			EWP	EWTh-1 EWTh-2 EWZr	EWTh-3	EWP	EWTh-1 EWTh-2 EWZr	EWTh-3
0.26	0.010	up to 15	b	up to 15	up to 15	b	up to 15	up to 15	b
0.51	0.020	5-20	b	5-15	5-20	b	10-20	5-20	10-20
1.02	0.040	15-80	b	10-60	15-80	10-80	20-30	20-60	20-60
1.59	1/16	70-150	10-20	50-100	70-150	50-150	30-80	60-120	30-120
2.38	3/32	150-250	15-30	100-160	140-235	100-235	60-130	100-180	60-180
3.18	1/8	250-400	25-40	150-210	225-325	150-325	100-180	160-250	100-250
3.97	5/32	400-500	40-55	200-275	300-400	200-400	160-240	200-320	160-320
4.76	3/16	500-750	55-80	250-350	400-500	250-500	190-300	290-390	190-390
6.35	1/4	750-1000	80-125	325-450	500-630	325-630	250-400	340-525	250-525

a. All values are based on the use of argon as the shielding gas. Other current values may be employed depending on the shielding gas, type of equipment, and application.
b. These combinations are not commonly used.

SHIELDING GASES

The inert gases, argon and helium, are used for gas tungsten arc welding. Neon, xenon, and krypton, the other inert gases, are not employed for welding because of their scarcity and relatively high cost. Of the reactive gases, only hydrogen and nitrogen have found limited use. Hydrogen is added to argon or helium in small quantities for mechanized welding of stainless steel. Nitrogen is sometimes added to argon for the joining of copper and copper alloys.

Argon

Argon (Ar) is a heavy monatomic gas with an atomic weight of 40. It is obtained from the atmosphere by liquefaction of air. After argon is refined to purities on the order of 99.99 percent, it may be stored and transported as a liquid at temperatures below −184° C (−300° F). Depending on the volume of use, argon may be supplied as a liquid or as a compressed gas. Because of the economics of liquid distribution, liquid argon for bulk storage can be purchased at much lower prices than cylinder argon, but equipment must be installed to vaporize the gas for distribution.

Argon is used more extensively than helium because of the following advantages:

(1) Smoother, quieter arc action

(2) Lower arc voltage at any given current value and arc length

(3) Greater cleaning action in the welding of such materials as aluminum and magnesium with alternating current

(4) Lower cost and greater availability

(5) Lower flow rates for good shielding

(6) Better cross-draft resistance

(7) Easier arc starting

The lower arc voltage characteristic of argon is particularly helpful in the manual welding of thin material because the tendency for excessive melt-thru is lessened. This same characteristic is advantageous in vertical or overhead welding since the tendency for the base metal to sag or run is decreased.

Helium

Helium (He) is the lightest monatomic gas, having an atomic weight of four. It is separated from natural gas. Welding grade helium is refined to a purity of at least 99.99 percent. Although some helium is distributed as a liquid, it is more commonly transported as a gas in high pressure cylinders. Special railroad cars and trailers are used for transporting gaseous helium, as they are for argon.

Because of its greater thermal conductivity, helium requires higher arc voltages and energy inputs than argon. Consequently, the greater power of the helium arc can be advantageous for the joining of metals of high conductivity and for high-speed mechanized applications. Also, helium is used more often than argon for welding heavy plate. Mixtures of argon and helium are useful when some balance between the characteristics of both is desired.

Characteristics of Argon and Helium

The chief factor influencing shielding effectiveness is the gas density. Argon is approximately one and one-third times as heavy as air and ten times as heavy as helium. Argon, after leaving the torch nozzle, forms a blanket over the weld area. Helium, because it is lighter, tends to rise in a turbulent fashion around the nozzle. Experimental work has consistently shown that to produce equivalent shielding effectiveness, the flow of helium must be two to three times that of argon. The same general relationship is true for mixtures of argon and helium, particularly those high in helium content.

The important characteristics are the voltage-current relationships of the tungsten arc in argon and in helium that are illustrated in Fig. 3.7. It

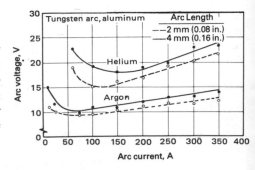

Fig. 3.7—Voltage-current relationship with argon and helium shielding

is apparent that, for equivalent arc lengths and welding currents, the arc voltage obtained with helium is appreciably higher than that with argon. Since heat in the arc is roughly measured by the product of current and voltage (arc power), the use of helium yields much higher available heat than does argon. The higher available heat favors the use of helium over argon at the higher current levels which are used for welding thick materials and those having high thermal conductivity or relatively high melting temperatures.

However, it should be noted that as the current decreases, the curves pass through a minimum voltage at current levels approximately 90 amperes (A) apart, and then the voltage increases as the current decreases. For helium, this increase in voltage occurs in the range of 50 to 150 A where much of the welding of thin materials is done. Since the voltage increase for argon occurs below 50 A, the use of argon in the 50 to 150 A range provides the operator with more latitude in arc length to control the welding operation.

It is apparent that to obtain equal arc power, appreciably higher currents must be used with argon than with helium. Since undercutting with either gas will occur at about equal currents, helium will produce satisfactory welds at much higher speeds.

The other influential characteristic is that of arc stability. Both gases provide excellent stability with direct current power. With alternating current power, which is used almost exclusively on aluminum and magnesium, argon yields excellent arc stability and good cleaning action. On the other hand, the use of helium results in poor arc stability and poor cleaning action. Thus, with alternating current, argon finds wide acceptance over helium.

Argon-Hydrogen

Argon-hydrogen mixtures are employed in special cases, such as mechanized welding of light gage stainless steel tubing where the hydrogen does not cause adverse metallurgical effects. Increased welding speeds can be achieved in almost direct proportion to the amount of hydrogen added to argon because of the increased arc voltage. However, the amount of hydrogen that can be added varies with the metal thickness and type of joint for each particular application. Excessive hydrogen will cause porosity. Hydrogen concentrations up to 35 percent can be used on all thicknesses of stainless steel if a root opening of approximately 0.25 to 0.5 mm (0.010 to 0.020 in.) is used. The use of argon-hydrogen mixtures is limited to stainless steel, nickel-copper, and nickel base alloys because the hydrogen produces adverse effects on most other materials.

The most commonly used argon-hydrogen mixture contains 15 percent hydrogen. This mixture is used most often for welding tight butt joints in stainless steel up to 1.6 mm (0.062 in.) thick at speeds comparable to helium (50 percent faster than argon). It is also used for welding stainless steel beer barrels and tube to tubesheet joints in a variety of stainless steels and nickel alloys. A hydrogen content of five percent is sometimes preferred for manual welding to obtain cleaner welds.

Selection of Shielding Gas

No inflexible rule governs the choice for any particular application. Either argon, helium, or a mixture of argon and helium may be used successfully for most applications with the possible exception of the manual welding of extremely thin material, for which argon is essential. Argon generally provides an arc that operates more smoothly and quietly, is handled more easily, and is less penetrating than the arc obtained by the use of helium. In addition, the lower unit cost and the lower flow rate requirements of argon make argon preferable from an economic point of view. For these reasons, argon is usually preferred for most applications, except where the higher heat penetration characteristics of helium are required for welding thick sections of high heat conductivity materials, such as aluminum or copper. A guide to the selection of gases is provided in Table 3.3.

The final selection of the "best" shielding gas must be the result of considering and weighing such factors as the type and thickness of material, joint design and fixturing, welding parameters, production rate and economics, and weld quality requirements. The following general recommendations are made primarily on the basis of metal and process but also on the influence of material thickness and the differences between manual and mechanized welding. Additional information is given in the *Welding Handbook,* Section 4, 6th edition.

Aluminum. For manual tungsten arc welding of aluminum, argon employed with alternating current and high frequency stabilization is superior to helium with direct current. Argon provides better arc starting, better cleaning action, and superior weld quality in production work than does helium.

Higher welding speeds can be obtained using helium with dcsp (electrode negative), but greater skill is required to produce satisfactory welds. When this combination is used, porosity frequently develops at the root of the joint due to insufficient cleaning action and poor control of the penetration unless a deep-grooved copper backing strip is employed. In mechanized welding, the arc length is essentially constant, and so variation in heat input is not as significant as it is in manual welding. For a high speed type application, such as tube welding, helium with dcsp (electrode negative) can be employed. However, the weld quality may be poorer than when argon is used because

of the absence of cleaning action. The cost of using helium is somewhat greater since its flow rate must be necessarily higher than that of argon for adequate shielding. Argon-helium mixtures can be employed where a compromise between speed and quality is satisfactory.

Mild Steel. Manual GTAW is employed primarily on material up to 3.2 mm (1/8 in.) thick. For applications where this process is employed, high quality, good appearing welds are required. Therefore, to provide the operator with the greatest ease of manipulation and freedom from overheating the molten metal, argon is generally used. In the welding of thicker materials, either argon or argon-helium mixtures can be employed depending on joint penetration. For manual or mechanized arc spot welding of steel, argon is superior to helium or argon-helium mixtures because of much greater electrode life, better weld nugget contour, ease of starting, and lower shielding gas cost. For the gas tungsten arc welding of pipe, argon is recommended as the shielding and backing gas. The quality and contour of pipe welds depend largely on the operator's skill. Therefore, argon is preferred over helium because the molten weld pool is much easier to control. When argon is used, the inside bead is more uniform and the penetration is easily controlled in all positions. Argon is

Table 3.3—Selection of shielding gases and power for gas tungsten arc welding

| Material | Thickness[c] | Shielding gas[a] and power[b] used | |
		Manual	Machine
Aluminum and its alloys	Under 3.2 mm	Ar (achf)	Ar (achf) or He (dcsp)
	Over 3.2 mm	Ar (achf)	Ar-He (achf) or He (dcsp)
Carbon steel	Under 3.2 mm	Ar (dcsp)	Ar (dcsp)
	Over 3.2 mm	Ar (dcsp)	Ar-He (dcsp) or He (dcsp)
Stainless steel	Under 3.2 mm	Ar (dcsp)	Ar-He (dcsp) or Ar-H_2 (dcsp)
	Over 3.2 mm	Ar-He (dcsp)	He (dcsp)
Nickel alloys	Under 3.2 mm	Ar (dcsp)	Ar-He (dcsp) or He (dcsp)
	Over 3.2 mm	Ar-He (dcsp)	He (dcsp)
Copper	Under 3.2 mm	Ar-He (dcsp)	Ar-He (dcsp)
	Over 3.2 mm	He (dcsp)	He (dcsp)
Titanium and its alloys	Under 3.2 mm	Ar (dcsp)	Ar (dcsp) or Ar-He (dcsp)
	Over 3.2 mm	Ar-He (dcsp)	He (dcsp)

a. Ar-He contains up to 75% helium; Ar-H_2 contains up to 15% hydrogen.
b. Abbreviations: achf=alternating current, high frequency; dcsp=direct current, straight polarity (electrode negative).
c. 3.2mm= 1/8 in.

recommended for the mechanized welding of steel up to a thickness of 1.90 mm (14 gage). For welding thicker sheets, the addition of helium is recommended to provide adequate weld penetration.

Stainless Steel and High-Temperature Alloys. For stainless steel and high-temperature alloys, argon is recommended for the manual tungsten arc welding of a wide range of thicknesses in all positions. This choice is based primarily upon the need for molten weld pool control. In mechanized welding, the weld pool control effect is minimized and the penetration and welding speed factors are of great importance. In a great many production applications, such as those found in fabrication of jet engines, food equipment, and chemical apparatus, argon, argon-helium mixtures, argon-hydrogen mixtures, or helium may be used. Argon is generally preferred for welding materials up to about 1.9 mm (14 gage) where high welding speed is not required. On heavier materials, especially where high speed is important, argon-helium mixtures are preferred. In continuous welding operations, such as tube welding where welding speed and bead contour are paramount, helium or an argon-hydrogen mixture is recommended.

Copper, Nickel, and Their Alloys. Argon is recommended for the manual gas tungsten arc welding of thin copper, nickel, and copper-nickel alloys. Argon-helium mixtures have some advantage in the welding of greater thicknesses. In comparing argon with other gases, less operator skill is required to control the molten pool to obtain the desired weld penetration and bead smoothness. Helium is recommended for welding heavy sections because the higher arc voltage provides a hotter arc for welding these high melting-point, high-conductivity metals. Argon can be employed if auxiliary preheat is used. In mechanized operations, argon is preferred for welding these materials up to the thickness of about 1.6 mm (1/16 in.); for greater thicknesses of nickel or copper, helium or an argon-helium mixture is recommended.

Magnesium. For gas tungsten arc welding of magnesium, argon shielding is preferred because of the excellent cleaning action, the ease of puddle manipulation in manual welding, and the low gas flow rate. Alternating current, high frequency (achf) power is generally used in combination with argon to obtain good cleaning action and deep penetration. Helium can be used with dcrp (electrode positive) power at very low currents to weld materials thinner than 1.6 mm (1/16 in.), but a high level of operator skill is required. In general, gas tungsten arc welding is superior to consumable electrode welding because better quality welds can be made in magnesium.

Titanium. For manual gas tungsten arc welding of titanium thinner than 3.2 mm (1/8 in.), argon or helium can be employed as the shielding gas. Low flow rates of argon, such as 7 liters/min (15 ft³/h), and helium flow rates of about 19 liters/min (40 ft³/h) are sufficient even when employing recommended large diameter gas nozzles. High gas flow rates cause turbulence in the gas pattern resulting in excessive air mixing and consequent contamination of the weld. Good chill from a copper backing strip and hold-down clamps must be employed in conjunction with good gas shielding to minimize embrittlement. It is preferred to use a grooved backing strip with inert gas backing to prevent air from contaminating the underside of the weld. For the manual welding of thicker plate, helium is recommended to obtain good penetration. Very little welding has been done outside of a gas purge chamber on the heavy thicknesses.

If necessary, a gas trailing shield should be used to provide additional shielding to the weld during cooling. In mechanized welding, the same chill methods, backing gas, and flow of gases through the gas nozzle must be used as in manual operation. In addition, a trailing gas shield must be employed to prevent the contamination of the weld at the higher travel speeds normally used for mechanical operation. Gas flow rates in the gas nozzle and trailing shield should be controlled independently.

Wind and Draft

A crosswind or draft moving at more than 1.5 kilometers per hour (1 mi./h) can dis-

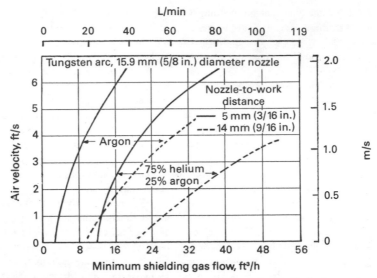

Fig. 3.8—Relationship of inert gas flow requirements to air cross draft velocity for gas tungsten arc welding

rupt the shielding gas coverage. Wind breaks or protective screens to prevent interference by the wind or draft are preferred to increasing the flow of shielding gas. Figure 3.8 shows the relationship of gas flow requirements to the cross draft velocity for argon and for 75 percent helium-25 percent argon mixture using two nozzle-to-work distances. Note that higher gas flow rates are required for the larger nozzle-to-work distance.

CONVENTIONAL WELDING SYSTEMS

Manual Welding

The required equipment for manual gas tungsten arc welding (Fig 3.2) consists of

(1) An electrode holder with gas passages and a nozzle to direct the shielding gas around the arc and a gripping mechanism to energize and hold a tungsten electrode

(2) A supply of shielding gas

(3) A flowmeter and gas pressure reducing regulator

(4) A power source

If the electrode holder is water cooled, a supply of cooling water is also required. This may come from the plant's water supply or a water circulator.

These basic components for manual welding are normally used in conjunction with a gas flow control valve and some means for conventionally switching the electric power on and off at the start and finish of welding. The individual components may differ considerably, depending upon power requirements, type of work, and details of the design.

Various types of electrode holders, cooled with air or water, are available to suit the different applications. Occasionally, they are fitted with gas control valves in the handles. All holders possess means for readily changing the electrodes and gas nozzles. Some are adjustable and permit the electrode angle to be changed with respect to the handle. Collets are available to accommodate the different sizes of electrodes. Electrode diameter depends upon the amount of welding current and electrode polarity. Gas nozzle size selection depends on electrode size, type of weld (butt, fillet), and weld area to be effectively shielded. Figure 3.9 shows a typical gas tungsten arc welding electrode holder (water cooled).

Gas flow regulating equipment generally consists of a single or dual stage pressure-reducing regulator and a gas measuring flowmeter, both of which may be incorporated into the same unit. Gas for shielding is supplied in cylinders or in bulk tanks for high usage.

Alternating or direct current power sources normally used for welding with covered electrodes are used with this process. Special dc and ac power sources that are designed specifically for gas tungsten arc welding are available with automatic means for controlling the flow of gas and water and the start and stop of welding. (For further details, refer to Chapter 1, "Arc Welding Power Sources.")

Semiautomatic Welding

Semiautomatic GTAW is done with equipment which controls only the filler metal feed. The advance of the welding is manually controlled. A semiautomatic electrode holder is similar to a manual electrode holder except that it is equipped with an attachment that brings the filler metal wire into the arc area. This attachment can be mounted on the electrode holder so that the filler metal can enter the arc from any direction. The filler metal is fed through a flexible conduit by a motor-driven wire feeder. Wire feed can be continuous or intermittent, depending on the particular system employed. The major advantages of semiautomatic GTAW are ease of operation and lower welder skill requirements.

Machine or Automatic Welding

Mechanized equipment is more complex but has the same basic components that are shown in Figs. 3.1 and 3.2. Depending upon the application, various other devices and controls may be required. These may include an arc voltage control for constantly checking and adjusting the electrode position (arc length) to maintain a uniform arc voltage. Electrode positioning devices, seam tracking devices, oscillators, and filler metal feeders also may be a part of an automatic installation.

The amount of automation or mechanization applied to gas tungsten arc welding depends on

(1) The quantity of identical welds
(2) The accessibility
(3) Quality control requirements

(4) The degree of perfection required in the weldment
(5) Available funding

Gas tungsten arc welding can be controlled with various devices that accept information about the desired process variable on punched tape or cards. In some devices, the input data are stored on a memory drum. Less sophisticated controls that use cam-actuated pressure or mechanical switches also are available. Automatic or machine welding can be controlled by commercially available programmed arc welding power sources. They will automatically initiate the arc, control the welding current, start and stop the travel, and terminate the arc. Programming of the welding current is also available.

Arc Spot Welding

Gas tungsten arc spot welding often is done manually with a pistol-like holder that has a vented, water cooled gas nozzle, a tungsten electrode that is concentrically positioned with respect to the gas nozzle, and a trigger switch for controlling the operation. Figure 3.10 shows the arrangement. Gas tungsten arc spot welding electrode holders are also available for automatic applications.

The configuration of the nozzle is varied to fit the contour of the weldment. Edge locating devices are normally used to prevent variations in the distance of the spot weld locations

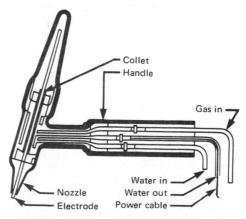

Fig. 3.9 — Section of typical gas tungsten arc welding electrode holder (water cooled)

Collet
Handle
Gas in
Water in
Water out
Power cable
Nozzle
Electrode

from the workpiece edge. The nozzle is often used to provide force against the workpiece to assure tight fit-up of the faying surfaces. This technique also provides control of electrode-to-work distance.

The high current density that is used in spot welding produces a magnetic focusing effect on the arc plasma. This effect is stronger when the electrode tip is tapered. If the angle at the apex of the taper is too small, molten metal will be expelled and a crater will be retained in the weld face.

Arc length is another important variable. If it is too long, the molten weld pool may overheat and undercut may be produced. If it is too short, the base metal may expand enough to contact the expanding electrode. This will contaminate the electrode and cause erratic arc action.

Spot welding may be done with either ac or dcsp (electrode negative). Power sources available for both types of current are discussed in Chapter 1.

Automatic sequencing controls are generally used for arc spot welding because of the relatively complex cycles used. The controls automatically establish the preweld gas and water flow, start the arc, time the arc duration, and provide the required postweld gas and water flow.

Penetration is controlled by adjustment of the amount of current and the length of time it flows. A reduction in the magnitude or time of current flow reduces the penetration and diameter of the spot weld. An increase produces the opposite effect. In some applications, multiple pulses of current are preferred to one long sustained pulse. Variations in the shear strength, nugget diameter, and penetration of the spot weld can be minimized with an accurate timer, an ammeter, a flowmeter, and tungsten electrodes that have precision machined tips.

The arc can be initiated by (1) the touch start system, which advances and retracts the tungsten electrode; (2) by the use of high frequency; or (3) by the use of the pilot arc start system where a low current arc is maintained between the tungsten electrode and nozzle. A typical pilot arc system is shown in Fig. 3.11.

The weld cycle timing should begin with the flow of welding current. The welding of a very thin metal will require very short welding times. A weld in 0.13 mm (0.005 in.) stainless steel would require a current rise to welding current in 1/2 cycle (measured on a 60 cycle per second time basis), hold for 4 cycles, and decay within 1/2 cycle. Electronic dc timers and cold cathode counting tubes are currently used.

The usual arc spot weld control consists of current and time control of the preweld, welding, and postweld operations. For material thicknesses over 1.3 mm (0.05 in.), welding rod (wire) should be added, if required, to overcome crater cracking or to improve nugget configuration. This may be accomplished by special wire feeders employing rate of feed control and sequence timing to introduce and retract the filler metal during the welding cycle.

AUXILIARIES AND ACCESSORIES

Gas Nozzles

Special gas nozzles are available with fine mesh wire screens through which the gas flows before it exits from the gas nozzle. Their use increases the effective length of the shielding gas column beyond the nozzle by reducing tur-

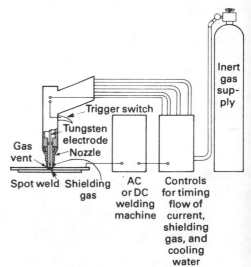

Fig. 3.10 — Gas tungsten arc spot welding arrangement

bulence in the gas stream as it emerges from the nozzle. This results in better gas shielding and permits electrode extension as much as 19 mm (3/4 in.) beyond the gas nozzle for increased visibility and accessibility. However, the special nozzles are more costly than standard nozzles.

Arc Starting Means

Retract starting can be used in automatic dc welding. The electrode is fed down until contact is made with the work, and it is then retracted to establish the arc.

High frequency starting can be used with dc or ac power sources for both manual and automatic applications. The high frequency generator superimposes a high voltage upon the welding circuit. These generators usually consist of a spark-gap oscillator that delivers a high voltage output at radio frequencies. The generator output is transferred from the spark-gap circuit to the electrode leg of the welding circuit by means of an air core transformer, as illustrated in Fig. 3.12. The high-frequency generator produces a series of closely spaced bursts of high voltage energy. This high voltage ionizes the gas between the electrode and the work. The ionized gas will then conduct welding current that initiates the welding arc.

The addition of high-frequency energy on the ac welding circuit also assists in reestablishing the arc at points of current reversal, i.e., twice every cycle. This technique of arc stabilization finds its greatest application in gas tungsten arc welding of aluminum and magnesium.

Because radiation from a high-frequency generator may disturb radio service, the use of this type of equipment is governed by regulations of the Federal Communications Commission. The user should follow the instructions of the manufacturer for the proper installation and use of high-frequency generating equipment.

Pilot arc starting may be used on dc welding apparatus. Because of its high reliability, the largest application of this starting technique is spot welding. The pilot arc is maintained between the welding electrode and an auxiliary electrode in the holder. As illustrated in Fig. 3.11, the arc is powered by a small auxiliary power source and provides conditions for initiating the welding arc in a manner analogous to the pilot light of a gas stove. The pilot arc is started by a scratch technique or by high-frequency energy.

Filler Metal Feed

Wire feeders are used to add filler metal

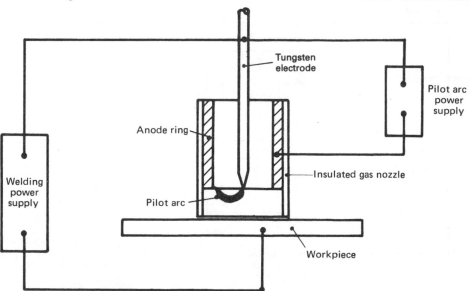

Fig. 3.11 — Pilot arc starting circuit used for gas tungsten arc spot welding

during machine and automatic welding. Either room temperature (cold) wire or preheated (hot) wire can be fed into the molten weld pool. Cold wire is fed into the leading edge and hot wire is fed into the trailing edge of the molten pool.

Cold Wire. The required system for the feeding of cold wire has three components:

(1) A wire drive mechanism

(2) A speed control

(3) A wire guide attachment to introduce the wire into the molten weld pool

The drive consists of a motor and gear train to power a set of drive rolls which push the wire. The control is essentially a constant speed governor which can be either a mechanical or an electronic device. The wire is usually guided from the drive mechanism to the attachment through a flexible conduit. An adjustable wire guide is attached to the electrode holder. It maintains the position at which the wire enters the weld and the angle of approach relative to the electrode, work surface, and the joint. In heavy duty applications, the wire guide is water cooled.

Wires in a range from 0.8 to 2.4 mm (1/32 to 3/32 in.) in diameter are fed at a constant rate.

Hot Wire. The equipment for a hot wire addition is similar to that for cold wire, except that the wire is electrically resistance heated to the desired temperature before it enters the molten weld pool. When using a preheated (hot) wire in machine and automatic gas tungsten arc welding in the flat position, the wire is fed mechanically to the weld pool through a holder from which inert gas flows to protect the hot wire from oxidation. This system is illustrated in Fig. 3.13.

Deposition rate is greater with hot wire than with cold wire and comparable to that in gas metal arc welding, as shown in Fig. 3.14. Normally, a mixture of 75% helium-25% argon is used to shield the tungsten electrode and the molten weld pool. The wire is electrically resistance heated by alternating current from a constant potential power source. The current flow is initiated when the wire contacts the weld surface. The wire is fed into the molten weld di-

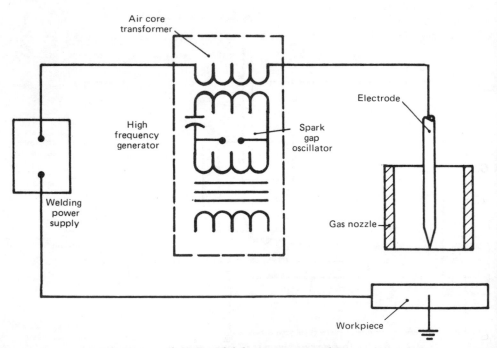

Fig. 3.12—High frequency arc starting

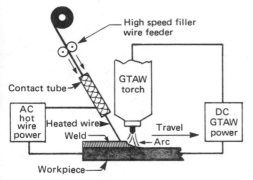

Fig. 3.13—Gas tungsten arc hot wire system

rectly behind the tungsten arc at a 40 to 60 degree angle with the tungsten electrode. Alternating current is used for heating the wire to avoid arc blow. With alternating current, the arc oscillates 30 degrees in the longitudinal direction when the heating current does not exceed 60 percent of the arc current. The oscillation is 120 degrees when the currents are equal. The amplitude of arc oscillation can be controlled by limiting the wire diameter to 1.2 mm (0.045 in.) so that the heating current will not exceed 60 percent of the arc current.

Preheated filler wire has been used successfully for the joining of carbon and low alloy steels, stainless steels, and alloys of copper and nickel. Preheating is not recommended for aluminum and copper because the low resistance of these filler wires requires high heating current, which results in excessive arc blow and uneven melting.

Gas Control Equipment

A combination regulator and flowmeter is widely used to control and measure the flow of shielding gas. High pressure in a cylinder or cylinder manifold is reduced to a safe working pressure. The lower pressure gas is metered through the flowmeter and controlled by manual adjustment of a throttle valve. The flow is indicated on the flowmeter tube or dial that is calibrated for the particular gas being metered.

In operations with high gas consumption, a centrally located cylinder manifold or a liquid storage tank can be installed to store gas. The gas

is piped from the storage containers to the welding stations. The pressure in the distribution line is regulated and individual flowmeters are mounted at each welding station.

When a shielding gas mixture must be used, standard proportions are commercially available in cylinders. Other desired mixtures can be obtained through the use of manually set mixers or automatic gas-ratio mixers, which can be operated from cylinders or bulk systems.

Traversing Mechanisms

Either the workpiece, the welding head, or the electrode holder can be moved, depending upon the size and type of work and the preference of the user. Traversing mechanisms maintain the position of the electrode with respect to the joint within close limits. Their speed ranges meet the requirements for a wide range of materials and thicknesses. They travel at a uniformly smooth rate without excessive vibration and maintain preset conditions accurately in repetitive operations.

Motor-driven carriages that run on tracks or directly on the workpiece are available in various sizes. They are used for straight line and contour welding. Units are also available for vertical and horizontal welding.

Side beam carriages are available for transporting welding heads and allied equipment in straight line operation. The carriage is supported

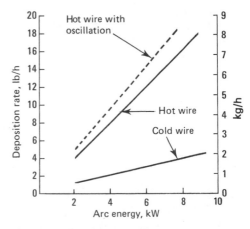

Fig. 3.14—Allowable deposition rates in steel for gas tungsten arc welding with cold and hot filler wire

on the vertical face of a flat track. The track can be mounted to the side of a beam, to construction brackets on a wall, or to posts from the floor.

Track sections are available in incremental lengths, and both systems can be extended in length. The various types of carriages are fitted with mechanical clutches so that the drive motor can be disengaged while the welding head is being positioned.

Welding head manipulators are available with many sizes of booms. They can be used for longitudinal welds and, in conjunction with a rotary weld positioner, for circumferential seams. Rotary weld positioners also are used for manual welding and with fixed welding head mountings.

Other Accessories

Oscillators are mechanical or electromagnetic devices used to impart oscillatory motion to the tungsten electrode or to the welding arc. They are used for surfacing, out-of-position welding, and other applications where weaving or widening the weld is beneficial. Commercially available mechanical and electromagnetic oscillators provide motions that are uniform, harmonic, or with dwell, as required.

Switches for foot control of current for manual gas tungsten arc welding are used to start the current flow, vary it as required during welding, and decay the current at the end of the weld. An on/off switch for starting and stopping the high frequency current may be included in the foot control.

Welding cable size is determined by the power source current rating, duty cycle, and total length of cable. See Chapter 2 for additional information.

MATERIALS

The welding arc operates in an atmosphere of inert gas to provide an intense source of heat that is clean, stable, and quiet. Argon and helium, most frequently used for shielding, provide an almost ideal environment in which to melt most metals. For these reasons, the gas tungsten arc welding process is admirably adapted for welding almost all metals or alloys that can be fused by an electric arc, that do not vaporize rapidly from the high heat of an arc, and that can be welded without cracking.

BASE METALS

Among the materials readily welded by this process are most grades of carbon, alloy, and stainless steel, aluminum and most of its alloys, magnesium and its alloys, copper, copper-nickel, phosphor-bronze, tin bronzes of various types, brasses, nickel, nickel-copper, nickel-chromium-iron, high-temperature alloys of various types, virtually all the hard-surfacing alloys, titanium, zirconium, gold, silver, and many others.

Details on the welding characteristics of specific metals and alloys can be found in *Welding Handbook*, Section 4, *Metals and Their Weldability*, 6th edition.[1] A condensed generalized treatment of the metallurgical aspects of welding can be found in Volume 1 of this edition.

The gas tungsten arc process is used to produce high quality welds in metals of thicknesses ranging from plate through sheet and down to foil. The process is especially suited to work requiring the utmost in quality or finish because of the precise heat control possible and the ability to weld either with or without filler metal.

Although the gas from the electrode holder can adequately shield the face side of the weld and adjacent base metal, it gives no protection to the root side of the joint. The root side of the joint will become hot enough to oxidize and this may cause a rough root surface, especially in thin

---

1. Volume 4 of the 7th edition, when published, will also contain this information.

gages. In many applications this is not important, but for high quality work, some protection is needed to obtain a smooth root surface and minimum contamination of the weld and base metals. This protection can consist of an inert gas blanket only or a metal backing strip used with or without inert-gas backing. Stainless steel and the super alloys benefit greatly by gas protection. The reactive metals titanium, zirconium, tantalum, and columbium require nearly perfect inert gas protection on the root surface. If they are not protected, their properties will be significantly altered by oxygen or nitrogen contamination, or both.

In general, best welding results are obtained with arcs operating on dcsp (electrode negative). The outstanding exceptions to this generality are in the welding of aluminum and magnesium where refractory surface oxides tend to prevent proper joining of molten joint edges. In these latter cases, alternating current arc power is generally used.

Cleanliness of both the weld joint areas and the filler metal is an important consideration when welding with the gas tungsten arc process. Oil, grease, shop dirt, paint, marking crayon, and rust or corrosion deposits all must be removed from the joint edges and metal surfaces to a distance beyond the heat-affected zone. Their presence during welding will lead to arc instability and contaminated welds. Depending upon the metallurgical response to these contaminants, welds can be porous, cracked, or contaminated.

FILLER METALS

Filler metals for joining a wide variety of metals and alloys are available for use with gas tungsten arc welding. Filler metals, if used, are most often similar, although not necessarily identical, to the metal that is being joined. When joining dissimilar metals, the filler metals will be different from one or both of the base metals.

Generally the filler metal composition is adjusted to allow for its use in the welded (cast) condition while matching the properties of the wrought base metal. To accomplish this, filler metals are produced to closer control on chemistry, purity, and quality than are base metals. Deoxidizers are frequently added to ensure weld soundness. Adjustments to filler metal chem-

istry are also made to allow for the operating characteristics of the welding process used. In the case of gas tungsten arc welding, there is little or no loss of deoxidizers as the filler metal is deposited directly into the molten puddle blanketed by the inert shielding gas. In the case of gas metal arc welding (GMAW) of steel, shielding gases may contain oxygen or carbon dioxide which can react with some of the deoxidizers and alloying elements of the consumable electrode before they reach the weld puddle.

The consumable electrode chemistry of GMAW often is enriched to offset these arc transfer losses. However, a solid bare wire (electrode) manufactured for the GMAW process is generally suitable for the GTAW process. Further modifications are made to some filler metal compositions to improve response to post-weld heat treatments.

The choice of a filler metal for any application is a compromise involving such factors as metallurgical compatibility, suitability for the intended operation, and cost. The tensile and impact properties, corrosion resistance, and electrical or thermal conductivities that are required in a particular weldment also must be considered. Thus, the alloy to be welded and the intended service generally determine the filler metal.

Table 3.4 lists the AWS filler metal specifications which are applicable for gas tungsten arc welding. These specifications establish filler metal classifications based on the mechanical properties or chemical compositions, or both, of each filler metal. They also set forth the conditions under which the filler metals must be tested. For some materials, radiographic standards of acceptability are given. Usability tests are also required by some of these specifications.

Appendices in the filler metal specifications provide useful background on the properties and uses of the filler metals within the various classifications. Manufacturers' catalogs often provide useful information on the proper use of their products. Brand name listings and addresses of vendors are shown in the latest edition of AWS A5.0, Filler Metal Comparison Charts.

Filler metals for gas tungsten arc welding are available in most alloys in the form of

Table 3.4—AWS specifications for filler metals suitable for gas tungsten arc welding

Specification number	Title
A 5.2	Iron and Steel Gas-Welding Rods
A 5.7	Copper and Copper Alloy Bare Welding Rods and Electrodes.
A 5.9	Corrosion-Resisting Chromium and Chromium-Nickel Steel Bare and Composite Metal Cored and Stranded Arc Welding Electrodes and Welding Rods
A 5.10	Aluminum and Aluminum Alloy Welding Rods and Bare Electrodes
A 5.13	Surfacing Welding Rods and Electrodes
A 5.14	Nickel and Nickel Alloy Bare Welding Rods and Electrodes
A 5.16	Titanium and Titanium Alloy Bare Welding Rods and Electrodes
A 5.18	Mild Steel Electrodes for Gas Metal Arc Welding
A 5.19	Magnesium-Alloy Welding Rods and Bare Electrodes
A 5.21	Composite Surfacing Welding Rods and Electrodes
A 5.24	Zirconium and Zirconium Alloy Bare Welding Rods and Electrodes

straight and cut lengths (rods), usually 914 mm (36 in.) long, for manual welding, and spooled or coiled continuous wire for mechanized welding. The filler metal diameters range from about 0.50 mm (0.020 in.) for fine and delicate work to about 4.76 mm (3/16 in.) for high current manual welding or surfacing.

Extra care must be exercised to keep the filler metals clean and free of all contamination in storage as well as in use. The hot end of the wire or rod should not be removed from the protection of the inert gas shield during the welding operation.

For a more detailed description of filler metals and their use with specific metals and welding processes, the reader is referred to Chapter 94, "Filler Metals," *Welding Handbook*, Section 5, 6th edition.[2]

GENERAL PROCESS APPLICATIONS

The gas tungsten arc welding process is used extensively for the welding of longitudinal seams of thin wall stainless steel and alloy pressure pipe and tubing on continuous strip mills, generally without filler metal. It is used for joining in all welding positions a broad spectrum of heavier wall metal and alloy pipe and tubing for the power generating, chemical, and petroleum industries, generally with filler metal. Aircraft and aerospace industries make extensive use of the process in manual, mechanized, and fully automated systems for joining components made of a wide variety of metals selected to meet critical strength-weight considerations. Jet engine casings and rocket motor cases are typical examples. It is one of the few processes permitting the rapid, satisfactory welding of such tiny or light-walled objects as transistor cases, instrument diaphragms, and delicate expansion bellows. The uses of GTAW in several major fields of industrial applications are further described in Chapters 85, 87, 91, and 92 of Section 5, *Applications of Welding Handbook*, 6th edition.

---

2. A chapter on filler metals is scheduled for Volume 4, 7th edition.

INDUSTRIAL PIPING

This process is being applied increasingly to industrial pipe welding and it is extensively used on almost all of the ferrous and nonferrous piping materials. The root passes of substantial quantities of carbon steel piping, especially those for critical application, are welded with the GTAW process. For practical reasons, the thinnest section that can be hand welded is approximately 0.8 mm (1/32 in.). The maximum thickness that may be welded is limited by available equipment and will vary for different metals. For pipe wall thicknesses over 6.4 to 9.5 mm (1/4 to 3/8 in.) (depending on the base metal), it is generally more economical to complete the pipe weld with another welding process, such as shielded metal arc or gas metal arc welding, after the gas tungsten arc root pass has been made.

AIRCRAFT

From the standpoint of man-hours, the gas tungsten arc welding process is a major welding method employed in the airframe industry.

Manual GTAW equipment is generally of the conventional transformer-rectifier type of power supply, usually with a saturable reactor or magnetic amplifier type of control. The electric control of the machine enables the welder to vary the current as required when welding by means of a foot control. Standard manual GTAW electrode holders are used with argon or mixtures of argon and helium for shielding.

Significant advances in automatic GTAW equipment have been made to meet the exacting standards required in the aircraft industry. The greatest progress has been made in the control of the various welding parameters. Present requirements dictate controlled up-and-down slope of current, delayed weld travel, delayed wire start, and advanced wire stop. In addition, weld variable changes must be programmed to meet the requirements of advanced design configurations. General practice is to instrument the machine variables of current, voltage, weld speed, and wire feed. Recording volt and ampere meters are often employed to monitor weld settings.

Automatic gas tungsten arc welding is applicable to material as thin as 0.076 mm (0.003 in.) foil. For thicknesses up to 0.5 mm (0.02 in.), it is not generally practical to add filler metal by means of wire. Therefore, a flange of about twice the metal thickness may be turned up on each piece to be groove welded. This flange is melted to provide the filler metal for the joint. This procedure has proved to be successful for the welding of face sheets in the fabrication of stainless steel brazed honeycomb aircraft panels. The welds are sometimes roll planished to improve their mechanical properties and to reduce weld thickness to the same thickness as the adjacent material. When welding butt joints in material over 0.5 mm (0.02 in.) thick, it may be necessary to add filler rod.

Automatic gas tungsten arc welding is especially suitable for girth and longitudinal groove welds that require minimum distortion, narrow weld beads and heat-affected zones, and consistent results. Automatic welding also saves considerable time per welded piece because of its ability to operate at regulated high speeds. Its limitations are (1) that consistent work alignment must be maintained, and (2) that additional tooling costs are incurred.

HEAT EXCHANGERS

Virtually all of the tube-to-tubesheet welding is done with automated gas tungsten arc welding equipment, with or without the addition of filler metal. Two types of tube-to-tubesheet GTAW machines are used. The first type, which makes welds at the face of the tubesheet, is mounted on the tubesheet with the tubesheet in the horizontal position, face up. The welding torch is mounted on a pantograph fixture that orbits the welding electrode around the weld joint. The second type of tube-to-tubesheet welding machine has an extendable rotary quill that permits making welds inside the hole from the face of the tubesheet. Recessed welds and boxed header welds are made with this equipment.

MACHINE CIRCUMFERENTIAL WELDING

The automatic GTAW systems presently in use for the joining of in-place tubing and piping

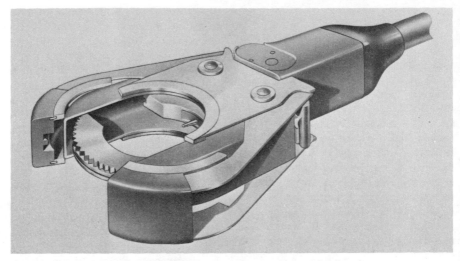

Fig. 3.15—Automatic gas tungsten arc tube welding head

Fig. 3.16—Automatic circumferential pipe welding machine with arc voltage control and filler wire feeding capability

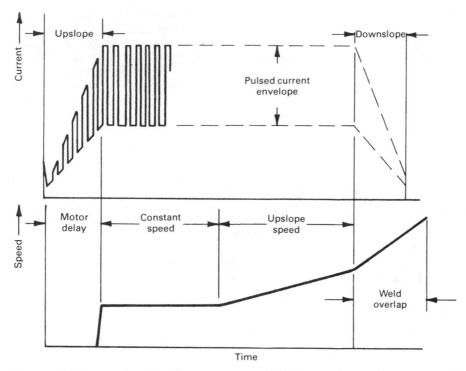

Fig. 3.17—Pulsed weld current and travel speed program for automatic gas tungsten arc welding of tubing

may be divided into two categories: those with capability of feeding filler metal, and those without. An automatic circumferential tube welding device. without filler wire feed is shown in Fig. 3.15. Butt joints in tubing are made in place with this type of equipment. An automatic circumferential pipe welding machine having a filler wire feeder, electrode holder oscillation, and arc voltage control is shown in Fig. 3.16.

Successful automatic gas tungsten arc welding of hydraulic and pressure tubing and fuel lines is performed using precision electronic and mechanical controls. A very important improvement has resulted from the use of pulsed current. The relation of pulsed current and travel speed during one weld program is shown in Fig. 3.17. Pulsing of the current begins at arc ignition. The travel start is delayed and the current pulse magnitude increases gradually until it reaches the operating level and complete penetration of the butt joint is achieved. Prior to arc extinction, the

pulse magnitude decreases gradually. The periods of gradual increase and decrease are called respectively upslope and downslope. Time between pulses, time at high current level, and

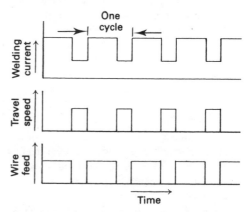

Fig. 3.18—Step-pulsed gas tungsten arc welding sequence

time at low current level may be varied to suit the individual application.

With pulsed current, satisfactory penetration and weld face contour can be maintained for the welding of tubing in a fixed position. The pulsed arc agitates the molten metal and thus minimizes porosity. Compared with a steady arc, the pulsing arc increases the penetration with less heat input to the joint than with steady dc. The pulsating current produces a solidification pattern different from the normal one. This may improve certain weld joint mechanical properties.

For the root pass of groove welds in pipe requiring the addition of filler metal, the variables of arc voltage, welding current, wire feed, and travel speed are most critical to obtain the fine balance of energy input needed for adequate penetration and fusion. Figure 3.18 illustrates "step-pulsed" GTA welding which includes the pulsation of filler metal, welding current, and travel speed. The technique of "step-pulsed" welding produces a series of controlled overlapping fusion zones that achieve a level of consistency and quality that cannot be duplicated by the most skilled welder. The subsequent passes can likewise be applied without the inherent constrictions of movement and control confronted by a welder.

PROCESS VARIABLES

WELDING CURRENT

A comparison of welding current output wave patterns produced by various types of ac and dc power supplies reveals a great variety of built-in ripples or pulses which affect process repeatability and interchangeability of the power supplies. For example, when commercial ac power supplies were first used for automatic welding of aluminum, it was noted that the magnitudes of the positive and negative pulses were changing during welding and so was weld quality. Wave balancing controls of pulsation in ac welding were developed to solve these problems.

Similar problems were found in dc welding. Direct current welding power sources free of ripple and with one percent regulation accuracy were developed for critical welding applications, such as in-place tube and pipe welding or aluminum plate welding without a backing strip.

ARC VOLTAGE

The voltage drop between the tip of the tungsten electrode and the work is influenced by the type of welding current and shielding gas used. The arc voltage is dependent on the shielding gas and the distance from the electrode to the work. Helium produces higher voltages than argon for the same arc gap (Fig. 3.7). Therefore, helium gas is used to obtain deeper penetration, particularly when welding aluminum alloys with dcsp (electrode negative). During manual GTAW, the operator is concerned with maintaining a controlled arc length (i.e., arc voltage) to make an acceptable weld.

The voltmeter on the power supply indicates the total of the voltage drops in the cables, connections, tools, electrode collet, tungsten electrode, and the arc. The voltage representing most nearly the potential drop across the arc should be measured between the electrode holder and the work.

Electrode tip geometry also affects arc voltage, as indicated in Fig. 3.19. It can be seen that, at the same tip-to-work distance, the arc voltage is higher with a sharper cone tip on the electrode. The record of the tungsten tip geometry is also important for transferability of GTAW process variables from machine to machine for a given job.

A common method of arc voltage control during automatic welding is to move the tungsten electrode electromechanically to maintain a preset arc voltage. The voltage is held constant by a reversible gear motor which raises or lowers the electrode to maintain the desired arc length. The basic arrangement of an automatic GTA voltage control is given in Fig. 3.20. In such a

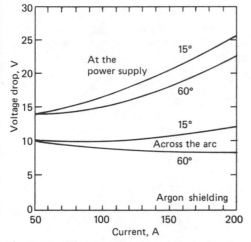

Fig. 3.19—Effects of tungsten electrode tip taper angle on the volt-ampere characteristics of the arc

electrode often encountered with voltage controlled heads is partially due to resistance changes which occur during the warm-up period of the welding circuit.

Another method of arc control is to move the tungsten electrode at a preset distance from the work and adjust the current, as needed, to maintain a constant potential or a constant arc power. These systems vary from simple manual adjustment to elaborate magnetic and eddy current devices.

TRAVEL SPEED

For a given current and voltage, the travel speed determines the amount of energy that is delivered per unit length of weld. Changes in energy per unit length have a strong effect on the shape of the weld.

Increasing the speed without changing the current reduces both the penetration and width of the weld; further reduction occurs if the current is decreased. Decreasing the speed without changing the current increases both the penetration and width of the weld, and further increase occurs if the current is increased. Increasing or decreasing both the speed and current together will maintain the penetration and width of the weld.

system, the measured arc voltage consists of two distinct voltage drops: the drop due to the arc itself and the drop in the tungsten electrode. The voltage drop in the tungsten electrode begins to rise immediately after arc initiation due to resistance heating. As the voltage drop increases, the arc voltage control (AVC) will shorten the arc length to maintain the sum at a constant value. The characteristic "diving" of the tungsten

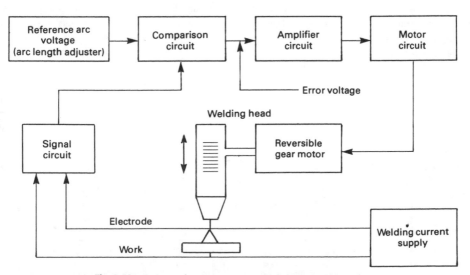

Fig. 3.20—Automatic gas tungsten arc voltage control system

The above rules on the direction of changes generally hold for all metals. The amount of the changes in weld penetration and width for a given change in speed and current depends on the composition and thickness of the metal being welded and on the joint configuration.

TUNGSTEN ELECTRODE CONDITIONING

The shape of the tungsten electrode tip is an important process variable in gas tungsten arc welding. A good selection for quick arc initiation and arc stability is a two percent thoriated tungsten electrode with a flat tip of the smallest diameter recommended for the current. However, tungsten electrodes are not available in a very large assortment of diameters, 1, 1.6, 2.4, and 3.2 mm (0.04, 1/16, 3/32, and 1/8 in.) being the common diameters. For this reason, shaping or tapering of the electrodes to a small diameter is a recommended practice. Table 3.5 is a guide for electrode tip preparation for a range of sizes with recommended current ranges.

During the welding operation, the accurately ground tip of a tungsten electrode is at a temperature probably in excess of 3000° C (5500° F), and the positively charged gaseous atoms and molecules, including metal vapors, are attracted by the electrode to produce the phenomenon of cathodic cleaning. As the current decays at the end of the weld, metallic vapors continue to flow and the electrode tip can be contaminated during this period. Certain elements in metals vaporize more readily than others and, by condensing on the tungsten tip, they can create problems in starting the arc again.

Another type of electrode contamination is caused by poor gas shielding during or after the welding operation. The worst contamination can be caused by weld puddle eruptions or splatter created by gas entrapment, surface impurities, or a constituent in the metal itself. While the metal vapor condensate and oxidation contamination may be corrected by a quick restart, by touch restart, or by light cleaning with abrasive paper, redressing of the electrode tip is required after splatter contamination.

WELD JOINT DESIGN

Proper joint design and preparation is very important in gas tungsten arc welding applications. There are several requirements and recommendations for a good design of any gas tungsten arc welded joint.

First, it must be formed by compatible materials. If filler metal is needed for metallurgical aspects or for weld reinforcement, it must be compatible with the base materials. Ideally, an automatic gas tungsten arc welding operation should perform without a need for operator dexterity and judgment in adding the welding rod (wire) to the weld puddle.

An important requirement, especially for automatic welding, is to have good fit-up. When two edges are brought together for a butt joint, they must be together without gapping. A very common joint is where one part is inserted into another, forming a self-locking joint. The amount of clearance between the parts that can be tolerated is a function of a number of factors. Pulsed dc welding techniques permit greater clearance between parts than with steady cur-

Table 3.5—Tungsten electrode tip shapes and current ranges						
					DCSP (electrode negative)	
Electrode diameter,		Diameter at tip,		Included angle,	Constant current range,	Pulsed current range,
mm	in.	mm	in.	degrees	A	A
1.02	0.040	0.125	0.005	12	2-15	2-25
1.02	0.040	0.25	0.010	20	5-30	5-60
1.59	0.062	0.5	0.020	25	8-50	8-100
1.59	0.062	0.8	0.030	30	10-70	10-140
2.38	0.093	0.8	0.030	35	12-90	12-180
2.38	0.093	1.1	0.045	45	15-150	15-250
3.18	0.125	1.1	0.045	60	20-200	20-300
3.18	0.125	1.5	0.060	90	25-250	25-350

rent. Furthermore, pulsed current techniques help to join parts which have unequal thicknesses of material in the immediate weld area. However, in general, joint geometries with matching thicknesses, included angles, and root faces are recommended for both manual and automatic applications.

Finally, a good joint design provides for easy cleaning. A small amount of oil, grease, moisture, or other foreign matter between two accurately placed parts can cause weld metal porosity in any metal. The ferrous alloys are much less critical to porosity formation, particularly when the weld is achieved by pulsed dc.

WELDING PROCEDURES

Most industrial gas tungsten arc welding operations are governed by predetermined welding procedures. Establishing these procedures may be quite involved depending upon the application and specification requirements. The welding procedure may reference a welding schedule for specific information. The welding schedule may be a chronological linking of welding and related activities which must be followed in their respective order for the purpose of producing the desired weld.

FIXTURES AND TOOLING

Fixtures and tooling can affect the size, shape, and uniformity of a weld cross section. Care should be taken in the selection of the materials from which they are made and in the design of the contacts between the work and the holding devices. A decision not to use fixturing usually means that the resulting weld distortion can be tolerated or it can be corrected by straightening operations. Some functions of a weld fixture are

(1) Locate parts relative to the assembly

(2) Maintain alignment during welding without excessive restraint (should not promote weld cracking)

(3) Control distortion in the weldment

Stainless steels are good backing materials for argon-shielded GTAW with one exception. When touch arc starting is used in the welding of thin metals, there is danger of welding the work to the edges of the groove in the backing strip.

Copper is predominantly used for weld back-up because it will not weld to thin metals. The

fast rates of weld cooling that can be obtained with copper are equally useful to control the heat from a helium-shielded arc. Rough, uneven root penetration is usually caused by a groove in the copper strip that is too narrow. To avoid magnetic arc blow, materials which can be magnetized by the arc current should not be used for fixturing near the welding arc.

ELECTRODE AND NOZZLE

The following recommendations with respect to the tungsten electrode and its gas shield should be followed to maintain weld quality and to promote welding economy in any specific application:

(1) The appropriate current (type and magnitude) should be selected for the electrode size to be used. Too great a current will cause excessive melting, dripping, or volatilization of the electrode tip. Too small a current will cause cathode bombardment and erosion due to the low operating temperature and resulting arc instability.

(2) The electrode should be properly broken and ground tapered by following the supplier's suggested procedures. Improper breakage may cause a jagged end or a bent electrode which usually results in a poorly shaped arc and electrode overheating.

(3) The electrodes should be carefully handled and kept as clean as possible. To obtain maximum cleanliness, they should be stored in their original packages until used.

(4) The shielding gas flow should be maintained, not only when welding, but also after

Table 3.6—Recommended lens shades for various welding current ranges

Shade no.	Welding current, A
6	Up to 30
8	30 to 75
10	75 to 200
12	200 to 400
14	Above 400

the arc is broken and until the electrode cools. When the electrode is properly cooled, the arc end will appear bright or polished. When improperly cooled, it may oxidize and have a colored film on the arc end. Unless removed, the oxidized tungsten tip may affect the weld quality on subsequent welds. All connections, both gas and water, should be checked for tightness.

(5) The electrode extension within the gas shielding pattern should be kept to a minimum. It is generally dictated by the application and equipment. Minimum extension will insure protection of the electrode by the gas, even at low gas flow rates.

SHIELDING GAS

A good practice is always to purge the gas system of air a few minutes before starting to weld after a period of downtime. Slight discoloration of areas near the weld and at the end of the weld may indicate improper inert gas coverage during or after gas tungsten arc welding. Backing gas (gas shield on the root side of the joint), leading and trailing gas coverage,

Table 3.7—Troubleshooting guide for gas tungsten arc welding

Problem	Cause	Remedy
Wasteful electrode consumption	1. Improper shielding (resulting in oxidation of electrode)	1. Clean nozzle; bring nozzle closer to work; step up gas flow.
	2. Operating on reverse polarity	2. Employ larger electrode or change to straight polarity.
	3. Improper size electrode for current required	3. Use larger electrode.
	4. Excessive heating in holder	4. Ground finish electrodes; change collet; check for improper collet contact.
	5. Contaminated electrode	5. Remove contaminated portion— erratic results will continue as long as contamination exists.
	6. Electrode oxidation during cooling	6. Keep gas flowing after stoppage of arc for at least 10-15 seconds. Rule: 1 second for each 10 amperes.
Erratic arc	1. Base metal is dirty, greasy	1. Use appropriate chemical cleaners, wire brush, or abrasives.
	2. Too narrow joint	2. Open joint groove; bring electrode closer to work; decrease voltage.
	3. Electrode is contaminated	3. Remove contaminated portion of electrode.
	4. Too large diameter electrode	4. Use ground finish-electrodes; change collet; check for improper collet contact.
	5. Arc too long	5. Bring holder closer to work to shorten arc.

or welding in a controlled atmosphere chamber will minimize weld quality problems caused by this process variable.

RECORDING INSTRUMENTS

Instrumentation for monitoring and recording welding process variables has become increasingly important with the mechanization of welding systems. Cost of commercially available equipment depends on the number of variables to be monitored, complexity of the instrumentation, and the precision required. Increased demands for quality control measures require incorporating these recording instruments in the welding systems to compare information with known standards. Recorders are used to provide accurate data on welding current

and voltage, weld travel speed, shielding gas flow, backing gas flow, gas moisture content, and weld related temperature values.

SAFE PRACTICES

In welding with the gas tungsten arc process, the same precautions and safe practices should be observed that apply to any other electric welding operation. The operator should be properly protected from the rays of the arc. This requires suitable fire resistant clothing to cover all exposed skin surfaces and a welder's helmet with the proper shade of glass to protect the eyes and the face. The shade of the glass lens depends on the intensity of the arc. Table 3.6 lists the recommended lens shades for different current ranges.

Table 3.7 (cont.)—Troubleshooting guide for gas tungsten arc welding

Problem	Cause	Remedy
Porosity	1. Entrapped gas impurities (hydrogen, nitrogen, air, water vapor)	1. Blow out air from all lines before striking arc; remove condensed moisture from lines; use welding grade (99.99%) inert gas.
	2. Possible use of old acetylene hoses	2. Use only new hoses. Acetylene impregnates hose.
	3. Gas and water hoses interchanged	3. Never interchange water and gas hoses. (Color coding helps.)
	4. Oil film on base metal	4. Clean with chemical cleaner not prone to break up in arc; DO NOT WELD WHILE BASE METAL IS WET.
Tungsten contamination of workpiece	1. Contact starting with electrode	1. Use high frequency starter; use copper striker plate.
	2. Electrode melting and alloying with base metal	2. Use less current or larger electrode; use thoriated or zirconium-tungsten electrode.
	3. Shattering of electrode by thermal shock	3. Make certain electrode ends are not slivered or cracked when using high current values. Use embrittled tungsten to facilitate easy and clean breakage.

(Reproduced by permission of Chas. A. Bennett Co., Inc., from *Welding: Principles & Practice*, by Sacks.)

WELDING PROBLEMS AND REMEDIES

There are a number of welding problems that may develop while setting up or operating a GTAW operation. The solution will require careful evaluation of the material, the fixturing, the welding equipment, and the procedures. Some problems that may be encountered and possible remedies are listed in Table 3.7.

WELD QUALITY

DISCONTINUITIES AND DEFECTS

Discontinuities are interruptions in the typical structure of a weldment, and they may occur in the base metal, weld metal, and heat-affected zones. Discontinuities that do not satisfy the requirements of a code or specification under which a weldment is being fabricated are classed as defects and, as such, are cause for rejection.

There are many types and sizes of discontinuities that do not qualify as defects. However, because defects may impair the performance of a weldment and also may be the cause of catastrophic failure, they are unacceptable.

The appearance of a weld does not necessarily indicate its quality. Discontinuities can be visible to surface inspection or they may be internal so that special techniques will be required for their detection. Discontinuities can be peculiar to the welding process employed, or they can be general types which are encountered with a wide variety of welding processes. Finally, discontinuities can be peculiar to the metallurgical response of the alloy to the melting and subsequent cooling cycles involved in welding operations.

One discontinuity found only in gas tungsten arc welding is tungsten inclusions. Discontinuities encountered in the general class of gas shielded arc welding processes are oxidation of weld pool and preceding joint surfaces due to loss of inert gas shielding. Discontinuities typical of arc welding of all types include inadequate joint penetration, incomplete fusion, and undercut. Discontinuities related to the metallurgical response of alloy systems to welding include cracking (hot or cold), embrittlement, or softening with loss of joint strength. Many, but by no means all, discontinuities can be avoided by careful welding.

PROBLEMS AND CORRECTIONS

Tungsten Inclusions

Particles and chunks of tungsten from the electrode can be found buried in a weld when improper welding procedures are used with the GTAW process. Typical causes are

(1) Contact of electrode tip with molten weld pool

(2) Contact of filler metal with hot tip of electrode

(3) Contamination of the electrode tip by spatter from the weld pool

(4) Exceeding the current limit for a given electrode, size, or type

(5) Extension of electrodes beyond their normal distances from the collet (as with long nozzles) resulting in overheating of the electrode

(6) Inadequate tightening of the holding collet or electrode chuck

(7) Inadequate shielding gas flow rates or excessive wind drafts resulting in oxidation of the electrode tip

(8) Defects such as splits or cracks in the electrode

(9) Use of improper shielding gases such as argon-oxygen or argon-CO_2 mixtures that are used for gas metal arc welding

Corrective steps are obvious once the causes are recognized and the welder is adequately trained.

Lack of Shielding

Discontinuities related to the loss of inert gas shielding are the tungsten inclusions previously described, porosity, oxide films and inclusions, incomplete fusion, cracking, and loss of joint strength and weld toughness. The extent to which they occur is strongly related to the characteristics of the metal being welded. For example, the properties of titanium, aluminum, nickel, and high strength steel alloys are seriously impaired with loss of inert gas shielding. Good shielding practices and the particular metal alloy requirements for shielding are described in detail in the *Welding Handbook,* Section 4, 6th edition. Spot checks of gas shielding effectiveness can often be made in advance of production welding by producing a stationary weld nugget or spot and continuing gas flow until the weld has cooled to a low temperature. A bright, silvery spot will be evident if shielding is effective.

General Weld Defects

Weld undercut from excessive travel speed or arc length is a defect common to all arc welding processes. Melt-thru, cold laps, incomplete penetration, overlaps, poor weld contour, and inadequate weld throat thickness are some of the additional defects which can be encountered. Corrective measures are similar for most processes. They include the proper balancing of heat and filler metal input rates to the weld joint.

SUPPLEMENTARY READING LIST

Chase, T.F. and Savage, W.F. Effect of anode composition on tungsten-arc characteristics. *Welding Journal* 50 (11), Nov. 1971, pp. 467-s—473-s.

Chihoski, R.A. Rationing of power between the gas tungsten-arc and electrode. *Welding Journal* 49 (2), Feb. 1970, pp. 69-s—82-s.

Dick, N.T. Tube welding by the pulsed TIG method. *Metal Construction & British Welding Journal* 5 (3), Mar. 1973, pp. 85-89.

Gas Tungsten-Arc Welding of Titanuim Piping and Tubing, D10.6 (latest ed.). Miami: American Welding Society.

"Gas Tungsten-Arc Welding (TIG Welding)," *Metals Handbook,* Vol. 6, 8th ed. Metals Park, Ohio: American Society for Metals, 1971, pp. 113-138.

Goodman, I.S., Ehringer, H.J., and Hackman, R.L. New gas tungsten-arc welding electrode. *Welding Journal* 42 (7), July 1963, pp. 567-570.

Grist, F.J. Improved, lower cost aluminum welding with solid state power source. *Welding Journal* 54 (5), May 1975, pp. 348-357.

GTA torch current control eliminates tungsten inclusions. *Metal Progress* 107 (2), Feb. 1975, p. 57.

Leitner, R.E., McElhinney, G.H., and Pruitt, E.L. Investigation of pulsed GTA welding variables. *Welding Journal* 52 (9), Sept. 1973, pp. 405-s—410-s.

McElrath, T., and Gorman, E.F. Argon-hydrogen shielding gas mixtures for tungsten-arc welding. *Welding Journal* 36 (1), Jan. 1957, pp. 28-35.

Multiple electrode TIG welding. *Welding & Metal Fabrication* 43 (4), May 1975, pp. 269-271.

Niles, R.W., and Pfender, E. Influence of the efficiency of the GTAW process. *Welding Journal* 54 (1), Jan. 1975, pp. 25-s—32-s.

Normando, N.J. Manual pulsed GTA welding. *Welding Journal* 52 (9), Sept. 1973, pp. 566-573.

Nuclear seals: perfect job for auto TIG. *Welding Design & Fabrication* 47 (8), Aug. 1974, pp. 35-38.

Petrie, T.W. and Pfender, E. Influence of the cathode tip on temperature and velocity fields in a gas tungsten-arc. *Welding Journal* 49 (12), Dec. 1970, pp. 588-s—596-s.

Rager, D.D. Direct current, straight polarity gas tungsten-arc welding of aluminum. *Welding Journal* 50 (5), May 1971, pp. 332-341.

Recommended Practices for Gas Tungsten-Arc Welding, C5.5 (latest ed.). Miami: American Welding Society.

Saenger, J.F. Gas tungsten-arc hot wire welding —a versatile new production tool. *Welding Journal* 49 (5), May 1970, pp. 363-371.

Specification for Tungsten Arc-Welding Electrodes, A5.12 (latest ed.). Miami: American Welding Society.

Spiller, K.R. Positional TIG welding of aluminum pipe. *Welding & Metal Fabrication* 43 (10), Dec. 1975, pp. 733-737, 746.

Troyer, W.E. Programming and pulsing the tungsten arc. *Metal Progress* 107 (3), Mar. 1975, pp. 99-106.

Wiley, R. Automatic TIG tames tough job. *Welding Design & Fabrication* 48 (5), May 1975, pp. 54-55.

4

Gas Metal Arc Welding

Prepared by a committee consisting of

A.B. CRICHTON, *Chairman*
  Union Carbide Corporation
J.R. CRISCI
  Department of the Navy
N.A. FREYTAG
  The Budd Company

V. HASKEN
  John Deere Company
R.B. HITCHCOCK
  E.I. DuPont de Nemours & Co.
G.P. YANOK
  Aluminum Company of America

4

Gas Metal Arc Welding

FUNDAMENTALS OF THE PROCESS

DEFINITION AND GENERAL DESCRIPTION

Gas metal arc welding (GMAW) is an electric arc welding process which produces coalescence of metals by heating them with an arc established between a continuous filler metal (consumable) electrode and the work. Shielding of the arc and molten weld pool is obtained entirely from an externally supplied gas or gas mixture, as shown in Fig. 4.1. The process is sometimes referred to as MIG or CO_2 welding.

When GMAW was first developed, it was considered to be fundamentally a high current density, small diameter, bare metal electrode process using an inert gas for arc shielding. Its primary application was for welding aluminum. As a result, the term MIG (Metal Inert Gas) was used and is still the most common reference for the process. Subsequent developments include operation at low current densities and pulsed direct current, application to a broader range of materials, and the use of reactive gases (particularly CO_2) or gas mixtures. This latter development has led to the formal acceptance of the term gas metal arc welding (GMAW) for the process because both inert and reactive gases are used.

GMAW is operated in semiautomatic, machine, and automatic modes. It is utilized particularly in high production welding operations. All commercially important metals such as carbon steel, stainless steel, aluminum, and copper can be welded with this process in all positions by choosing the appropriate shielding gas, electrode, and welding conditions.

ARC POWER AND POLARITY

The vast majority of GMAW applications require the use of direct current reverse polarity (electrode positive). This type of electrical connection yields a stable arc, smooth metal transfer, relatively low spatter loss, and good weld bead characteristics for the entire range of welding currents used.

Direct current straight polarity (electrode negative) is seldom used, since the arc can become very unstable and erratic even though the electrode melting rate is higher than that achieved with dcrp (electrode positive). When employed, dcsp (electrode negative) is used in conjunction with a "buried" arc or short circuiting metal transfer. Penetration is lower with straight polarity than with reverse polarity direct current.

Alternating current has found no commercial acceptance with the GMAW process for two reasons: (1) The arc is extinguished during each half cycle as the current reduces to zero, and it may not reignite if the cathode cools sufficiently; and (2) Rectification of the reverse polarity cycle promotes the erratic arc operation. A more detailed discussion on the effect of polarity on metal transfer can be found in Chapter 2, *Welding Handbook,* Volume 1, 7th edition.

METAL TRANSFER

Filler metal can be transferred from the electrode to the work in two ways: (1) when the electrode contacts the molten weld pool thereby establishing a short circuit, which is known as short circuiting transfer (short circuiting arc welding); and (2) when discrete drops are moved across the arc gap under the influence of gravity or electromagnetic forces. Drop transfer can be either globular or spray type.

Shape, size, direction of drops (axial or nonaxial), and type of transfer are determined by a number of factors. The factors having the most influence are

(1) Magnitude and type of welding current
(2) Current density
(3) Electrode composition
(4) Electrode extension
(5) Shielding gas
(6) Power supply characteristics

Axially directed transfer refers to the movement of drops along a line that is a continuation of the longitudinal axis of the electrode. Nonaxially directed transfer refers to movement in any other direction.

Short Circuiting Transfer

Short circuiting arc welding uses the lowest range of welding currents and electrode diameters associated with GMAW. Typical current ranges for steel electrodes are shown in Table 4.1. This type of transfer produces a small, fast freezing weld pool that is generally suited for the joining of thin sections, for out-of-position welding, and for the filling of large root openings. When weld heat input is extremely low, plate distortion is small. Metal is transferred from the electrode to the work only during a period when the electrode is in contact with the weld pool. There is no metal transfer across the arc gap.

The electrode contacts the molten weld pool at a steady rate in a range of 20 to over 200 times each second. The sequence of events in the transfer of metal and the corresponding current and voltage are shown in Fig. 4.2. As the wire touches the weld metal, the current increases. It would continue to increase if an arc did not form, as shown at E in Fig. 4.2. The rate of current

increase must be high enough to maintain a molten electrode tip until filler metal is transferred. Yet, it should not occur so fast that it causes spatter by disintegration of the transferring drop of filler metal. The rate of current increase is controlled by adjustment of the inductance in the power source. The value of inductance required depends on both the electrical resistance of the welding circuit and the temperature range of electrode melting. The open circuit voltage of the power source must be low enough so that an arc cannot continue under the existing welding conditions. A portion of the energy for arc maintenance is provided by the inductive storage of energy during the period of short circuiting.

As metal transfer only occurs during short circuiting, shielding gas has very little effect on this type of transfer. Spatter can occur. It is usually caused by either gas evolution or electromagnetic forces on the molten tip of the electrode.

Globular Transfer

With a positive electrode (dcrp), globular transfer takes place when the current density is relatively low, regardless of the type of shielding gas. However, carbon dioxide (CO_2) shielding yields this type of transfer at all usable welding currents. Globular transfer is characterized by a drop size of greater diameter than that of the electrode.

Globular, axially directed transfer can be achieved in a substantially inert gas shield without spatter. The arc length must be long enough to assure detachment of the drop before it contacts the molten metal. However, the resulting weld is likely to be unacceptable because of lack of fusion, insufficient penetration, and excessive reinforcement.

Carbon dioxide shielding always yields nonaxially directed globular transfer. The departure from axial motion is due to an electromagnetic repulsive force acting upon the bottom of the molten drop, as shown in Fig. 4.3. Flow of electric current through the electrode generates several forces that act on the molten tip. The most important of these are pinch force *(P)* and anode reaction force *(R)*. The magnitude of the pinch force is a direct function of welding current and wire diameter, and is usually responsible for drop detachment. With CO_2 shielding, the wire electrode is melted by the arc heat conducted through the molten drop. The electrode tip is not enveloped by the arc plasma. The molten drop grows until it detaches by short circuiting or gravity, as R is never overcome by P. High speed photography shows that the arc moves over the area between the molten drop and workpiece, while R tends to support the drop. Spatter can therefore be very severe.

Spray Transfer

In a gas shield of at least 80 percent argon or helium, filler metal transfer changes from globular to spray type as welding current increases for a given size electrode. For all metals, the change takes place at a current value called the globular-to-spray transition current. Table 4.2 lists transition currents for various metal electrodes.

Spray type transfer has a typical fine arc column and pointed wire tip associated with it. Molten filler metal transfers across the arc as fine droplets. The droplet diameter is equal to or less than the electrode diameter. The metal spray is axially directed. The reduction in droplet size is also accompanied by an increase in the rate of droplet detachment, as illustrated in Fig. 4.4. Metal transfer rate may range from less than 100 to several hundred droplets per second as the electrode feed rate increases from approximately 42 to 340 mm/s (100 to 800 in./min).

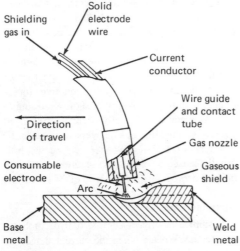

Fig. 4.1—Gas metal arc welding process

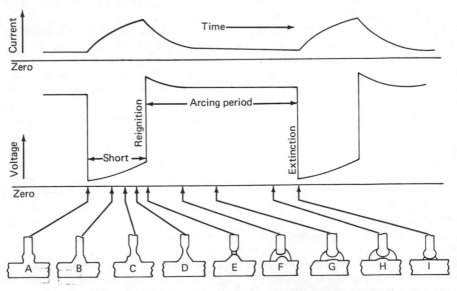

Fig. 4.2—Oscillograms and sketches of short circuiting arc metal transfer

Table 4.1—Typical current ranges for gas metal arc welding of steel with short circuiting transfer

| Wire diameter, | | Welding current, A[a] | | | |
| mm | in. | Flat position | | Vertical and overhead positions | |
		Minimum	Maximum	Minimum	Maximum
0.76	0.030	50	150	50	125
0.89	0.035	75	175	75	150
1.14	0.045	100	225	100	175

a. DCRP (electrode positive)

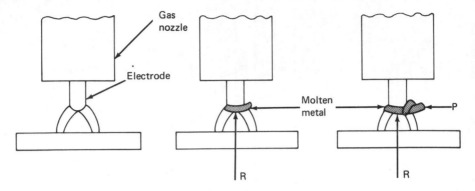

Fig. 4.3—Nonaxial globular transfer

The mechanisms of axial spray transfer appear to be chiefly influenced by both the electromagnetic forces on the molten electrode tip and the arc plasma. The action of the latter is at the electrode tip. The electromagnetic forces arise from the interaction of the welding current and the coaxial magnetic field that it produces at the electrode tip. The electromagnetic pinch force squeezes on the molten portion of the electrode and accelerates it towards the work.

Another influence on spray transfer is the vaporization of metal at the electrode tip.

When CO_2 is used for shielding, a globular-to-spray transition does not occur. Drops become smaller as the current increases, but they are not all axially directed; and there is much more spatter than with an inert gas shield. The spatter is minimized if the welding conditions are adjusted so that the tip of the electrode and the arc are located in a cavity below the surface

Table 4.2—Globular-to-spray transition currents for a variety of electrodes

Wire electrode type	Wire electrode diameter		Shielding gas	Minimum spray arc current, A
	mm	in.		
Mild steel	0.76	0.030	98% argon-2% oxygen	150
Mild steel	0.89	0.035	98% argon-2% oxygen	165
Mild steel	1.14	0.045	98% argon-2% oxygen	220
Mild steel	1.59	0.062	98% argon-2% oxygen	275
Stainless steel	0.89	0.035	99% argon-1% oxygen	170
Stainless steel	1.14	0.045	99% argon-1% oxygen	225
Stainless steel	1.59	0.062	99% argon-1% oxygen	285
Aluminum	0.76	0.030	Argon	95
Aluminum	1.14	0.045	Argon	135
Aluminum	1.59	0.062	Argon	180
Deoxidized copper	0.89	0.035	Argon	180
Deoxidized copper	1.14	0.045	Argon	210
Deoxidized copper	1.59	0.062	Argon	310
Silicon bronze	0.89	0.035	Argon	165
Silicon bronze	1.14	0.045	Argon	205
Silicon bronze	1.59	0.062	Argon	270

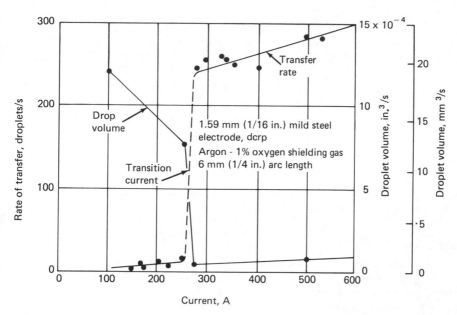

Fig. 4.4— *Variation in volume and transfer rate of drops with welding current (steel electrode)*

of the base metal. Under this condition, most of the nonaxially directed droplets are trapped in the weld pool.

The spray arc is best adapted for welding relatively thick parts because of the high welding currents associated with it. The minimum base metal thickness limit cannot be well defined. It will depend on the type of metal being welded and the welding conditions. Thicknesses down to 3 mm (1/8 in.) can be welded in some metals, such as aluminum.

PROCESS VARIATIONS

In addition to the three basic types of metal transfer which characterize the GMAW process, there are several unique variations of significance.

Pulsed Spray Welding

Pulsed spray welding is a variation of the GMAW process that is capable of all-position welding at higher energy levels than short circuiting arc welding. The power source provides two current levels: a steady "background" level, which is too low to produce spray transfer; and

a "pulsed peak" current, which is superimposed upon the background current at a regulated interval, as shown in Fig. 4.5. The pulse peak is well above the transition current, and usually one drop is transferred during each pulse. The combination of the two levels of current produces a steady arc with axial spray transfer at effective welding currents below those required for conventional spray arc welding. Because the heat input is lower, this variation in operation is capable of welding thinner sections than are practical with the conventional spray transfer.

Arc Spot Welding

Gas metal arc spot welding is a method of joining analogous to resistance spot welding and riveting. A variation of continuous gas metal arc welding, the process fuses two pieces of sheet metal together by penetrating entirely through one piece into the other. No joint preparation is required other than cleaning of the overlap areas. The welding gun remains stationary while a spot weld is being made. Mild steel, stainless steel, and aluminum are commonly joined by this method.

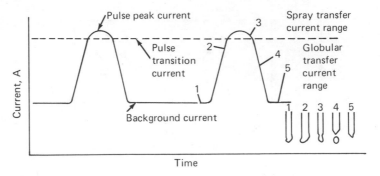

Fig. 4.5—Pulsed-spray arc welding current characteristic

Electrogas Welding

The electrogas (EG) variation of the GMAW process is a fully automatic, high deposition rate method for the welding of butt, corner, and T-joints in the vertical position. The electrogas variation essentially combines the mechanical features of electroslag welding (ESW) with the GMAW process. Water-cooled copper shoes span the gap between the pieces being welded to form a cavity for the molten metal. A carriage is mounted on a vertical column; this combination provides both vertical and horizontal movement. Welding head, controls, and electrode spools are mounted on the carriage. Both the carriage and the copper shoes move vertically upwards as welding progresses. The welding head may also be oscillated to provide uniform distribution of heat and filler metal.

This method is capable of welding metal sections of from 13 mm (1/2 in.) to more than 51 mm (2 in.) in thickness in a single pass. Deposition rates of 16 to 20 kg (35 to 45 lb) per hour per electrode can be achieved. Chapter 7, "Electroslag and Electrogas Welding," covers this process in detail.

EQUIPMENT

Gas metal arc welding equipment consists of a welding gun, a power supply, a shielding gas supply, and a wire-drive system which pulls the wire electrode from a spool and pushes it through a welding gun. Also, a source of cooling water may be required for the welding gun. In passing through the gun, the wire becomes energized by contact with a copper contact tube, which transfers current from a power source to the arc. While simple in principle, a system of accurate controls is employed to initiate and terminate the shielding gas and cooling water, operate the welding contactor, and control the electrode feed speed as required. The basic features of GMA welding equipment are shown in Fig. 4.6. The GMAW process is used for semiautomatic, machine, and automatic welding. Semiautomatic GMAW is often referred to as manual welding.

WELDING GUNS

Welding guns for GMAW are available for manual manipulation (semiautomatic welding) and for machine or automatic welding. Because the electrode is fed continuously, a welding gun

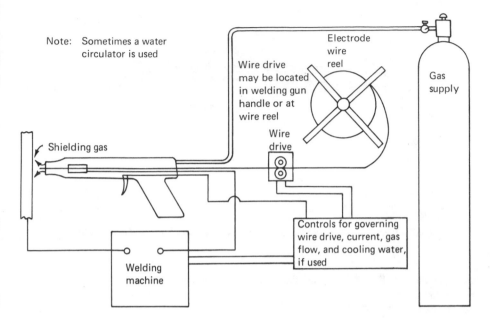

Note: Sometimes a water circulator is used

Electrode wire reel

Wire drive may be located in welding gun handle or at wire reel

Gas supply

Shielding gas

Wire drive

Controls for governing wire drive, current, gas flow, and cooling water, if used

Welding machine

Fig. 4.6—Diagram of gas metal arc welding equipment

must have a sliding electrical contact (contact tube) to transmit the welding current to the electrode. The gun must also have a gas passage and a nozzle to direct the shielding gas around the arc and molten weld pool. Cooling is required to remove the heat generated within the gun, and also heat radiated from the welding arc and the molten weld metal. Shielding gas or internal circulating water, or both, are used for cooling. An electrical switch is needed to start and stop the welding current, electrode feed system, and shielding gas flow.

Semiautomatic Guns

Semiautomatic, hand held guns are usually in the form of a pistol, or they are shaped similar to an oxyacetylene torch, with electrode wire fed through the barrel or handle. Typical curved and pistol gun designs are shown in Figs. 4.7 and 4.8. In some versions of the pistol design, where the most efficient cooling is desirable, water is directed through passages in the gun to cool both the contact tube and the metal shielding gas nozzle. The curved gun utilizes a curved current-carrying body at the front end through which the shielding gas is brought to the nozzle. Such a

gun, designed for small diameter wires, is extremely flexible and maneuverable. It is particularly suited for welding in tight, hard-to-reach corners and other confined areas. The service lines, consisting of a power cable, water hose, and gas hose, enter at the handle and rear barrel section of the gun. Guns are equipped with metal nozzles of various internal diameters to insure adequate gas shielding. The orifice usually varies from approximately 10 to 22 mm (3/8 to 7/8 in.) depending upon welding requirements. The nozzles are usually threaded to facilitate replacement.

Current input into the electrode is accomplished in the gun by a threaded or clamped copper-base alloy contact tube. The electrode is fed through the tube. Positive electrical contact is obtained by various means. To obtain optimum welding current input into the electrode, the contact tubes are supplied in various hole sizes, requiring the selection of the proper tube for a given electrode size and type.

The conventional pistol type holder is used also for arc spot welding applications where filler metal is required. The heavy nozzle of the holder is slotted to exhaust the gases away from the spot. The pistol grip handle permits easy

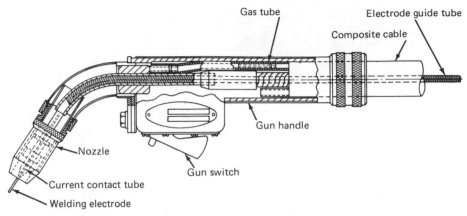

Fig. 4.7—Typical semiautomatic gas-cooled, curved-neck gas metal arc welding gun

manual loading of the holder against the work. The welding control is designed to regulate the flow of cooling water and the supply of shielding gas. It is also designed to prevent the wire from freezing to the weld by timing the weld over a a preset interval.

Air-cooled guns are available for applications where water is not readily obtainable as a cooling medium. These guns are generally available for service up to 600 A, intermittent duty, with CO_2 shielding gas. However, they are usually limited to 200 A with argon or helium shielding. The holder is generally of a pistol-type

construction and its operation parallels the water-cooled type. One exception is that the water service line and the water-cooled power cable are replaced by individual gas and power lines, or a composite shielding gas-power cable assembly. Three general types of air-cooled guns are available:

(1) A gun that has the electrode wire fed to it through a flexible conduit from a remote wire feeding mechanism. The conduit is generally in the 3.7 m (12 ft) length range due to the wire feeding limitations of a push-type system (Figs. 4.7 and 4.8). Steel wires of 0.89 to 2.38 mm (0.035

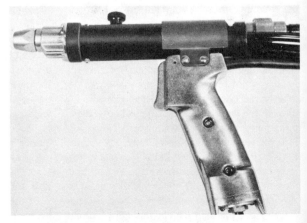

Fig. 4.8—Typical semiautomatic gas metal arc welding guns

Fig. 4.9—Gas metal arc welding guns with self-contained electrode wire drives and wire supply

to 3/32 in.) diameter and aluminum wires of 1.19 to 3.18 mm (3/64 to 1/8 in.) diameter can be fed with this arrangement.

(2) A gun that has a self-contained wire feed mechanism and electrode wire supply. The wire supply is generally in the form of a 102 mm (4 in.) diameter, 0.45 to 1.1 kg (1 to 2-1/2 lb) spool. This type of gun employs a pull-type wire feed system, and it is not limited by a 3.7 m (12 ft) flexible conduit. Typical guns are shown in Fig. 4.9. Wire diameters of 0.76 to 1.19 mm (0.030 to 3/64 in.) are normally used with this type of gun.

(3) A pull-type gun that has the electrode wire fed to it through a flexible conduit from a remote spool. This incorporates a self-contained wire feeding mechanism. It can also be used in a push-pull type feeding system. The system permits the use of flexible conduits in lengths up to 15 m (50 ft) or more from the remote wire feeder. Figure 4.10 shows two typical pull-type guns.

Aluminum and steel electrodes with diameters of 0.76 to 1.59 mm (0.030 to 1/16 in.) can be used with these types of feed mechanisms.

Water-cooled guns for manual GMAW are similar to gas-cooled types with the addition of water cooling ducts. The ducts circulate water around the contact tube and the gas nozzle. Water cooling permits the gun to operate continuously at rated capacity and at lower temperatures. Water-cooled guns are used for applications requiring 200 to 750 A. The water in and out lines to the gun add weight and reduce maneuverability of the gun for welding.

The selection of air- or water-cooled guns is based on the type of shielding gas, welding current range, materials, weld joint design, and existing shop practice. Air-cooled guns are heavier than water-cooled guns of the same welding current capacity. However, air-cooled guns are easier to manipulate to weld out-of-position and in confined areas.

Fig. 4.10—Pull type gas metal arc welding guns for use with a remotely located push wire drive system

Machine Welding Guns

Machine type guns, with capacities up to 1200 A, employ the same basic design principles and features used in semiautomatic gas metal arc welding guns. The machine welding gun is generally water-cooled since the service conditions are more severe. They are generally operated for long periods of arc time. The gun is mounted directly below the electrode wire drive head which feeds the electrode wire through the gun. The electrode is fed by feed rolls into a guide assembly which supports and protects the electrode during the welding operation. Typical machine GMAW guns are shown in Fig. 4.11. Electrode diameters of 0.76 to 6.4 mm (0.030 to 1/4 in.) are normally used with these guns.

WIRE FEEDERS AND WELDING CONTROLS

Wire Feeders

The wire electrode drive mechanisms contained in hand-held (semiautomatic) guns consist of small motors and drive rolls. These motors are usually electric or compressed air-powered with adjustable speed. In push-pull systems, the two drive motors are synchronized to avoid damage to the wire electrode. Wire feeders for semiautomatic gas metal arc welding vary considerably in individual design. Unit speed controls may be mechanical, electromechanical, or electronic. Most wire feeders are now designed for use with constant voltage power sources. The welding current is adjusted by increasing

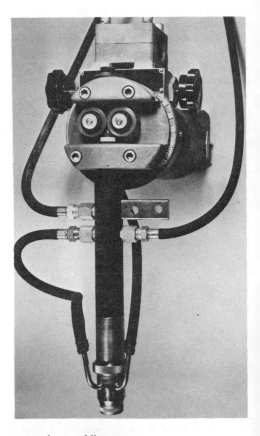

Fig. 4.11—Machine-type gas metal arc welding guns

or decreasing electrode speed for a given setting of the power source. With constant current power sources, a voltage-sensing circuit is used to maintain the desired arc length by varying the electrode speed.

The wire feed motor is usually a dc type, and it provides the power for driving the electrode wire through the gun to the work. The wire feed is held constant for the majority of GMAW applications. Therefore, most feed motors are shunt wound or permanent magnet types. Occasionally, a variable speed motor will be necessary if a constant current type of power source is used. This type of motor can be a series wound or one of the above types. Its speed will vary as the control unit increases or decreases the wire feed speed to maintain constant arc length (voltage).

Welding Control Unit

The welding control unit may be a separate package for remote operation, or it may be integrated with the wire feed drive unit. The main function of the control unit is to regulate the speed of the wire feed motor. Motor speed regulation is usually accomplished with an electronic governor in the control unit. Electrode feed speed is manually set by the operator to obtain the desired welding current from a constant voltage power source. If a constant current power source is used, the control unit varies the electrode feed rate so that a preset arc voltage (length) is maintained. The control also regulates the starting and stopping of electrode feed upon the appropriate manual or automatic switch operation.

In addition, the control units will contain several of the following:

(1) An electrode feed jogging switch to feed the electrode through the unit when not welding

(2) A shielding gas purging switch for manual control of shielding gas flow

(3) Electrode speed or arc voltage adjustment

(4) A braking system to prevent electrode stubbing into the molten weld pool when welding is stopped

(5) Timers for preweld and postweld gas and water flow

(6) A water pressure switch to insure coolant flow

(7) A meter to indicate the load on the wire drive motor

One of several arc starting systems may also be included in the control circuitry. One type is a slow speed start in which the electrode advances slowly toward the work until the arc starts. Then the electrode speeds up. Another type is a retract start in which the electrode is touched to the work and then retracted to draw an arc. With constant voltage power sources, the arc can be started by just feeding the electrode to the work.

Most controls for automatic welding contain provisions for operation from an operator's station, from limit switches, or from weld duration timers. They are so arranged that their circuits can be properly oriented with the fixture controls in order that full weld programming can be achieved. Gas flow regulating equipment and power souces are the same as those used for semiautomatic equipment.

AUTOMATIC WELDING EQUIPMENT

Automatic welding equipment installation is effectively used when the work can be easily brought to the welding station, or when a large amount of welding must be done. Weld travel speed and weld quality can be greatly increased because arc travel is automatically controlled.

Basically, all the equipment is similar to that used for machine welding except for additional automation which includes beams, carriages mounted on tracks, and specially built positioners and fixtures. The welding gun will likely be a machine type, as shown in Fig. 4.11.

POWER SOURCES

The GMAW process uses power sources similar to those used with other continuous electrode feed welding processes, such as flux cored and submerged arc welding. The process requires a source of direct current, which may be supplied by a transformer-rectifier or a motor-generator power source. The power source rating depends on the amperage range required for the applications. Some applications may require anywhere from 15 to 1200 A. Power source ratings are based on either a 60 percent or 100 percent duty cycle. The various types of arc welding power sources and their characteristics are covered in Chapter 1.

Constant Voltage Power Supply

The arc voltage is established by setting the output voltage on the power supply. The power source will supply the necessary amperage to melt the welding electrode at the rate required to maintain the preset voltage (or relative arc length). The speed of the electrode drive is used to control the average welding current. This characteristic is generally preferred for the welding of all metals. The use of this type of power supply in conjunction with a constant wire electrode feed results in a self-correcting arc length system.

Constant Current Power Supply

With this type, the welding current is established by the appropriate setting on the power supply. Arc length (voltage) is controlled by the automatic adjustment of the electrode feed rate. This type of welding is best suited to large diameter electrodes and machine or automatic welding, where very rapid change of electrode feed rate is not required. Most constant current power sources have a drooping volt-ampere output characteristic. However, true constant current machines are available.

Constant current power sources are not normally selected GMAW because of the greater control needed for electrode feed speed. The systems are not self-regulating.

Pulsed Direct Current Power Supply

This type of power source pulses the dc output from a low background value to a high peak value, as described previously. Because the average power is lower, pulsed welding current can be used to weld thinner sections than those that are practical with steady dc spray transfer.

POWER SOURCE VARIABLES

There are several power source adjustments necessary for the production of the best possible welding conditions for a particular application. The continuous control of the arc voltage is critically important. Some control of the volt-

ampere slope characteristic and the inductance is also beneficial. Refer to Chapter 1 for further details.

Voltage

Arc voltage is the electrical potential between the electrode and the workpiece. The voltage indicated by the power source meter is commonly referred to as the arc voltage and, consequently, a direct measure of the arc length. This is incorrect because there are many locations in the welding circuit where a drop in voltage occurs other than across the arc. These voltage drops are a function of cable size and length, conduction efficiency of power carrying connections, condition of the current contact tube in the gun, and electrode extension. If they are kept to a minimum, the voltage reading on the power supply can approach closely the true arc voltage. It is generally impractical in production work to measure true arc voltage, which would have to be measured between the electrode tip and the work.

As the arc voltage is increased, the length of the arc increases. The arc length per volt varies primarily with the composition of the shielding gas. As a rule, the optimum arc voltage is dependent upon a variety of factors, including metal thickness, the type of joint, the position of the welding, electrode size, gas composition, and the type of weld. Typical arc voltages for welding various metals with some of the most commonly used shielding gases are shown in Table 4.3.

Slope

Figure 4.12 illustrates the output volt-ampere characteristics for a dc power source at two particular settings. The slope of each curve is referred to as the "slope" of the power source. The slope of the power source, as specified by the manufacturer, is measured at its output terminals. It is not the total slope of the arc system. Anything which adds impedence to the welding circuit increases both the slope and the voltage drop at a given welding current. Various components of the welding circuit, such as power cables, connections, loose terminals, and dirty contacts, add to the slope. Therefore, slope is best measured at the arc in a welding system.

Table 4.3—Typical arc voltages for gas metal arc welding of various metals[a]

Metal	Drop transfer 1.6 mm (1/16 in.) diameter electrode					Short circuiting transfer 0.9 mm (0.035 in.) diameter electrode			
	Argon	Helium	25% Ar-75% He	Ar-O$_2$ (1-5% O$_2$)	CO$_2$	Argon	Ar-O$_2$ (1-5% O$_2$)	75% Ar-25% CO$_2$	CO$_2$
Aluminum	25	30	29	—	—	19	—	—	—
Magnesium	26	—	28	—	—	16	—	—	—
Carbon steel	—	—	—	28	30	17	18	19	20
Low alloy steel	—	—	—	28	30	17	18	19	20
Stainless steel	24	—	—	26	—	18	19	21	—
Nickel	26	30	28	—	—	22	—	—	—
Nickel-copper alloy	26	30	28	—	—	22	—	—	—
Nickel-chromium-iron alloy	26	30	28	—	—	22	—	—	—
Copper	30	36	33	—	—	24	22	—	—
Copper-nickel alloy	28	32	30	—	—	23	—	—	—
Silicon bronze	28	32	30	28	—	23	—	—	—
Aluminum bronze	28	32	30	—	—	23	—	—	—
Phosphor bronze	28	32	30	23	—	23	—	—	—

a. Plus or minus approximately ten percent. The lower voltages are normally used on light material and at low amperage; the higher voltages with high amperage on heavy material.

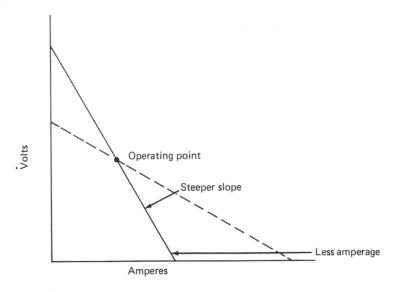

Volts

Operating point

Steeper slope

Less amperage

Amperes

Fig. 4.12—Effect of changing slope on short circuiting amperage

Current

The main use of slope in a GMAW system is to control the magnitude of the short-circuit current. This is the current available from the power source on the short circuit between the electrode and workpiece. The short circuit current determines the amount of pinch force available on the electrode. Pinch force is the term used to describe the squeezing force on a current carrying conductor due to the current flowing through it. In any conductor, the pinch force is proportional to the square of the current flowing through it. This pinch force will eventually cause the molten electrode tip to "neck down" and finally separate from the solid electrode material.

Slope controls the short circuit current and the maximum pinch force on the electrode by changing the volt-ampere characteristics of the power source. The steeper the slope of the volt-ampere curve, the smaller are the short circuit current and the pinch force. The flatter the slope, the higher are the short circuit current and pinch force. When used in combination with a voltage adjustment, it is possible to use slope adjustment to vary the slope of the volt-ampere curve as it passes through a given operating point. Thus, the short circuiting amperage can be changed, as shown in Fig. 4.12.

The short circuit current is important since the resultant pinch force affects the way a molten drop detaches from the electrode tip. This, in turn, affects the arc stability. When little or no slope is present in the power supply circuit, the short circuit current will rise to a very high level. At high currents the pinch force is very high, and a violent parting of the molten drop from the electrode tip takes place. The high pinch effect will violently squeeze the metal aside and "clear" the short circuit. This causes excessive weld spatter.

When the short circuit current is limited to low values by a steep slope, the electrode can carry the full current, and the short circuit will not clear itself. The wire electrode then piles up on the workpiece, or it will freeze to the weld pool occasionally and melt in two.

When the short circuit current is at the correct value, the parting of the molten drop from the electrode tip is smooth with very little spatter. Most constant voltage power supplies are equipped with a means of changing the slope of the volt-ampere characteristic. They may be stepped or continuously adjustable to provide the correct short circuit current for the application. Some power sources have a fixed slope which has been predetermined for a given type of gas metal arc welding.

Inductance

When the load changes on a power supply, the current takes a finite time to attain its new level. The circuit parameter primarily responsible for such a time lag is the inductance. The rate of current rise is determined by the inductance of the power supply, as shown in Fig. 4.13. Because the pinch force on the molten metal at the electrode tip increases with current, the rate of the pinch force buildup is also affected by the inductance in the circuit.

For short circuiting arc welding, inductance can be added to control the rate of current rise to minimize spatter. Inductance also influences arc time. Increasing the inductance will increase the arc time and decrease the frequency of short circuiting. The increased arc time produces a more fluid puddle and a flatter, smoother weld bead. For each particular electrode feed rate, there is a best value of inductance.

In spray arc welding, the addition of some inductance to the power supply will produce a softer arc start without reducing the final amount of welding current available. Spatter is held to a minimum when conditions of both adequate current and correct rate of current rise exist. Too much inductance in the welding circuit will cause erratic arc starting.

SHIELDING GAS EQUIPMENT

Regardless of the type of gas supply system used, constant pressure and flow of the gas must be maintained. The purposes of gas pressure regulators are to reduce the pressure from the source to a working pressure and to maintain a constant delivery pressure. In addition, the regulator must be adjustable to provide gas at a desired pressure within its operating range.

Flowmeters are used to control the rate of

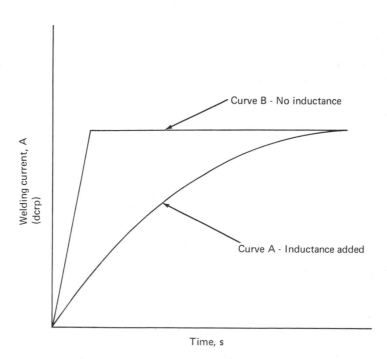

Curve B - No inductance

Welding current, A (dcrp)

Curve A - Inductance added

Time, s

Fig. 4.13—Change in welding current rise due to inductance

gas flow to the welding gun. They may have single or dual flow rates for a particular gas, calibrated in liters per minute or cubic feet per hour (CFH). The inlet gas pressure to the flowmeter is specified by the manufacturer, and the pressure regulator must be set accordingly. Gas flow is adjusted with a valve at the flowmeter outlet.

Combination pressure regulator-flow meter units are available for some gases. The delivery pressure to the flowmeter is adjusted and locked by the manufacturer. It should not be changed except for adjustment after repair.

Shielding gases are supplied either in gas form in high pressure cylinders or in liquid form and stored in tanks with vaporizers. Cylinders may be manifolded for increased storage capacity at the welding station.

Proportioners are available for mixing gases such as argon and carbon dioxide. Some gases, such as argon and oxygen, can be mixed using a wye connection and equal delivery pressures.

Cylinder, regulator, and flowmeter connections for various gases consist of different sizes to eliminate the possibility of a mismatched regulator and cylinder. Hose connections are also different.

The Compressed Gas Association has a complete set of specifications for cylinder valves and regulator inlet connections that are not interchangeable. Regulators and flowmeters should only be used for the gases for which they are designed.

Additional information is contained in Chapter 3, since the gases and equipment used for GMAW is also used for gas tungsten arc welding. For information on the best and most economical method of handling specific gases, the gas suppliers and equipment manufacturers should be consulted.

MATERIALS

The consumables used in GMAW are electrodes and shielding gases. The chemical composition of the electrode, in combination with the shielding gas, will influence the weld metal composition which determines the chemical and mechanical properties of the weldment. Factors influencing the selection of the shielding gas-wire electrode combination are

(1) Base metal composition
(2) Base metal mechanical properties
(3) Base metal condition and cleanliness
(4) Type of service or applicable specification requirement
(5) Welding position
(6) Type of filler metal transfer (spray, globular, or short circuiting)

ELECTRODES

The electrodes for gas metal arc welding are usually quite similar or identical in composition to those used for welding with most other bare electrode processes. As a rule, the compositions of the electrode and the base metal are as nearly alike as practicable, commensurate with good welding characteristics and weld properties. In some cases, this involves very little modification from the base metal composition. In other cases, obtaining satisfactory welding and weld metal characteristics requires an appreciable composition change, perhaps the use of an electrode of completely different composition. For example, the electrodes that are most satisfactory for welding manganese bronze, a copper-zinc alloy, are either aluminum bronze or copper-manganese-nickel-aluminum alloys. Somewhat similarly, although not to the same degree, the electrodes that are most suitable for welding the higher strength aluminum and steel alloys are usually quite different in composition from the base metal on which they are to be used. This is because some alloys, such as 6061, that are quite satisfactory or desirable as base metals are unsuitable as weld metal. Accordingly, electrode alloys are designed to produce the desired weld metal properties and the acceptable operating characteristics.

Whatever modifications are made in the

composition of electrodes, deoxidizers or other scavenging elements are nearly always added. This is done to minimize porosity in the weld or to assure desired mechanical properties in the weld by the reactions with oxygen, nitrogen, or hydrogen that may be in the shielding gas or may accidentally reach the metal from the surrounding atmosphere. The addition of appropriate deoxidizers in the right quantity, most essential in electrodes that are to be used with an oxygen containing shielding gas, is desirable in most other cases. In arc spot welding, a large amount of deoxidizer is usually provided to compensate for the excessive loss that results from a long welding arc and the large pool of molten metal. The deoxidizers most frequently used in steel electrodes are manganese, silicon, and aluminum. Titanium and silicon are the principal ones used in nickel alloys. Copper alloys may be deoxidized with titanium, silicon, or phosphorus depending on the type and desired end results.

The electrodes used for GMAW are quite small in diameter compared to those used for other types of welding. Wire diameters of 1.02 to 1.59 mm (0.045 to 1/16 in.) are about average. However, electrode diameters as small as 0.5 mm (0.020 in.) and as large as 3.18 mm (1/8 in.) are sometimes used. Because of the small sizes of the electrode and the comparatively high currents used for GMAW, the melting rates of the electrodes are very high. The rates range from approximately 40 to 340 mm/s (100 to 800 in./min) for all metals except magnesium, which reaches speeds up to 590 mm/s (1400 in./min). Because of this rapid melting, the electrodes always must be provided as long, continuous strands of suitably tempered wire that can be fed smoothly and continuously through the welding equipment. The wires are normally provided on conveniently sized spools or in coils. Spools are more common because of the tendency for small size wires to snarl unless supported by a suitable frame.

Because of its relatively small size, the electrode has a high surface-to-volume ratio. Therefore, any drawing compounds, oil, or other foreign matter on or worked into the surface of the electrode tend to be in high proportion relative to the amount of metal present. These foreign materials may cause weld metal defects, such as porosity and cracking. Consequently, the electrode should be clean and free from contaminants.

In addition to welding, the GMAW process is widely used for surfacing where the overlayed weld deposit provides desirable corrosion resistance or other properties. These overlays are normally applied to carbon steel. Such overlays must be carefully engineered and evaluated for satisfactory results. Weld metal dilution with the base metal is a function of arc characteristics and technique. Multiple layers are usually required to obtain suitable deposit chemistry at the surface. Considerable weld overlay is executed automatically where dilution, bead width, thickness, and overlaps are controlled. Additional information on surfacing is contained in Chapter 14, "Surfacing."

The GMAW process is used for spot welding, and electrode selection is based on metallurgical and mechanical considerations. Dilution between the filler and base metals can be complex. Therefore the spot welds should be evaluated carefully by suitable tests. Additional information on filler metals is contained in Chapter 94, "Filler Metals," of the *Welding Handbook,* Section 5, 6th edition.[1] The types of electrodes generally recommended and used for the welding of various metals and alloys are discussed in Section 4, *Metals and Their Weldability,* of the *Welding Handbook,* 6th edition.[2]

SHIELDING GASES

When molten, most metals combine with the basic elements in air, oxygen and nitrogen, to form metal oxides and nitrides. Contamination of the weld metal can result in low strength, low ductility, and excessive weld defects such as porosity and lack of fusion.

The primary purpose of the shielding gas in GMAW is to protect the molten weld metal from contamination and damage by the surrounding atmosphere. However, several other factors

1. Chapter 94 is scheduled for revision in Vol. 4 of the 7th ed.
2. Section 4 is scheduled for revision in the 7th ed.

affect the choice of a shielding gas. Some of these factors are

(1) Arc and metal-transfer characteristics during welding

(2) Penetration, width of fusion, and shape of reinforcement

(3) Speed of welding

(4) Undercutting tendency

All the above factors influence the finished weld and the overall result. Cost also must be considered.

Table 4.4 lists the principal gases used in gas metal arc welding. It is readily noted that most of these are mixtures of gases. Properties of pure shielding gases, used primarily for gas tungsten arc welding, are described in Chapter 2, Volume 1, of the *Welding Handbook,* 7th edition.

Argon and helium, used most frequently for GMAW of nonferrous metals, are completely inert. The selection of one or the other, or mixtures of the two in various combinations, can be made so that the desirable metal transfer, penetration, bead geometry, and other weld characteristics can be obtained.

Helium and CO_2 have higher thermal conductivities than argon. For a given welding current and arc length, the arc voltage and, therefore, heat input are higher with a more thermally conductive gas. Arc energy (plasma) is more consistently dispersed. As shown in Fig. 4.14, bead contour and penetration patterns will be different.

Although the pure inert gases protect the weld metal from reaction with air, they are not suitable for all welding applications. By mixing controlled quantities of reactive gases with them, a stable arc and substantially spatter free metal transfer are obtained simultaneously. Reactive gases and mixtures of such gases provide other types of arcs and metal transfer. Only a few reactive gases have been successfully used either alone in combination with inert gases for welding. These reactive gases include oxygen, nitrogen, and carbon dioxide. Although hydrogen and nitrogen have been considered as additives to control the amount of the joint penetration, they are recommended only for a limited number of highly specialized applications where their presence will not cause porosity or embrittlement of the weld metal. As a rule, it is not practical to use the reactive gases alone for arc shielding. Carbon dioxide is the outstanding

Table 4.4—Shielding gases and gas mixtures for gas metal arc welding

Shielding gas	Chemical behavior	Remarks
Argon	Inert	Virtually all metals except steels
Helium	Inert	Aluminum, magnesium and copper alloys for greater heat input and to minimize porosity
Ar+He (20-80% to 50-50%)	Inert	Aluminum, magnesium, and copper alloys for greater heat input to minimize porosity (better arc action than 100% helium)
Nitrogen		Greater heat input on copper (Europe)
Ar+25-30% N_2		Greater heat input on copper (Europe); better arc action than 100% nitrogen
Ar+1-2% O_2	Slightly oxidizing	Stainless and alloy steels; some deoxidized copper alloys
Ar+3-5% O_2	Oxidizing	Carbon and some low alloy steels
CO_2	Oxidizing	Carbon and some low alloy steels
Ar+20-50% CO_2	Oxidizing	Various steels, chiefly short circuiting arc
Ar+10% CO_2+ 5% O_2	Oxidizing	Various steels (Europe)
CO_2+20% O_2	Oxidizing	Various steels (Japan)
90% He+7.5% Ar+2.5% CO_2	Slightly oxidizing	Stainless steels for good corrosion resistance, short circuiting arc
60 to 70% He +25 to 35% Ar+4 to 5% CO_2	Oxidizing	Low alloy steels for toughness, short circuiting arc

exception. It is suitable alone, mixed with inert gas, or mixed with oxygen for welding a variety of carbon and low alloy steels. Carbon dioxide shielding is inexpensive. All the other gases except nitrogen are used chiefly as small additions to one of the inert gases (usually argon).

Nitrogen has been used alone, or mixed with argon, for welding copper. The most extensive use of nitrogen, however, is in Europe, where little or no helium is available.

Argon and Helium

Argon and helium, used most frequently for gas metal arc welding of nonferrous metals, are completely inert. Although the two gases are equally inert, they differ in other properties. These differences are reflected in their effects on metal transfer through the arc, joint penetration, weld bead shape, undercut, and other weld variables. The selection of argon, helium, or a mixture of the two to shield a particular metal is made so that the desired effects will be obtained.

Helium has a higher thermal conductivity than argon. For any given arc length and current, the arc voltage is higher with helium than with argon shielding. The difference is shown in Fig. 4.15. Consequently, more heat is produced at any given current with a shield of helium than with one of argon. This tends to make helium preferable for use in the welding of thick metals, especially those of high heat conductivity such as aluminum and copper alloys. Conversely, argon is preferable for use in the welding of the light gages of metal and metals of low heat conductivity because it produces lower arc energy. This is especially true for welding in other than the flat position.

Helium is a very light gas with an atomic weight just over 4. Argon is approximately 10 times as heavy as helium and approximately 1-1/3 times as heavy as air. Its atomic weight is approximately 40. The heavier a gas is, the more effective it is at any given flow rate for arc shielding. Largely because of its weight, argon tends to form a blanket over the weld area after leaving the welding nozzle. Helium tends to rise in turbulent fashion and disperse from the weld region because it is lighter than air. Therefore, helium shielding generally requires higher flow rates than when shielding with argon.

Weld reinforcement and penetration patterns differ with argon and helium shielding, or mixtures of the two. Welds made with helium usually exhibit wider reinforcement than welds made with argon. Welds made with argon generally are more deeply penetrated at the center than at the edges, as shown in Fig. 4.14. Helium is added to argon to increase the penetration while retaining the desirable metal transfer characteristic of argon.

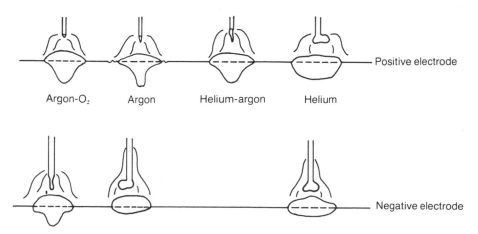

Fig. 4.14—Bead contour and penetration patterns for several shielding gases

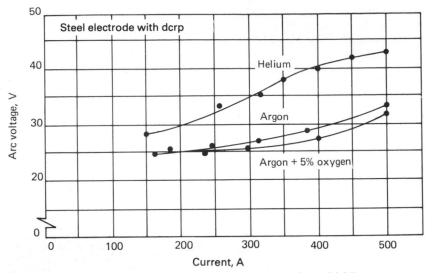

Fig. 4.15—Change in arc voltage with current using three shielding gases

Bead shape and penetration are greatly influenced by metal transfer characteristic. Spray transfer tends to produce relatively deep penetration along the centerline of the bead and relatively shallow penetration at the edges because of the plasma jet effect. Globular and short circuiting transfer tend to produce broader, shallower penetration. As a rule, spray transfer is obtained more readily in argon than in helium. In either gas, the shape of the bead cross section varies with the polarity of the electrode. Welds made with reverse polarity (electrode positive) exhibit more acceptable shapes than those made with straight polarity (electrode negative). See Fig. 4.14.

Oxygen and Carbon Dioxide Additions to Argon and Helium

Spray Transfer. Although the pure inert gases are often essential or preferable for use in the welding of nonferrous metals, they do not always provide the most satisfactory operational characteristics for welding ferrous metals. There is a tendency with pure argon shielding for the filler metal to draw away from, or not flow out to the fusion line or toe of the weld on carbon steels and most low alloy steels. The metal transfer tends to be somewhat erratic and spat-

tery when all iron base alloys are welded, especially with helium. The use of helium or argon-helium mixtures usually fails to improve the situation. For steels, the addition of oxygen or carbon dioxide to argon stabilizes the arc, promotes favorable metal transfer, and minimizes spatter. At the same time, these additions change the cross section of the weld bead and promote the wetting and flow of the weld metal along the edges of the weld. This change is shown in Fig. 4.16. Undercut is reduced or eliminated with the change in bead cross section; bead penetration is decreased; and porosity is often reduced.

Additions of oxygen or carbon dioxide to either argon or helium change the operating characteristics with dc straight polarity (electrode negative). The arc is stabilized; metal transfer is improved; and joint penetration is increased. The change for additions of less than one percent is generally not enough for satisfactory operation with a bare electrode. However, when welding carbon steel with an argon-oxygen mixture that contains five percent oxygen, the electrode tip tapers and the metal transfers in a stream of fast moving drops. The weld reinforcement becomes less convex, and the penetration shape of the bead cross section

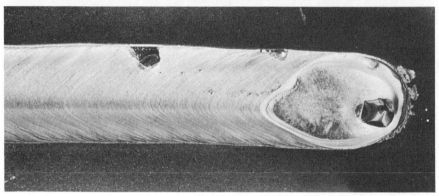

Fig. 4.16—Weld beads of steel deposited with pure argon (top) and 98% argon-2% oxygen shielding (bottom)

approaches that obtained with reverse polarity (electrode positive). At the same time, there is a marked reduction in the melting rate of the electrode.

The amount of oxygen or carbon dioxide required to produce a noticeable change in the arc stability or metal transfer in gas metal arc welding is quite small; even 0.5 percent oxygen is sufficient. However, additions of from 1 to 5 percent oxygen and 3 to 10 percent carbon dioxide are more common. The amount added depends on joint geometry, welding position, base metal composition, and welding technique. Oxygen or carbon dioxide additions to helium or argon-helium mixtures are used on occasion for short circuiting and pulsed dc operation. Normally, such additions are made only to argon

to achieve good axially directed spray transfer and improved wetting.

The addition of oxygen or carbon dioxide to an inert gas, which causes the shielding gas to become oxidizing, may cause porosity in some ferrous metal unless it is counteracted by the addition of suitable deoxidizers in the electrode. It may also cause some loss of certain alloying elements, such as chromium, vanadium, aluminum, titanium, manganese, and silicon. Consequently, electrodes that are to be used with an oxidizing shielding gas must contain deoxidizers to offset the effects of the oxygen. Some wires are designed for fast, inexpensive, and acceptable quality welding; others are designed for porosity free welding that meets the most rigid code requirements.

Short Circuiting Transfer. Gases used in welding with short circuiting metal transfer often differ from those used in welding with spray transfer. For example, argon-carbon dioxide mixtures that contain 20 to 50 percent carbon dioxide are frequently used to weld steel with short circuiting transfer, but they are seldom used with spray transfer. Argon or argon-helium mixtures are employed for welding most non-ferrous metals. Reactive gases or mixtures of inert and reactive gases are used for the welding of steels.

The polyatomic or "high voltage" gases are used more frequently in gas mixtures for short circuiting welding than for spray transfer welding to increase heat input and improve wetting between the filler metal and base metal. Sometimes the percentage of reactive gas used must be restricted to control gas-metal reactions that may be harmful metallurgically. Argon-carbon dioxide mixtures perform satisfactorily for the welding of stainless steels, but they may increase the carbon content of the weld metal and reduce its corrosion resistance, especially in multiple pass welds. Consequently, a less reactive mixture of 90 percent helium-7.5 percent argon-2.5 percent carbon dioxide is generally used to achieve adequate corrosion resistance and minimize oxidation of the weld. With this mixture, both the helium and the carbon dioxide increase the arc energy for a given current; the carbon dioxide also improves the arc stability. As a result, better wetting and better weld shape are achieved.

Similarly, mixtures of 60 to 70 percent helium-25 to 35 percent argon-4 to 5 percent carbon dioxide are used for the welding of low alloy steels when notch toughness is important. Percentage reduction of carbon dioxide to minimal levels increases the Charpy V-notch energy absorption capacity of the weld metal to one that is comparable to that achieved with axially directed spray transfer welding with a 98 percent argon-2 percent oxygen mixture. The flow rates required for short circuiting welding are frequently about one-half of those for spray transfer welding. The lower flow rates can be used because of the lower welding current and smaller weld puddle.

Carbon Dioxide

Although argon and helium are used for gas metal arc welding of most metals, carbon dioxide has become widely used (along with argon-oxygen mixtures) for the welding of mild steels. Higher welding speed, better joint penetration, sound deposits with good mechanical properties, and lower costs have led to the extensive use of carbon dioxide. With a bare electrode, axially directed spray transfer cannot be achieved with pure carbon dioxide. The transfer is either short circuiting, globular, or nonaxially directed spray.

The chief drawback of carbon dioxide shielding is the rather harsh arc and excessive spatter. Spatter is reduced by the use of a very short, uniform arc length with the tip of the electrode below the adjacent work surface. Proper adjustment of the power supply inductance also minimizes spatter. If good control of the arc is maintained, the amount of spatter is kept within tolerable bounds, and sound weld deposits are achieved with proper procedures.

· Sound welds can be made consistently if the electrode is designed for use with carbon dioxide and contains the appropriate balance of deoxidizers. Carbon dioxide oxidizes at welding temperatures. However, when the carbon contents of a steel electrode and the base metal are less than 0.07 percent, the weld metal carburizes. On the basis of manganese and silicon losses, a shield of carbon dioxide may be regarded as equivalent to a shield of inert gas containing about 10 percent oxygen. As a result, the surface of carbon dioxide shielded welds is usually heavily oxidized. Despite this oxidizing condition, porosity is not a problem when a suitable deoxidized electrode and a reasonably short arc length are used. Sound welds can be readily reproduced in carbon and some low alloy steels with a shield of carbon dioxide gas. When good impact properties are essential, argon-CO_2 mixtures are employed.

Shielding Gas Selection

It has been demonstrated that the choice of a shielding gas depends on the metal to be welded, section thickness, process variation, quality requirements, metallurgical factors, and cost. Argon, helium, and argon-helium mixtures are generally used with nonferrous metals. Argon-oxygen, argon-carbon dioxide, or argon-helium mixtures, and also pure carbon dioxide are employed for ferrous metals. The application needs, therefore, determine shielding gas selection. General applications are given in Tables 4.5 and 4.6. For detailed information on gas selection for particular metals, the appropriate chapter in Volume (Section) 4 of the *Handbook* should be consulted.

PROCESS VARIABLES

A list of some of the variables that affect weld penetration and bead geometry is given below.

(1) Welding amperage
(2) Arc voltage
(3) Travel speed
(4) Electrode extension
(5) Electrode inclination
(6) Electrode size
(7) Weld joint position

Knowledge and control of these variables are essential if welds of good quality are to be consistently obtained. Also, shielding gas selection influences weld penetration and bead geometry, as discussed previously.

The importance of one variable adjustment over another differs. However, regardless of the particular application, the adjustment of variables is strongly influenced by (a) composition, metallurgical structure, and thickness of the base metal; (b) electrode composition; (c) welding position; (d) quality requirements; and (e) the quantity of completed weldments required.

WELDING AMPERAGE

When all other variables are held constant, the welding amperage varies with the electrode speed or melting rate in a nonlinear relation. As the electrode feed speed is varied, the welding amperage will vary in a like manner with a constant voltage power source. Figure 4.17 shows the typical melting rates for four sizes of carbon steel electrodes. At the lower amperage range for each electrode size, the curve is nearly linear. However, at higher welding amperages, particularly with small diameters, the melting rate curve becomes nonlinear, progressively increasing at a higher rate as welding amperage increases. This change can be attributed to resistance heating of the electrode extension beyond the contact tube. Figure 4.17 also shows that when the diameter of the electrode is increased at any electrode feed speed, a higher welding amperage is required. Each type of wire (e.g., steel, aluminum, etc.) has different melting rate characteristics.

With all other variables held constant, an increase in welding amperage (electrode feed speed) will result in

(1) Increasing the depth and width of the weld penetration
(2) Increasing the deposition rate
(3) Increasing the size of the weld bead

ARC VOLTAGE

No specific values of arc voltage are consistently appropriate for normal production welding, although typical values are shown in Table 4.3. Trial runs are necessary to adjust the arc voltage if it is to produce the most favorable filler metal transfer and weld bead appearance. These trial runs are essential because arc voltage is dependent upon a variety of factors, including metal thickness, the type of joint, the position of welding, electrode size, shielding gas composition, and the type of weld. From any specific value of arc voltage, a voltage increase tends to flatten the weld bead and increase the fusion zone width. Reduction in voltage results in a narrower weld bead with a high crown and deeper penetration. Excessively high voltage may cause porosity, spatter, and undercutting;

Table 4.5—Selection of gases for gas metal arc welding with spray transfer

Metal	Shielding gas	Advantages
Aluminum	Argon	0-25 mm (0-1 in.) thick: Best metal transfer and arc stability; least spatter.
	75% Helium-25% argon	25-76 mm (1-3 in.) thick: Higher heat input than argon.
	90% Helium-10% argon	Over 76 mm (3 in.) thick: Highest heat input; minimizes porosity.
Magnesium	Argon	Excellent cleaning action.
Carbon steel	Argon-3-5% oxygen	Good arc stability; produces a more fluid and controllable weld pool; good coalescence and bead contour; minimizes undercutting; permits higher speeds, compared with argon.
	Carbon dioxide	High-speed mechanized welding; low-cost manual welding.
Low alloy steel	Argon-2% oxygen	Minimizes undercutting; provides good toughness.
Stainless steel	Argon-1% oxygen	Good arc stability; produces a more fluid and controllable weld pool, good coalescence and bead contour; minimizes undercutting on heavier stainless steels.
	Argon-2% oxygen	Provides better arc stability, coalescence, and welding speed than 1% oxygen mixture for thinner stainless steel materials.
Copper, nickel, and their alloys	Argon	Provides good wetting; good control of weld pool for thickness up to 3.2 mm (1/8 in.).
	Helium-argon	Higher heat inputs of 50 and 75% helium mixtures offset high heat conductivity of heavier gages.
Reactive metals (Ti, Zr, Ta)	Argon	Good arc stability; minimum weld contamination. Inert gas backing is required to prevent air contamination on back of weld area.

excessively low voltage may cause porosity and overlap at the weld edges.

TRAVEL SPEED

Travel speed is the linear rate at which the arc is moved along the weld joint. With all other conditions held constant, weld penetration is a maximum at some travel speed. The penetration will decrease when the travel speed is changed, and the weld bead will either become narrower or wider.

When the travel speed is decreased, the filler metal deposition per unit length increases; and a large, shallow weld pool is produced. The welding arc impinges on this pool rather than the base metal as the arc advances. This limits penetration but produces a wide weld bead.

As the travel speed is increased, the thermal energy transmitted to the base metal

Table 4.6—Selection of gases for gas metal arc welding with short circuiting transfer

Metal	Shielding gas	Advantages
Carbon steel	Argon- 20-25% CO_2	Less than 3.2 mm (1/8 in.) thick: high welding speeds without melt-through; minimum distortion and spatter: good penetration.
	Argon- 50% CO_2	Greater than 3.2 mm (1/8 in.) thick: minimum spatter; clean weld appearances; good weld pool control in vertical and overhead positions.
	CO_2	Deeper penetration; faster welding speeds; minimum cost.
Stainless steel	90% Helium- 7.5% argon- 2.5% CO_2	No effect on corrosion resistance; small heat-affected zone; no undercutting; minimum distortion; good arc stability.
Low alloy steel	60-70% Helium- 25-35% argon- 4-5% CO_2	Minimum reactivity; good toughness; excellent arc stability, wetting characteristics, and bead contour; little spatter.
	Argon- 20-25% CO_2	Fair toughness; excellent arc stability, wetting characteristics, and bead contour; little spatter.
Aluminum, copper, magnesium, nickel, and their alloys	Argon and argon- helium	Argon satisfactory on sheet metal; argon-helium preferred on thicker sheet metal.

from the arc is decreased. Therefore, melting of the base metal is slowed and occurs nearer to the surface of the base metal. Thus, penetration and bead width are decreased. As travel speed is increased further, there is a tendency toward undercutting along the edges of the weld bead because there is insufficient deposition of filler metal to fill the path melted by the arc.

ELECTRODE EXTENSION

The electrode extension is the distance between the last point of electrical contact, usually the end of the contact tube, and the end of the electrode, as shown in Fig. 4.18. As this distance increases, so does the electrical resistance of the electrode. Resistance heating causes the electrode temperature to rise. Then, less welding current is required to melt the electrode at a given feed rate.

There is a need to control electrode extension, because too long an extension results in excess weld metal being deposited with low arc heat. This will cause poor weld bead shape and shallow penetration. Also, as the contact tube-to-work distance increases, the arc becomes less stable. Good electrode extension is from 6 to 13 mm (1/4 to 1/2 in.) for short circuiting transfer and from 13 to 25 mm (1/2 to 1 in.) for other types of metal transfer.

ELECTRODE POSITION

As with all arc welding processes, the position of the welding electrode with respect to the weld joint affects the weld bead shape and penetration. The effects are greater than those of arc voltage or travel speed. Electrode position is described by the relationships of the electrode axis with respect to (1) the direction of travel, (2) the travel angle, and (3) the angle

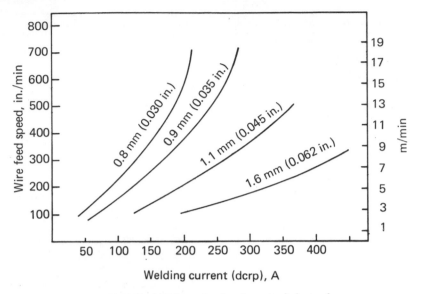

Fig. 4.17—Melting rate of carbon steel electrodes

between the axis and the adjacent work surface (work angle). When the electrode points opposite from the direction of travel, it is called the backhand welding technique. When the electrode points in the direction of travel, it is called the forehand welding technique.

The effect of the electrode position with respect to the direction of travel is shown in Fig. 4.19. When the electrode is changed from the perpendicular to the forehand technique with all other conditions unchanged, the penetration decreases and the weld bead becomes wider and flatter. Maximum penetration is obtained in the flat position with the backhand technique at

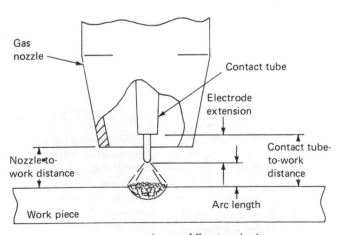

Fig. 4.18—Gas metal arc welding terminology

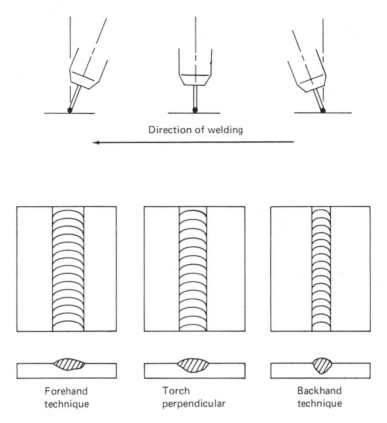

Fig. 4.19—Effect of electrode position and welding technique

a travel angle of about 25 degrees from the perpendicular. Backhand technique also produces a more convex, narrower bead; a more stable arc; and less spatter on the workpiece.

When producing fillet welds in the horizontal position, the electrode should be positioned about 45 degrees to the vertical member (work angle). For all positions, the electrode travel angle normally used is in the range of 5 to 15 degrees for good control of the molten weld pool.

ELECTRODE SIZE

Each electrode diameter of a given composition has a usable amperage range. The welding amperage range is limited by undesirable effects, such as the absence of wetting at very low values, and also spatter, porosity, and poor bead appearance with excessively high values. The electrode melting rate is a function of current density. If two wires of different diameters are operated at the same amperage, the smaller will have the higher melting rate.

Penetration is also a function of current density. For example, a 1.14 mm (0.045 in.) diameter electrode will produce deeper penetration than a 1.59 mm (1/16 in.) diameter electrode when it is used at identical amperage. However, the weld bead profile will be wider with the larger electrode. The reverse is also true when a small weld bead profile is specified. Since smaller diameter wires are more costly on a weight basis, for each application there is a wire size that will give minimum cost welds.

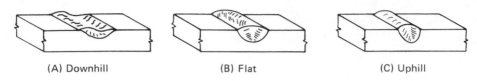

(A) Downhill   (B) Flat   (C) Uphill

Fig. 4.20—Effect of work inclination on weld bead shape

WELDING POSITION

Most spray type GMAW is done in the flat position, while pulsed and short circuiting GMAW can be used in all positions. Fillet welds made in the flat position with spray transfer are usually more uniform, less likely to have unequal legs and convex profiles, and less susceptible to undercutting than similar fillet welds made in the horizontal position.

To overcome the pull of gravity on the weld metal in the vertical and overhead positions of welding, small diameter electrodes are usually used with either short circuiting metal transfer or spray transfer with pulsed direct current. Electrode diameters of 1.14 mm (0.045 in.) and under are best suited for out-of-position welding. The low heat input allows the molten pool to freeze quickly. Vertical-down welding is usually effective on sheet metal in the vertical position.

When welding is done in the flat position, the inclination of the work surface with respect to the horizontal will influence the weld bead shape, penetration, and travel speed. The effect is the same whether the welding gun moves or the work moves. In the case of circumferential welding, the work rotates under the welding gun and inclination is obtained by moving the welding gun either direction from top dead center.

By positioning the work with the weld axis at 15 degrees to the horizontal and welding downhill, weld reinforcement can be decreased under welding conditions that would produce excessive reinforcement when the work is in the flat position. Also when traveling downhill, speeds can usually be increased. At the same time, penetration is lower, which is beneficial for welding sheet metal.

Downhill welding affects the weld, as shown in Fig. 4.20. The weld puddle tends to flow toward the electrode and preheats the base metal, particularly at the surface. This produces an irregularly shaped fusion zone, called a "secondary wash." As the angle of declination increases, the middle surface of the weld is depressed, penetration decreases, and the width of the weld increases.

Uphill welding affects the fusion zone contour and the weld surface, as illustrated in Fig. 4.20. The force of gravity causes the weld puddle to flow back and lag behind the electrode. The edges of the weld lose metal, which flows to the center. As the angle of inclination increases, reinforcement and penetration increase, and the width of the weld decreases. The effects are exactly the opposite of those produced by downhill welding. When higher welding currents are used, the maximum angle decreases.

WELDING PROCEDURES

The welding procedures for GMAW are similar to those for other arc welding processes. Adequate fixturing and clamping of the work are required with adequate accessibility for the welding gun. Fixturing must hold the work rigid to minimize distortion from welding. It should be designed for easy loading and unloading. Good connection of the work lead (ground) to the workpiece or fixturing is required. Location of the connection is important, particularly when welding ferromagnetic materials such as steel. Generally the best direction of welding is away from the work lead connection. The position of the electrode with respect to the weld joint is important in order to obtain the desired joint penetration, fusion, and weld bead geometry. Electrode positions for automatic GMAW are similar to those used with submerged arc welding.

When complete joint penetration is required, some method of weld backing will help to control it. A backing strip, backing weld, or copper backing bar can be used. Backing strips and backing welds usually are left in place. Copper backing bars are removable. For additional information on fixturing, grounding, electrode position, weld backing, and other factors in preparation for welding, refer to Chapters 2 and 6.

The assembly of the welding equipment should be done according to the manufacturer's directions. All gas and water connections should be tight; there should be no leaks. Aspiration of water or air into the shielding gas will result in erractic arc operation and contamination of the weld. Porosity may also appear.

The gun nozzle size and the shielding gas flow rate should be set according to the recommended welding procedure for the material and joint design to be welded. Joint designs that require long nozzle-to-work distances will need higher gas flow rates than those used with normal nozzle-to-work distances. The gas nozzle should be of adequate size to provide good gas coverage of the weld area. When welding is done in confined areas or in the root of thick weld joints, small size nozzles are used.

The gun contact tube and electrode feed drive rolls are selected for the particular electrode composition and diameter, as specified by the equipment manufacturer. The contact tube will wear with usage, and must be replaced periodically if good electrical contact with the electrode is to be maintained and heating of the gun is to be minimized.

Electrode extension is set by the distance between the tip of the contact tube and the gas nozzle opening. The extension used is related to the type of GMAW, short circuiting or spray type transfer. It is important to keep the electrode extension (nozzle-to-work distance) as uniform as possible during welding. Therefore, depending on the application, the contact tube may be inside, flush with, or extending beyond the gas nozzle.

The electrode feed rate and welding voltage are set to the recommended values for the electrode size and material. With a constant voltage power source, the welding current will be established by the electrode feed rate. Trial bead welds should be made to established proper voltage (arc length) and feed rate values. Other variables, such as slope control or inductance, or both, should be adjusted to give good arc starting and smooth arc operation with minimum spatter. The optimum settings will depend on the equipment design and controls, electrode material and size, shielding gas, weld joint design, base metal composition and thickness, welding position, and welding speed.

Volume (Section) 4, *Metals and Their Weldability,* contains information on the procedures for gas metal arc welding of various metals and alloys for which the process is suitable.

WELD QUALITY

High quality welds are produced by this welding process when proper welding procedures are used. The absence of flux or electrode covering eliminates slag inclusions in the weld. Some dross formation may occur when highly deoxidized steel electrodes are used, and it should be removed before the next weld bead or pass is made.

Inert gas shielding provides excellent protection of the weld area from oxygen and nitrogen contamination. Hydrogen is virtually eliminated as a cause of cracking in the weld and heat-affected zones of low alloy steels. On the other hand, the process permits low cost welding of carbon steels with the use of inexpensive CO_2 gas shielding.

Weld defects may occur with GMAW welding when the process variables, materials, or welding procedures are improper. Some of the defects specifically related to this process, their probable causes, and recommended corrective actions are given in Table 4.7.

PROCESS SELECTION

There are many factors that influence the selection of a welding process, with each factor assuming different degrees of importance depending upon the industry and application involved. In most cases, welding cost is of primary concern. Other factors, such as production rate, operator skill, available equipment, quality requirements, and material composition and thickness, influence the decision to select the GMAW process. The following discussion on process selection compares this process to several other welding processes.

WELD COSTS

Welding costs are usually derived by considering three cost categories: materials, labor, and overhead.

Materials

These items are considered the consumables of the process; they include the wire electrode and shielding gas. Wire costs per unit weight are dependent on the electrode diameter, type of alloy, size of package, and quantity purchased. Small diameter wire electrodes, such as those used for gas metal arc and gas tungsten arc welding, are more costly than the larger diameter wires used for submerged arc welding. Flux cored electrodes are usually more costly than solid wire electrodes because they are fabricated wire electrodes containing flux.[3]

Shielding gas is another consumable expense associated with GMAW. Inert gas or gas mixtures are used for both ferrous and nonferrous materials, but they are higher priced than carbon dioxide (CO_2), which has found wide use for carbon and low alloy steel fabrication. Liquefied gas distribution systems within a plant offer considerable savings over individual high-pressure gas cylinders when they can be justified for volume production. Submerged arc welding, which does not require a gas shield, uses a granular flux distributed over the weld zone immediately prior to welding. On the average, every pound of wire deposited consumes a pound of flux.

Material costs are also influenced by the inherent efficiency of each process in depositing metal. Estimates have been made of the deposition efficiency for each of the major consumable electrode processes, as shown in Table 4.8. These values reflect the electrode losses in such forms

---

3. Flux cored electrodes are described in Chapter 5.

**Table 4.7—Weld defects, their possible causes,
and corrective actions**

Possible causes	Corrective actions
Weld metal cracks	
A. Too high a weld depth-to-width ratio	Increase the arc voltage or decrease the welding current to widen the weld bead and decrease the penetration.
B. Too small a weld bead (particularly fillet and root beads)	Decrease the travel speed to increase the cross section of the bead.
C. Rapid cooling of the crater at the end of a weld	Use a taper power control to reduce the cooling rate.
	Fill craters adequately.
	Use a back step welding technique to end the weld on top of a finished bead.
Inclusions	
A. Use of multiple pass, short circuiting arc welding (slag type inclusions)	Remove any glossy dross islands from the weld bead before making subsequent passes.
B. High travel speeds (film type inclusions)	Reduce the travel speed.
	Use a more highly deoxidized electrode.
	Increase the arc voltage.

as metal vapor, slag, spatter, and unusable electrode (stub) losses.

Labor

Generally, direct labor is the dominant cost factor in any welding operation. The labor costs of welders or welding operators are a direct function of the travel speed of the process and the operating factor or duty cycle involved. The operating factor is the ratio of the actual arc time to the total time. There is considerable difference between welding processes with respect to travel speeds. For instance, shielded metal arc welding can possibly be done at a maximum of approximately 20 mm/s (50 in./min), while gas metal arc welding can be done at welding speeds of 65 to 85 mm/s (150 to 200 in./min), or more.

The operating factor is reduced by any operation that decreases arc time, such as frequent electrode replacement and the removal of fused slag or spatter. All three operations are associated with shielded metal arc welding, while submerged arc welding requires only fused slag removal.

Depending on the application and shielding gas used, only the removal of spatter may be required with GMAW. In general, the use of this process results in one of the highest operating factors.

Overhead

Items such as ancillary facilities, services, and utilities necessary for production welding are included in the overhead. The amortization of capital equipment used for welding is also considered an overhead expense. The power source and associated equipment used for shielded metal arc welding are the least complicated, and therefore the least expensive of all the arc welding processes. Both semiautomatic gas metal arc and flux cored arc welding require an electrode wire feeder and a more elaborate delivery system (gun with shielding gas flow). Consequently, the amortization expenses are greater. Submerged arc welding and the automatic versions of gas metal arc and flux cored arc welding equipment, with their associated controls and guidance fixtures, increase the overhead even more.

Table 4.7 (continued)—Weld defects, their possible causes, and corrective actions

Possible causes	Corrective actions
	Porosity
A. Inadequate shielding gas coverage	Increase the shielding gas flow to displace all air from the weld area.
	Decrease the shielding gas flow to avoid turbulence and entrapment of air in the gas.
	Remove spatter from the interior of the gas nozzle.
	Eliminate drafts (from fans, open doors, etc.) blowing into the welding arc.
	Use a slower travel speed.
	Reduce the nozzle-to-work distance.
	Hold the gun at the end of the weld until the molten crater solidifies.
B. Electrode contamination	Use only clean and dry electrode wire.
	Eliminate pickup of lubricant on electrode in the wire feeder or conduit.
C. Workpiece contamination	Remove all grease, oil, rust, paint, and dirt from work surfaces before welding.
	Use a more highly deoxidizing electrode.
D. Arc voltage too high	Reduce operating voltage.
E. Excess nozzle-to-work distance	Reduce electrode extension.

ADAPTABILITY

A process is sometimes selected on the basis of not only how well it performs in a given application but also on how well it will adapt to a variety of conditions. GMAW may be used with all of the major commercial metals including carbon steels, alloy steels, stainless steels, aluminum, magnesium, copper, titanium, zirconium, and nickel alloys. A wide range of joint thicknesses can be welded with the process. For instance, when operated in the short circuiting mode (low voltage), 1 to 3 mm ($1//32$ to $1/8$ in.) thick material can readily be joined in any welding position. When spray transfer mode or drop transfer mode with CO_2 shielding is used, thicknesses of 2 mm ($1/16$ in.) and greater are routinely welded. A 6 to 8 mm ($1/4$ to $5/16$ in.) single pass fillet weld is the maximum practical size in the horizontal position for most applications. When steel joints thicker than 13 mm ($1/2$ in.) are to be welded in production, submerged arc welding or flux cored arc welding is generally a more desirable process.

The versatility of GMAW is well illustrated by its ability to be used either semiautomatically, where the electrode is fed automatically through a manually guided and positioned welding gun, or automatically, where the entire process is fully controlled. Gas metal arc welding can also be adapted to surfacing applications whereby a particular metal is welded to the surface of another to enhance its corrosion, wear, or hardness properties.

Table 4.7 (continued)—Weld defects, their possible causes, and corrective actions

Possible causes	Corrective actions
Incomplete fusion	
A. Weld zone surfaces not free of film or scale	Clean all groove faces and weld zone surfaces of any mill scale or impurities prior to welding.
B. Insufficient heat input	Increase the electrode feed speed and the arc voltage.
	Reduce the travel speed.
C. Too large a weld puddle	Reduce arc weaving to produce a more controllable weld pool.
D. Improper welding technique	When using a weaving technique, dwell momentarily on the groove faces.
	Keep the electrode directed at the leading edge of the weld pool.
E. Improper joint design	Maintain the included angle of the groove joint large enough to allow access to the bottom of the groove using proper electrode extension and arc characteristics.
	Change the groove design to a "J" or "U" type.

OPERATOR SKILL

One of the chief advantages of GMAW is that, in general, it does not require the degree of operator skill that is essential to gas tungsten arc or shielded metal arc welding. Skill and adequate training are prerequisites to quality GMAW, but the welder need not be concerned with controlling the important variable of arc length. The semiautomatic nature of GMAW (automatic electrode feed), combined with a constant potential power supply, creates the desirable condition of almost constant arc length with nominal gun-to-work distance variations. This feature is not present with shielded metal arc or gas tungsten arc welding where the operator's positioning of the electrode is directly related to arc length and voltage.

ADVANTAGES AND LIMITATIONS

The decision as to whether GMAW should be used in preference to another arc welding process can best be made by reviewing the advantages and limitations of the process.

Advantages

Some of the advantages of gas metal arc welding compared to other processes are as follows:

(1) It overcomes the restriction of limited electrode length with shielded metal arc welding.

(2) Welding can be done in all positions, which is a limitation on the use of submerged arc welding.

(3) Welding speeds are higher than those with shielded metal arc welding because of the continuous electrode feed, absence of slag, and higher filler metal deposition rates.

(4) Deeper penetration is possible than with shielded metal arc welding, which permits the use of smaller size fillet welds for equivalent strengths.

**Table 4.7 (continued)—Weld defects, their possible causes,
and corrective actions**

Possible causes	Corrective actions
Lack of penetration	
A. Improper joint preparation	Joint design must be adequate to provide access to the bottom of the groove while maintaining proper nozzle-to-work distance and arc characteristics.
	Reduce root face height.
	Provide or increase the root opening in butt joints.
B. Improper welding technique	Position the electrode at the proper travel angle to achieve maximum penetration.
	Keep the arc on the leading edge of the weld pool.
C. Inadequate heat input	Increase electrode feed to obtain higher welding current.
	Maintain proper nozzle-to-work distance.
Excessive melt-through	
A. Excessive heat input	Reduce the electrode feed rate and arc voltage.
	Increase the travel speed.
B. Improper joint preparation.	Reduce excessive root opening.
	Increase root face height.

Limitations

This process has some limitations on its applications compared to shielded metal arc welding.

These are

(1) The welding equipment is more complex, more costly, and less portable.

(2) It is more difficult to weld in hard-to-reach places because the welding gun must be close to the joint.

(3) The welding arc must be protected against strong air drafts that will disperse the shielding gas. This limits outdoor applications.

(4) Weld metal cooling rates are higher than

Table 4.8—Electrode deposition efficiency for several welding processes

Process	Efficiency, %
Shielded metal arc	65
Flux cored arc	82
Gas metal arc	95
Submerged arc	100

with the welding processes that deposit a slag blanket over the weld metal. High cooling rate will alter the metallurgical and mechanical properties of the weld joint in most metals.

GAS METAL ARC SPOT WELDING

INTRODUCTION

The essential differences between conventional GMAW and gas metal arc spot welding are that there is no movement of the gun during spot welding, and the weld time is a relatively short period of no more than a few seconds.

The process is used to spot weld two overlapped sheets together by penetrating through one sheet and into the other, and to tack weld other types of joints together. The process can be used in the same manner as resistance spot welding. Mild steel, stainless steel, and aluminum are commonly spot welded with the process.

The weld is made by placing the welding gun on the joint so the spot weld will be made at the proper location. The gun is held motionless. When the trigger is depressed, shielding gas flow is initiated. After a preflow interval, the electrode feed is energized and the arc starts. At the completion of a preset weld time, the welding current and electrode feed are stopped. Finally, gas flow stops automatically. Because of the inherent differences in the weld shape of conventional welds and spot welds, the effects of the variables involved should be reviewed so that a better understanding of the process will be obtained.

WELDING VARIABLES

Welding Current

The welding current is a direct function of the electrode feed rate and has a major influence on weld metal penetration. Penetration increases approximately as the square of the current with CO_2 spot welding. Increasing the penetration will generally produce a larger diameter weld at the interface between sheets and, consequently, greater weld shear strength. Virtually all gas metal arc spot welding is carried out with direct current reverse polarity, although straight polarity can be used under special conditions.

Welding Time

The duration of welding current has an important effect on penetration and weld size at the interface. Penetration increases with weld time at a decreasing rate to a limiting value which depends on the current, voltage, and electrode diameter. Also, the weld reinforcement height increases with weld time.

Control of the weld time is best provided with an electronic timer. Even with the most consistent timers, however, the actual weld time can vary considerably because an arc is not always initiated the instant the advancing electrode touches the work surface. Sometimes the electrode "stubs," causing the wire to melt and form a momentary arc, which is often of short duration. Stubbing of the wire electrode can occur several times before an arc is finally established. Unfortunately, during this period little or no heating of the base metal is realized. Inconsistent weld time can be overcome by use of a control that does not time the weld period until the arc voltage reaches a preset value.

Arc Voltage

The arc voltage is indirectly a measure of the arc length and, consequently, has a major bearing on the shape of the arc spot. If the welding amperage is maintained constant, an increase in arc voltage will cause the weld diameter at the interface to increase and the reinforcement height and weld metal penetration to decrease slightly. If the arc voltage is too high, heavy spatter is likely to occur. On the other hand, insufficient arc voltage may result in the formation of a depression in the center of the reinforcement and inadequate fusion at the edge of the weld.

Electrode Size

The electrode size is selected on the basis of the part thickness being welded. As in conventional GMAW, there is an overlap of the appropriate electrode sizes that may be used for a particular joint thickness. The capacity of the electrode wire feeder must also be considered for electrode size selection.

Electrode Extension

The electrode extension (contact tube-to-work distance) is maintained constant in arc spot welding by a spacer nozzle that is positioned on the work. Since an increase in electrode extension reduces the current at the same voltage, the weld metal penetration will also decrease. Normally, a 13 mm (1/2 in.) extension is used for most applications. Small variations from this value have only slight influence on the weld shape and quality.

Shielding Gas

The shielding gases for arc spot welding are selected by using the same criteria as those employed for continuous welding. Arc characteristics, filler metal transfer, weld penetration, weld shape, materials joined, and cost are all important factors to be considered in shielding gas selection. The shielding gas performs the same functions as in continuous GMAW, but a flow rate of 25 percent to 50 percent of that normally used is satisfactory. Preflow and postflow gas timing controls should be provided on production equipment to permit purging the gas nozzle before and after each weld.

JOINT DESIGN

Gas metal arc spot welding is applicable to all the materials welded with the process. Many weld joint types are used, with slight modification of the shielding gas nozzle design, including lap, fillet (inside corner or edge of a lap joint), corner (outside corner), and plug welds. For best results, the top member should be no thicker than the bottom member in lap joints, where a melt-through technique is employed. If the top member must be thicker than the bottom one, plug welding should be used. In plug welding, care must be taken to insure complete penetration into the base plate and also into the side walls of the top member. Lack of fusion is a common problem in this type of joint. A copper backing block may be required on the back side of an arc spot weld to prevent excessive penetration when the bottom sheet is thinner than the top.

Fit-up of mating parts in gas metal arc spot welding is not as critical as with resistance spot welding, where intimate contact and high forces are essential. With lap joints, the sheets are forced together by loading from the welding gun. Interface gaps of up to 1.6 mm (1/16 in.) or one-half the top sheet thickness, whichever is smaller, have little effect on the joint strength. With larger gaps, the weld deposit may sink below the top sheet surface, and cause a reduction in shear strength.

WELDING POSITION

Vertical and overhead arc spot welds can be made successfully on sheets up to 1.3 mm (0.05 in.) thick. To weld out of position, the short circuiting mode of metal transfer must generally be used. Spot welding of heavier gage material is usually restricted to the flat position because of the influence of gravity on the molten weld pool.

WELD QUALITY

All of the above welding variables have an influence to varying degrees on the quality of the weld. They should be controlled as much as possible. But in addition, good practice suggests that the resultant weld quality should be monitored directly. Unfortunately, the shape and size of the weld reinforcement is not a reliable indication of the weld size.

One inspection technique involves examining the back side of the joint for slight melt-through or a bump indicating adequate penetration. However, when the back side is inaccessible or the gage combination is not favorable, periodic destructive testing of welded coupons is the only dependable way to assess weld quality. This may be done with the tension-shear test[4] to determine the spot weld strength, or by metallographic examination to determine weld size and soundness. The consistency of weld strength or size in arc spot welding is generally not as good as with resistance spot welding. This is especially true with aluminum.

---

4. See Chapter 5 of the *Welding Handbook,* Vol. 1, 7th ed., and AWS C1.1, Recommended Practices for Resistance Welding, for information on the tension-shear test method.

NARROW GAP WELDING

Narrow gap welding is a modification of conventional automatic GMA welding where square grooves or V-grooves with extremely small included angles are utilized in thick sections of ferrous metals. Root openings of approximately 6 to 10 mm (1/4 to 3/8 in.) are used, regardless of the material thickness to be joined. This narrow gap requires the use of specially designed welding torch accessories: (a) water-cooled contact tubes and (b) nozzles that introduce shielding gases from the plate surface. All positioned welds are made from one side of the plate.

Welding is usually done with two small diameter wire electrodes in tandem, being fed through one or two contact tubes. Axial spray transfer with dcrp is most commonly used, although pulsed power can be employed. Each electrode is oriented so that a weld bead is directed toward each sidewall. Travel speeds must be high, resulting in low heat input and small weld puddles which are easily controlled in out-of-position welding. Heat-affected zones are narrow.

The necessary use of low heat input in thick materials can cause sidewall or inter-bead lack of fusion, the most often encountered defect when narrow gap welding. Careful attention to every element of the welding schedule, particularly wire electrode placement, is necessary to minimize the recurrence of this defect. Interpass cleaning of all slag islands is also necessary if slag entrapment defects are to be avoided.

Narrow gap welding has certain advantages and disadvantages over other arc welding processes. The advantages are
(1) Lower residual stresses and distortion
(2) Improved as-welded joint properties
(3) Better economy
The disadvantages are
(1) More prone to defects
(2) Difficulty of removing defects when detected
For thicknesses over 50 mm (2 in.), this process is competitive with other automatic arc welding processes used for heavy plate welding.

SUPPLEMENTARY READING LIST

Aldenhoff, B.J., Stearns, J.B., and Ramsey, P.W. Constant potential power sources for multiple operation gas metal arc welding. *Welding Journal* 53 (7), July 1974, pp. 425-429.

Butler, C.A., Meister, R.P., and Randall, M.D. Narrow gap welding—a process for all positions. *Welding Journal* 48 (2), Feb. 1969, pp. 102-108.

Kiyolara, M., *et al.* On the stabilization of GMA welding of aluminum. *Welding Journal* 56 (3), Mar. 1977, pp. 20-28.

Lesnewich, A. MIG Welding with Pulsed Power. Bulletin 170. New York: Welding Research Council, 1972.

Liptak, J.A. Instabilities in gas metal arc welding and their causes. *Welding Journal* 48 (5), May 1969, pp. 396-401.

Manz, A.F. Inductance vs slope control for gas metal arc power. *Welding Journal* 48 (9), Sept. 1969, pp. 707-712.

MIG Welding Handbook. New York: Union Carbide Corporation, Linde Div., 1974.

Morris, R.W. Application of multiple electrode gas metal arc welding to structural steel fabrication. *Welding Journal* 47 (5), May 1968, pp. 379-385.

Recommended Safe Practices for Gas Shielded Arc Welding, AWS A6.1. Miami: American Welding Society, 1966.

Shackleton, D.N., and Lucas, W. Shielding gas mixtures for high quality mechanized GMA welding of Q & T steels. *Welding Journal* 53 (12), Dec. 1974, pp. 537-s—547-s.

The Procedure Handbook of Welding, 12th ed. Cleveland, Ohio: Lincoln Electric Company, 1973.

5

Flux Cored Arc Welding

Prepared by a committee consisting of

L. J. PRIVOZNIK, *Chairman*
   Westinghouse Electric Corp.

J. E. BALL
   Airco Welding Products

J. B. CHRISTOFFERSON
   Chicago Bridge & Iron Co.

P. A. KAMMER*
   Teledyne McKay

L. MOTT
   Hobart Brothers Co.

R. QUINN
   Caterpillar Tractor Co.

R. S. SABO
   Lincoln Electric Co.

C. R. ZIMMERMAN
   Chemetron Corp.

A. J. ZVANUT
   Stoody Co.

*Now with Gorham International, Inc.

5

Flux Cored Arc Welding

FUNDAMENTALS OF THE PROCESS

Currently increasing in use for welding ferrous metals, flux cored arc welding (FCAW) is an arc welding process which produces coalescence of metals by heating them with an arc between a continuous filler metal (consumable) electrode and the work. Shielding is provided by a flux contained within the tubular electrode and may be supplemented by an externally supplied gas.

The feature that distinguishes the FCAW process from other arc welding processes is the enclosure of fluxing ingredients within a continuously fed electrode. Both the particular operating characteristics of the process and the resulting weld properties begin with a consideration of this feature.

Flux cored arc welding offers two major process variations that differ in the method used to shield the arc and weld pool from atmospheric contamination (oxygen and nitrogen). One type, self-shielded FCAW, protects the molten metal to some extent through the decomposition and vaporization of the flux core by the heat of the arc. The other type, gas shielded FCAW, makes use of a protective gas flow in addition to the flux core action to shield the arc and the weld pool. With both methods, the electrode core material provides a relatively thin slag covering to protect the solidifying weld metal.

There is a third variant of the process known as FCAW-electrogas. Because of the similarity of this FCAW process variation to certain other single pass, vertical-up welding process is also used in machine welding where this volume in Chapter 7, "Electroslag and Electrogas Welding."

Flux cored arc welding is usually applied as semiautomatic welding, where the gun is hand held and manipulated by the welder. The process is also used in machine welding, where the operator continuously monitors the operation during mechanized travel. Automatic welding is also applied, as in the automotive industry, where light gage body parts and casings are welded with push-button control.

PRINCIPAL FEATURES

The benefits of flux cored arc welding are achieved by combining two general features: (1) the productivity of continuous welding, and (2) the metallurgical benefits that can be derived from a flux. In this respect FCAW possesses some of the characteristics of shielded metal arc welding (SMAW), gas metal arc welding (GMAW), and submerged arc welding (SAW). Consequently, in certain applications each of the two FCAW process types offers certain trade-off benefits over the related processes. This is most notably true with respect to the SMAW process.

Combined features, as well as those that distinguish the two major versions of the pro-

154

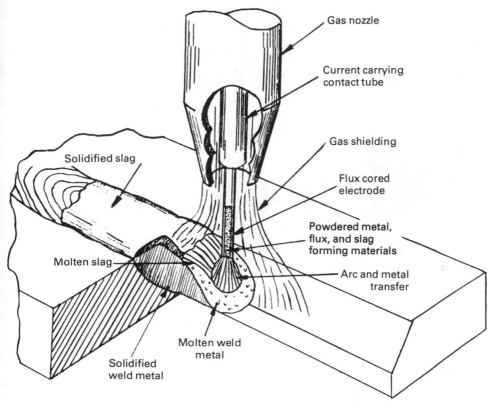

Fig. 5.1—Gas shielded flux cored arc welding

cess, are shown in Fig. 5.1 which illustrates the gas shielded version, and in Fig. 5.2 which illustrates the self-shielded type. Both figures illustrate the melting and deposition of matched quantities of filler metal and flux, together with the formation of the slag covering the weld metal.

In the gas shielded method, shown in Fig. 5.1, the shielding gas (usually but not exclusively carbon dioxide) protects the molten metal from the oxygen and nitrogen of the air by forming an envelope around the arc and over the weld pool. Little need exists for denitrification of the weld metal because the nitrogen of the air is mostly excluded. Although most of the air is excluded, some oxygen is present in the atmosphere. It may be present as an additive to argon or from dissociation of CO_2 to form carbon monoxide and oxygen. The compositions of

the electrodes are designed to tolerate small amounts of oxygen in the shielding gas. Thus, flux cored electrodes are designed specifically for self-shielding, for auxiliary gas shielding, or for use with or without gas shielding.

In the self-shielded method shown in Fig. 5.2, some shielding is obtained from vaporized flux ingredients which displace the air and provide protection to the molten weld pool during welding. Since in its transfer through the arc, the molten filler metal is outside the flux, this method does not rely as much on shielding to help achieve weld metal soundness as does the gas shielded method. Instead, the emphasis is on the addition of deoxidizing and denitrifying constituents to the filler metal and flux. This explains why self shielded electrodes can operate in strong air currents sometimes encountered when welding outdoors.

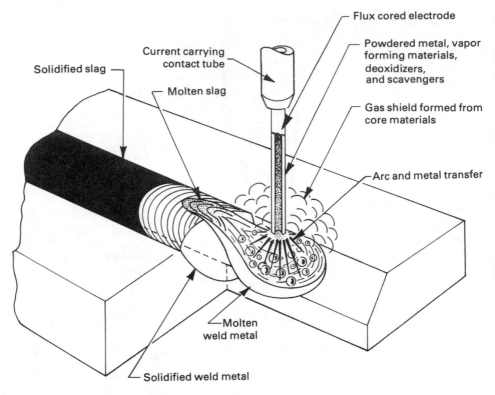

Flux cored electrode

Powdered metal, vapor forming materials, deoxidizers, and scavengers

Current carrying contact tube

Gas shield formed from core materials

Solidified slag

Molten slag

Arc and metal transfer

Molten weld metal

Solidified weld metal

Fig. 5.2—Self-shielded flux cored arc welding

One feature of the self-shielded method is the use of long electrode extensions. Electrode extension is the length of unmelted electrode extending beyond the end of the contact tube during welding. Self-shielded electrode extensions of 19 to 95 mm (3/4 to 3-3/4 in.) are generally used, depending on the application.

Increasing the electrode extension increases the resistance heating of the electrode. This preheats the electrode and lowers the voltage drop across the arc. At the same time, the welding current is decreased, which lowers the heat available for melting the base metal. The resulting weld bead is narrower and shallower. This makes the process suitable for welding light gage material and for bridging gaps due to poor fit-up. If the arc length (voltage) and welding current are maintained (higher voltage setting and electrode feed rate), deposition rate is increased by longer electrode extension.

In contrast, the gas shielded method is suited to the production of narrow, deeply penetrating welds. Shorter electrode extensions and higher welding currents are used. For fillet welding, narrower welds with larger throat lengths are possible when compared to SMAW. The electrode extension principle cannot be equally applied to the gas shielded method because of adverse effects on the shielding.

PRINCIPAL APPLICATIONS

Application of the two methods of the FCAW process overlap. However, the specific characteristics of each method make each one suitable for different operating conditions. The process is used to weld carbon and low alloy steels, stainless steels, and cast irons. It is also used for arc spot welding of lap joints in sheet and plate.

The method of FCAW used depends on the type of electrodes available, the mechanical property requirements of the welded joints, and the joint designs and fit-up. Generally, the self-shielded method can be used for some applications that could be done by the shielded metal arc welding process. The gas shielded method can be used for some applications that could be welded by the gas metal arc welding process. The advantages and disadvantages of the FCAW process must be compared to those of other applicable processes when it is evaluated for a specific application.

Higher productivity, compared to shielded metal arc welding, is the chief appeal of flux cored arc welding for many applications. This generally translates into lower overall costs per pound of metal deposited in joints that offer continuous welding and easy FCAW gun and equipment accessibility. Here, the advantages are higher deposition rates, higher operating factors, and higher deposition efficiency (no stub loss).

Consequently, semiautomatic FCAW has found wide application in shop fabrication, maintenance, and field erection work. It has been used to produce weldments conforming to the ASME Boiler and Pressure Vessel Code, the rules of the American Bureau of Shipping, and the AWS Structural Welding Code. The latter permits its use for producing selected pre-qualified joints.

In the field of stainless steel welding, both self-shielded and gas shielded FCAW have been used for general fabrication, surfacing, joining dissimilar metals, and repair of castings.

The major disadvantages, compared to the SMAW process, are the higher cost of the equipment, the relative complexity of the equipment in setup and control, and the restriction on operating distance from the electrode wire feeder. Compared to any process, FCAW (especially the self-shielded type) generates large volumes of welding fumes which, except in field work, require suitable exhaust equipment. Compared to the slag-free GMAW process, the need for removing slag between passes is an added labor cost. This is especially true in root bead welding where the GMAW process is sometimes preferred.

EQUIPMENT

SEMIAUTOMATIC EQUIPMENT

The basic equipment for self-shielded and gas shielded flux cored arc welding is similar, the major difference being the provision of a system for supplying and metering gas to the arc of the gas shielded electrode, as shown in Fig. 5.3. The recommended power source is the dc, constant voltage type similar to those used for gas metal arc welding. The power supply should be capable of operating at the maximum current required for the specific application. At present, most semiautomatic applications use less than 650 A. The controls should be capable of voltage adjustments in increments of one volt or less. Constant current (dc) power sources of adequate capacity, with appropriate controls and wire feeders, may also be used.[1]

The purpose of the wire feed control is to supply the continuous electrode to the welding arc at a constant preset rate. The rate at which the electrode is fed into the arc determines the welding amperage that a constant voltage power source must supply. If the electrode feed rate is changed, the welding machine will automatically change the welding current to maintain a preset arc voltage. Electrode feed rate may be governed by either mechanical or electronic means.

---

1. See Chapter 1 for additional information on arc welding power sources.

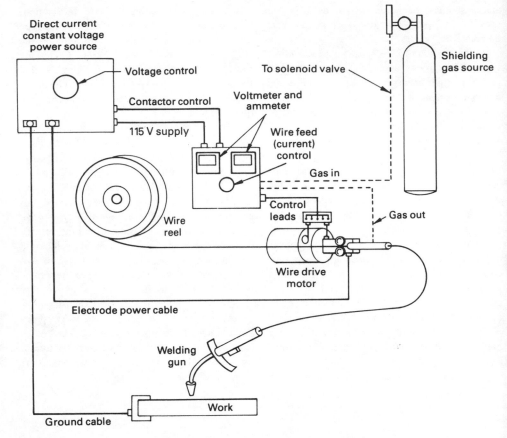

NOTE: Gas shielding is used only with flux cored electrodes that require it.

Fig. 5.3—Typical semiautomatic flux cored arc welding equipment

Flux cored electrodes require the use of drive rolls that will not flatten or otherwise distort them. Various grooved and knurled surfaces are used to advance the electrode. Some wire feeders have a single pair of drive rolls, while others have two pairs with at least one roll of each pair driven.

Typical semiautomatic welding guns are shown in Fig. 5.4 and Fig. 5.5. They are designed for handling comfort, ease of manipulation, and durability. The guns provide internal contact with the electrode to conduct the welding current. Welding current and elec-

trode feed are actuated by a switch mounted on the gun.

Welding guns may be either gas-cooled or water-cooled. Gas-cooled guns are favored since they are more compact, lighter in weight, and require less maintenance than water-cooled guns. Capacity ratings range up to 600 A, continuous duty. Guns may have either straight or curved nozzles. The curved nozzle can vary from 40° to 60°. In some applications, the curved nozzle enhances flexibility and ease of electrode manipulation.

Some self-shielded flux cored electrodes

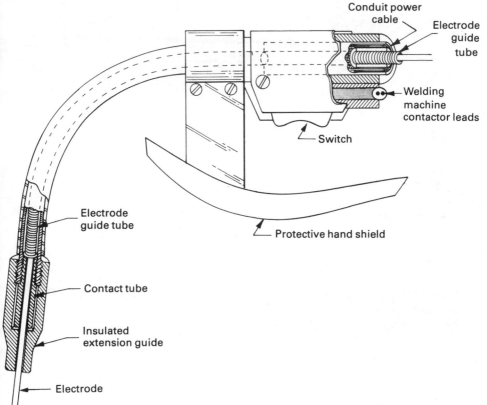

Fig. 5.4—Gun for semiautomatic self-shielded flux cored arc welding

require a specific minimum electrode extension to develop proper shielding. Welding guns for these electrodes generally have guide tubes with an insulated extension guide to support the electrode and assure a minimum electrode extension. Details of a self-shielded electrode nozzle showing the insulated guide tube are illustrated in Fig. 5.6.

AUTOMATIC EQUIPMENT

Figure 5.7 shows the equipment layout of an automatic flux cored arc welding installation. For automatic operation, a dc constant voltage power source designed for 100 percent duty cycle is recommended. The size of the power source is determined by the current required for the work to be performed. Because large electrodes, high electrode feed rates and long welding times may be required, electrode feeders necessarily have higher capacity drive motors and heavier duty components than similar equipment for semiautomatic operation.

A typical nozzle assembly for automatic self-shielded flux cored arc welding appears in Fig. 5.8. Figure 5.9 shows two typical nozzle assemblies for automatic gas shielded flux cored arc welding. Nozzle assemblies for the latter may be designed for side shielding or for concentric shielding of the electrode. Side shielding permits welding in narrow, deep grooves and minimizes spatter buildup in the nozzle. Nozzle

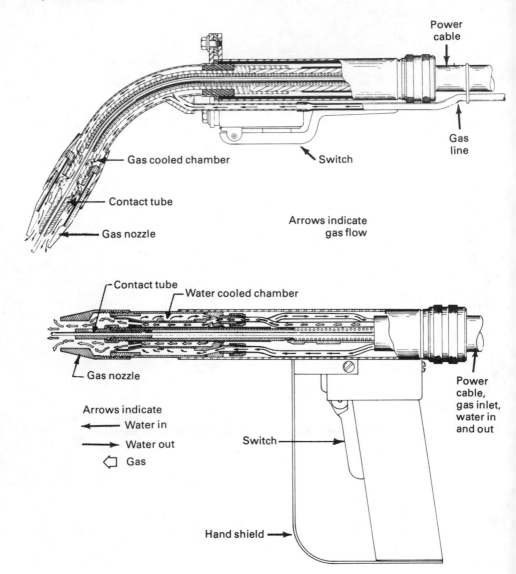

Fig. 5.5—Guns for semiautomatic gas shielded flux cored arc welding

assemblies may be air- or water-cooled. In general, air-cooled nozzle assemblies are preferred for operation up to 600 A. Water-cooled nozzle assemblies are recommended for welding currents above 600 A. For higher deposition rates with gas shielded electrodes, tandem welding guns may be used, as shown in Fig. 5.10.

For large scale surfacing, increased productivity can be obtained from automatic multiple-electrode oscillating equipment. Such installations may include a track-mounted manipulator supporting a multiple-electrode oscillating welding head with individual electrode feeders and a track-mounted, power driven turning roll, in addition to power supply, electronic controls, and electrode supply system. Figure 5.11 illustrates the operating details of a six-electrode oscillating system for self-shielded surfacing of a vessel shell with stainless steel.

FUME EXTRACTORS

As a result of safety and health requirements for controlling air pollution, several manufacturers have introduced gas shielded welding guns equipped with integral fume extractors. A fume extractor usually consists of an exhaust nozzle that encircles the gun nozzle. The nozzle is ducted to a filter canister and an exhaust pump. The aperture of the fume extracting nozzle is located at a sufficient distance behind the tip of the gun nozzle to draw in the fumes

rising from the arc without disturbing the shielding gas flow.

The chief advantage of this fume extraction system is that it is in close proximity to the fume source wherever the welding gun is used. In contrast, a portable fume exhaust would not generally be positioned as close to the fume source. It would also require repositioning of the exhaust hood for each significant change in welding location.

One disadvantage of the fume extractor system is that the added weight and bulk make semiautomatic welding more cumbersome for the welder. If not properly installed and maintained, fume extractors can cause welding problems by disturbing the gas shielding. In a well-ventilated welding booth, a fume extractor-welding gun combination may not be necessary.

MATERIALS

BASE METALS WELDED

Most steels that are weldable with the SMAW, GMAW, or SAW processes are readily welded using the FCAW process.[2] Examples of these steels include the following:

(1) Mild steel, structural and pressure vessel grades, such as ASTM A36, A515, and A516.

(2) High strength, low alloy structural grades, such as ASTM A440, A441, A572, and A588.

(3) High strength quenched and tempered alloy steels, such as ASTM A514, A517, and A533.

(4) Chromium-molybdenum steels, such as 1-1/4% Cr-1/2% Mo and 2-1/4% Cr-1% Mo.

(5) Corrosion resistant wrought stainless steels, such as AISI Types 304, 309, 316, 347, 410, 430, and 502; also cast stainless steels such as ACI Types CF3 and CF8.

(6) Nickel steels, such as ASTM A203.

(7) Abrasion resistant alloy steels when welded with filler metal having a yield strength less than that of the steel being welded.

2. See Chapters 61, 62, 63, 64, and 65, *Welding Handbook*, Section 4, 6th ed., for information on materials welded by FCAW.

ELECTRODES

Flux cored arc welding owes much of its versatility to the wide variety of ingredients that can be included in the core of the round electrode. The electrode consists of a low carbon steel or alloy steel sheath surrounding a core of fluxing and alloying materials. The composition of the flux core will vary according to the electrode classification and to the particular manufacturer of the electrode.

Most flux cored electrodes are made by passing steel strip through rolls that form it into a U-shaped cross section. The formed strip is filled with a measured amount of granular core material (flux). The filled shape is then closed by closing rolls that shape it round and tightly compress the core material. The round tube is next pulled through drawing dies that reduce its diameter and further compress the core. The electrode is drawn to final size, and finally it is wound on spools or in coils. Other methods of manufacture are also used.

Manufacturers generally consider the precise composition of their cored electrodes to be proprietary information. By proper selection

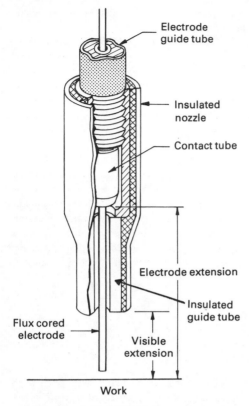

Fig. 5.6—Self-shielded electrode nozzle

(2) Promote weld metal soundness by shielding the molten metal from oxygen and nitrogen in the air

(3) Scavenge impurities from the molten metal by use of fluxing reactions

(4) Produce a slag cover to protect the solidifying weld metal from the air, and to control the shape and appearance of the bead in the different welding positions for which the electrode is suited

(5) Stabilize the arc by providing a smooth electrical path to reduce spatter and facilitate the deposition of uniformly smooth, properly sized beads

Table 5.1 lists most of the elements commonly found in the flux core, their sources, and the purposes for which they are used.

In mild and low alloy steel electrodes, a proper balance of deoxidizers and, in the case of self-shielded electrodes, denitrifiers must be maintained to provide a sound weld deposit with adequate ductility and toughness. Deoxidizers, such as silicon and manganese, combine with oxygen to form stable oxides. This helps to control the formation of carbon monoxide which causes porosity, and also the loss of alloying elements through oxidation. The denitrifiers, such as aluminum, combine with nitrogen and tie it up as stable nitrides. This prevents nitrogen porosity and the formation of other nitrides which might be harmful.

of the ingredients in the core (in combination with the composition of the sheath), it is possible to

(1) Produce welding characteristics ranging from high deposition rates in the flat position to proper fusion and bead shape in the overhead position

(2) Produce electrodes for various gas shielding mixtures and for self shielding

(3) Vary alloy content of the weld metal from mild steel for certain electrodes to high alloy stainless steel for others

The primary functions of the flux core ingredients are to

(1) Provide the mechanical, metallurgical, and corrosion resistant properties of the weld metal by adjusting the chemical composition

CLASSIFICATIONS OF ELECTRODES

Mild Steel Electrodes

Most mild steel FCAW electrodes are classified according to the requirements of AWS A5.20, Specification for Mild Steel Electrodes for Flux Cored Arc Welding. The classification system follows the general pattern for electrode classification. It may be explained by considering a typical designation, E70T-1.

The prefix "E" indicates an electrode, as in other electrode classification systems. The first number refers to the minimum as-welded tensile strength, in 10 ksi units. In this example, the number "7" indicates that the electrode has a minimum tensile strength of 70 ksi. The second number indicates the welding positions for

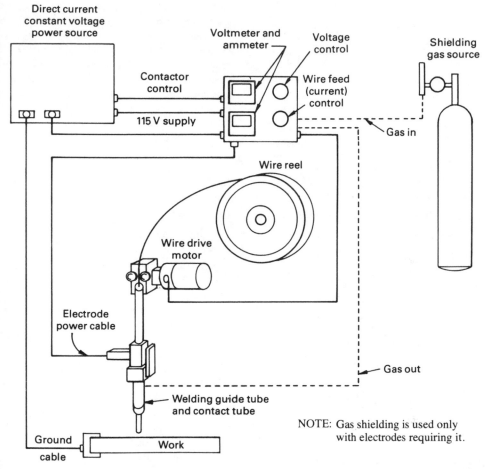

Direct current constant voltage power source

Voltmeter and ammeter

Voltage control

Shielding gas source

Contactor control

Wire feed (current) control

115 V supply

Gas in

Wire reel

Wire drive motor

Electrode power cable

Gas out

Welding guide tube and contact tube

NOTE: Gas shielding is used only with electrodes requiring it.

Ground cable

Work

Fig. 5.7—Typical flux cored automatic arc welding equipment

which the electrode is designed. Here the "0" means that the electrode is designed for flat and horizontal positions. However, small sizes of the classification may be suitable for vertical or overhead positions, or both.

The letter "T" indicates that the electrode is of tubular construction (a flux cored electrode). The suffix number (in this example "1") places the electrode in a particular grouping built around the chemical composition of deposited weld metal, method of shielding, and suitability of the electrode for single or multiple pass welds.

Mild steel FCAW electrodes are classified on the basis of whether or not carbon dioxide is required as a separate shielding gas, the type of current, their usability for welding position and number of weld passes, and the chemical composition and as-welded mechanical properties of deposited weld metal. Electrodes are designed to produce weld metals having specified chemical compositions and mechanical properties when the welding and testing are done to the specification requirements.

Electrodes are produced in standard sizes ranging from 1.58 to 3.97 mm (1/16 to 5/32 in.)

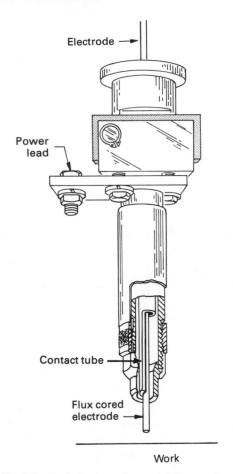

Fig. 5.8—Typical automatic arc welding gun for self-shielded electrodes

manufacturer's recommended range may be used for out-of-position welding.

Mild steel electrodes are basically designed to produce weld metals having a minimum tensile strength of 414 MPa (70 000 psi), with and without carbon dioxide shielding. In addition, minimum yield and impact strengths as well as minimum ductility may be required. Some electrodes are intended for single pass welding in the flat and horizontal positions.

Additional information on mild steel flux cored electrodes is given in Chapter 94, "Filler Metals," *Welding Handbook,* Section 5, 6th Edition,[3] and AWS A5.20, Specification for Mild Steel Electrodes for Flux Cored Arc Welding.[4]

Low Alloy Steel Electrodes

Flux cored electrodes are commercially available for welding low alloy steels.[5] The electrodes are designed to produce deposited weld metals having chemical compositions and mechanical properties similar to those produced by low alloy steel SMAW electrodes. They are generally used to weld low alloy steels of similar chemical composition.

Some electrode classifications are designed for welding in all positions. Others are designed for flat and horizontal positions only, although vertical and overhead welding may be done by using the proper size and low welding current. The manufacturer's recommendations should be followed.

Most low alloy steel FCAW electrodes are designed for the gas shielded method using either a *1* or a *5* flux core formulation and CO_2. They generally produce weld metals having Charpy V-notch impact strengths of 27 J (20 ft•lb) at -18° C (0° F) or below. A few nickel steel electrodes with *4* or *8* formulations are available for self-shielded FCAW. Charpy V-notch impact requirements for deposited weld metal are generally 27 J (20 ft•lb) at -18° C (0° F). Thus for low temperature applications, the *1* or *5* flux

diameter. Special sizes may also be available. Weld properties may vary appreciably depending on electrode size and current used, plate thickness, joint geometry, preheat and interpass temperatures, surface conditions, base metal composition and admixture with the deposited metal, and shielding gas. Many electrodes are designed primarily for welding in the flat and horizontal positions. They may be used in other positions if the proper choice of welding current and electrode diameter are made. Selected electrode with diameters below 2.38 mm (3/32 in.) and welding currents on the low side of the

3. Chapter 94 is scheduled for revision in *Welding Handbook,* Vol. 4, 7th ed.
4. The latest revision of AWS A5.20 should be consulted.
5. AWS Specification A5.29 for low alloy steel flux cored electrodes is in preparation. Publication is expected in 1978.

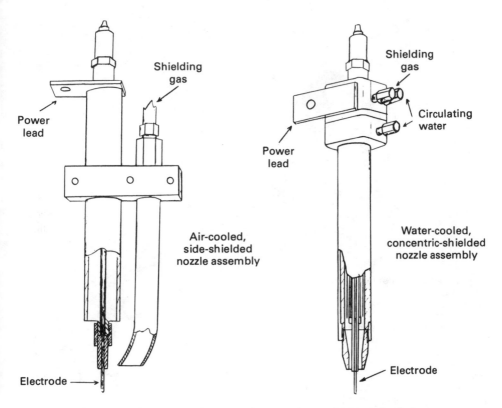

Fig. 5.9—Typical automatic arc welding guns for gas shielded electrodes

core formulations are best. Additional information may be found in Chapter 94, "Filler Metals," *Welding Handbook,* Section 5, 6th edition.[3]

Electrodes for Surfacing

Flux cored electrodes are produced for certain types of surfacing applications, such as restoring usable service parts and hardfacing. Such electrodes possess many of the advantages of the electrodes used for joining, but there is less standardization of weld metal analyses and performance characteristics. Literature from various manufacturers should be consulted for details on flux cored surfacing electrodes.

Most flux cored surfacing electrodes deposit iron base alloys which may be either ferritic, martensitic, or austenitic in structure. The electrodes are designed to produce a surface with

corrosion resistance, wear resistance, toughness, or antigalling properties. They may be used to restore worn parts to original dimensions.

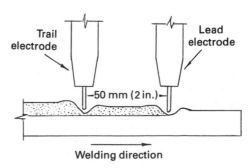

Fig. 5.10—Automatic tandem arc welding with two gas shielded flux cored electrodes

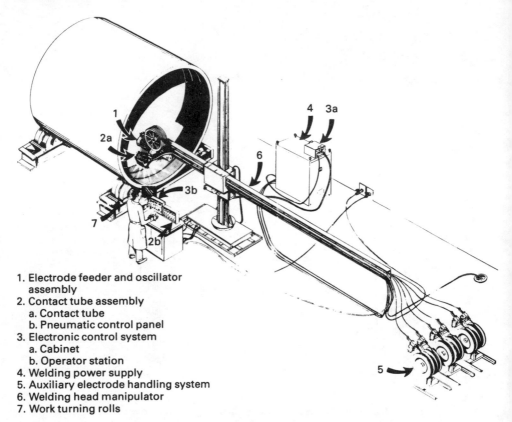

1. Electrode feeder and oscillator
   assembly
2. Contact tube assembly
   a. Contact tube
   b. Pneumatic control panel
3. Electronic control system
   a. Cabinet
   b. Operator station
4. Welding power supply
5. Auxiliary electrode handling system
6. Welding head manipulator
7. Work turning rolls

Fig. 5.11 — A typical multiple weave surfacing installation

Stainless Steel Electrodes

The classification system of AWS A5.22, Specification for Flux Cored Corrosion-Resisting Chromium and Chromium-Nickel Steel Electrodes,[6] prescribes requirements for flux cored corrosion resisting chromium and chromium-nickel steel electrodes. These electrodes are classified on the basis of the chemical composition of the deposited weld metal and the shielding medium to be employed during welding. Table 5.2 identifies the shielding designations used in the classification and indicates the respective current characteristics.

With the EXXXT-1 classifications using

carbon dioxide shielding, there is some minor loss of oxidizable elements and some increase in carbon content. With the EXXXT-3 classifications, which are used without external shielding, there is some loss of oxidizable elements and pickup of nitrogen which may be significant. Low welding currents coupled with long arc lengths (high arc voltages) may cause excessive nitrogen pickup. Nitrogen stabilizes the austenite and, therefore, it may reduce the ferrite content of the weld metal.

The requirements of the EXXXT-3 classifications are different from those of the EXXXT-1 classifications because shielding with a flux system alone is not as effective as shielding with both a flux system and a separately applied external shielding gas. The EXXXT-3 deposits,

6. The latest revision of AWS A5.22 should be consulted.

	Table 5.1—Common core elements in flux cored electrodes	
Element	Usually present as	Purpose in weld
Aluminum	Metal powder	Deoxidize and denitrify
Calcium	Minerals such as fluorspar (CaF_2) and limestone ($CaCO_3$)	Provide shielding and form slag
Carbon	Element in ferroalloys such as ferromanganese	Increase hardness and strength
Chromium	Ferroalloy or metal powder	Alloying to improve creep resistance hardness, strength and corrosion resistance
Iron	Ferroalloys and iron powder	Alloy matrix in iron base deposits, alloy in nickel base and other nonferrous deposits
Manganese	Ferroalloy such as ferromanganese or as metal powder	Deoxidize; prevent hot shortness by combining with sulfur to form MnS; increase hardness and strength; form slag
Molybdenum	Ferroalloy	Alloying to increase hardness strength, and in austenitic stainless steels to increase resistance to pitting-type corrosion
Nickel	Metal powder	Alloying to improve hardness, strength, toughness and corrosion resistance
Potassium	Minerals such as potassium bearing feldspars and silicates and in frits	Stabilize the arc and form slag
Silicon	Ferroalloy such as ferrosilicon or silicomanganese; mineral silicates such as feldspar	Deoxidize and form slag
Sodium	Minerals such as potassium-bearing feldspars and silicates and in frits	Stabilize the arc and form slag
Titanium	Ferroalloy such as ferrotitanium; in mineral, rutile	Deoxidize and denitrify; form slag; stabilize carbon in some stainless steels
Zirconium	Oxide or metal powder	Deoxidize and denitrify
Vanadium	Oxide or metal powder	Increase strength

Table 5.2—Shielding designations and welding current characteristics
for stainless steel flux cored electrodes

AWS designations[a] (all classifications)	External shielding medium	Current and polarity
EXXXT-1	CO_2	dcrp[b] (electrode positive)
EXXXT-2	Ar + 2% O	dcrp[b] (electrode positive)
EXXXT-3	None	dcrp[b] (electrode positive)
EXXXT-G	None specified	Not specified

a. The classifications are given in AWS A5.22, Specification for Flux Cored Corrosion-Resisting Chromium and Chromium-Nickel Steel Electrodes. The letters "XXX" stand for the chemical composition (AISI Type) such as 308, 316, 410, and 502.

b. Direct current reverse polarity.

therefore, usually have a higher nitrogen content than the EXXXT-1 deposits. This means that to control the ferrite content of the weld metal, the chemical compositions of the EXXXT-3 deposits must have different Cr/Ni ratios than those of the EXXXT-1 deposits.

The mechanical properties of deposited weld metal are specified for each classification. Properties include minimum tensile strength and ductility. Radiographic soundness requirements are also specified.

Although welds made with electrodes meeting the AWS specification requirements are commonly used in corrosion or heat resisting applications, it is not practical to require electrode qualification tests for corrosion or scale resistance on welds or weld metal specimens. Special tests which are pertinent to an intended application should be established by agreement between the electrode manufacturer and the user.

SHIELDING GASES

Carbon Dioxide

Carbon dioxide (CO_2) is the most widely used shielding gas in flux cored arc welding. Low cost and deep weld penetration are two advantages of CO_2 gas shielding. It gives a globular type of metal transfer.

Carbon dioxide is relatively inactive at room temperature. When it is heated to high temperature by the welding arc, CO_2 dissociates to form carbon monoxide (CO) and oxygen (O), as indicated by the chemical equation

$$2CO_2 \rightleftharpoons 2CO + O_2$$

Thus, the arc atmosphere contains a considerable amount of oxygen to react with elements in the molten metal. The oxidizing tendency of CO_2 shielding gas has been recognized in developing flux cored electrodes. Deoxidizing materials are added to the core of the electrode to compensate for the oxidizing effect of the CO_2.

In addition, molten iron reacts with CO_2, producing iron oxide and carbon monoxide in a reversible reaction:

$$Fe + CO_2 \rightleftharpoons FeO + CO$$

At red heat temperatures, some of the carbon monoxide dissociates to carbon and oxygen as indicated by

$$2CO \rightleftharpoons 2C + O_2$$

The effect of a CO_2 shielding atmosphere on the carbon content of the weld metal is unique. Depending upon the original carbon contents of the base metal and the electrode, the CO_2 atmosphere can behave as either a carburizing or decarburizing medium. The carbon content of the weld metal may be either increased or decreased depending upon the carbon contributed by the electrode and the base metal. If the carbon content of the weld metal is below approximately 0.05 percent, the molten weld pool will tend to pick up carbon from the CO_2 shielding atmosphere. On the other hand, if the carbon content of the weld metal is greater than approximately 0.10 percent, the molten weld pool will show a tendency to lose carbon. The loss of carbon may be attributed to the formation of carbon monoxide (CO) because of the oxidizing characteristics of CO_2 shielding gas at high temperatures.

When this reaction occurs, the carbon monoxide can be trapped in the weld metal as porosity. This tendency is minimized by an adequate level of the deoxidizing elements in the core of the electrode. Oxygen will react with the deoxidizing elements in preference to carbon in the steel. This reaction results in formation of solid oxide compounds that float to the surface of the molten weld pool where they form part of the slag covering.

Gas Mixtures

Gas mixtures used in flux cored arc welding combine the separate advantages of two or more gases. The higher the percentage of inert gas in mixtures with CO_2 or oxygen, the higher will be the transfer efficiencies of the deoxidizers contained in the core. Argon is capable of protecting the molten weld pool at all welding temperatures. Its presence in sufficient quantities in a shielding gas mixture results in less oxidation than occurs with 100 percent CO_2 shielding.

The mixture commonly used in FCAW is 75 percent argon-25 percent carbon dioxide. Weld metal deposited with this mixture generally has higher tensile and yield strengths than weld metal deposited with 100 percent CO_2 shielding. When welding with this mixture, the electrode transfer approaches a spray type. The Ar-CO_2 mixture is mainly used for out-of-position welding because of better arc characteristics and greater operator appeal.

The use of shielding gas mixtures with high percentages of inert gas for electrodes designed for CO_2 shielding may cause an excessive buildup of manganese, silicon, and other deoxidizing elements in the weld metal. The higher alloy content of the weld metal will change its mechanical properties. Therefore, electrode manufacturers should be consulted for the mechanical properties of weld metal obtained with specific shielding gas mixtures. If data are not available, tests should be made to determine the mechanical properties for the particular application.

Shielding gases are dispensed from cylinders, manifolded cylinder groups, or from bulk tanks which are piped to individual welding stations. Regulators and flowmeters are used to control pressure and flow rates. Since regulators can freeze up when withdrawal of CO_2 gas from storage tanks is high, heaters are available to prevent this problem. Additionally, by manifolding several tanks together, flow rates from individual tanks can be reduced. Welding grade gas purity is required because small amounts of moisture can result in porosity. The dew point of shielding gases should be below $-40°$ C ($-40°$ F).

PROCESS CONTROL

WELDING CURRENT

Welding current is proportional to electrode feed rate for a specific electrode diameter, composition, and electrode extension. The relationship between electrode feed rate and welding current for typical mild steel gas shielded electrodes, self-shielded mild steel electrodes, and self-shielded stainless steel electrodes is presented in Figs. 5.12, 5.13, and 5.14 respectively. A constant voltage power source of the proper size is used to melt the electrode at the rate that maintains the preset output voltage (arc length). If the other welding variables are held constant for a given diameter of electrode, changing the

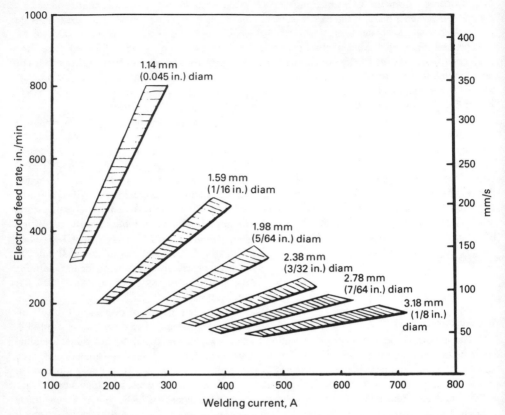

Fig. 5.12—Electrode feed rate vs welding current range for E70T-1 steel electrodes with CO_2 shielding

welding current will have the following major effects:

(1) Increasing current increases electrode deposition rate.

(2) Increasing current increases penetration.

(3) Excessive current produces convex weld beads with poor appearance.

(4) Insufficient current produces large droplet transfer and excessive spatter.

(5) Insufficient current can result in pickup of excessive nitrogen and also porosity in the weld metal when welding with self-shielded flux cored electrodes.

As welding current is increased or decreased by changing electrode feed rate, power supply output voltage should be changed to maintain the optimum relationship of arc voltage to cur-

rent. For a given electrode feed rate, measured welding current varies with the electrode extension. As the electrode extension increases, welding current will decrease, and vice versa.

ARC VOLTAGE

Arc voltage and arc length are closely related. The voltage shown on the meter of the welding power supply is the sum of the voltage drops throughout the welding circuit. This includes the drop through the welding cable, the electrode extension, the arc, the workpiece and the ground cable. Therefore, arc voltage will be proportional to the meter reading provided all other circuit elements (including their temperatures) remain constant.

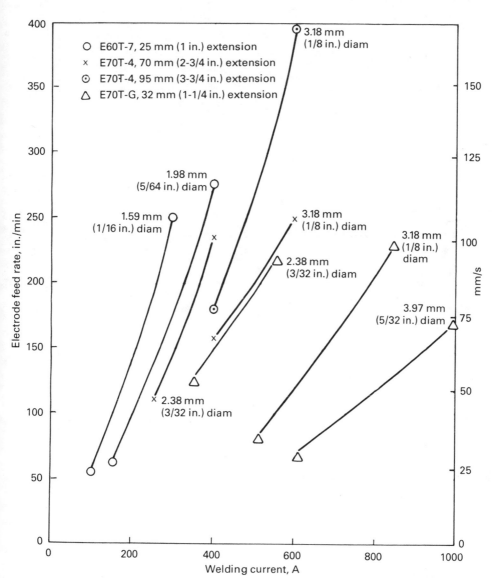

Fig. 5.13—Electrode feed rate vs welding current for self-shielded mild steel electrodes

The appearance, soundness, and properties of welds made with flux cored electrodes can be affected by the arc voltage. Too high an arc voltage (too long an arc) will result in excessive spatter and wide, irregularly shaped weld beads. With self-shielded electrodes, too high an arc voltage will result in excessive nitrogen pickup.

With mild steel electrodes, this may cause porosity. With stainless steel electrodes, it will reduce the ferrite content of the weld metal, and this in turn may result in cracking. Too low an arc voltage (too short an arc) will result in narrow convex beads with excessive spatter and reduced penetration.

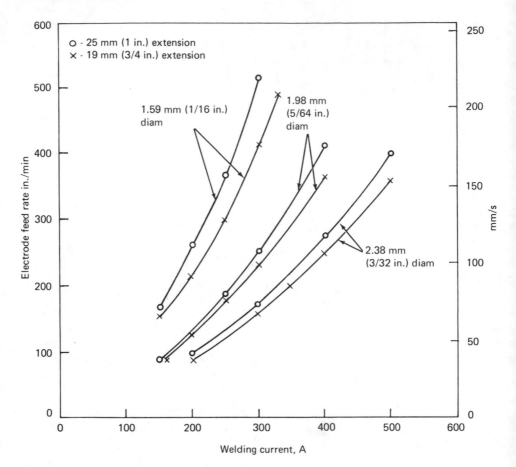

Fig. 5.14 — Electrode feed rate vs welding current for self-shielded E308T-3 stainless steel electrodes

ELECTRODE EXTENSION

The unmelted electrode that extends beyond the contact tube during welding (electrode extension) is resistance heated in proportion to its length, assuming other variables remain constant. As explained earlier, electrode temperature affects arc energy, electrode deposition rate, and weld penetration. It also can affect weld soundness and arc stability.

In effect, the use of electrode extension principle as an operating factor in FCAW introduces a new variable that must be held in balance with the shielding conditions and the related welding variables. For example, the melting and activation of the core ingredients must be consistent with that of the containment tube, as well as with arc characteristics. Other things being equal, too long an extension produces an unstable arc with excessive spatter. Too short an extension may cause excessive arc length at a particular voltage setting. With gas shielded electrodes, it may result in excessive spatter buildup in the nozzle that can interfere with the gas flow. Poor shielding gas coverage may cause weld metal porosity and excessive oxidation.

Most manufacturers recommend an extension of 19 to 38 mm (3/4 to 1-1/2 in.) for gas shielded electrodes and from approximately 19 to

95 mm (3/4 to 3-3/4 in.) for self-shielded types, depending on the application. For optimum settings in these ranges, the electrode manufacturer should be consulted.

TRAVEL SPEED

Travel speed influences weld bead penetration and contour. Other factors remaining constant, penetration at low travel speeds is greater than that at high travel speeds. Too low travel speeds at high currents can result in overheating of the weld metal. This will cause a rough appearing weld with the possibility of mechanically trapping slag. Too high travel speeds will result in an irregular, ropy appearing bead.

SHIELDING GAS FLOW

For gas shielded electrodes, gas flow rate is a variable affecting weld quality. Inadequate flow will result in poor shielding of the transferring filler metal and the weld metal, resulting in weld porosity and oxidation. Excessive gas flow can result in turbulence and mixing with air. The effect on the weld quality will be the same as inadequate flow. Either extreme will increase weld metal impurities. Correct gas flow will depend on the type and the diameter of the gun nozzle, distance of the nozzle from the work, and air movements in the immediate region of the welding operation.

DEPOSITION RATE AND EFFICIENCY

Deposition rate in any welding process is the weight of material deposited per unit of time. Deposition rate is dependent on welding variables such as electrode diameter, electrode composition, electrode extension, and welding current. Deposition rate versus welding current for various diameters of gas shielded and self-shielded mild steel electrodes and self-shielded stainless steel electrodes are presented in Figs. 5.15, 5.16, and 5.17 respectively.

Deposition efficiencies of FCAW electrodes will range from 85 to 90 percent for those used with gas shielding and 80 to 87 percent for self-shielded electrodes. Deposition efficiency is the ratio of weight of metal deposited to the weight of electrode consumed.

ELECTRODE ANGLE

The angle at which the electrode is held during welding determines the direction in which the arc force is applied to the molten weld pool. When welding variables are properly adjusted for the application involved, the arc force can be used to oppose the effects arising from the force of gravity. In the FCAW and SMAW process, the arc force is used not only to help shape the desired weld bead, but also to prevent the slag from running ahead of and becoming entrapped in the weld metal.

When making groove and fillet welds in the flat position, gravity tends to cause the molten weld pool to run ahead of the weld. To counteract this, the electrode is held at an angle to the vertical with the electrode tip pointing backwards toward the weld, i.e., away from the direction of travel. This travel angle, defined as the *drag angle*, is measured from a vertical line in the plane of the weld axis, as shown in Fig. 5.18(A).

The proper drag angle depends on the FCAW method used, the base metal thickness, and the position of welding. For the self-shielded method, drag angles should be about the same as those used with shielded metal arc welding electrodes. For flat and horizontal positions, drag angles will vary from approximately 20° to 45°. Larger angles are used for thin sections. As material thickness increases, the drag angle is decreased to increase penetration. For vertical-up welding, the drag angle should be 5° to 10°.

With the gas shielded method, the drag angle should be small, usually 2° to 15°, but not more than 25°. If the drag angle is too large, the effectiveness of the shielding gas will be lost.

When fillet welds are made in the horizontal position, the weld pool tends to flow both in the direction of travel and at right angles to it. To counteract the latter, the electrode should point at the bottom plate close to the corner of the joint and, in addition to the drag angle, a *work angle* of 40° to 50° from the vertical member should be used. Figure 5.18(B) shows the electrode offset and the work angle used for horizontal fillets.

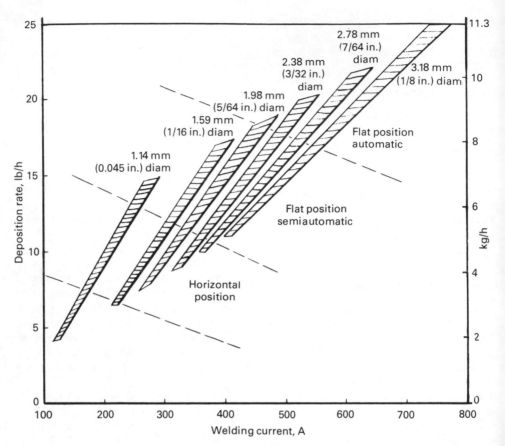

Fig. 5.15—Deposition rate vs welding current for E70T-1 mild steel electrodes with CO_2 shielding

JOINT DESIGNS AND WELDING PROCEDURES

The joint designs and welding procedures appropriate for flux cored arc welding will depend on whether the gas shielded or self-shielded method is used. However, all the basic joint types can be welded by either method. All the basic weld groove shapes to which shielded metal arc welding is commonly applied can be welded by both of the FCAW methods.

There may be some differences in specific groove dimensions for a specific joint between the two methods of FCAW and between the FCAW and SMAW processes. Because FCAW electrode formulations, purpose and operating characteristics differ between classifications, the values of their welding procedure variables may also differ.

FOR GAS SHIELDED ELECTRODES

In general, joints can be designed to take advantage of the penetration achieved by high current densities. Narrower grooves with smaller groove angles, narrower root openings, and

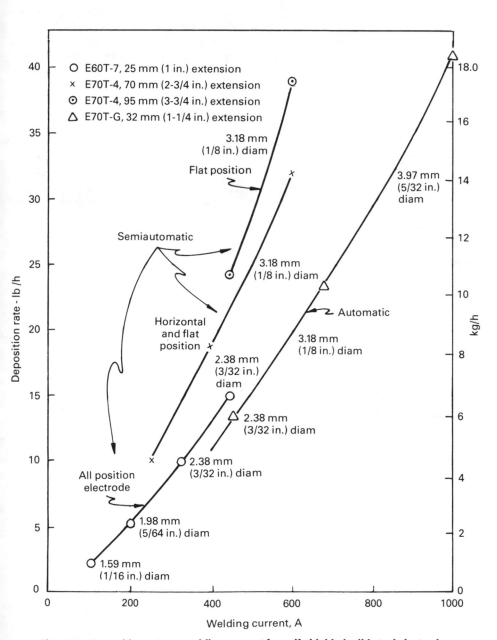

Fig. 5.16 — Deposition rate vs welding current for self-shielded mild steel electrodes

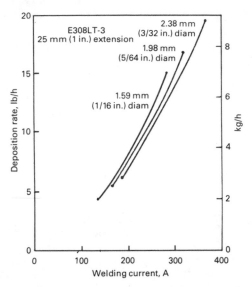

Fig. 5.17—Deposition rate vs welding current for self-shielded E308LT-3 stainless steel electrodes

and horizontal positions, can be reduced in size by approximately 2 to 3 mm (1/16 to 1/8 in.) when design specifications and quality control standards permit.

Adequate CO_2 gas shielding is required to obtain sound welds. Flow rates required depend on nozzle size, draft conditions, and electrode extension. Welding in still air requires flow rates in the range of 14 to 19 liters/min (30 to 40 ft³/h). When welding in moving air or when electrode extension is longer than normal, flow rates of up to 26 liters/min (55 ft³/h) may be needed. Flow rates for side shielding nozzles are generally the same or slightly higher than those for concentric nozzles. Nozzle openings must be maintained free from excessive adhering spatter.

larger root faces than are practical with SMAW can be used with the gas shielded method.

For the basic butt joint designs, the following points should be considered:

(1) The joint should be designed so that a constant electrode extension can be maintained when welding succeeding passes in the joint.

(2) The joint should be designed so that the root is accessible, and any necessary electrode manipulation during welding is easily done.

Groove angles for various metal thicknesses are properly designed when they provide welding accessibility for the appropriate gas nozzle and electrode extension. Side shielding nozzles for automatic welding permit better accessibility into narrower joints and also the use of smaller groove angles than do concentric nozzles. Sound welds that meet the quality requirements of several fabrication codes can be obtained with the proper welding procedure. Typical joint designs and welding procedures for gas shielded FCAW of carbon steel are given in Table 5.3. Welding power may vary with the electrode manufacturer. Because of increased penetration compared to SMAW, fillet welds made with gas shielded electrodes, in the flat

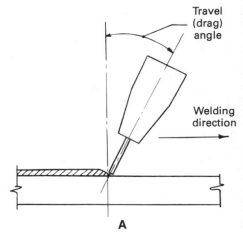

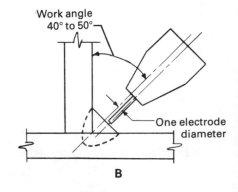

Fig. 5.18—Welding electrode positions

Table 5.3—Typical gas shielded flux cored arc welding procedures for carbon steel (E70T-1 electrodes)

Joint design	Thickness, T		Root opening, R		Total passes	Electrode diameter		Welding power, dcrp. (ep)		Production rate	
	mm	in.	mm	in.		mm	in.	V	A	m/h	ft/h
				Flat position groove welds (semiautomatic)							
	3	1/8	1.6	1/16	1	1.98	5/64	28	325	61	200
	13	1/2	6	1/4	2	2.38	3/32	32	450	9	30
	13	1/2	0	0	2	2.38	3/32	30	480	11	35
	25	1	0	0	6	2.78	7/64	32	525	3	9
	16	5/8	5	3/16	3	2.78	7/64	32	525	7	23
	25	1	5	3/16	6	2.78	7/64	32	525	4	12
	16	5/8	3	1/8	3	2.78	7/64	32	525	8	27
	25	1	3	1/8	6	2.78	7/64	32	525	4	12
	25	1	0	0	6	2.78	7/64	32	525	8	26
	51	2	0	0	20	2.78	7/64	32	525	2	7
	25	1	0	0	4	2.78	7/64	32	500	7	21
	51	2	0	0	12	2.78	7/64	32	500	2	5

Table 5.3 (continued)—Typical gas shielded flux cored arc welding procedures for carbon steel (E70T-1 electrodes)

Joint design	Thickness, T		Root opening, R		Total passes	Electrode diameter		Welding power, dcrp. (ep)		Production rate	
	mm	in.	mm	in.		mm	in.	V	A	m/h	ft/h
Horizontal position groove weld (semiautomatic)											
	13	1/2	3	1/8	6	1.98	5/64	28	350	5	17
	25	1	3	1/8	18	1.98	5/64	28	350	2	5
Vertical position groove welds (semiautomatic)											
	10	3/8	0	0	2	1.14	0.045	22	180	6	20
	13	1/2	0	0	3	1.14	0.045	22	180	4	13
	13	1/2	2.4	3/32	4	1.14	0.045	22	180	3	10
	25	1	2.4	3/32	9	1.14	0.045	22	180	1	3
	25	1	1.6	1/16	6	1.14	0.045	22	180	2	7
	51	2	1.6	1/16	16	1.14	0.045	22	180	0.6	2
Vertical position fillet weld (semiautomatic)											
	3	1/8	0	0	1	1.14	0.045	22	180	55	180
	13	1/2	0	0	2	1.14	0.045	22	180	4	12

Table 5.3 (continued)—Typical gas shielded flux cored arc welding procedures for carbon steel (E70T-1 electrodes)

Joint design	Thickness, T		Root opening, R		Total passes	Electrode diameter		Welding power, dcrp. (ep)		Production rate	
	mm	in.	mm	in.		mm	in.	V	A	m/h	ft/h
	Horizontal position fillet weld (semiautomatic)										
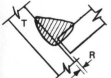	5	3/16	0	0	1	1.98	5/64	28	350	55	180
	13	1/2	0	0	3	2.78	7/64	30	450	9	30
	Flat position groove welds (automatic)										
	6	1/4	3	1/8	1	2.38	3/32	28	475	37	125
	13	1/2	3	1/8	1	2.78	7/64	28	550	24	80
	19	3/4	0	0	3	3.18	1/8	30	750	14	45
	38	1-1/2	3	1/8	9	3.18	1/8	30	750	3	11
	19	3/4	0	0	4	2.38	3/32	31	450	13	44
	51	2	0	0	10	3.18	1/8	30	750	3	9
	Flat position fillet weld (automatic)										
	13	1/2	0	0	1	3.18	1/8	29	700	16	54
	25	1	0	0	5	3.18	1/8	30	750	4	14

In moving air, curtains must be provided around the welding area to avoid loss of gas shielding. When the gas is obtained from a liquid supply system, precautions must be taken to prevent freezing of the pressure regulator. A dual cylinder setup with a heater attachment will prevent freezing at the regulator. Freezing can also be prevented by using a manifold system. Where practical, bulk gas systems are preferred.

FOR SELF-SHIELDED MILD STEEL ELECTRODES

The basic joint types that are suitable for the gas shielded FCAW and SMAW processes are also suitable for self-shielded FCAW. Although the general shapes of weld grooves are similar to those used for shielded metal arc welding, specific groove dimensions may differ. The differences are largely due to the higher deposition rates and shallower penetration.

Electrode extension introduces another welding procedure variable that may influence joint design. When using a long electrode extension for making flat position groove welds without backing, adequate root penetration must be considered. Welding of the first pass in the groove may best be accomplished by the use of the SMAW process for better control of fusion and penetration. Similarly, in grooves with backing, the root opening must be sufficient to permit complete fusion by globular metal transfer.

Typical joint designs and procedures for welding carbon steel by the use of self-shielded electrodes are given in Table 5.4. The welding current and voltage may vary for a specific electrode size and classification from different manufacturers. Depending on the joint spacing and technique used for root pass welding, back gouging and welding may be required when no backing strip is used.

When welding in the flat position, techniques similar to those used with low hydrogen covered electrodes are followed. When making vertical welds on plates 19 mm (3/4 in.) and thicker, the root pass may be deposited vertically

down for joints without backing and vertical-up for joints with backing. Subsequent passes are deposited vertical-up, using a technique similar to that used for low hydrogen covered electrodes.

FOR SELF-SHIELDED STAINLESS STEEL ELECTRODES

Typical joint designs and welding procedures for self-shielded stainless steel electrodes are given in Table 5.5. These electrodes, in their present state of development, are limited to butt welds in the flat position, fillet welds in the flat or horizontal position, and surfacing in the flat or horizontal position. If welding of stainless steels in other positions is required, the electrode manufacturers should be consulted for their recommendations for the application. In general, joint geometry for butt welds should be approximately the same as that used for shielded metal arc welding. When applying a surfacing weld on carbon or low alloy steels, special precautions must be taken to control dilution during the initial surfacing passes.

EDGE PREPARATION AND FIT-UP TOLERANCES

Edge preparation for welding with flux cored electrodes can be done by oxyfuel gas cutting, plasma arc cutting, or machining, depending on the type of base metal and joint design required. Fit-up tolerances for welding will depend on the

(1) Level of quality required for the joint
(2) Method of welding (gas shielded or self-shielded; automatic or semiautomatic)
(3) Thickness of the base metal welded
(4) Type and size of electrode
(5) Position of welding

In general, mechanized and automatic flux cored arc welding require adherence to close tolerances in both joint preparation and fit-up. Welds made with semiautomatic equipment can accept somewhat wider tolerances.

Table 5.4—Typical self-shielded flux cored arc welding procedures for carbon steel

Joint design	Plate thickness, T		Root opening		Total passes	Electrode diameter		Welding power, dc		Production rate[b]		Electrode extension	
	mm	in.	mm	in.		mm	in.	A	V (p)[a]	m/h	ft/h	mm	in.
Flat position groove welds (semiautomatic)													
	3.4	0.14	4	5/32	1	2.38	3/32[c]	300	29 (+)	34	110	70	2-3/4
	10	3/8	10	3/8	2	3.18	1/8[c]	500	33 (+)	12	39	70	2-3/4
	13	1/2	10	3/8	3	3.18	1/8[c]	500	32 (+)	8	25	70	2-3/4
	25	1	10	3/8	6	3.18	1/8[c]	550	36 (+)	4	14	95	3-3/4
Weld by SMAW	13	1/2	2	3/32	2	2.38	3/32[c]	350	29 (+)	9	30	70	2-3/4
	76	3	2	3/32	26	3.18	1/8[c]	550	36 (+)	0.8	2.5	95	3-3/4
	10	3/8	10	3/8	2	3.18	1/8[c]	500	32 (+)	6	21	70	2-3/4
	32	1-1/4	10	3/8	7	3.18	1/8[c]	550	36 (+)	3	9	95	3-3/4
Vertical position groove welds (semiautomatic)													
Vertical down	2.7	0.105	3	1/8	1	1.98	5/64[d]	250	20 (−)	38	125	25	1
	6	1/4	5	7/32	3	1.98	5/64[d]	350	25 (−)	15	50	25	1

See page 184 for notes.

Table 5.4 (continued)—Typical self-shielded flux cored arc welding procedures for carbon steel

Joint design	Plate thickness, T mm	in.	Root opening mm	in.	Total passes	Electrode diameter mm	in.	Welding power, dc A	V (p)[a]	Production rate[b] m/h	ft/h	Electrode extension mm	in.
Vertical position groove welds (semiautomatic)
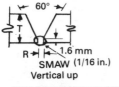	8	5/16	2	3/32	1	1.59	1/16[d]	150	18 (−)	5	15	25	1
	25	1	2	3/32	1	1.59	1/16[d]	195	21 (−)	0.6	2	25	1
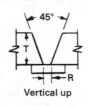	10	3/8	5	3/16	2	1.59	1/16[d]	170	19 (−)	2	7	25	1
	25	1	5	3/16	6	1.98	5/64[d]	180	19 (−)	0.5	1.8	25	1
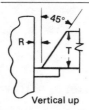	10	3/8	6	1/4	1	1.59	1/16[d]	170	19 (−)	2	6	25	1
	38	1-1/2	6	1/4	4	1.98	5/64[d]	180	19 (−)	0.3	0.9	25	1
Vertical position fillet weld (semiautomatic)
	6	1/4	0	0	1	1.59	1/16[d]	130	18	6	20	25	1
	16	5/8	0	0	1	1.59	1/16[d]	185	21	3	9	25	1
	38	1-1/2	0	0	4	1.59	1/16[d]	190	21	6	21	25	1
Horizontal position groove welds (semiautomatic)
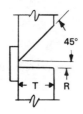	8	5/16	5	3/16	3	2.38	3/32[c]	300	28 (+)	6	21	70	2-3/4
	32	1-1/4	5	3/16	16	3.18	1/8[c]	400	29 (+)	1	4	70	2-3/4

Joint design labels: SMAW (1/16 in.) Vertical up (60°, R, 1.6 mm); Vertical up (45°, R); Vertical up (45°, R, T); Vertical up; 45° (T, R).

See page 184 for notes.

Table 5.4 (continued)—Typical self-shielded flux cored arc welding procedures for carbon steel

Joint design	Plate thickness, T mm	in.	Root opening mm	in.	Total passes	Electrode diameter mm	in.	Welding power, dc A	V (p)[a]	Production rate[b] m/h	ft/h	Electrode extension mm	in.
Horizontal position groove welds (semiautomatic)													
	19	3/4	2	3/32	6	2.38	3/32[c]	300	28 (+)	3	10	70	2-3/4
	38	1-1/2	2	3/32	12	3.18	1/8[c]	400	29 (+)	2	5	70	2-3/4
Overhead position groove and fillet welds (semiautomatic)													
	2.7	0.105	0	0	1	1.59	1/16[d]	150	18 (−)	30	100	25	1
	19	3/4	0	0	6	1.59	1/16[d]	180	19 (−)	2	6	25	1
Weld by SMAW													
	8	5/16	1.6	1/16	2	1.59	1/16[d]	150	18 (−)	5	16	25	1
	25	1	1.6	1/16	8	1.59	1/16[d]	170	19 (−)	0.6	2	25	1
Horizontal position fillet welds (semiautomatic)													
	2.7	0.105	0	0	1	1.98	5/64[d]	235	20 (−)	41	135	25	1
	5	3/16	0	0	1	2.38	3/32[d]	335	21 (−)	35	115	25	1
	6	1/4	0	0	1	2.38	3/32[c]	325	29 (+)	30	100	25	1
	25	1	0	0	5	3.18	1/8[c]	450	29 (+)	5	17	70	2-3/4

See page 184 for notes.

Table 5.4 (continued)—Typical self-shielded flux cored arc welding procedures for carbon steel

Joint design	Plate thickness, T mm	in.	Root opening mm	in.	Total passes	Electrode diameter mm	in.	Welding power, dc A	V (p)[a]	Production rate[b] m/h	ft/h	Electrode extension mm	in.
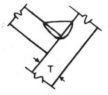 **Flat position fillet welds (semiautomatic)**													
	8	5/16	0	0	1	2.38	3/32[c]	350	30 (+)	30	100	70	2-3/4
	25	1	0	0	4	3.18	1/8[c]	580	27 (+)	9	30	95	3-3/4
Flat position groove welds (automatic)													
	1.2	0.05	0	0	1	2.38	3/32[e]	425	26 (+)	396	1300	25	1
	5	3/16	0	0	1	3.97	5/32[e]	950	27 (+)	175	575	32	1-1/4
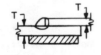 **Horizontal position fillet welds (automatic)**													
	1.2	0.05	0	0	1	2.38	3/32[e]	475	26 (+)	396	1300	25	1
	5	3/16	0	0	1	3.97	5/32[e]	900	26 (+)	152	500	32	1-1/4
	1.5	0.06	0	0	1	2.38	3/32[e]	425	26 (+)	305	1000	25	1
	5	3/16	0	0	1	3.97	5/32[e]	875	27 (+)	160	525	32	1-1/4

a. (p)- Polarity: + electrode positive; − electrode negative
b. Production rate at 100 percent operator factor
c. E70T-4 electrode
d. E60T-7 electrode
e. E70T-G electrode

Table 5.5 — Typical self-shielded flux cored arc welding procedures for stainless steels using stainless steel electrodes

Joint design	Weld size, T mm	Weld size, T in.	Root opening, R mm	Root opening, R in.	Total passes	Electrode diameter mm	Electrode diameter in.	Welding power dcrp (ep) A	Welding power dcrp (ep) V	Production rate m/h	Production rate ft/h	Electrode extension mm	Electrode extension in.
						Flat position groove welds							
	6	1/4	3	1/8	1	2.38	3/32	300	27.5	18	60	25	1
	10	3/8	3	1/8	2	2.38	3/32	300	27.5	10	33	25	1
	13	1/2	5	3/16	2	2.38	3/32	300	27.5	8	27	25	1
	19	3/4	5	3/16	4	2.38	3/32	300	27.5	2	7	25	1
	22	7/8	10	3/8	6	2.38	3/32	300	27.5	2	7	25	1
	32	1-1/4	10	3/8	8	2.38	3/32	300	27.5	2	5	25-32	1 to 1-1/4
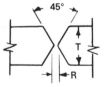	13	1/2	3	1/8	2	2.38	3/32	300	27.5	9	30	25	1
	76	3	3	1/8	25	2.38	3/32	300	27.5	0.5	1.5	25-32	1 to 1-1/4
	10	3/8	10	3/8	3	2.38	3/32	300	27.5	6	20	25	1
	32	1-1/4	10	3/8	8	2.38	3/32	300	27.5	2	5	25-32	1 to 1-1/4

Table 5.5 (continued)—Typical self-shielded flux cored arc welding procedures for stainless steels using stainless steel electrodes

Joint design	Weld size, T mm	in.	Root opening, R mm	in.	Total passes	Electrode diameter mm	in.	Welding power dcrp (ep) A	V	Production rate m/h	ft/h	Electrode extension mm	in.
						Flat position fillet weld							
	10	3/8	0	0	1	2.38	3/32	300	27.5	15	50	25	1
	19	3/4	0	0	3	2.38	3/32	300	27.5	4	12	25	1
						Horizontal position fillet weld							
	3	1/8	0	0	1	1.59	1/16	185	24	27	90	13	1/2
	10	3/8	0	0	1	2.38	3/32	300	27	15	50	25	1

WELD QUALITY

The quality of welds that can be produced with the FCAW process depends on the type of electrode used, the method (gas shielded or self-shielded), condition of the base metal, weld joint design, and welding conditions. Particular attention must be given to each of these factors to produce sound welds with the best mechanical properties.

The impact properties of mild steel weld metal may be influenced by the FCAW method used. Self-shielded electrodes are generally highly deoxidized types that may produce weld metal with relatively low notch toughness. However, some self-shielded electrodes have good impact properties. Gas shielded electrodes are available that meet the Charpy V-notch strength requirements of the AWS electrode specifications, but some of them will not. Therefore, notch toughness requirements should be considered before selecting the method and the specific electrode for an application.

A few mild steel electrodes are designed to tolerate some amount of mill scale and rust on the base metals. Some deterioration of weld quality should be expected when welding dirty materials. When these electrodes are used for multipass welding, cracking may occur in the weld metal.

In general, sound welds can be produced in mild and low alloy steels by the FCAW process that will meet the requirements of several construction codes. Particular attention must be given to all the factors affecting weld quality to meet the code requirements.

When less stringent requirements are needed, advantages of higher welding speeds and currents can be utilized. Minor defects that are not objectionable from the design and service standpoints may be present in the welds.

Flux cored arc welds can be produced in stainless steels with qualities equivalent to those of gas metal arc welds. Welding position and arc length are significant factors when self-shielded electrodes are used. Out-of-position welding procedures should be carefully evaluated with respect to weld quality. Excessive arc length generally causes nitrogen pickup in the weld metal. Because nitrogen is an austenite stabilizer, absorption of excess nitrogen into the weld may increase the susceptibility to microfissuring.

Low alloy steels are generally welded with the gas shielded method using *1* or *5* electrode core formulations when good low temperature

toughness is required. The combination of gas shielding and proper flux formulation generally produces sound welds with good mechanical properties and notch toughness. Self-shielded electrodes containing nickel for good strength and impact properties as well as aluminum as a deoxidizer are also available. In general, the composition of the electrode should be similar to that of the base metal.

SUPPLEMENTARY READING LIST

Bishel, R.A. Flux-cored electrode for cast iron welding. *Welding Journal* 52 (6), June 1973, pp. 372-381.

Cary, Howard. Match wire to the job. *Welding Engineer* 55 (4), Apr. 1970, pp. 44-46.

Hinkel, J.E. Long stickout welding—a practical way to increase deposition rates. *Welding Journal* 47 (3), Mar. 1968, pp. 869-874.

Hoitomt, M., and Lee, R.K. All-position production welding with flux-cored gas-shielded electrodes. *Welding Journal* 51 (11), Nov. 1972, pp. 765-768.

Metals Handbook, Vol. 6, 8th ed. Metals Park, Ohio: American Society for Metals, 1971, pp. 24-45.

Method for Sampling Airborne Particulates Generated by Welding and Allied Processes, AWS F1.1. Miami: American Welding Society, 1976.

The Procedure Handbook of Arc Welding, 12th

ed. Cleveland, Ohio: The Lincoln Electric Company, 1973.

Safety in Welding and Cutting, ANSI Z49.1 Miami: American Welding Society, 1973.

Specification for Mild Steel Electrodes for Flux-Cored Arc Welding, AWS A5.20. Miami: American Welding Society, 1969.

Specification for Flux Cored Corrosion-Resisting Chromium and Chromium-Nickel Steel Electrodes, AWS A5.22. Miami: American Welding Society, 1974.

Wick, W.C., and Lee, R.K. Welding low-alloy steel castings with the flux-cored process. *Welding Journal* 47 (5), May 1968, pp. 394-397.

Zvanut, A.J., and Farmer, H.N., Jr. Self-shielded stainless steel flux cored electrodes. *Welding Journal* 51 (11), Nov. 1972, pp. 775-780.

6

Submerged Arc Welding

Prepared by a committee consisting of

N. S. WAMACK, *Chairman*
  Combustion Engineering, Inc.
R. E. CLOTFELTER
  Chicago Bridge & Iron Co.
L. COLAROSSI
  Pittsburgh-Des Moines Steel Co.
L. W. MOTT
  Hobart Brothers Co.

R. S. SABO
  The Lincoln Electric Co.
G. D. UTTRACHI
  *Linde Division, Union
  Carbide Corp.*
S. W. WISMER
  Westinghouse Electric Corp.

6
Submerged Arc Welding

FUNDAMENTALS OF THE PROCESS

DESCRIPTION

Submerged arc welding (SAW) produces coalescence of metals by heating them in an arc (or arcs) maintained between a bare metal electrode (or electrodes) and the work. The arc is shielded by a blanket of granular, fusible material placed over the welding area. Filler metal is obtained from the electrode and sometimes from a supplementary welding rod or other metallic addition.

The feature that distinguishes submerged arc welding from other arc welding processes is the granular material that covers the welding area. By common usage this material is termed a flux, although it performs several important functions in addition to those strictly associated with a fluxing agent. Flux plays a central role in achieving the high deposition rates and weld quality that characterize the SAW process in joining and surfacing applications.

PRINCIPLES OF OPERATION

Basically, in submerged arc welding, the end of a continuous bare wire electrode is inserted into a mound of flux that covers the area or joint to be welded. An arc is initiated, causing the base metal, electrode, and flux in the immediate vicinity to melt. The electrode is advanced in the direction of welding and mechanically fed into the arc while flux is steadily added. The melted base metal and filler metal flow together to form a molten pool in the joint. At the same time, the melted flux floats to the surface to form a protective slag cover. Unmelted flux is reclaimed for reuse.

The production of a submerged arc weld is illustrated in Fig. 6.1. Here it can be seen that the unmelted, nonconductive flux acts as an air shield, a heat insulator, and an effective radiation shield. Not so obvious are the reactions that occur between the conductive molten flux (slag) and the weld metal, and the effect of dissociated (ionized) flux constituents on the arc plasma.

Reactions between the flux and molten weld metal influence the cleanliness, properties, and sometimes the composition of the weld metal. Flux composition influences the stability, temperature, and heat distribution of the arc plasma. The molten slag cover influences weld bead appearance, promotes slow cooling, and continues to provide protection from atmospheric contamination to the weld metal as the metal cools.

The various properties and functions of the flux make it possible to employ the high welding currents for which the process is noted. The significance of high currents is shown in Fig. 6.2, where corresponding deposition rates are given for some variations of the submerged arc welding process. Comparable data are given for shielded metal arc welding.

Butt joints in thicknesses up to approximately 19 mm (3/4 in.) can be made by sub-

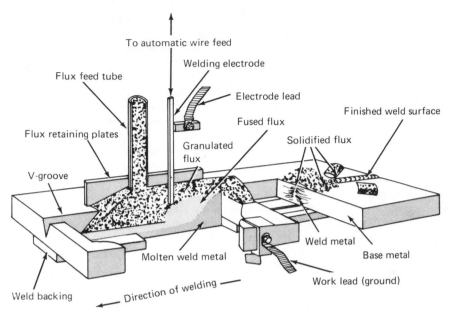

Flux feed tube

To automatic wire feed

Welding electrode

Electrode lead

Flux retaining plates

Fused flux

Finished weld surface

Granulated flux

Solidified flux

V-groove

Weld metal

Molten weld metal

Base metal

Weld backing

Direction of welding

Work lead (ground)

Fig. 6.1—The submerged arc welding process

merged arc welding in a single pass with two electrodes. Joint designs may vary from square grooves to relatively small V-grooves, smaller than those used in shielded metal arc welding. Some form of weld backing is usually required. In these types of joints, the base metal composition can greatly influence the composition and properties of the weld metal. The solidification pattern of large single pass welds shows a columnar grain structure. Even so, good mechanical properties are obtained with single pass welds.

Using multiple pass welding techniques, the filler metal and flux have a greater influence on weld metal composition and properties than in single pass welding. Individual weld beads show finer grain structures, and the annealing effect of successive passes helps to reduce shrinkage stresses. Complete interpass slag removal is important. Some, but not all, slags are self-peeling. Undercutting, which entraps slag, should be avoided.

Welds made by this process have good ductility, uniformity, and density. Good impact strengths are obtainable, but specific procedures and techniques must be evaluated in combination with the electrode and flux to obtain high impact values. Good corrosion resistance, depending on the requirements, can be obtained by proper selection of the electrode and flux. With correct procedure, mechanical properties at least equal to those of the base metal are consistently obtained.

GENERAL METHODS

There are three general methods by which the process can be applied: semiautomatic welding, automatic welding, and machine welding. Each of these methods requires that the work be positioned so that the flux and molten weld pool will remain in place until they have solidified. This condition can be satisfied in either the flat position or the horizontal fillet welding position. Many types of fixtures, such as carriages, turning rolls, tilting turntables, and manipulators are available for positioning of workpieces.

Semiautomatic welding utilizes a hand-held welding gun, through the nozzle of which both flux and electrode are delivered. The electrode is driven by a wire feeder. Flux may be supplied by gravity from a small hopper mounted on the gun, or it may be force fed through a hose. This method of welding offers the flexibility of manual guidance using small electrodes and moderate travel speeds.

Automatic welding is done with equipment that performs the welding operation without the need of a welding operator to continually monitor and adjust the controls. The cost of completely self-regulating equipment is justified on the basis of the high production rates obtainable with it. Applications may be found in the automotive industry or in the manufacture of semi-finished steel products, such as various types of piping produced by pipe mills.

Machine welding employs equipment that performs the complete welding operation but requires constant monitoring by a welding operator to start and stop welding and to adjust the controls, when needed. Of the three methods, machine welding is the most widely used. The major techniques and process variations discussed in this chapter are based on the use of machine and automatic welding.

In machine welding, arc initiation and prompt transition to a steady-state operation are achieved with the aid of equipment designed for that purpose. Steady-state operation requires that the electrode be melted at the same rate that it is being fed into the arc. This equilibrium is achieved through arc length control. Since electrode melting is current related and arc

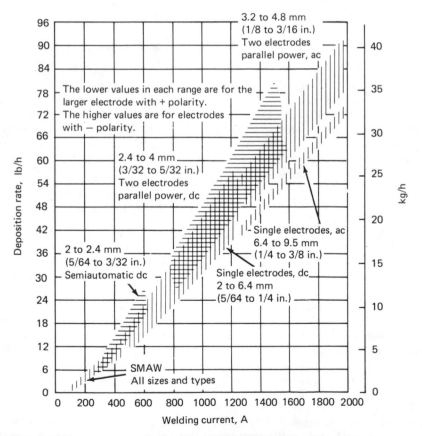

Fig. 6.2—Approximate deposition rates of some submerged arc process variations compared to shielded metal arc welding

length is voltage related, the system must be capable of continuously correcting a voltage change with either a current or a feed rate change, or both.

During machine welding, assuming the work and equipment are properly set up, the operator may need to make occasional adjustments. These may result from changes in seam tracking, workpiece or joint geometry, flux deposition, amperage, or other conditions that may vary.

The following conditions should be evaluated when semiautomatic, machine, and automatic submerged arc welding are under consideration: (1) production volume, (2) weld length, (3) joint thickness, (4) weld accessibility, (5) tooling and fixturing needs, (6) weld quality, and (7) weld appearance.

PROCESS VARIATIONS

There are several variations of machine and automatic submerged arc welding that will permit higher deposition rates with good control of the weld bead size and penetration. Various multiple electrode systems that use one or more power sources with different types of circuit connections are available. For example, two electrodes can be positioned in tandem so that their arcs will produce a single molten weld pool. In this configuration, the arcs may be operated from the same power source by connecting them either in series or in parallel, or they may be operated from separate power sources. In the latter case, one system uses a dc source for the lead arc and an ac source for the trail arc. Another system uses two ac power sources with an adjustable phase-shift control to adjust the interaction between the two ac arcs.

Multiple arcs are not always arranged with their electrodes precisely in tandem. For surfacing applications, the electrodes are positioned transverse or at some angle to the direction of travel. Triangular or rectangular patterns may also be used. Likewise, tandem arcs do not always feed a single weld pool. For single pass welding of large fillet welds in the horizontal position, two electrodes are spaced approximately 75 to 125 mm (3 to 5 in.) apart to produce separate weld pools, since a single large pool cannot be contained.

Various multiple arc configurations may be used to control the weld profile and increase the deposition rates over single arc operation. Weld deposits may range from wide beads with shallow penetration for surfacing to narrow beads with deep penetration for thick joints. Part of this versatility is derived from the use of ac arcs.

The principles which favor the use of ac to minimize arc blow in single arc welding are often applied in multiple arc welding to create a favorable arc deflection. The current flowing in adjacent electrodes sets up interacting magnetic fields that can either reinforce or diminish each other. In the space between the arcs, these magnetic fields are used to produce forces that will deflect the arcs (and thus distribute the heat) in directions beneficial to the intended welding application.

Various types of power sources and related equipment are designed and manufactured especially for multiple arc welding. These relatively sophisticated machines are intended for high production on long runs of repetitive type applications.

To a lesser extent, special power sources are also designed for single electrode submerged arc welding. Single electrode power sources are basically the same as those used for other arc welding processes, except that higher power ratings are generally required.

EQUIPMENT

The basic elements of SAW equipment consist of (1) a wire feeder to drive the electrode through the contact tube of a welding gun to the work, (2) a welding power source to energize the electrode at the contact tube, (3) a means of storing and distributing the flux as required, and (4) a means of traversing the weld joint. Equipment is available for vacuum removal, cleaning, and storing flux from the joint for reuse.

POWER SOURCES

Although coordination among all elements of the electrical and mechanical subsystems is necessary for a successful operation, the power source determines most limits within which the SAW process can operate.

Welding power may be provided by a constant current dc rectifier or generator, a constant voltage dc rectifier or generator, or a constant current ac transformer. The terms "constant voltage" and "constant potential" are used interchangeably with respect to this type of power source. The general characteristics of power sources and the methods of connecting them for multiple arc applications are given in Chapter 1.

The power sources should provide the high amperages, at 100% duty cycle, required by most submerged arc welding installations. Although most welding is done in a range from 400 to 1500 A, currents as high as 4000 A at 55 V, or as low as 150 A at 18 V may be used.

Accordingly, it is economical to relate the size of the power source to the range of intended applications, thus ensuring the availability of adequate welding current. Criteria for the selection of the type of current employed will be discussed later. This section describes the characteristics of each type of power source with reference to the submerged arc welding application.

Constant Current (DC) Types

Constant current power sources of the rectifier or motor-generator type are widely. used for dc welding. If the current required exceeds the output of a single machine, two or more rectifiers or generators of the same type can be connected in parallel.

To achieve a steady-state operation with automatic wire feeders and constant current power sources, it is necessary to maintain a reasonably constant arc voltage (arc length) to obtain an almost constant welding current. This is done by monitoring the arc voltage to control the electrode feed speed. The control circuit simply increases the speed of the wire feed motor when the arc voltage (arc length) increases above a set value, and reduces the speed when the voltage decreases. Because of the inertia of this system, there are always some minor voltage and current fluctuations.

Constant Voltage (DC) Types

Another approach to wire feeding and current control for submerged arc welding combines a constant speed wire feeder and a constant voltage power source. The output of the constant voltage type differs from that of the constant current type in that the characteristic volt-ampere curve is nearly flat. Short circuit current is therefore very high. Some constant voltage units of either the rectifier or motor-generator type are built so that the slope of the characteristic curve may be varied to suit the application. This sloping volt-ampere characteristic (2 to 3 V per 100 A) limits the high amperage fluctuations that can occur. Where large amperage excursions can occur, the slope should be increased to limit them.

For certain applications, constant voltage units have several advantages over the conventional constant current types. The constant voltage system uses a constant speed wire feed drive that operates independently from the welding circuit. Arc length control is faster and more precise. Because of the nearly flat volt-ampere characteristic, a small change in arc voltage (arc length) causes a relatively large amperage change which immediately increases or decreases the melting rate of the electrode. The response is almost instantaneous.

A high short circuit current is provided by constant voltage machines, and this makes arc initiation easier than with constant current types. Where it is desirable to fill the crater at the end of the weld, the wire feed speed can be gradually reduced. This, in turn, reduces the welding current as the crater fills.

The control circuits used with constant potential power sources are simple because the arc is inherently self-regulating and can easily be started and stopped.

Constant voltage power is the most advantageous for welding steel sheet of 3.6 mm (0.14 in.) thickness or less. The uniformity of the voltage permits higher welding speeds than those that can be obtained with constant current power. With power supplies having only modestly sloping volt-ampere characteristics (3 V per 100 A), the electrode size should be selected so that the current density is in excess of A/mm² (40 000 A/in.²) for optimum deposition rates.

The advantages of constant voltage are not as evident when thicker materials are being welded. The power source for these welds should be adjusted to provide a more drooping characteristic.

Constant potential power sources are available with a large amount of inductance added to the dc welding circuit. The large inductance reduces wide current excursions and minimizes arc outages which are normally encountered at low current densities of approximately 25 to 100 A/mm² (15 000 to 60 000 A/in.²). This system enables constant voltage units to be employed for most applications where dc submerged arc welding is useful. Constant voltage power sources are available in capacities up to 1200 A continuous rating.

Alternating Current Types

Alternating current is generally supplied by heavy duty welding transformers of 1000, 1500, or 2000 A capacity, equipped with motor operated reactors and primary contactors. These transformers, or those of smaller capacity with the same open circuit voltage, may be connected in parallel to obtain output power greater than that of a single unit. They must be connected so their output voltages and currents will be in phase. Transformers for multiple wire, multiple power welding should be supplied with taps so that the necessary circuit connections can be made.

For submerged arc welding with ac power, the open-circuit voltage should be a minimum of 80 volts to ensure reignition of the arc during current reversals. Since variations in the primary power supply may affect the quality of the weld, a constant input primary voltage should be maintained.

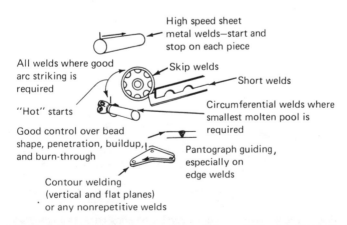

High speed sheet metal welds—start and stop on each piece

All welds where good arc striking is required

Skip welds

Short welds

"Hot" starts

Circumferential welds where smallest molten pool is required

Good control over bead shape, penetration, buildup, and burn-through

Pantograph guiding, especially on edge welds

Contour welding (vertical and flat planes) or any nonrepetitive welds

Fig. 6.3—Typical applications for dc submerged arc welding (single electrode)

POWER SOURCE SELECTION

Welding with dc power provides better control of the weld bead shape, depth of penetration, and welding speed. In addition, arc starting is much easier than with ac power. Direct current is preferred where fast, accurate arc striking is essential; where close arc control is needed; and where difficult contours are to be followed at high travel speeds.

Direct current, reverse polarity (electrode positive) provides the best control of the bead shape because the arc is very stable and because the molten weld pool is reasonably small. **Direct current also produces better weld penetration.**

Highest filler metal deposition rates are obtained with dc straight polarity (electrode negative) but penetration is low. Alternating current produces penetration somewhere between the two modes of dc. Some applications where dc is best for SAW are shown in Fig. 6.3.

Alternating current minimizes arc blow. This assumes increasing importance at welding currents above 900 A. Alternating current is usually preferred for the trail arc or arcs for multiple wire, multiple power welding. Two arcs approximately 102 mm (4 in.) or less apart are deflected by each other's magnetic field. Two dc arcs of like polarity flare together, whereas arcs of unlike polarity flare apart. If one ac and one dc arc are employed, their deflections can be controlled. This combination is the easiest to use when the application varies from day to day. When two ac arcs are used, the deflections can be directed so that both the arc blow and lagging of the arc at high welding speeds can be counteracted.

Alternating current may be used where arc blow cannot otherwise be controlled, or for short welds where grounding of the work is a problem. Typical applications for ac power are shown in Fig. 6.4.

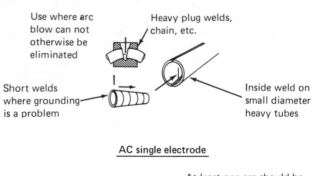

Use where arc blow can not otherwise be eliminated

Heavy plug welds, chain, etc.

Short welds where grounding is a problem

Inside weld on small diameter heavy tubes

AC single electrode

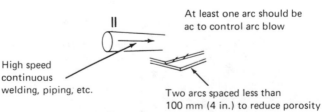

At least one arc should be ac to control arc blow

High speed continuous welding, piping, etc.

Two arcs spaced less than 100 mm (4 in.) to reduce porosity

Multiple electrode

Fig. 6.4—Typical applications for ac submerged arc welding (single and multiple electrodes)

WIRE FEEDING SYSTEMS

Equipment for feeding the electrode wire in SAW is similar to that used for gas metal arc and flux cored arc welding. The electrode is stored in a coil or drum which may be mounted on the welding head (mechanized welding) or remotely for semiautomatic welding. The electrode is driven through the welding gun to the arc by electric motor powered drive rolls.

There are two methods used for controlling electrode feed speed. With a constant voltage power source, the drive motor speed is set to furnish the electrode at the required constant speed. With a constant current power source, the electrode feed speed (drive motor speed) varies as necessary to maintain constant arc length. The drive motor speed control senses arc voltage and automatically changes the motor speed and, thus, electrode speed to compensate for arc length variations. If the arc length decreases, the electrode speed decreases to allow the electrode to melt off to the set length (voltage). Then the speed increases again. If the arc length increases, the electrode speed increases momentarily to decrease the arc length.

FLUX FEED

The welding flux is deposited just ahead of or concentric with the arc from a flux delivery tube as welding progresses along the joint. The granular flux is stored in a hopper mounted on the welding head and gravity fed through the tube. With some semiautomatic welding guns, the flux is force fed from a remote hopper or tank.

SAFETY PRECAUTIONS

Since the end of the electrode and the welding zone are completely covered at all times during the actual welding operation, the weld is made without the sparks, spatter, smoke, or flash commonly observed in other arc welding processes. No protective shields or helmets are necessary, but safety glasses should be worn as routine eye protection. The glasses may be tinted for protection against flash when the arc is inadvertently initiated without flux cover.

Since submerged arc welding can produce some fumes and gases hazardous to health, it is common practice to provide adequate ventilation, especially where welding is done in confined areas. For details on safety, refer to ANSI Standard Z49.1, Safety in Welding and Cutting, which is available from the American Welding Society.

Welding operators and others who may be associated with SAW equipment operation should be thoroughly familiar with the manufacturer's operating instructions. Particular attention should be paid to the precautionary information contained in the operating manual.

MATERIALS

BASE METALS

Submerged arc welding is used for joining many ferrous and nonferrous metals and alloys. It is also used to apply cladding to base metals to provide corrosion resistance or other properties. The general classes of base metals welded are

(1) Carbon steels up to 0.29% carbon

(2) Heat treated carbon steels (normalized or quenched and tempered)

(3) Low alloy steels, quenched and tempered, up to 690 MPa (100 000 psi) yield strength

(4) Chromium-molybdenum steels (1/2% to 9% Cr and 1/2% to 1% Mo)

(5) Austenitic chromium-nickel stainless steels

(6) Nickel and nickel alloys (solid solution types)

The alloy compositions in the above classes that can be submerged arc welded depend on the availability of suitable electrodes and fluxes. Electrodes and many flux-electrode combinations classified by the American Welding Society for carbon steels and low alloy steels are generally available. Information on electrodes and fluxes for welding other base metals is available from the manufacturers. Industrial applications that employ submerged arc welding for joining various base metals are discussed in applicable chapters of Section 5 of this *Handbook,* 6th edition.

ELECTRODES

Submerged arc electrodes are available for producing such types of weld metal as carbon steel, low alloy steel, high carbon steel, special alloy steel, stainless steel, nonferrous alloys, and special alloys for surfacing applications. These electrodes are supplied as bare solid wire and as composite electrodes (similar to flux cored arc welding electrodes).

Electrodes are normally packaged in the form of coils or drums ranging in weight from 11 to 454 kg (25 to 1000 lb). Large electrode packages are economical for increasing operating efficiency and eliminating end-of-coil waste.

Steel electrodes are usually copper coated, except for welding corrosion resisting materials, or for certain nuclear applications. The copper coating helps to obtain good shelf life and decreases contact tube wear. Electrodes are generally packaged to ensure long shelf life when stored indoors under normal conditions.

Submerged arc welding electrodes normally vary in size from 1.6 to 6.4 mm (1/16 to 1/4 in.) in diameter. General guidelines for amperage range selection are presented in Table 6.1. The wide amperage ranges are typical of submerged arc welding.

Carbon Steel Electrodes and Fluxes

The AWS Specification A5.17 prescribes the requirements for electrodes and fluxes for submerged arc welding of carbon steels. The electrodes are classified on the basis of their chemical composition, as manufactured. The fluxes are classified on the basis of weld metal properties obtained when used with specific electrodes. The chemical composition requirements for electrodes should be obtained from the latest revision of AWS Specification A5.17. Reference should be made to this specification and also to Chapter 94, "Filler Metals," in Section 5 of the *Welding Handbook,* 6th edition, for additional information.[1]

Low Alloy Steel Electrodes and Fluxes

Alloy steel weld metal is produced by the use of solid alloy steel electrodes, fluxes containing the alloying elements, or composite electrodes where the core contains the alloying elements. Alloy steel and composite electrodes are normally used with a neutral flux. Alloy-bearing fluxes are generally used with a carbon steel electrode. Electrode-flux combinations are available for many commonly used carbon steel alloys. They are used for the welding of low alloy, high strength steels, and for surfacing.

---

1. Chapter 94 is scheduled for revision in Vol. 5, 7th ed., of the *Welding Handbook.*

Table 6.1—Submerged arc welding electrode amperage ranges		
Electrode diameter		Current range
mm	in.	A
1.6	1/16	150-400
2.0	5/64	200-600
2.4	3/32	250-700
3.2	1/8	300-900
4.0	5/32	400-1000
4.8	3/16	500-1100
5.6	7/32	600-1200
6.4	1/4	700-1600

AWS Specification A5.23 prescribes the requirements for solid and composite electrodes and fluxes for welding low alloy steels. The fluxes are classified on the basis of weld metal properties obtained when used with specific electrodes. The chemical composition requirements for the electrodes or deposits, or both, and other information, should be obtained from the latest revision of the specification.

Stainless Steel Electrodes

AWS Specification A5.9 covers filler metals for welding corrosion or heat resisting chromium and chromium-nickel steels when used with submerged arc and other welding processes. This specification includes steels in which chromium exceeds 4 percent and nickel does not exceed 50 percent of the composition. Bare electrodes are classified on the basis of their chemical composition, as manufactured, and composite electrodes on the basis of chemical analysis of a fused sample. The American Iron and Steel Institute numbering system is used for these alloys.

Nickel and Nickel Alloy Electrodes

Nickel and nickel alloy electrodes in wire form are available for submerged arc welding. They are used for welding wrought and cast forms of commercially pure nickel and various nickel alloys. AWS Specification A5.14 covers nickel and nickel alloy filler metals for submerged arc welding and for other processes. As in the other filler metal specifications, the electrodes are classified on the basis of their chemical compositions, as manufactured. For specific information on the properties and use of these electrodes, reference should be made to the latest revision of AWS Specification A5.14.

FLUXES

The flux shields the molten weld pool from the atmosphere by covering the metal with molten slag (flux); it cleans the molten weld pool; it modifies the chemical composition of the weld metal; and it influences the shape of the weld bead and its mechanical properties. Fluxes are granular mineral compounds mixed according to various formulations. They are produced by several different manufacturing methods. Based on manufacturing method, the different types of fluxes are (1) fused, (2) bonded, (3) agglomerated, and (4) mechanically mixed.

Fused Fluxes

To manufacture a fused flux, the raw materials are dry mixed and then melted in an electric furnace. After melting and any final additions, the furnace charge is poured and cooled. Cooling may be accomplished by shooting the melt through a stream of water or by pouring it onto large chill blocks. In either case, the result is a glassy appearing product which is then crushed, screened for sizing, and packaged.

Fused fluxes have the following advantages:
(1) Good chemical homogeneity
(2) Easy removal of the fines without affecting the flux composition
(3) Normally nonhygroscopic, which simplifies handling, storage, and welding problems
(4) Permit recycling through the feeding and recovery systems without a significant change in particle size or composition

The main disadvantage of fused fluxes is the difficulty of adding deoxidizers and ferroalloys during manufacture without segregation or extremely high losses. The high temperatures associated with melting the raw ingredients cause the problem.

Bonded Fluxes

To manufacture a bonded flux, the raw materials are powdered, dry mixed, and bonded with either potassium silicate, sodium silicate, or a mixture of the two. This part of the operation is somewhat similar to the manufacture of the covering materials for shielded metal arc electrodes. After bonding, the wet mix is pelletized and baked at a relatively low temperature. The pellets are then broken up, screened to size, and packaged.

The advantages of bonded fluxes are as follows:

(1) The addition of deoxidizers and alloying elements is easy. Alloying elements can be added either as ferro-alloys, or as elemental metals, to produce alloys not readily available as electrodes, or to adjust weld metal composition.

(2) Permit a thicker layer of flux when welding.

(3) Color identification.

Bonded fluxes have the disadvantages of

(1) A tendency to absorb moisture like shielded metal arc electrodes

(2) Possible gas evolution from the molten slag

(3) Possible change in flux composition due to segregation or removal of fine mesh particles

Agglomerated Fluxes

The agglomerated fluxes are manufactured in much the same manner as bonded fluxes except that a ceramic binder is used instead of a silicate binder. The ceramic binder requires drying at relatively high temperatures that may limit the use of deoxidizers and ferro-alloys, as in the case of the fused fluxes.

Mechanically Mixed Fluxes

To produce a mechanically mixed flux, the manufacturer or user may mix two or more fused, bonded, or agglomerated fluxes in any ratio necessary to yield the desired results.

Mechanically mixed fluxes have the advantage that several commercial fluxes may be mixed for highly critical or proprietary welding operations.

Mechanically mixed fluxes have the disadvantage of

(1) Segregation of the various fluxes during shipment, storage, and handling

(2) Segregation of the various fluxes in the feeding and recovery systems during the welding operation

(3) Inconsistency in the flux from mix to mix

Particle Size and Distribution

Flux particle sizes and size distribution are important because of their influence on feeding and recovery, amperage level, and weld bead smoothness and shape. As amperage requirements increase, the average particle size should be decreased. At the same time, the percentage of small particles should be increased. If the amperage is too high with a given particle size, the arc may be unstable and cause ragged, uneven bead edges. Coarse particle size fluxes are normally preferred when rusty steel is welded because they are less dense and allow generated gases to escape more easily.

To provide a quantitative measure, flux manufacturers usually mark their packages with sizing information presented in the form of two mesh numbers. The numbers represent the largest and smallest particle sizes present when standardized screens are used for measuring them. The first number identifies the mesh (screen) through which essentially all of the particles will pass, and the second number identifies the mesh through which most or all of the particles will not pass.

The largest and smallest particle size numbers do not provide all the information that may be needed for uniform weld reproducibility. For instance, a flux that is designated as having a mesh 20 × 200 does not indicate whether the flux is coarse with some fines, or is fine with some coarse. All that is known is the range.

For some large users who wish to know accurately the "fineness" or "coarseness" of a flux, it is necessary to weigh and screen a sample through various mesh screens. The flux that is retained on each screen is weighed. The amount of flux retained on each screen can be expressed as a percentage of the total sample. It is common practice to use only five or six mesh screens when determining particle distribution. Standard wire cloth sieves used to determine particle size are covered by ASTM Specification E11.

Flux Usage

When a relatively fine flux is used, the flux feeding and recovery systems must be considered. If the flux is too fine, it will pack and not feed properly. If a fine flux or a flux with a small amount of fine particles is recovered by a vacuum system, the fine particles may be trapped by the system. Only the coarser particles will be returned to the feeding system for reuse, which may cause welding problems.

In applications where low hydrogen considerations are important, fluxes must be kept dry. Fused fluxes do not contain chemically bonded H_2O, but the particles can hold surface moisture. Bonded fluxes contain chemically bonded H_2O and can hold surface moisture as well. The bonded fluxes may need to be controlled in the same manner as low-hydrogen shielded metal arc electrodes. The user should follow the directions of the flux manufacturer for specific baking procedures.

When alloy-bearing fluxes are used, it is necessary to maintain a fixed ratio between the flux and the electrode melted for consistent weld metal composition. This ratio is actually determined by the welding procedure variables. For example, deviation from an established volt-ampere relation will change the alloy content of the weld metal by changing the flux-electrode melting ratio.

Fluxes are also identified as basic, acid, or neutral. The basicity or acidity of a flux is related to the ease with which the component oxides of the flux ingredients dissociate into a metallic cation and an oxygen anion. Oxides which dissociate easily are called basic, while those which dissociate to only a small degree are called acidic. A neutral flux is one that does not readily oxidize the alloying elements or add significant alloy to the weld.

Often, the basicity, acidity, or neutrality of a flux is referred to as the ratio of CaO or MnO to SiO_2. Fluxes having ratios greater than one are called basic, near to one are neutral, and less than one are acidic. Basic fluxes are used when resistance to brittle fracture is a welding criterion.

GENERAL PROCESS APPLICATIONS

With proper selection of equipment, SAW can be applied to a wide range of industrial applications. The high quality of welds, the high deposition rates, the deep penetration, and the adaptability to automatic operation make the process particularly suitable for fabrication of large weldments. It is used extensively in ship and barge building, railroad car fabrication, pipe manufacturing, and the fabrication of structural members where long welds are required. Automatic SAW installations are used to manufacture mass produced assemblies joined with repetitive short welds.

The process can be used to weld materials ranging from 1.5 mm (0.06 in.) thick sheet to very thick, heavy weldments. Submerged arc welding is not suitable for all metals and alloys. It is widely used for welding carbon steels, low alloy structural steels, and stainless steels. Some high strength structural steels, high carbon steels, maraging steels, and copper alloys can be joined by SAW. However, better joint properties are usually obtained with these metals by the use of a process with lower heat input to the base metal, such as gas metal arc welding.

Materials for which the process is not suitable include cast iron, austenitic manganese steels, high-carbon tool steels, aluminum, and magnesium. Fluxes are not available for aluminum or magnesium.

Submerged arc welding can be used to weld butt joints in the flat position, fillet welds in the flat and horizontal positions, and for surfacing in the flat position. With special tooling and fixturing, lap and butt joints can be welded in the horizontal position.

OPERATING VARIABLES

Knowledge and control of the operating variables in submerged arc welding are essential if high production rates and welds of good quality are to be obtained consistently. These variables, in the approximate order of their importance, are

(1) Welding amperage
(2) Type of flux and particle distribution
(3) Welding voltage
(4) Welding speed
(5) Electrode size
(6) Electrode extension
(7) Type of electrode
(8) Width and depth of the layer of flux

The operator should know how the variables affect the welding action and what changes he should make to them. The success of the operation depends on their proper selection and control.

WELDING AMPERAGE

Welding amperage is the most influential variable because it controls the rate at which the electrode is melted, the depth of penetration, and the amount of base metal melted. If the amperage is too high at a given travel speed, the depth of fusion or penetration will be too great. The resulting weld may have a tendency to melt through the metal being joined. High amperage also leads to waste of electrodes in the form of excessive reinforcement. This overwelding increases weld shrinkage and usually causes greater distortion. If the amperage is too low, inadequate penetration or incomplete fusion may result. The effect of amperage variation is shown in Fig. 6.5.

Some application rules to remember concerning welding amperage are

(1) Increasing amperage increases penetration and melting rate.
(2) Excessively high amperage produces a digging arc; an undercut; or a high, narrow bead.
(3) Excessively low amperage produces an unstable arc.

ARC VOLTAGE

Arc voltage adjustment varies the length of the arc between the electrode and the molten weld metal. If the arc voltage increases, the arc length increases; if the arc voltage decreases, the arc length decreases.

The arc voltage has little effect on the electrode deposition rate which is determined mainly by the welding amperage. The voltage principally determines the shape of the weld bead cross section and its external appearance. Figure 6.6 illustrates this effect.

Increasing the arc voltage with constant amperage and travel speed will

(1) Produce a flatter and wider bead
(2) Increase flux consumption
(3) Tend to reduce porosity caused by rust or scale on steel
(4) Help bridge excessive root opening when fit-up is poor
(5) Improve pickup of alloying elements from the flux when they are present

Excessively high arc voltage will

(1) Produce a widebead shape that is subject to cracking
(2) Make slag removal difficult in groove welds
(3) Produce a concave shaped fillet weld that may be subject to cracking
(4) Increase undercut along the edge(s) of fillet welds

Lowering the arc voltage produces a "stiffer" arc which improves penetration in a deep weld groove and resists arc blow. An excessively low voltage produces a high, narrow bead and causes difficult slag removal along the bead edges.

TRAVEL SPEED

With any combination of welding amperage and voltage, the effects of changing the travel speed conform to a general pattern. If the travel speed is increased (1) power or heat input per unit length of weld is decreased; (2) less filler metal is applied per unit length of weld,

Semiautomatic welding
2.4 mm (3/32 in.) wire, 35 V, 10 mm/s
(24 in./min)

Fully automatic welding
5.6 mm (7/32 in.) wire, 34 V, 13 mm/s
(30 in./min)

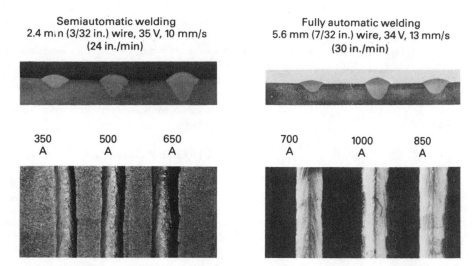

| 350 A | 500 A | 650 A | | 700 A | 1000 A | 850 A |

Fig. 6.5—Effect of amperage variation on weld bead shape and penetration

and consequently less weld reinforcement results; and (3) penetration decreases. Thus, the weld bead becomes smaller, as shown in Fig. 6.7.

Weld penetration seems to be affected more by travel speed than by any variable other than current. This is true except for excessively slow speeds when the molten weld pool is beneath the welding electrode. Then the penetrating force of the arc is cushioned by the molten pool. Excessive speed may cause undercutting.

Travel speed can be adjusted to control weld size and penetration within limits. In these respects, it is interrelated with the current and the type of flux. Excessively high travel speeds decrease fusion between the filler metal and the base metal and increase tendencies for undercut, arc blow, porosity, and uneven bead shape. Relatively slow travel speeds give time for gases to be forced out of the molten metal as it solidifies. This reduces the tendency for poros-

Semiautomatic welding
2.4 mm (3/32 in.) wire, 500 A, (24 in./min)

Fully automatic welding
5.6 mm (7/32 in.) wire, 850 A, (30 in./min)

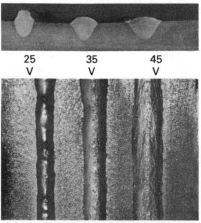

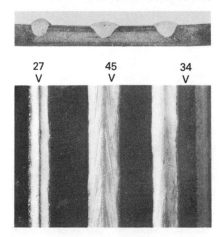

| 25 V | 35 V | 45 V | | 27 V | 45 V | 34 V |

Fig. 6.6—Effect of arc voltage variations on weld bead shape and penetration

ity. Excessively slow speeds produce (1) a convex bead shape that is subject to cracking; (2) excessive arc exposure, which is uncomfortable for the operator; and (3) a large molten pool that flows around the arc, resulting in a rough bead and slag inclusions.

ELECTRODE SIZE

Electrode size affects the weld bead shape and the depth of penetration at a fixed amperage, as shown in Fig. 6.8. Small electrodes are used with semiautomatic equipment to provide flexibility of movement. They are also used for multiple electrode, parallel power equipment. Where poor fit-up is encountered, a larger electrode is better than smaller ones for bridging large root openings.

Electrode size also influences the deposition rate. At any given amperage setting, a small diameter electrode will have a higher current density and a higher deposition rate than a larger diameter electrode. However, a larger diameter electrode can carry more current than a smaller electrode, so the larger electrode can ultimately produce a higher deposition rate at higher amperage. If a desired electrode feed rate is higher (or lower) than the feed motor can maintain, changing to a larger (or smaller) size electrode will permit the desired deposition rate.

For a given electrode size, a high current density results in a "stiff" arc that penetrates into the base metal. On the other hand, a lower current density in the same size electrode results in a "soft" arc that is less penetrating. The electrode size also affects the arc starting characteristics. Arc ignition is easier with smaller electrodes.

ELECTRODE EXTENSION

At current densities above 125 A/mm² (80 000 A/in.²), electrode extension becomes one of the important variables. At high current densities, resistance heating of the electrode between the contact tube and the arc can be utilized to increase the electrode melting rate. The longer the extension, the greater the amount of heating and the higher the melting rate. This resistance heating is commonly referred to as I²R heating.

In developing a procedure, an electrode extension of approximately eight times the electrode diameter is a good starting point. As the procedure is developed, the length is modified to achieve the optimum electrode melting rate with fixed amperage.

Increased electrode extension effectively adds a resistance element in the welding circuit and consumes some of the energy previously

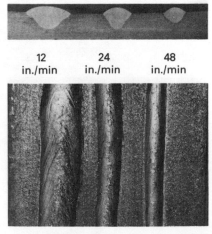

Semiautomatic welding
2.4 mm (3/32 in.) wire, 500 A, 35 V

| 12 in./min | 24 in./min | 48 in./min |

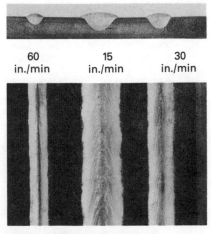

Fully automatic welding
5.6 mm (7/32 in.) wire, 850 A, 34 V

| 60 in./min | 15 in./min | 30 in./min |

Fig. 6.7—Effect of travel speed variation on weld bead shape and penetration

supplied to the arc. With lower voltage across the arc, bead width and penetration decrease (see Fig. 6.6). Because lower arc voltage increases the convexity of the bead, the bead shape will be different from one made with a normal electrode extension. Therefore, when the electrode extension is increased to take advantage of the high electrode melting rate, the voltage setting on the machine should be increased to maintain proper arc length.

The condition of the contact tube can affect the effective electrode extension. Contact tubes should be replaced at predetermined intervals to insure consistent welding conditions.

Deposition rates can be increased some 25 percent to 50 percent by using long electrode extensions with no change in welding amperage. With single electrode, automatic SAW, the deposition rate may approach that of the two wire method with two power sources.

However, changing to long extension welding is somewhat similar to changing from reverse to straight polarity dc power (positive to negative electrode). The increase in deposition rate is accompanied by a decrease in penetration. Thus, changing to long electrode extension is not recommended when deep penetration is needed. On the other hand, when melt-through is a problem which might be encountered with sheet gages, increasing the electrode extension will be beneficial. However, as electrode extension increases, it may become more difficult to maintain the electrode tip in the correct position with respect to the joint.

Suggested maximum electrode extensions for solid steel electrodes for SAW are

(1) For 2.0, 2.4, and 3.2 mm (5/64, 3/32, and 1/8 in.) electrodes, 75 mm (3 in.)

(2) For 4.0, 4.8, and 5.6 mm (5/32, 3/16, and 7/32 in.) electrodes, 125 mm (5 in.)

TYPE OF FLUX OR ELECTRODE

The various types of flux and electrodes were discussed earlier with respect to mechanical properties of the weld deposit. The flux and electrodes also affect other operational aspects of making the weld. For example, an electrode with a low electrical conductivity, such as stainless steel, can be heated by the welding current

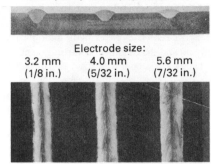

Fully automatic welding
600 A, 30 V, 13 mm/s (30 in./min)

Electrode size:
3.2 mm (1/8 in.)　　4.0 mm (5/32 in.)　　5.6 mm (7/32 in.)

Fig. 6.8—Effect of electrode size on weld bead shape and penetration

with normal electrode extension. The effect is similar to long electrode extension. For the same size electrode and amperage, the melting rate of a stainless steel electrode will be higher than that of a carbon steel electrode.

The various fluxes also affect the weld to some extent because of their different physical properties. The fluxes can cause considerable variations in bead shape, cleaning action, tendency to undercut, freezing pattern, tolerance for contaminants, and weld appearance. The manufacturer's literature should be consulted for information regarding these properties.

WIDTH AND DEPTH OF FLUX

The width and depth of the layer of granular flux influence the appearance and soundness of the finished weld as well as the welding action. If the granular layer is too deep, the arc is too confined and a rough weld with a rope-like appearance is likely to result. The gases generated during welding cannot readily escape, and the surface of the molten weld metal is irregularly distorted. If the granular layer is too shallow, the arc will not be entirely submerged in flux. Flashing and spattering will occur. The weld will have a poor appearance, and it may be porous. Figure 6.9 shows the effects of proper depth and too shallow a depth of flux on weld bead surface appearance.

An optimum depth of flux exists for any set of welding conditions. This depth can be established by slowly increasing the flow until the welding arc is submerged, and flashing no longer occurs. The gases will then puff up quietly around the electrode, sometimes igniting.

Dams are sometimes used to confine the flux to the joint area. Flux dams that are too close together interfere with the normal lateral flow of weld metal. This results in a weld reinforcement that is narrow, steep-sided, and poorly faired into the base metal.

During welding, the unfused granular flux can be removed a short distance behind the welding zone after the fused flux has solidified. However, under certain conditions, it may be best not to disturb the flux until the heat from welding has been almost evenly distributed throughout the section thickness.

Fused flux should not be forcibly loosened while the weld metal is at a high temperature. If allowed to cool, the fused material will often detach itself. Then it can be brushed away with little effort. Sometimes a small section is forcibly removed for quick inspection of the weld surface appearance.

It is important that no foreign material be picked up when reclaiming the flux. To prevent this, a space approximately 300 mm (12 in.) wide should be cleaned on both sides of the weld joint before the flux is laid down. If the recovered flux contains fused pieces, it should be passed through a screen with openings no larger than 3.2 mm (1/8 in.) to remove the coarse particles.

The flux is thoroughly dry when packaged by the manufacturer. If it is exposed to high humidity, it should be dried by baking it before it is used. Any moisture will cause porosity in the weld. The manufacturer's recommendation should be followed.

MULTIPLE ELECTRODE OPERATIONS

Multiple electrode welding is used to increase deposition rates. Because of the high heat input, multiple electrode welding may also reduce the solidification rate of the weld metal and give more time for gases to escape. Welding may be done with either a single power source feeding two or more electrodes, or with a separate power source for each electrode, or a combination of both. Examples of welding with one power source are multiple electrode, parallel power and two electrode, series power techniques. When separate power sources are used, the practice is called the multiple electrode, multiple power technique.

Multiple Electrode, Parallel Power

The multiple electrode, parallel power technique consists of two or more electrodes fed through electrically connected contact tubes supplied from a single power source. This technique results in deposition rates ranging to more than double that of a single electrode. It is used

Correct depth	Too shallow

19 mm (3/4 in.) depth 16 mm (5/8 in.) plate Result: Smooth top, sound weld structure	6 mm (1/4 in.) depth 16 mm (5/8 in.) plate Result: Gas pockets in weld; open arcing occurred

Fig. 6.9—Effect of proper and improper depth of flux on weld appearance

to increase travel speed on welds where high deposition rate is a major consideration, for example, large, flat position fillet welds. The two arcs flare together, causing back blow at the front arc and forward blow at the trailing arc. Because of the parallel connection, arc blow cannot be modified with different types of current for the electrodes.

This is the simplest form of multiple electrode welding and usually results in approximately a 50 percent increase in speed over a single arc. Figure 6.10 indicates some applications for this method. If more than two electrodes are used, they may be arranged in a triangular or rectangular pattern, which may be skewed in a manner calculated to best suit the joint and desired weld deposit.

Two Electrode, Series Power

The two electrode, series power technique is used for high deposition rates with a minimum of penetration into the base metal. When the electrodes are in a position transverse to the direction of travel, a wide deposit results with minimum dilution from the base metal. The technique is particularly valuable for surfacing with corrosion resistant or wear resistant alloys.

Each electrode in series arc welding operates independently, each having its own feed motor and voltage control. The power supply cable is connected to one welding head and the return power cable is connected to the second welding head instead of to the workpiece; thus, the two electrodes are in series. The welding current travels from one electrode to the other through the molten weld pool and surrounding material. The electrodes are usually positioned at an angle of 45 degrees to each other, and transverse to the direction of travel. The distance between the point of intersection of the electrodes and the surface of the work is the most important factor for control of the shape and quality of the deposited metal.

The magnetic field surrounding the arc stream affects the shape of the weld. The electrodes in series arc welding are of opposite polarity, which creates a force that tends to spread the arcs away from each other.

Either ac or dc can be used, depending upon the application. Alternating current is preferred for carbon steel, and direct current is best for nonferrous applications. When direct current is used, penetration is slightly deeper for the positive electrode than for the negative electrode. When a multiple pass, wide deposit is made, this difference can be minimized by positioning the welding heads so that the positive electrode is located over the previously deposited metal. A deposit that has fairly uniform penetration into the base metal can be produced by this means.

Multiple Electrode, Multiple Power

Multiple electrode, multiple power arcs (two or more electrodes, each with separately controlled power source) have a speed advantage on both V-groove and square groove welds. Two electrodes in tandem usually increase the welding speed and rate of deposition at least 100 percent over the single electrode.

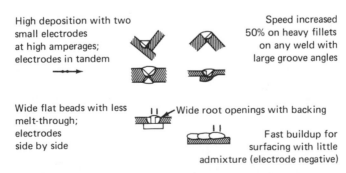

High deposition with two small electrodes at high amperages; electrodes in tandem

Speed increased 50% on heavy fillets on any weld with large groove angles

Wide flat beads with less melt-through; electrodes side by side

Wide root openings with backing

Fast buildup for surfacing with little admixture (electrode negative)

Fig. 6.10—Typical applications of two electrodes, parallel power welding

To produce a good bead shape at high speeds, it is necessary to have control over the magnetic arc blow that occurs when two arcs are operating close together. Forward arc blow can be used to govern bead shape and reduce undercutting. Some means of choosing the proper balance between back blow and forward blow must be available. Where dc power is used for all electrodes, proper balance might be obtained by the interchange of electrode polarities and positions, by the relationship of current and voltage values, or by arc spacing. However, at best only a compromise can be obtained because of the limited polarity relationships between the electrodes.

When ac is introduced on one of the electrodes, there is a regular cycle change in polarity relationship between the ac arc and the dc arc. Each arc then moves in accordance with the momentary field surrounding it. Adjustment of the welding conditions noted above, but principally the welding amperage and voltage, will place the trailing arc in a position pointing forward for a sufficient part of the cycle to obtain the desired weld.

When ac power is supplied to two electrodes independently, there are three ways in which the transformers can be connected to a three phase line for serving the two electrodes. These are (1) the closed delta, (2) the open delta, and (3) the Scott connection. Details of the closed delta and Scott connections are found in Chapter 1, "Arc Welding Power Sources." With these connections, there is not only the regular cycle change of polarity, but also a displacement of the time at which the difference in polarity occurs between the arcs. Thus, these power supply connections provide more versatile control over the direction of arc blow.

A Scott connection is the most economical to set up and the simplest to use. The amount of control is adequate for many applications. The welding currents between adjacent arcs are displaced by approximately 90 degrees. The ground currents in the workpiece are of reasonable magnitude to properly influence the result. This connection is widely used with two electrode, multiple power welding for single pass welds in plates, and for deep groove multiple pass welds in heavy-walled pressure vessels.

An open-delta connection is also economical to set up. It generally requires a reactor in the ground circuit to reduce the magnitude of the ground current, and thus bring the phase relation between the arc currents close to 90 degrees. A closed-delta connection provides complete control over the phase relation between the arc currents and the magnitude of the ground current. This connection is used to provide the best control of arc deflection, weld penetration, and weld contour.

A third electrode, supplied by a transformer properly phased with the other transformers, can increase welding speed even more. The speed can be as much as 50 percent greater than with two electrodes.

The necessity for considering the spacing between the arcs, the relation among the various welding currents and voltages, and the position and diameters of the electrodes makes the multiple electrode technique most suitable for long runs of repetitive joints. Precise welding procedures can be established for a particular joint type, base metal, and thickness from welding conditions previously developed with like or approximately similar weldments and fixturing.

Direct current power may be used with two electrodes for light gage weldments. The polarities, spacing, and welding conditions for the individual electrodes are varied to produce the desired weld. Where heavier material is to be welded, at least one electrode should be operated with alternating current. Where maximum control of the amount and direction of arc blow is necessary, such as in deep grove multiple pass welds, alternating current should be supplied to each electrode. The welding transformers can be connected in delta or Scott methods.

TYPES OF WELDS

Submerged arc welding is used for making groove, fillet, and plug welds, and for surfacing. Groove and fillet welds are most easily made in the flat position because the molten weld pool and the flux are most easily contained in this position. However, simple techniques are available for producing these types of welds in the horizontal welding position. Surfacing and plug welding are done in the flat position.

Welds made by this process may be classified with respect to (1) type of joint, (2) type of groove, (3) welding method (semiautomatic or machine), (4) welding position (flat or horizontal), (5) single or multiple pass deposition, (6) single or multiple electrode operation, and (7) single or multiple power supply (series, parallel, or separate connections).

Since it would be impractical to attempt data tabulation for such a large array of possible weld types, a few of the simpler types have been selected for illustration in Tables 6.2 through 6.5. The data given in these tables are intended only as a guide to show some typical power, travel speed, and electrode consumption conditions.

GROOVE WELDS

Groove welds are commonly made in butt joints ranging from 1.2 mm (0.05 in.) sheet metal to thick plate. The greater penetration inherent in submerged arc welding permits square groove joints 13 mm (1/2 in.) or more in thickness to be completely welded from one side, provided some form of backing is used to support the molten metal (see Tables 6.2 and 6.3 for typical data). Single pass welds up to 7.9 mm (5/16 in.) and two pass welds up to 15.9 mm (5/8 in.) thick are made in steel with a square groove butt joint, no root opening, and a weld backing.

With multiple pass welding, using single or multiple electrodes, plate ranging from 38 to 610 mm (1-1/2 to 24 in.) can be welded. Welds in thick material, when deposited from both sides, can utilize V-or U-grooves on one or both sides of the plate.

A form of groove welding of butt joints in the horizontal position is known as three o'clock welding from the analogy to the clock position. This type of weld may be made from both sides of the joint simultaneously, if desired. In most cases, the electrodes are positioned at a 10 to 30 degree angle above the horizontal. To support the flux and molten metal, some form of sliding support (or a proprietary moving belt) is used.

FILLET WELDS

Using a single electrode, fillet welds up to 9.5 mm (3/8 in.) in throat size can be made in the horizontal position with one pass. Larger single-pass, horizontal position fillet welds may be made by the use of multiple electrodes. Welds larger than 7.9 mm (5/16 in.) are usually made in the flat position by positioning the work (see Tables 6.4 and 6.5 for some typical examples). Fillet welds made by the submerged arc welding process can have greater penetration than those made by shielded metal arc welding, thereby exhibiting greater strength for the same size weld.

PLUG WELDS

Submerged arc welding is used effectively to make high quality plug welds. The electrode is positioned in the center of the hole and remains in this position until the weld is complete. The time required is dependent on welding amperage and hole size. Because of the deep penetration obtained with this process, it is essential to have adequate thickness in the weld backing.

SURFACING WELDS

Both single and multiple electrode submerged arc welding methods are used to provide a base metal with special surface properties. The purpose may be to repair or reclaim worn equipment that is otherwise serviceable or to impart

Table 6.2—Typical welding conditions for single electrode, machine submerged arc welding of steel plate using one pass (square groove)

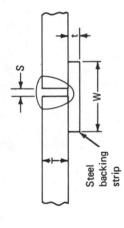

Plate thickness T		Root opening S		Current	Voltage	Travel speed		Electrode diam		Electrode consumption		Backing strip			
												t, min		W, min	
mm	in.	mm	in.	A^a	V^a	mm/s	in./min	mm	in.	kg/m	lb/ft	mm	in.	mm	in.
3.6	10 ga.	1.6	0-1/16	650	28	20	48	3.2	1/8	0.104	0.070	3.2	1/8	15.9	5/8
4.8	3/16	1.6	1/16	850	32	15	36	4.8	3/16	0.194	0.13	4.8	3/16	19.0	3/4
6.4	1/4	3.2	1/8	900	33	11	26	4.8	3/16	0.248	0.20	6.4	1/4	25.4	1
9.5	3/8	3.2	1/8	950	33	10	24	5.6	7/32	0.357	0.24	6.4	1/4	25.4	1
12.7	1/2	4.8	3/16	1100	34	8	18	5.6	7/32	0.685	0.46	9.5	3/8	25.4	1

a. DC, reverse polarity (electrode positive)

Table 6.3.—Typical welding conditions for single electrode, two pass submerged arc welding of steel plate (square groove)

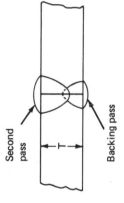

Second pass

Backing pass

Plate thickness T		Second pass						Backing pass						Electrode consumption	
		Current, A[a]	Voltage V[a]	Travel speed		Electrode diam.		Current A[a]	Voltage V[a]	Travel speed		Electrode diam			
mm	in.			mm/s	in./min	mm	in.			mm/s	in./min	mm	in.	kg/m	lb/ft
				Semiautomatic welding											
3.6	10 ga.	325	27	21	50	1.6	1/16	250	25	21	50	1.6	1/16	0.104	0.070
4.8	3/16	350	32	19	46	1.6	1/16	300	29	19	46	1.6	1/16	0.131	0.088
6.4	1/4	375	33	18	42	1.6	1/16	325	34	18	42	1.6	1/16	0.158	0.106
9.5	3/8	475	35	12	28	2.0	5/64	425	33	12	28	2.0	5/64	0.268	0.18
12.7	1/2	500	36	9	21	2.0	5/64	475	34	9	21	2.0	5/64	0.417	0.28
15.9	5/8	500	37	7	16	2.0	5/64	500	35	7	16	2.0	5/64	0.640	0.43
				Machine welding											
6.4	1/4	575	32	20	48	4.0	5/32	475	29	20	48	4.0	5/32	0.164	0.11
9.5	3/8	850	35	14	32	4.0	5/32	500	33	14	32	4.0	5/32	0.343	0.23
12.7	1/2	950	36	11	27	4.8	3/16	700	35	11	27	4.8	3/16	0.506	0.34
15.9	5/8	950	36	9	22	4.8	3/16	900	36	9	22	4.8	3/16	0.745	0.50

a. DC, reverse polarity (electrode positive)

Table 6.4—Typical welding conditions for fillet welds in steel made in the flat position using semiautomatic submerged arc welding

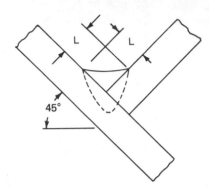

Weld leg, L		Current, A	Voltage V	Travel speed		Electrode diam		Electrode comsumption	
mm	in.			mm/s	in./min	mm	in.	kg/m	lb/ft
4.8	3/16	360[a]	30	21	50	1.6	1/16	0.070	0.047
6.4	1/4	425[b]	45	14	33	2.0	5/64	0.209	0.14
7.9	5/16	450[b]	47	10	23	2.0	5/64	0.328	0.22
9.5	3/8	450[b]	47	7	16	2.0	5/64	0.477	0.32
12.7	1/2	450[b]	47	4	9	2.0	5/64	0.849	0.57

(a) DC, reverse polarity (electrode positive).
(b) DC, straight polarity (electrode negative).

Table 6.5—Typical welding conditions for horizontal fillet welding of steel plate by submerged arc welding

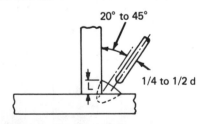

Weld leg, L		Current, A[a]	Voltage, V[a]	Travel speed		Electrode diam, d		Electrode comsumption	
mm	in.			mm/s	in./min	mm	in.	kg/m	lb/ft
3.2	1/8	400	24	27	64	3.2	1/8	0.045	0.03
4.8	3/16	500	26	18	42	4.0	5/32	0.089	0.06
6.4	1/4	650	30	14	32	4.0	5/32	0.164	0.11
7.9	5/16	700	33	10	24	4.0	5/32	0.253	0.17

(a) DC, reverse polarity (electrode positive).

desired properties to surfaces of original equipment. The high deposition rates achieved by submerged arc welding make this process well suited to large area surfacing applications.

To give ample treatment to this special form of welding, the subject of surfacing, including submerged arc surfacing, is presented separately in Chapter 14 of this volume.

WELDING PROCEDURES

Choice among various techniques for submerged arc welding is influenced by many factors. The effects of process variables have been discussed earlier. Other factors affecting processing technique include joint design, fit-up, fixturing, backing methods, preheat, postheat, and other physical considerations. The quality of preweld operations must be considered if the high production benefits of submerged arc welding are to be realized.

JOINT DESIGN AND EDGE PREPARATION

Types of joints used in submerged arc welding include chiefly butt, T-, and lap joints, although edge and corner joints can also be welded. The principles of joint design and methods of edge preparation are similar to those of most other arc welding processes. Typical welds include fillet, square groove, single and double V-groove, and single and double U-groove welds.

Joint designs, especially for plate welding, often call for a root opening of 0.8 to 1.6 mm (1/32 to 1/16 in.) to prevent angular distortion or cracking due to shrinkage stresses. However, a root opening that is larger than that required for proper welding will increase welding time and resultant costs. This is true for both groove and fillet welding.

Edge preparation may be done by any of the thermal cutting methods or by machining. The accuracy of edge preparation is important, especially for machine or automatic welding. For example, if a joint designed with a 6.4 mm (1/4 in.) root face were actually produced with a root face that tapered from 7.9 to 3.2 mm (5/16 to 1/8 in.) along the length of the joint, the weld might be acceptable because of excessive melt-thru

or lack of penetration. In such a case, the capability of the cutting equipment, as well as the skill of the operator, should be checked and corrected.

JOINT FIT-UP

Joint fit-up is an important part of the assembly or subassembly operations, and it can materially affect the quality, strength, and appearance of the finished weld. When welding plate thicknesses, the deeply penetrating characteristics of the submerged arc process emphasize the need for close control of fit-up. Uniformity of joint alignment and of the root opening must be obtained.

A poor fit-up may or may not be due to careless work. Discrepancies in layout, edge preparation, or forming can produce a condition in which the fitter, at best, can only effect a compromise between undersizing and oversizing. Where a fit-up problem exists, the discrepancies should be checked and corrective action taken to maintain low welding costs.

WELD BACKING

Submerged arc welding creates a large volume of molten weld metal which remains fluid for an appreciable period of time. This molten metal must be supported and contained until it has solidified at the root of the weld.

There are several methods commonly used to support molten weld metal when complete joint penetration is required. They are (1) backing strips, (2) backing weld, (3) copper backing bars, (4) flux backing, and (5) backing tapes. In the first two methods, the backing may become a part of the completed joint. The other three methods employ temporary backing, which is removed after the weld is completed. Methods (1) and (2) may or may not leave the backing in place, depending on the design requirements of the joint.

In many joints, the root face is designed to be thick enough to support the first pass of the weld. This method may be used for butt welds (partial joint penetration), for fillet welds, and for plug or slot welds. Supplementary backing or chilling is sometimes used. It is most important that the root faces of groove welds be tightly butted at the point of maximum penetration of the weld.

Backing Strip

In this method, the weld penetrates into and fuses with a backing strip, which temporarily or permanently becomes an integral part of the assembly (see Table 6.2).

Backing strips should be of metal that is compatible with the metal being welded. If the design permits, the joint may be so located that a part of the structure forms the backing (see Fig. 2.6, pp. 61). It is important that the contact surfaces be clean and close together; otherwise porosity and leakage of molten weld metal may occur.

Backing Weld

In a weld metal backed joint, the backing pass is usually made with some other process, such as flux cored arc welding. This backing pass forms a support for subsequent SAW passes made from the opposite side (see Fig. 2.7, pp. 61). Manual or semiautomatic welds can be used as backing for submerged arc welds when alternate backing methods are not convenient because of inaccessibility, poor joint penetration or fit-up, or difficulty in turning the weldment.

When shielded metal arc welding the backing in low carbon steels, in either the flat or overhead position, low hydrogen electrodes, such as E7016 or E7018, are recommended. However, E7028 or E7027 electrodes may be used for flat position welds, and E6010 or E6011 electrodes may be used for overhead welds. The E6012 and E6013 electrodes, and their iron powder equivalents, are not recommended because they tend to cause porosity in the finished weld joint.

The weld backing may remain as a part of the completed joint if it is of suitable quality. If necessary, the weld may be removed by oxygen or arc gouging, by chipping, or by machining after the submerged arc weld has been made. It is then replaced by a permanent submerged arc weld.

Copper Backing

With some joints, a copper backing bar is used to support the molten weld pool. Copper is used because of its high thermal conductivity, which prevents the weld metal from fusing to the backing bar. Where it is desirable to reinforce the underside of the weld, the backing bar may be grooved to the desired shape of the reinforcement. The backing bar must have enough mass to prevent it from melting beneath the arc, thus contaminating the weld with copper. Caution must be used to prevent copper pickup in the weld caused by harsh arc starts. Sometimes water is passed through the interior of the copper backing bar to keep it cool, particularly for high production welding applications.

The copper backing bar does not become a permanent part of the weld. In some welding applications, copper backing is designed to slide so a relatively short length can be used in the vicinity of the arc and molten weld pool. In still other applications, the copper backing may be in the form of a rotating wheel.

Flux Backing

Flux, under moderate pressure, is sometimes used as backing material for submerged arc welds. Usually, loose granular flux is placed in a trough on a thin piece of flexible sheet ma-

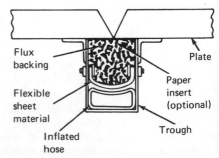

Fig. 6.11—A method of supporting flux backing for submerged arc welding

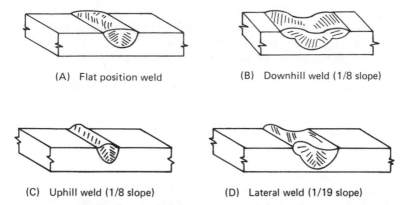

(A) Flat position weld   (B) Downhill weld (1/8 slope)

(C) Uphill weld (1/8 slope)   (D) Lateral weld (1/19 slope)

Fig. 6.12—Effect of work inclination on weld bead shape in 13 mm (1/2 in.) plate

terial. Beneath the flexible sheet, there is an inflatable rubberized canvas fire hose. The hose is inflated to no more than 35 to 70 kPa (5 to 10 psi) to develop moderate flux pressure on the backside of the weld. This concept is shown in Fig. 6.11.

FIXTURING

The main purpose of fixturing is to hold a workpiece assembly in proper alignment during handling and welding. Some assemblies may require stiffening fixtures to maintain their shape. In addition, some type of clamping or fixturing may be required to hold the joint alignment for welding and to prevent warpage and buckling from the heat of welding.

For assemblies that are inherently rigid, tack welding alone may suffice. Heavy section thicknesses in themselves offer considerable restraint against buckling and warpage. In intermediate cases, a combination of tack welding, fixturing, and weld sequencing may be required. For joints of low restraint in light gage materials, clamping is needed. Clamping bars maintain alignment and remove heat to reduce or prevent warpage. Tack welds are usually necessary.

Fixtures also include the jigs and tooling used to facilitate the welding operation. Weld seam trackers and travel carriages are used to guide machine or automatic welding heads. Turning rolls can be used to rotate cylindrical workpieces during fit-up and welding. Rotating turntables with angular adjustment are used to posi-

tion weld joints in the most favorable position for welding. Manipulators with movable booms are used to position the welding head, and sometimes the welding operator, for hard-to-reach locations.

INCLINATION OF WORK

The inclination of the work during welding can affect the weld bead shape, as shown in Fig. 6.12. Most submerged arc welding is done in the flat position. However, it is sometimes necessary or desirable to weld with the work slightly inclined so that the weld progresses downhill. For example, in high speed welding of 1.3 mm (0.050 in.) steel sheet, a better weld results when the work is inclined 15 to 18 degrees, and the welding is done downhill. Penetration is less than when the sheet is in a horizontal plane. The angle of inclination should be decreased as plate thickness increases to increase penetration.

Downhill welding affects the weld, as shown in Fig. 6.12 (B). The weld pool tends to flow under the arc and preheat the base metal, particularly at the surface. This produces an irregularly shaped fusion zone, called a "secondary wash." As the angle of inclination increases, the middle surface of the weld is depressed, penetration decreases, and the width of the weld increases.

Uphill welding affects the fusion zone contour and the weld surface, as illustrated in Fig. 6.12 (C). The force of gravity causes the weld pool to flow back and lag behind the welding electrode. The edges of the base metal melt and flow to the middle. As the angle of inclination increases, reinforcement and penetration increase, and the width of the weld decreases. Also, the larger the weld pool, the greater the penetration and center buildup. These effects are exactly the opposite of those produced by downhill welding. The limiting angle of inclination when welding uphill with currents up to 800 amperes is about 6 degrees, or a slope of approximately one in ten. When higher welding currents are used, the maximum workable angle decreases. Greater inclination than approximately 6 degrees makes the weld uncontrollable.

Lateral inclination of the workpiece produces the effects shown in Fig. 6.12(D). The limit for a lateral slope is approximately 3 degrees or 1 in 20. Permissible lateral slope varies somewhat, depending on the size of the weld puddle.

GROUNDING THE WORK

The method of attachment and the location of the ground (work connection) are important considerations in submerged arc welding since they can affect the arc action, the quality of the weld, and the speed of welding. A poor ground location can cause or increase arc blow which may cause porosity and poor bead shape. However, experimenting is often needed because it is not always possible to predict the effect of ground location. Generally, the best direction of welding is away from the ground (work connection).

In welding long seams, there may be a tendency for the welding current to change slowly. This can happen because the path and electrical characteristics of the circuit change as the weld progresses. A more uniform weld can frequently be obtained by attaching a work lead (ground) to both ends of the object being welded.

When the longitudinal seam of light gage cylinders is welded in a clamping fixture with copper backing, it is usually best to connect the work lead on the bottom of the cylinder itself at the start end. If this is not possible, then the work lead should be attached to the fixture at the start end. It is undesirable to connect the work lead to the start end of a copper backing bar because the welding current will enter or leave the work at the point of best electrical contact, not necessarily at the starting end. This current sets up a magnetic field around some length of the backing bar, causing arc blow.

When grounding is through a sliding shoe, two or more shoes should always be used. This will prevent interruptions of current in case electrical contact with one shoe is lost. Preload, tapered roller bearings are excellent for rotating connections, and they give better performance than sliding brushes.

The electrode and work leads (cables) should be kept as close together as possible, and free of coils. Welding cables should not be hung on or wrapped around steel objects, particularly when alternating current is used. With alternating current, a substantial inductive voltage loss (in the range of 5 to 10 V) can be induced in the welding cables if they are not kept close together.

ARC LENGTH CONTROL

Arc length control by either voltage or current control methods have considerable bearing on the success of the application. Arc length control has two functions: (1) control of the arc energy and the electrode at the instant of arc striking, so that the electrode does not freeze to the work or melt back to the contact tube when the arc is established; and (2) maintenance of the correct arc length after the arc is established, even though there may be variations in the distance from the contact tube to the work.

Arc length control when striking the arc is the more critical one of the two requirements. The reasons for this are the high demands placed on the power source and the importance of producing sound weld metal at the start. Although various arc starting methods exist, satisfactory arc striking requires the best arc control method available. Less stringent demands are made on the control to maintain the proper arc length as long as no radical changes in contour are encountered. In many cases, the workpiece design will limit the choice of the arc starting method.

With one voltage control method, the wire feeder retracts the electrode to start the arc. This is the best method to use for starting starts, for low electrode current density, or for welding at high speeds. The speed of the electrode feed is regulated for sudden changes in the work surface. This method is recommended whenever close control of the bead shape is essential.

Voltage control without the ability to retract the electrode is another method used. Arc striking may be done by one or more of the methods described later. The control may be set to govern arc length during welding by controlling the electrode speed in the forward direction only. This is satisfactory for high current density welding when repetitive starting is no problem. It is probably the simplest method, but it may not have the sensitivity required for certain types of work.

Current control of arc length depends on the high short circuit current characteristics of the power source to start the arc. The amperage increases or decreases during welding with a corresponding increase or decrease in electrode melting rate to maintain the preset arc voltage. A constant voltage power source and a constant speed wire feeder are used.

ARC STARTING METHODS

The method used to start the arc in a particular application will depend on such factors as the time required for starting relative to the total setup and welding time, the number of pieces to be welded, and the importance of starting the weld at a particular place on the joint. There are six methods of starting.

Fuse Ball Start

A tightly rolled ball of steel wool about 10 mm (3/8 in.) in diameter is positioned in the joint directly beneath the welding electrode. The welding electrode is lowered onto the steel wool until the ball is compressed to approximately one-half its original height. The flux is then applied and welding is started. The steel wool ball creates a current path to the work, but it melts away rapidly while creating an arc.

Sharp Wire Start

The welding electrode, protruding from the contact tube, is snipped with a pair of bolt cutters. This forms a sharp, chisel-like config-

uration at the end of the wire. The electrode is then lowered until the end just contacts the workpiece. The flux is applied and welding is commenced. The chisel point melts away rapidly to start the arc.

Scratch Start

The welding electrode is lowered until it is in light contact with the work, and the flux is applied. Next, the welding current is applied immediately after the carriage is started. The motion of the carriage prevents the welding wire from fusing to the workpiece.

Molten Flux Start

Whenever there is a molten puddle of flux, an arc may be started by simply inserting the electrode into the puddle and applying the welding current. This method is regularly used in multiple-electrode welding. When two or more welding electrodes are separately fed into one weld pool, it is only necessary to start one electrode to establish the weld pool. Then the other electrodes will start when they are fed into the molten pool.

Wire Retract Start

Retract arc starting is one of the most positive methods, but the welding equipment must be designed for it. It is cost effective when frequent starts have to be made and when starting location is important.

Normal practice is to move the electrode down until is just contacts the workpiece. Then the end of the electrode is covered with flux, and the welding current is turned on. The low voltage between the electrode and the work signals the wire feeder to withdraw the tip of the electrode from the surface of the workpiece. The arc strikes as this action takes place. As the arc voltage builds up, the wire feed motor quickly changes direction to feed the welding electrode toward the surface of the workpiece. Electrode feed speeds up until the electrode melting rate and arc voltage stabilize to the preset values.

If the workpiece is light gage metal, the electrode should make only light contact, consistent with good electrical contact. The welding head should be rigidly mounted. The end of the

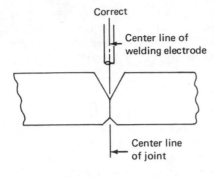

(A) Welding electrode directly
over joint center line

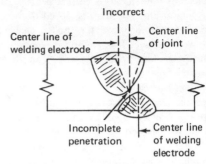

(B) Electrode not held to center line
results in incomplete penetration

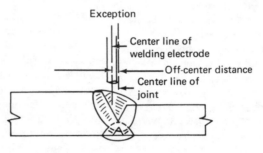

(C) Off-center alignment is sometimes
required when butt-welding
plates of different thickness

Fig. 6.13—Effect of electrode location with respect to the weld groove

electrode must be clean and free of fused slag. Wire cutters are often used to snip off the tip of the electrode (preferably to a point) before each weld is made. The electrode size should be chosen to permit operation with high current densities since higher current density permits easier starting.

High Frequency Start

This method requires special equipment but requires no manipulation by the operator other than closing a starting switch. It is particularly useful as a starting method for intermittent welding or for welding at high production rates where a large number of starts are required.

When the welding electrode approaches to within approximately 1.6 mm (1/16 in.) above the workpiece, a high frequency, high voltage

generator in the welding circuit causes a spark to jump from the electrode to the workpiece. This spark produces an ionized path through which the welding current can flow, and the welding action begins. This technique is commonly used for arc starting.

ARC TERMINATION

In one system, the travel and the electrode feed will stop at the same time when the "stop weld" button is pushed. A second type of system will stop the travel, but the electrode continues to feed for a controlled length of time. A third type of system reverses the direction of travel for a controlled amount of time while welding continues. The latter two systems are used to fill the weld crater.

ELECTRODE POSITION

In determining the proper position of the welding electrode, three factors must be considered:

(1) The alignment of the welding electrode in relation to the joint.

(2) The angle of tilt in the lateral direction, that is, the tilt in a plane perpendicular to the joint (work angle).

(3) The forward or backward direction in which the welding electrode points (travel angle). Forward is the direction of travel. Hence, a forward pointing electrode is one that makes an acute angle with the finished weld. A backward pointing electrode makes an obtuse angle with the finished weld.

Most submerged arc welds are made with the electrode axis in a vertical position. Pointing the electrode forward or backward becomes important when multiple arcs are being used, when surfacing, and when the workpiece cannot be inclined. Pointing the electrode forward results in a weld configuration similar to downhill welding; pointing the electrode backward results in a weld similar to uphill welding. Pointing the electrode forward or backward does not affect

the weld configuration as much as uphill or downhill positioning of the workpieces.

When butt joints are welded between equal thicknesses, the electrode should be aligned with the joint center line, as shown in Fig. 6.13(A). Improper alignment may cause lack of joint penetration as shown in Fig. 6.13(B). When unequal thicknesses are butt welded, the electrode must be located over the thick section to melt it at the same rate as the thin section. Figure 6.13(C) shows this requirement.

When horizontal fillet welding, the electrode is aligned, as shown in Table 6.5. The centerline of the electrode should be aligned below the root of the joint and toward the horizontal piece at a distance equal to one-fourth to one-half of the electrode diameter. The greater distance is used when making larger sizes of fillet welds. Careless or inaccurate alignment may cause undercut in the vertical member or produce a weld with unequal legs.

When horizontal fillet welds are made, the electrode is tilted between 20 and 45 degrees from the vertical (work angle). The exact angle is determined by either or both of the following factors:

(1) Clearance for the welding gun nozzle,

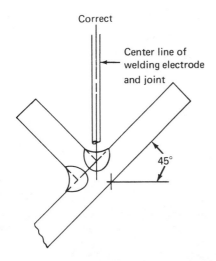

(A) Alignment for fillet welds in flat position

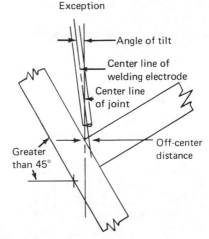

(B) When more than the usual amount of penetration is required

Fig. 6.14—Electrode positions for fillet welds in the flat position

especially when structural sections are being welded to plate.

(2) The relative thicknesses of the members forming the joint. If the possibility of melting through one of the members exists, it is necessary to direct the electrode toward the thicker member.

Normally, horizontal fillet welding should be done with the welding electrode positioned perpendicular to the axis of the weld.

When fillet welding is done in the flat position, the electrode axis is normally in a vertical position and bisects the angle between the work pieces, as shown in Fig. 6.14(A). When making positioned fillet welds where greater than normal penetration is desired, the electrode and the workpieces are positioned, as shown in Fig. 6.14(B). The electrode is positioned so that its center line intersects the joint near its center. The electrode may be tilted to avoid undercutting.

RUN-ON AND RUN-OFF TABS

Where a weld starts and finishes at the abrupt ends of a workpiece, it is sometimes necessary to provide a means of supporting the weld metal and molten slag so that it does not spill off. Tabs, such as those shown in Fig. 2.11, (p. 67), are the most commonly used method. An arc is started on a run-on tab that is tack welded to the start end of the weld, and it is stopped on a run-off tab at the finish end of the weld. The tabs are large enough so that the weld metal on the work itself is properly shaped at the ends of the joint. When the tabs are prepared, the groove should be similar to the one being welded, and the tabs must be wide enough to support the flux.

A variation of the tab is a copper dam that holds the flux which, in turn, supports the weld metal at the ends of the joint.

CIRCUMFERENTIAL WELDS

Circumferential welds differ from those made in the flat position because of the tendency for the molten flux and weld metal to flow away from the arc. To prevent spillage or distortion of the bead shape, welds must solidify as they pass the 6 or 12 o'clock positions. Figure 6.15 illustrates the bead shapes that result from various electrode positions with respect to the 6 and 12 o'clock positions.

Too little displacement on an outside weld or too much displacement on an inside weld produces deep penetration and a narrow, very convex bead shape. Undercutting may also take place. Too much displacement on an outside weld or too little displacement on an inside weld produces a shallow, concave bead.

The flux is granular and will spill off of small diameter work if it is not contained. If spilling of flux occurs, the arc is uncovered and poor quality welds result. One method of overcoming this is to use a nozzle assembly that

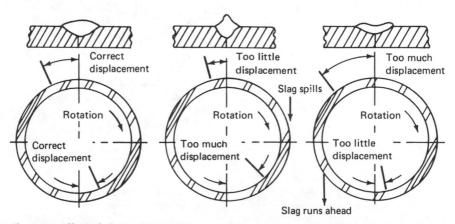

Fig. 6.15—Effect of electrode positions on weld bead shape when circumferential welding

pours the flux concentric with the arc; this gives it less chance to spill. A wire brush or some other flexible heat-resisting material can be attached to the nozzle assembly so that it rides the work ahead of the arc and contains the flux.

Regardless of electrode position, if the molten pool is too big for the diameter of work, the molten weld metal will spill simply because it cannot freeze fast enough. Bead size, as measured by the volume of deposited metal per unit length of weld, depends on the amperage and travel speed used. Lower amperages and higher travel speeds will reduce the size of the bead.

SLAG REMOVAL

On multiple pass welds, slag removal becomes very important because no subsequent passes should be made if slag is present. The factors that are particularly important in dealing with slag removal are bead size and bead shape. Smaller beads tend to cool more quickly and slag adherence is reduced. Flat to slightly convex beads that blend evenly with the base metal make slag removal much easier than very concave or undercut beads. For this reason, a decrease in voltage will improve slag removal in narrow grooves. On the first pass of two-pass welds, a concave bead that blends smoothly to the top edges of the joint is much easier to clean than a convex bead that does not blend well.

PREHEAT AND INTERPASS TEMPERATURES

Preheat and interpass temperature control may be required with some base metals to produce acceptable welds for the application. Preheat temperature and control of welding heat input may be particularly important with some alloy steels to produce a metallurgical structure with required mechanical properties and good resistance to cracking. They may also be important for the control of shrinkage and distortion of the weldment caused by internal stresses set up from expansion and contraction during welding.

The requirements depend primarily on the chemical composition, heat treatment, and thickness of the base metal. Information on the subject for various base metals is given in the appropriate chapter of Volume (Section) 4 of the *Welding Handbook*. In special cases, the metal producer's recommendations should be followed.

WELD QUALITY

Submerged arc deposited weld metal is usually clean and free of injurious porosity because of the excellent protection afforded by the blanket of molten slag. When porosity does occur, it may be found on the weld bead surface or beneath a sound surface. Various factors that may cause porosity are

(1) Contaminants in the joint
(2) Electrode contamination
(3) Insufficient flux coverage
(4) Contaminants in the flux
(5) Entrapped flux at the bottom of the joint
(6) Segregation of constituents in the weld metal
(7) Excessive travel speed
(8) Slag residue from tack welds made with covered electrodes

As with other welding processes, the base metal and electrode must be clean and dry. High travel speeds and associated fast weld metal solidification do not provide time for gas to escape from the molten weld metal. The travel speed can be reduced, but other solutions should be investigated first to avoid higher welding costs. Porosity from covered electrode tack welds can be avoided by using electrodes that will not leave a porosity causing residue. Several recommended ones are E6010, E6011, E7016, and E7018 classes.

Used flux should be discarded if the flux recovery equipment cannot remove solid contaminants. The flux should be kept clean and dry. When exposed to excessive humidity, it should be baked to drive off the moisture.

Flux entrapment can be avoided by properly backing butt welds so that all the flux will be melted and displaced by molted weld metal. When the joint is to be only partially penetrated, the weld bead should clear the weld joint backing by at least 4 mm (5/32 in.). Also, no root opening should be used.

Sometimes problems arise in weld metal with cracking, coarse grain structure, and segregation caused by the relatively slow solidification of a large mass of molten weld metal. This large mass of molten metal may not have a favorable cross section for proper progressive solidification as welding proceeds. Deeply penetrating weld beads sometimes have bulbous shapes which are narrower at the surface than at the midpoint. This bead shape is susceptible to cracking in the center because it is the last area to solidify. Microshrinkage voids tend to develop here.

Changes in welding procedures can be made to reduce the weld bead penetration and the amount of base metal dilution. The bead shape can be changed to produce one that is least prone to cracking for the particular weld joint.

SUPPLEMENTARY READING LIST

Butler, C.A., and Jackson, C.E. Submerged arc welding characteristics of the $CaO-TiO_2-SiO_2$ system. *Welding Journal* 46 (10), Oct. 1967, pp. 448-s—456-s.

Hinkel, J.E., and Forsthoefel, F.W. High current density submerged arc welding with twin electrodes. *Welding Journal* 55 (3), Mar. 1976, pp. 175-180.

Jackson, C.E. Fluxes and Slags in Welding. Bulletin 190. New York: Welding Research Council, Dec. 1973.

Keith, R.H. Weld backing comes of age. *Welding Journal* 54 (6), June 1975, pp. 422-430.

Kubli, R.A., and Sharav, W.B. Advancements in submerged arc welding of high impact steels. *Welding Journal* 40 (11), Nov. 1961, pp. 497-s—502-s.

Lewis, W.J., Faulkner, G.E., and Rieppel, P.J. Flux and filler wire developments for submerged arc welding HY-80 steel. *Welding Journal* 40 (8), Aug. 1961, pp. 337-s—345-s.

McKeighan, J.S. Automatic hard facing with mild steel electrodes and agglomerated alloy fluxes. *Welding Journal* 34 (4), Apr. 1955, pp. 301-308.

Metals Handbook, Vol. 6, 8th Ed. *Welding and Brazing*. Metals Park, Ohio: American Society for Metals, 1971, pp. 46-77.

Patchett, B.M. Some influences of slag composition on heat transfer and arc stability. *Welding Journal* 53 (5), May 1974, pp. 203-s—210-s.

Renwick, B.G., and Patchett, B.M. Operating characteristics of the submerged arc process. *Welding Journal* 55 (3), Mar. 1976, pp. 69-s—76-s.

Submerged Arc Welding Handbook. New York: Union Carbide Corp., Linde Div., 1974.

The Procedure Handbook of Arc Welding, 12th. ed. Cleveland, Ohio: Lincoln Electric Co., 1973.

Troyer, W., and Mikurak, J. High deposition submerged arc welding with iron powder joint-fill. *Welding Journal* 53 (8), Aug. 1974, pp. 494-504.

Uttrachi, G.D., and Messina, J.E. Three-wire submerged arc welding of line pipe. *Welding Journal* 47 (6), June 1968, pp. 475-481.

Wittstock, G.G. Selecting submerged arc fluxes for carbon and low alloy steels. *Welding Journal* 55 (9), Sept. 1976, pp. 733-741.

Wilson, R.A. A selection guide for methods of submerged arc welding. *Welding Journal* 35 (6), June 1956, pp. 549-555.

7

Electroslag and Electrogas Welding

Prepared by a committee consisting of

J.R. HANNAHS, *Chairman*
Bowser-Morner Testing Laboratories
D.R. AMOS
Westinghouse Electric Corporation
W.P. BENTER, JR.
U.S. Steel Corporation
L.N. FARINGHY
Airco Welding Products

J.E. NORCROSS
Welding Industry Consultant
B.L. SHULTZ
General American Transportation Corp.
R.C. SHUTT
The Lincoln Electric Co.
R.B. SMITH
Union Carbide Corporation

W.E. TROYER
Hobart Brothers Company

7
Electroslag and Electrogas Welding

INTRODUCTION

Single-pass welding of plates, castings, and forgings is the most desirable technique from the standpoint of cost and residual welding stresses. Electroslag (ESW) and electrogas (GMAW-EG or FCAW-EG) processes are capable of making single pass welds in plate thicknesses.

Electroslag and electrogas welding have many similarities. Both processes are done in the vertical position between copper shoes (dams) and usually in a single pass. Machine welding equipment is used to deposit high quality weld metal at high deposition rates. Angular distortion is usually absent in butt welds with both processes. Mechanically, the electrogas equipment is similar to that used for the conventional method of electroslag welding.

The main difference between electroslag and electrogas welding is how the energy is produced for welding. Electroslag welding uses the electrical resistance of a molten slag bath to convert electrical energy to thermal energy. The source of the slag is a granular flux used with the process. In electrogas welding, heat generation is by an electric arc between the electrode and the molten weld pool. Both processes shield the weld pool, one using molten slag and the other a gas

ELECTROSLAG WELDING

FUNDAMENTALS OF THE PROCESS

Definition and General Description

Electroslag welding (ESW) is a welding process producing coalescence of metals with molten slag which melts the filler metal and the surfaces of the work to be welded. The molten weld pool is shielded by the molten slag, which moves along the full cross section of the joint as welding progresses. The process is initiated by an arc which heats the flux and melts it to form the slag. The arc is then extinguished, and the conductive slag is maintained in a molten condition by its resistance to electric current passing between the electrode and the work.

Usually a square groove joint is positioned so that the axis or length of the weld is vertical or near vertical. Except for circumferential welds, there is no manipulation of the work once welding has started. Electroslag welding is a machine welding process, and once started, it continues to completion. Since no arc exists, the welding action is quiet and spatter free. Extremely high metal deposition rates allow the welding of very thick sections in one pass. A high quality weld deposit is the result of the nature of the melting and solidification during welding. There is no angular distortion of the welded plates.

Principles of Operation

The process is initiated by starting an electric arc between the electrode and the joint bottom. Granulated welding flux is then added and melted by the heat of the arc. As soon as a sufficiently thick layer of molten slag (flux) is formed, all arc action stops and the welding current passes from the electrode through the slag by electrical conduction. Welding is started in a sump or on a starting tab to allow the process to stabilize before the welding reaches the work.

Heat generated by the resistance of the molten slag to the passage of the welding current is sufficient to fuse the edges of the workpiece and the welding electrode. The interior temperature of the bath is in the vicinity of 1925° C (3500° F). The surface temperature is approximately 1650° C (3000° F). Melted electrode and base metal collect in a pool beneath the molten slag bath and slowly solidify to form the weld. There is progressive solidification from the bottom upward, and there is always molten metal above the solidifying weld metal.

Run-off tabs are required to allow the molten slag and some weld metal to extend beyond the top of the joint. Both starting and run-off tabs are usually removed flush with the ends of the joint.

Process Variations

There are two variations of electroslag welding that are in general use. One variation uses a wire electrode with a nonconsumable guide (contact) tube to direct the electrode into the molten slag bath. This variation will be referred to as the "conventional method." The other variation is similar to the first, except that a consumable guide (contact) tube is used. This variation will be called the "consumable guide method." With the conventional method, the welding head moves progressively upward as the weld is deposited. With the consumable guide method, the welding head remains stationary at the top of

the joint, and both the guide tube and the electrode are progressively melted by the molten flux.

Conventional Method. Figure 7.1 illustrates this method of electroslag welding. One or more electrodes are fed into the joint, depending on the thickness of the material being welded. The electrodes are fed through nonconsumable wire guides which are maintained 50 to 75 mm (2 to 3 in.) above the molten flux. Horizontal oscillation of the electrodes may be used to weld very thick materials.

Water-cooled copper shoes (dams) are normally used on both sides of the joint to contain the molten weld metal and slag bath. The shoes are attached to the welding machine and move vertically with the machine. Vertical movement of the welding machine is consistent with the electrode deposition rate. Movement may be either automatic or controlled by the welding operator.

Vertical movement of the shoes exposes the weld surfaces. There is normally a slight reinforcement on the weld, which is shaped by a groove in the shoe. The weld surfaces are covered with a thin layer of slag. This slag consumption must be compensated for during welding by the addition of small amounts of flux to the molten slag bath. Fresh flux is normally added manually. Flux-cored wires may be used to supply flux to the bath.

The conventional method of electroslag welding can be used to weld plates ranging in thickness from approximately 13 to 500 mm (1/2 to 20 in.). Thicknesses from 19 to 460 mm (3/4 to 18 in.) are most commonly welded. One oscillating electrode will successfully weld up to 120 mm (5 in.), two electrodes up to 230 mm (9 in.), and three electrodes up to 500 mm (20 in.). With each electrode, the process will deposit from 11 to 20 kg (25 to 45 lb) of filler metal per hour. The diameter of the electrode used is generally 3.2 mm (1/8 in.). Electrode metal transfer efficiency is almost 100 percent. The normally large weld made by electroslag welding will consume approximately 2.3 kg (5 lb) of flux for each 45 kg (100 lb) of deposited weld metal.

Consumable Guide Method. Figure 7.2 shows the consumable guide method of ESW. In

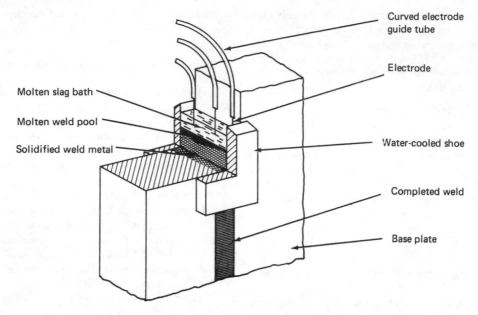

Curved electrode guide tube

Electrode

Molten slag bath

Molten weld pool

Solidified weld metal

Water-cooled shoe

Completed weld

Base plate

Fig. 7.1—Nonconsumable guide method of electroslag welding (three electrodes)

this method, filler metal is supplied by both an electrode and its guiding member. The electrode wire is generally directed to the bottom of the joint by a guide tube extending the entire joint length (height). Welding current is carried by the guide tube, and it melts off just above the surface of the slag bath. Thus, the welding machine does not move vertically, and stationary or nonsliding shoes are used.

As with the conventional method, one or more electrodes may be used; and they may oscillate horizontally in the joint. Since the guide tube carries electrical current, it may be necessary to insulate it from the joint side walls (base plate) and shoes. A flux coating may be provided on the outside of the consumable guide to insulate it and to help replenish the slag bath. Other forms of insulation include doughnut shaped insulators, fiber glass sleeves, and tape.

As welding proceeds and the slag bath rises, the consumable guide melts and becomes part of the weld metal. The consumable guide provides approximately 5 to 10 percent of the filler metal. On short welds, shoes may be the same length as the joint. Several sets of shoes may be required for longer joints. As the metal solidifies, a set of shoes is removed from the joint and then placed above the next set. This "leap frog" pattern is repeated until the weld is complete.

The consumable guide method can be used to weld sections of unlimited thickness. When using stationary electrodes, each electrode will weld approximately 63 mm (2.5 in.) of plate thickness. One oscillating electrode will successfully weld up to 130 mm (5 in.), two oscillating electrodes up to 300 mm (12 in.), and three oscillating electrodes up to 450 mm (18 in.). Weld lengths up to 9 m (30 ft) have been routinely accomplished with a single stationary electrode. Control problems may occur on long weld lengths. Therefore, if the required amount of oscillation cannot be properly controlled, additional electrodes may be added and oscillation stopped.

EQUIPMENT

The equipment for both ESW process methods is the same except for the design of the electrode guide tubes and the requirements for

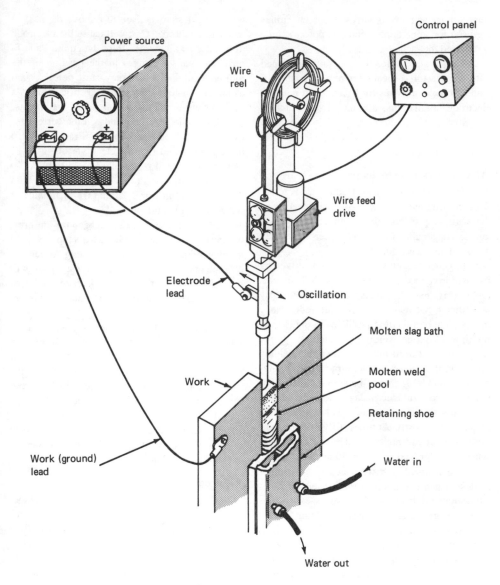

Fig. 7.2—Consumable guide method of electroslag welding

vertical travel. Major components of electro-slag welding equipment are

(1) Power source
(2) Wire feeder and oscillator
(3) Electrode guide tube
(4) Welding controls
(5) Travel carriage
(6) Retaining shoes (dams)

Power Sources

Power sources are typically of the constant voltage transformer-rectifier type with ratings of 750 or 1000 A, dc, at 100 percent duty cycle. They are similar to those used for submerged arc welding. Load voltages generally range from 30 to 55 V; therefore, the minimum open-circuit voltage of the power source should be 60 V. Constant voltage ac power sources of similar ratings are used for some applications. A power source is required for each electrode.

The power source is generally equipped with a contactor, a means for remote control of output voltage, a means for balancing in multiple electrode installations, a main power switch, a range control, an ammeter, and a voltmeter. Additional information on welding power sources is given in Chapter 1.

Wire Feeders and Oscillators

The function of a wire feed device is to deliver the wire electrode at a constant speed from the wire supply through the guide tube to the molten slag bath. The wire feeder is usually mounted on the welding head.

In general, each wire electrode is driven by its own drive motor and feed rolls. A dual gearbox to drive two electrodes from one motor may be utilized, but it does not provide redundancy in the event of a feeding problem. In the case of multiple electrode welding, the failure of one wire drive unit need not shut down the welding operation if a corrective measure can be accomplished quickly. It should be stressed, however, that for successful electroslag welding, it is vital to avoid a shut down because weld repair at the restart can be costly. At times, fifty or more continuous operating hours are demanded of these wire drives for heavy, long weldments.

The motor-driven wire feeders are similar in design and operation to those used for other continuous electrode welding processes, such as gas metal arc and submerged arc welding. Figure 7.3 shows a typical wire feed unit. The feed rolls normally consist of a geared pair, the driving force thus being applied by both rolls. Configuration of the roll groove may vary, depending on whether a solid or cored electrode is used. With solid wire, care must be taken that the wire feeds without slipping but is not squeezed too tightly so that it is knurled. Knurled wire can create the effect of a file and wear the components between the feed rolls and the weld. Rolls with an oval groove configuration have been found to perform best for both types of wire without danger of crushing a metal cored electrode.

Frequently a wire straightener, either of the simple three roll design or the more complex revolving design, is used to remove the cast in the wire electrode. Cast will cause the electrode to wander as it emerges from the guide. This, in turn, can cause changes in the position of the molten weld pool which may cause defects, such as lack of fusion. Cast in the electrode causes a more severe problem on large heavy weldments.

Electrode speed depends on the current required for the desired deposition rate, and also the diameter and type of electrode being used. Generally a speed range of 17 to 150 mm/s (40 to 350 in./min) is entirely suitable for use with 2.4 mm (3/32 in.) or 3.2 mm (1/8 in.) diameter metal cored or solid electrodes.

Electrodes should be packaged for uniform, uninterrupted feeding. The wire supply should allow feeding with minimal driving torque requirements so that binding or wire stoppage will not take place. The wire package must be of adequate size to complete the entire weldment without stopping.

Electrode oscillation devices are required when the joint thickness exceeds approximately 70 mm (2-1/4 in.) per electrode. Oscillation of the electrode guide tube(s) can be done by the use of motor-operated mechanical drive mechanisms, such as a lead screw or a rack and pinion. The drive must be adjustable for travel distance, travel speed, and adjustable delay at the end of each stroke. Control of oscillation movement is generally done using electronic circuitry.

Electrode Guide Tube

Conventional Method. With this method, the nonconsumable wire guide tube, or "snorkel," guides the electrode from the wire feed rolls into the molten slag bath. It also functions as an electrical contact to energize the electrode. The exit end of the tube is positioned close to the molten slag, and it will deteriorate with time.

Guide tubes are generally made of beryllium copper alloy and supported by two narrow, rectangular bars brazed to them. Beryllium copper is used because it retains reasonable strength at elevated temperatures. The tubes are wrapped with insulating tape to prevent short circuiting to the work.

To feed the electrodes vertically into the molten slag, the guides must be curved and narrow enough to fit into the joint root opening.

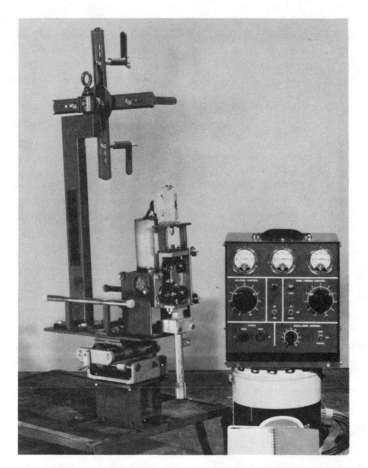

Fig. 7.3—Typical wire feed unit (consumable guide method)

They are generally less than 13 mm (1/2 in.) in diameter. To overcome cast in the electrode, integral wire straightening can be designed into the tubes.

Consumable Guide Method. The guide tube is made of a steel that is compatible with the base metal and slightly longer than the joint to be welded. It is commonly 16 mm (5/8 in.) in outside diameter and 3.2 to 4.8 mm (1/8 to 3/16 in.) inside diameter. Smaller diameters are necessary for welding sections less than 19 mm (3/4 in.) thickness.

The guide tube is attached to a copper alloy support tube, which mounts to the welding head. Welding current is transmitted from the copper tube to the steel tube, and then to the electrode.

For welds more than 600 to 900 mm (2 to 3 ft) long, it is necessary to insulate the guide tubes to prevent short circuiting with the work. The entire tube length may be coated with flux; or insulator rings, spaced 300 to 450 mm (12 to 18 in.) apart, are slipped over the tube and held in place with small weld buttons on the tube. The flux covering or insulator rings melt and help replenish the slag bath as the guide tube is consumed.

Retaining Shoes

The retaining shoes and the associated water circulation system are included in this category. The function of the shoes is to maintain the molten metal and the flux bath within the weld cavity. The shoes are fabricated of copper and generally include water passages at critical heat build-up points to prevent melting or overheating. Each shoe, shown in Fig. 7.2, generally has a cavity machined in the side toward the weld to provide for a slight reinforcement of weld metal.

The shoes may be cooled by a water circulation system or by tap water. Water circulators must have a 32 to 42 kJ/h (30 000 to 40 000 Btu/h) heat removal capacity. A recirculating system does not normally cause condensation on the shoes. Tap water frequently is at a lower temperature than ambient air, causing condensation on the shoes. If the condensation runs down the shoes and collects in the starting tabs prior to starting the weld, weld porosity will likely occur. Condensation on the inside of the shoes during welding will evaporate because of the advancing slag bath. Thus, it is best to turn on the tap water just before starting the weld.

With conventional welding, water-cooled shoes are mounted on the welding head and travel upward as welding progresses. With the consumable guide method, the shoes do not move. However, they can be repositioned, leap frog fashion, as welding progresses upward. Sometimes, the shoes are not water-cooled, but they must be massive to avoid meltings. The shoes are clamped in place, usually by wedges against U-shaped bridges (strongbacks) across the joint, or by large C-clamps on short welds made by the consumable guide method.

Welding Head and Controls

In electroslag welding, the weldments are relatively large and heavy. Therefore, it is more convenient to establish a single location for the power source and then use long cables, a remote control, and a portable welding head at the weld joint location. The control boxes are generally lightweight and contain a minimum of components. The controls interconnect between the power source and wire feeders on the welding head.

The welding head will include the wire feeder, electrode supply, wire guide tube(s), electrical connections to the guide tubes, and a means of attaching it to the work. It may also contain provisions for multiple electrode operation and an electrode oscillation drive unit. Where portability is required, the wire feeder and electrode supply may be located a short distance away from the welding head, as in semiautomatic gas metal arc welding.

Electroslag welding controls consist of a console mounted near the welding head which contains the following component groups:

(1) Ammeters; voltmeters; a remote control for each of the power sources, such as a switch to operate the weld contactor; and remote voltage control in the form of either a manual rheostat or a motorized rheostat which is triggered by a reversing switch.

(2) A speed control for each of the wire drive motors. This control will also jog and reverse the wire drive. On some consoles, the power contactor and wire drive can be activated by a common switch.

(3) Oscillator controls to move the guides back and forth in the weld joint. Adjustable limit switches control the established length of stroke. Two timers control dwell duration at each end of the stroke.

(4) Control for the vertical rise of the welding head. The type of control depends on whether rise is activated manually or automatically. An electric eye sensor may be used with the rise device for automatic control. The sensor is aimed at a point below the top of the containment shoe and adjusted to detect the top of the molten slag bath. When the bath rises above the point, the rise drive is activated and moves the welding head and shoe up until the bath is no longer detectable. A continuous incremental rise is obtained automatically in this manner.

In a given joint, if the plate thickness, root opening, number of electrodes, and electrode feed speed are known, then the vertical rate of rise may be computed approximately. This speed may be set on a variable-speed motor by the welding operator. As welding progresses, minor adjustments may be made to keep the molten slag bath and liquid metal pool within the containment shoes.

(5) Alarm systems for equipment or system malfunctions.

MATERIALS

Base Metals

Many types of carbon steels can be electroslag welded in production, such as AISI 1020, AISI 1045, ASTM A36, ASTM A441, and ASTM A515. They generally can be welded without postweld heat treatment.

In addition to carbon steels, other steels are successfully electroslag welded. They include AISI 4130, AISI 8620, ASTM A302, HY80, austenitic stainless steels, ASTM A514, ingot iron, and ASTM A387. Most of these steels require special electrodes and a grain refining post weld heat treatment to develop required weld or weld heat-affected zone properties.

Consumables

The composition of electroslag weld metal is determined by base metal and filler metal compositions and their relative dilution. The filler metal and consumables utilized in electroslag welding include the electrode, the flux, and in the case of consumable guide welding, a consumable guide tube and its insulation. The welding consumables can be effectively utilized to control the final chemical composition and mechanical properties of the weld metal.

Electrodes. There are two types of electrodes used with the electroslag welding process, solid and metal cored electrodes. The solid electrodes are more widely used. Various chemical compositions are available with each type of electrode to produce desired weld metal mechanical properties.

Metal cored electrodes permit adjustment of filler metal composition for welding alloy steels through alloy additions (ferroalloys) in the core. They also provide a means for replenishing flux in the molten bath. The metal tube is low carbon steel. The use of metal cored electrodes may result in excessive buildup of slag in the bath when the core is composed entirely of flux (flux cored).

In the electroslag welding of carbon steels and high strength, low alloy steels, the electrodes usually contain less carbon than the base metal, similar to other welding processes. Weld metal strength and toughness is achieved by

alloying with manganese, silicon, and other elements. This approach reduces the tendency towards weld metal cracking in steels containing up to 0.35 percent carbon.

The electrode wire compositions used for welding higher alloy steels normally match the base metal compositions. Higher alloy steels generally develop their mechanical properties by a combination of chemical composition and heat treatment. Usually it will be necessary to heat-treat an electroslag weld in a higher alloy steel to develop the desired weld metal and heat-affected zone properties. Thus, the best approach is to have the weld metal composition match the base material composition so that both will respond to heat treatment to approximately the same degree.

When the electrode wire for electroslag welding is selected, its dilution with the base metal must be considered. In a typical electroslag weld, the dilution runs from 30 percent to 50 percent base metal. The amount of dilution or base metal melting is dependent upon the welding procedure. The filler metal and melted base metal thoroughly mix to provide a weld with almost uniform chemical composition throughout.

The most popular electrode sizes are 2.4 mm (3/32 in.) and 3.2 mm (1/8 in.) diameter. However, 1.6 to 4.0 mm (1/16 to 5/32 in.) diameter electrodes have been successfully used. Smaller diameter electrodes provide a higher deposition rate than do larger electrodes at the same welding amperage. From practical applications, it has been found that either 2.4 mm (3/32 in.) or 3.2 mm (1/8 in.) diameter electrodes provide the optimum combination of deposition rate, feedability, welding amperage ranges, and ability to be straightened.

Electrodes for electroslag welding are supplied in the forms of 27 kg (60 lb) coils, large spools, or large drums. Since it is essential that enough electrode be available to complete the entire weld joint, it has been found most practical and economical to use spools up to 340 kg (750 lb) or drums up to 340 kg (750 lb).

Flux. The flux is a major part of the successful operation of the electroslag welding process. Flux composition is of utmost importance since its characteristics determine how well the electroslag process operates. During the welding operation, the flux is melted into slag that transforms the electrical energy into thermal energy to melt the filler metal and base metal. Also, the slag (flux) must conduct the welding current, protect the molten weld metal from the atmosphere, and provide for a stable operation.

An electroslag welding flux must have several important characteristics. When molten, it must be electrically conductive and yet have adequate electrical resistance to generate sufficient heat for welding. However, if its resistance is too low, arcing will occur between the electrode and the slag bath surface. The molten slag viscosity must be fluid enough for good circulation to ensure even distribution of heat in the joint. A slag that is too viscous will cause slag inclusions in the weld metal, and one that is too fluid will leak out of small openings between the work and the retaining shoes. The melting point of the flux must be well below that of the metal being welded, and its boiling point must be well above the operating temperature to avoid losses that could change operating characteristics. The molten slag (flux) should be reasonably inert to reactions with the metal being welded, and it should be stable over a wide range of welding conditions and slag pool sizes. Solidified slag on the weld surfaces should be easy to remove, although some commercial fluxes may not have this characteristic. Electroslag flux in original unopened packages should be protected against moisture pickup by the packaging under normal conditions. Flux reconditioning may be necessary if the flux has been exposed to high humidity.

Fluxes for electroslag welding are usually combinations of complex oxides of silicon, manganese, titanium, calcium, magnesium, and aluminum, with some calcium fluoride always present. Special characteristics are achieved by variations in the composition of the flux.

Only a relatively small amount of flux is used during electroslag welding. An initial quantity of flux is required to establish the process. Flux solidifies as slag in a thin layer on the cold surfaces of the retaining shoes and on both weld faces. It is necessary to add flux to the molten pool during welding to maintain the required depth. Neglecting losses by leakage, the total flux used is approximately 0.5 kg (1 lb) for each 9 kg (20 lb) of deposited metal. However, as the plate thickness or weld length increases, the flux consumption approaches 0.5 kg. (1 lb) for each 36 kg (80 lb) of deposited metal.

Consumable Guide Tube. The primary function of the consumable guide tube is to provide support to the electrode wire from the welding head to the molten slag bath. The consumable guide melts periodically just at the top of the rising molten slag bath. Its use permits the welding head to be fixed in position at the top of the vertical seam. The electrode cable is attached to the guide tube. Welding current is conducted to the electrode, as it passes through the guide tube, and then into the molten slag bath.

Most consumable guide tubing is 16 mm (5/8 in.) or 13 mm (1/2 in.) outside diameter. The inside diameter of the tubing is normally determined by the size of the electrode wire being used. The amount of metal contributed to the weld by the guide tube is relatively small. Since the guide tube does become part of the weld, its composition should be compatible with the desired weld metal compostion.

When making short welds, a bare consumable guide tube can be used. However, for long welds, it must be insulated to prevent electrical contact with the base metal. A flux coating can be used to provide electrical insulation, and at the same time add flux to the slag bath. Other forms of insulation include doughnut shaped insulators, fiber glass sleeves, and tape. The form of insulation must not affect either the deposited weld metal or the operating characteristics of the flux.

Specifications for Consumables

AWS A5.25, Specification for Consumables Used for Electroslag Welding of Carbon and High Strength Low Alloy Steels, classifies electrodes and fluxes for ESW. Metal cored electrodes are classified on the basis of chemical analysis of weld metal taken from an undiluted ingot. The solid electrodes are classified on the basis of their chemical composition as manufactured. Since the consumable guide tube contributes only a small amount of filler metal to the joint, it does not affect the flux-electrode classification. However, the guide tubes must conform to AISI Specifications for 1008 to 1020 carbon steel tubing.

Metal cored electrodes deposit weld metals of low carbon steel and low alloy steels. The low alloy steel deposits contain small amounts of nickel and chromium, and either copper or molybdenum. Carbon is less than 0.15 percent. Solid electrodes are divided into three classes: medium manganese (approximately 1 percent), high manganese (approximately 2 percent), and special classes.

In the flux-electrode classification system, both types of electrodes are used in any combination with six fluxes. The fluxes are classified on the basis of the mechanical properties of a weld deposit made with a particular electrode and a specified base metal. The compositions of the fluxes are left to the discretion of the manufacturer. Two levels of tensile strength for flux-electrode combinations are specified: 415 to 550 MPa (60 to 80 ksi) and 485 to 655 MPa (70 to 95 ksi). For each level of strength, two of the three flux-electrode classifications must meet minimum toughness requirements as determined by the Charpy V-notch impact tests.

METALLURGICAL CONSIDERATIONS

During electroslag welding, heat is generated by the resistance to current flow as it passes from the electrode(s) through the molten slag into the weld pool. The molten slag, being conductive, is electromagnetically stirred in a vigorous motor action, which maintains uniform temperatures within the pool. Heat diffuses throughout the entire cross section being welded. Temperatures attained in the electroslag welding process are considerably lower than those in arc welding processes. However, the molten slag pool temperature must be higher than the melting range of the base metal for satisfactory welding.

The molten zone, both slag and weld metal, advances relatively slowly, usually approximately 13 to 38 mm/min (1/2 to 1-1/2 in./min). Generally the weld is completed in one pass. The basic differences between electroslag welds

and consumable electrode arc welds, which are made with a point source of heat and multiple passes in heavy plate, result in some different mechanical and metallurgical characteristics for electroslag welds compared to arc welds. Some of these are

(1) In the as-deposited condition, a generally favorable residual-stress pattern is developed in electroslag welded joints. The weld surfaces and heat-affected zones are normally in compression, and the center of the weld is in tension.

(2) Because of the symmetry of most vertical ESW butt welds (square groove joints welded in a single pass), there is no angular distortion in the weldment. There is slight distortion in the vertical plane caused by weld metal contraction, which can readily be adjusted for during joint fit-up.

(3) The weld metal stays molten long enough to permit some slag refining action, as in electroslag remelting. The progressive solidification allows gasses in the weld metal to escape and nonmetallic inclusions to float up and mix with the slag bath. High quality, sound weld deposits are generally produced by ESW.

(4) The circulating slag bath washes the groove faces and melts them in the lower portion of the bath. The weld deposit contains up to 50 percent admixed base metal, depending on welding conditions. Therefore, the composition of steel being welded and the amount melted significantly affect both the chemical composition of the weld metal and the resultant weld joint mechanical properties.

(5) The prolonged thermal cycle results in a weld metal structure that consists of large prior austenite grains that generally follow a columnar solidification pattern. The grains are oriented horizontally at the weld metal edges and turn to a vertical orientation at the center of the weld, as shown in Fig. 7.4(b). The microstructure of electroslag welds in low carbon steels generally consists of acicular ferrite and pearlite grains with proeutectoid ferrite outlining the prior austenite grains. It is very common to observe coarse prior austenite grains at the periphery of the weld and a much finer grained region near the center of the weld. This fine-grained region appears equiaxed

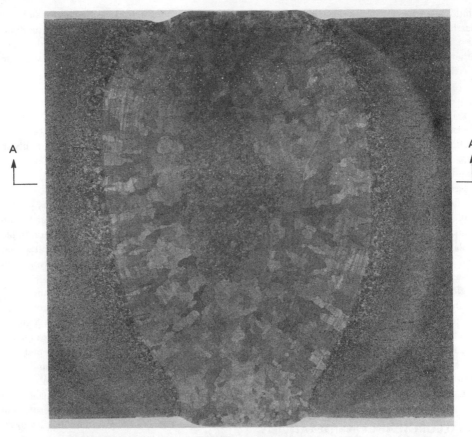

(a) Transverse section

(b) Longitudinal section at A-A

Fig. 7.4—Transverse and longitudinal section through a 100 mm (4 in.) thick electroslag weld

in a transverse cross section; however, longitudinal sections reveal its columnar nature, as shown in Fig. 7.4. Changes in weld metal composition, and to a lesser extent welding procedure, can markedly change the relative proportions of the coarse- and fine-grained regions to the extent that only one may be present.

(6) The relatively long time at high temperatures and the slow cooling rate after welding result in wide heat-affected zones with a relatively coarse grain region. The cooling rate is slow enough so that only relatively soft, high-temperature transformation products are formed. For most steels, this is an advantage, particularly if stress-corrosion cracking may be a problem.

Preheat and Postheat

Preheating is not required or generally used in electroslag welding. By its nature, the process is self-preheating in that a significant amount of heat is conducted into the workpieces and preheats them ahead of the weld. Also, because of the very slow cooling rate after welding, postheating is usually unnecessary.

Postweld Heat Treatment

Most applications of electroslag welding, particularly the welding of structural steel, require no postweld heat treatment. As discussed previously, as-deposited electroslag welds have a favorable residual stress pattern which is negated by a postweld heat treatment. Subcritical postweld heat treatments (stress relief) can be either detrimental or beneficial to mechanical properties, particularly notch toughness. They are generally not employed.

The properties of carbon and low alloy steel welds can be greatly altered by heat treatment. Normalizing removes nearly all traces of the cast structure of the weld and almost equalizes the properties of the weld metal and base metal. This may improve the resistance to brittle fracture initiation and propagation above certain temperatures as measured by the Charpy V-notch impact test.

Quenched and tempered steels are not usually joined by ESW. They must be heat-treated after welding to obtain adequate mechanical strength properties in the weld and heat-affected zones. Such a heat treatment is very difficult to do on large, thick structures.

MECHANICAL PROPERTIES

The mechanical properties of electroslag welds will depend on the type and thickness of base metal, electrode composition, electrode-flux combination, and the welding conditions. All of these will influence the chemical composition, metallurgical structure, and mechanical properties of the weldment. In general, electroslag welds are used in structures under static or fluctuating load conditions. One major concern is the notch toughness of the weld metal and heat-affected zones under service conditions, particularly at low temperatures. This must be carefully evaluated for the particular applications so that design requirements will be met, and the weldments will perform satisfactorily in service.

Typical mechanical properties of deposited weld metal in selected structural steels are given in Table 7.1, and for some low alloy steels in Table 7.2. The number and type of electrodes used for welding are given. Some variables that could affect the mechanical properties are unknown and, therefore, the data should not be used for design purposes.

GENERAL PROCESS APPLICATIONS

Structural

Probably the widest use of the ESW process is for structural applications. The electroslag process has many unique advantages that make it a highly desirable welding process. The high weld metal deposition rates, the low percentage of weld defects, and the fact that it is an automatic process are a few reasons for the use of electroslag welding. For thick sections, good mechanical properties, and more intricate configurations, electroslag welding is a low cost welding process if the weldments meet the design requirements and service conditions. However, if the welding process is stopped during the welding of a joint for any reason, the restart area must be carefully inspected for discontinuities. Those which are considered unacceptable for

Table 7.1—Typical mechanical properties of weld metal from electroslag welds in structural steels, as-welded (consumable guide method)

Base metal, ASTM	Thickness		Electrode		Yield strength		Tensile strength		Elong., % in 51 mm (2 in.)	Reduction of area, %	Impact strength[a]	
	mm	in.	Type (AWS)	No.	MPa	ksi	MPa	ksi			J	ft·lb
A441	25	1	EM13K-EW	1	344	49.9	523	75.8	28.0	59.0	23	17
A441	64	2-1/2	EM13K-EW	1	316	45.8	505	73.3	26.5	66.0	37	27
A36	152	6	EM13K-EW	2	317	46.0	548	79.5	28.5	52.8	–	–
A36	305	12	EM13K-EW	2	255	37.0	464	67.3	33.5	71.0	–	–
A572 Gr. 42	203	8	EM13K-EW	2	400	58.2	585	84.8	25.0	67.6	38	28
A572 Gr. 60	57	2-1/4	EH10Mo-EW	1	423	61.5	680	98.5	18.0	35.6	–	–

a. Impacts at −17.8° C (0° F)

Table 7.2—Typical mechanical properties of weld metal from electroslag welds in carbon and alloy steels

Base metal, ASTM	Thickness		Electrode		Heat treatment	Yield strength		Tensile strength		Elong., % in 51 mm (2 in.)	Charpy impact tests, −12.2° C (10° F) Notch location					
											WMFG[a]		WMCG[b]		BM HAZ[c]	
	mm	in.	Type	No.		MPa	ksi	MPa	ksi		J	ft·lb	J	ft·lb	J	ft·lb
A204-A	89	3-1/2	Mn-Mo	1	NT[d]	382	55.5	562	81.5	27	46	34	39	29	18	13
A515 GR70	38	1-1/2	Mn-Mo	1	SR[e]	358	52.0	579	84.1	26	61	45	35	26	10	7
A515 GR70	51	2	Mn-Mo	1	SR	469	68.1	587	85.2	23	63	46	29	21	7	5
A515 GR70	86	3-3/8	Mn-Mo	2	NT	396	57.5	537	78.0	29	46	34	45	33	30	22
A515 GR70	171	6-3/4	Mn-Mo	2	NT	313	45.5	512	74.3	31	33	24	29	21	16	12
A302-B	76	3	Mn-Mo-Ni	1	NT	393	57.0	565	82.0	28	72	53	71	52	86	63
A387-C	76	3	1-1/4 Cr-1/2 Mo	2	NT	320	46.5	503	73.0	29	95	70	103	76	78	57
A387-D	83	3-1/4	2-1/4 Cr-1 Mo	1	SR	396	57.5	565	82.0	25	63	46	68	50	65	48
A387-D	191	7-1/2	2-1/4 Cr-1 Mo	2	SR	551	80.0	658	95.5	20	84[f]	62	102[f]	75	113[f]	83

a. WMFG—Weld metal, fine grain
b. WMCG—Weld metal, coarse grain
c. BM HAZ—Base metal, heat-affected zone
d. NT—Normalized and tempered
e. SR—Stress relieved
f. Impacts at 10° C (50° F)

the application must be repaired by another acceptable welding process.

A common structural application of the electroslag process is the transition joint between different flange thicknesses, which is a type of butt joint. By using copper shoes designed for this type joint configuration, the varying thicknesses present no problem.

Another very common use of electroslag welding in the structural area is the welding of stiffeners in box columns and wide flanges. In all cases, the stiffener weld would be a T-joint. Figure 7.5 shows this application.

A very impressive use of electroslag welding is the joining of large wide flange beams. The web thickness is 51 mm (2 in.) and the height of the weld, or length of the flange, is 8.2 m (27 ft). The type weld is a T-joint. Each side of the web may be welded to a flange, either singly or simultaneously. Another common use of electroslag welding is in the splicing of flanges, which is a simple butt weld between plates of the same thickness.

Machinery

In the machinery area, manufacturers of large presses and machine tools work with large, heavy plates. Quite often the design requires plates that are larger than the mill can produce in one piece. Electroslag welding is used to splice two or more plates together.

Other machinery applications include kilns, gear blanks, motor frames, press frames, turbine rings, shrink rings, crusher bodies, and rims for road rollers. These parts are formed from plate and welded along a longitudinal seam.

Pressure Vessels

Pressure vessels for the chemical, petroleum, marine, and power generating industries are made in all shapes and sizes, with wall thicknesses from less than 13 mm (1/2 in.) to greater than 400 mm (16 in.). In current practice, plate may be rolled to form the shell of the vessel, and the longitudinal seam welded. In very large or thick walled vessels, the shell may be fabricated from two or more curved plates and joined by several longitudinal seam welds.

Steels used in pressure vessel construction are generally made with controlled rolling practices, or they are heat treated. Consequently, when welding these steels with a high heat input, such as with electroslag welding, the weld heat-affected zone does not have adequate mechanical properties. To improve the mechanical properties, weldments are given the required heat treatment.

Electroslag welding has also been used to join branch pipes to thick walled vessels, to weld lifting lugs to the vessel, and to make massive weld buildups around large diameter vessels.

Ships

Electroslag welding is used in the ship building industry for both in-shop and on-ship applications. Main hull section joining is done with the conventional method. Vertical welding of the side shell, from the bilge area to, but not including, the sheer strake may be done with an electroslag plate crawler. Plate thickness of approximately 25 mm (1 in.) are welded, and the length of welding is approximately 12 to 21 m (40 to 70 ft), depending on the size of the ship.

Figure 7.6 shows an electroslag plate crawler. A single electrode is fed into the joint through a guide tube. Once welding is initiated, the entire assembly is propelled upward by a serrated drive wheel that tracks the weld joint. The vertical speed of the plate crawler is set by remote control. Two water-cooled copper shoes slide along the joint to contain the molten slag and weld metal. The copper shoes are held tight against the plate by spring tension between the front and rear sections of the plate crawler.

The plate crawler may be used to weld plates ranging in thickness from 13 mm (1/2 in.) to 51 mm (2 in.). Normally either a square groove or single V-groove joint is used. Vertical travel speeds of up to 3 mm/s (7 in./min) are possible. Longitudinal and transverse hull stiffeners 13 to 51 mm (1/2 to 2 in.) thick, 300 to 900 mm (12 to 36 in.) long are welded with the consumable guide tube method.

Castings

Electroslag welding is often used to fabricate cast components. The metallurgical characteristics of a casting and an electroslag weld are similar, and both respond to post weld heat treatment in a similar way. Many large difficult-

Fig. 7.5—Electroslag welding a stiffener plate in a building column

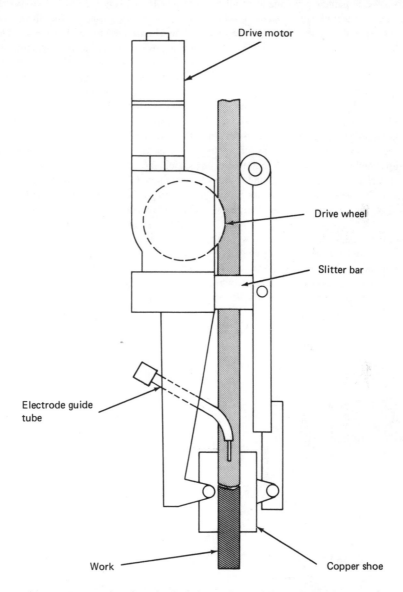

Fig. 7.6—Conventional method of electroslag welding using a plate crawler

to-cast components are now produced in smaller, higher quality units, and then electroslag welded together. Costs are reduced and the quality usually is improved. Compatible weld metal produces a homogeneous structure. Color match, machinability, and other desirable properties are thus produced.

Economics

To appreciate the true overall economy of electroslag welding, the first consideration is the cost of joint preparation. A square oxygen-cut joint is suitable preparation for the process in carbon steels. No elaborate joint matching or close fit-up is required. In welds of 75 mm (3 in.) or more in thickness, electroslag welding requires much less weld metal, and as much as 90 per-cent less flux, than a comparable submerged arc weld.

Once the pieces are in place for welding, the weld is completed without stopping if the process remains under control. Down or non-productive time, common in most arc welding processes, can range from 30 percent to 75 per-

cent. There is generally no down time with elec-troslag welding once the operation is started.

The deposition rate for ESW is approximately 16 to 20 kg/h (35 to 45 lb/h) per electrode. In very heavy plates, using three electrodes, 47 to 61 kg/h (105 to 135 lb/h) of weld metal can be deposited. Using a joint spacing of 29 mm (1-1/8 in.), the rate of welding is shown in Fig. 7.7. Heavy plates ranging from 76 to 305 mm (3 to 12 in.) in thickness are welded at speeds between 610 to 1220 mm/h (2 to 4 ft/h).

Another significant saving is achieved by the elimination of angular distortion and subsequent rework. Angular distortion can be a major factor in heavy multiple pass welding when it is done from one or from both sides.

Electroslag welding normally produces a high percentage of defect-free weldments which minimizes repair costs. Slag entrapment, porosity, and lack of fusion can be avoided in most cases. It must be clearly recognized that stoppage of a heavy electroslag weld in process can be very costly. To restart without producing a defect is difficult and may be impossible.

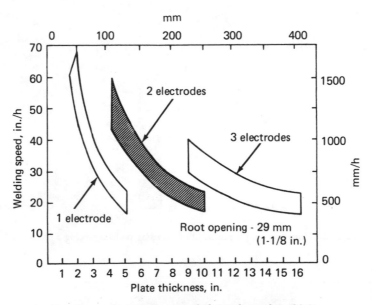

Fig. 7.7—Electroslag welding speeds for various plate thickness

PROCESS VARIABLES

Welding variables are those factors which affect the operation of the process, weld quality, and economics of the process. A smooth running process and a quality weld deposit result when all of the variables are in proper balance. In electroslag welding, it is essential that the effects of each variable be fully understood since they differ from those with the conventional arc welding processes.

Form Factor

The slow cooling rate and solidification pattern of an electroslag weld are similar to metal cooling in a mold. Heat is removed from the molten weld metal by the cool base metal and the retaining shoes which form the mold in ESW. Solidification begins at these cooler areas and progresses toward the center of the weld. However, since filler metal is added continuously and the joint fills during welding, solidification progresses from the bottom of the joint, as shown in Fig. 7.4.

The angle at which the grains meet in the center is determined by the shape of the molten weld pool. Weld pool shape can be expressed by the term "form factor." The form factor is the ratio of the weld pool width to its maximum depth. Width is the root opening plus the total penetration into the base plates. Depth is the distance from the top of the molten weld pool to its lowest level of the liquid-solid interface. Welds having a high form factor (wide width and shallow weld pool) tend to solidify with the grains meeting at an acute angle. Welds having a low form factor (narrow width and deep weld pool) tend to solidify with the grains meeting at an obtuse angle. Thus, the form factor indicates how the grains from opposite sides of the weld meet at the center.

The angle at which the grains meet in the center determines whether the weld will have a high or low resistance to hot center line cracking. If the dendritic grains meet at an obtuse (large) included angle, the cracking resistance will be low. However, if the angle is acute (small), the cracking resistance will be high. Therefore, maximum resistance to cracking is obtained with a high form factor.

The shape of the molten weld pool and resultant form factor is controlled by the welding variables. However, form factor alone does not control cracking. The base metal composition (especially the carbon content), the filler metal composition, and joint restraint also have a significant effect on cracking propensity.

Welding Amperage

Welding amperage and electrode feed rate are directly proportional and can be treated as one variable. Increasing the electrode feed speed increases the welding amperage and the deposition rate when a constant voltage power source is used.

As the welding amperage is increased, so is the depth of the molten weld pool. When welding with a 3.2 mm (1/8 in.) diameter electrode and below approximately 400 A, an increase in amperage also increases weld width. The net result is a slight decrease in form factor. However, when operating a 3.2 mm (1/8 in.) electrode above 400 A, an increase in amperage reduces weld width. Thus, the net effect of increasing the welding amperage is to decrease the form factor and thus lower the resistance to cracking.

Welding amperages of 500 to 700 A are commonly used with 3.2 mm (1/8 in.) electrodes. Metals or conditions prone to cracking may require a high form factor associated with welding amperages below 500 amperes.

Welding Voltage

Welding voltage is an extremely important variable. It has a major effect on the depth of fusion into the base metal and on the stable operation of the process. Welding voltage is the primary means for controlling depth of fusion. Increasing the voltage increases depth of fusion and the width of the weld. Depth of fusion must be somewhat more than necessary in the center of the weld to assure complete fusion at the edges where the chilling effect of the water-cooled shoes must be overcome.

Since an increase in welding voltage increases the weld width, it also increases the form factor and thereby increases cracking resistance.

The voltage must also be maintained within limits to assure stable operation of the process. If the voltage is low, short circuiting or arcing to the weld pool will occur. Too high a voltage may produce unstable operation because of slag spatter and arcing on the top of the slag bath. Welding voltages of 32 to 55 volts per electrode are used. Higher voltages are used with thicker sections.

Electrode Extension

The distance between the slag pool surface and the end of the guide (contact) tube, when the conventional welding method is used, is referred to as the dry electrode extension. There is no dry extension in the consumable guide method. Using constant voltage power and constant electrode feed speed, increasing the electrode extension will increase the resistance in the dry extension, which must be balanced by a greater length of electrode within the conductive bath. This causes the power supply to reduce its current output, which slightly increases the form factor. If the voltage setting is increased to restore the previous current level, the form factor will noticeably increase.

Electrode extensions of 50 to 75 mm (2 to 3 in.) are generally used. Extensions of less than 50 mm (2 in.) usually cause overheating of the guide tube. Those greater than 75 mm (3 in.) cause overheating of the electrode because of the increased electrical resistance. Hence, at long extensions, the electrode will melt at the slag bath surface instead of in the bath. This will result in instability and improper slag bath heating.

Electrode Oscillation

Plates up to 75 mm (3 in.) thick can be welded with a stationary electrode and high voltage. However, the electrode is usually oscillated horizontally across the plate thickness when the material exceeds 50 mm (2 in.). The oscillation pattern distributes the heat and helps to obtain better edge fusion. Oscillation speeds vary from 8 to 40 mm/s (20 to 100 in./min) with the speed increasing with the plate thickness. Generally, oscillation speeds are based on a traverse time of 3 to 5 seconds. Increasing the oscillation speed reduces the weld width and hence the form factor. Thus, the oscillation speed must be balanced with the other variables. A dwell period is used at each end of the oscillation travel to obtain complete fusion with the base metal

and to overcome the chilling effect of the retaining shoes. The dwell time may vary from 2 to 7 seconds.

Slag Bath Depth

A minimum slag bath depth is necessary so that the electrode will enter into the bath and melt beneath the surface. Too shallow a bath will cause slag spitting and arcing on the surface. Excessive bath depth reduces the weld width and hence the form factor. Slag bath circulation is poor with excessive depth, and slag inclusions may result. A bath depth of 38 mm (1-1/2 in.) is optimum, but it can be as low as 25 mm (1 in.) or as high as 51 mm (2 in.) without significant effect.

Number of Electrodes and Spacing

As the metal thickness per electrode increases, the weld width decreases slightly; but the weld pool depth decreases greatly. Thus, the form factor improves as the thickness of material increases for a given number of electrodes. However, a point is reached where the weld width at the cool retaining shoes is less than the root opening, and lack of edge fusion results. At this point, the number of electrodes must be increased. In general, one oscillating electrode can be utilized for sections up to 130 mm (5 in.) thick, and two oscillating electrodes up to 300 mm (12 in.) thick. Each additional oscillating electrode will accommodate approximately 150 mm (6 in.) of additional thickness. This applies to both the wire and consumable guide methods. If nonoscillating electrodes are used, each electrode will handle approximately 65 mm (2-1/2 in.) of plate thickness.

Root Opening

A minimum root opening is needed for sufficient slag bath size, good slag circulation, and, in the case of the consumable guide method, clearance for the guide tube and its insulation. Increasing the root opening does not affect the weld pool depth. However, it does increase the weld width and hence the form factor. Excessive root openings will require extra amounts of filler metal, which may not be economical. Also, excessive root openings may cause a lack of edge fusion. Root openings are generally in the range of 20 to 40 mm (3/4 to 1-1/2 in.), depending on the base metal thickness, the number of electrodes, and the use of electrode oscillation.

WELDING PROCEDURE

Joint Preparation

One of the major advantages of the electroslag process is the relatively simple joint preparation. It is basically a square groove joint, so the only preparation required is a straight edge on each groove face, which can be produced by thermal cutting, machining, etc. If sliding retaining shoes are used, the surfaces of the plate on either side of the groove must be reasonably smooth to prevent slag leakage and jamming of the shoes.

The joint should be free of any oil, heavy mill scale, or moisture, which is true for any welding process. However, the joint need not be as clean as would normally be required for other welding processes. Oxygen-cut surfaces should be free of adhering slag, but the slightly oxidized surface is not detrimental.

Care should be exercised to protect the joint prior to initiating the weld. Moisture bearing materials packed around the shoes, commonly known as "mud," will cause porosity. It should be completely dry before welding. Also, leaking water-cooled shoes can cause porosity or weld face defects.

Joint Fit-up

Prior to welding, components should be set up with the proper joint alignment and root opening. Rigid fixturing or strongbacks that bridge the joint should be used. Strongbacks are bridge-shaped plates that are welded to each component along the joint so that alignment during welding is maintained. They are designed with clearance to span fixed-position retaining shoes. After welding is completed, they are removed.

For the consumable guide method, imperfect joint alignment can be accommodated to some degree. Plates with large misalignments may be welded by special retaining shoes that are adaptable to the fit-up. Alternatively, the space between the shoes and the work is packed with refractory material or steel strips (the steel must be similar in composition to the base metal). Afterwards, the steel strips may be removed in

such a way as to blend the weld faces smoothly with the adjacent base metal.

Experience will dictate the proper root opening for each application. As welding progresses up the joint, the parts are drawn together by weld shrinkage. Therefore, the root opening at the top of the joint should be approximately 3 to 6 mm (1/8 in.) more than at the bottom to allow for this shrinkage. Factors influencing shrinkage allowance include material type, joint thickness, and joint length.

The use of proper root opening and shrinkage allowance is important for maintaining the dimensions of the weldment. However, if the root opening proves incorrect, the welding conditions may be varied to compensate for it, within limits. For example, if the initial root opening is too small, the wire feed speed may be lowered to reduce deposition rate. If the root opening is too large, the wire feed speed may be increased within good operating limits. Regardless of the type of base metal, the voltage should be increased to account for a wider root opening.

If the root opening is greater than the capability of the normal number of electrodes used, an additional electrode may be added if there is space available. In some cases, excessive root opening may be compensated for by oscillation of the electrode(s).

When a root opening is too small, the joint can fill too fast, causing weld cracks or lack of edge fusion. It is also possible for small root openings to close up from weld shrinkage and stop an oscillating guide tube from traversing.

Inclination of Work

The axis of the weld joint should be in the vertical or near vertical position. It may deviate as much as 10 degrees from vertical. With greater deviations, it becomes increasingly difficult to weld without slag inclusions and lack of edge fusion. Alignment of the guide tube can be a problem with the consumable guide method. Large insulators (flux rings) or flux coated guide tubes with spring clips are required to keep the tube aligned in the joint.

Grounding the Workpiece

A good electrical ground (work connection) is important because of the relatively high amounts of current used for the ESW process. Normally, two 4/0 welding cables are sufficient for each electrode. It is best to attach the work-lead directly under the sump, that is, below the electrode. In that location, the effect of a very strong magnetic field in the weldment on filler metal transfer will be minimized.

Spring type ground clamps are not recommended because they tend to overheat. A more positive connection, such as a "C" type clamp, is best.

Run-on and Run-off Tabs, and Starting Plate

Where full penetration is required for the full length of the joint, run-on and run-off tabs will be required. The run-on tabs, frequently referred to as starting tabs, are located at the bottom of the joint. They are used in conjunction with a starting plate to initiate the welding process. Generally, the tabs and starting plate are made from the same or similar material as the base metal. The starting tabs and plate form a sump in which the weld is started. In this case, the sump is removed and discarded after welding. The tabs and starting plate are the same thickness as the base plates.

Where the run-on tabs and starting plate are disposable, a tab is welded to the bottom of each base plate. The starting plate is welded across the bottom of the tabs to form the sump. The faces of the sump are flush with the base plate surfaces.

Copper sumps may be used, in which case water cooling is usually necessary. The arc is not started on the copper sump because it would melt through the water jacket. Normally, one or two small blocks of base metal are placed in the bottom of the copper sump, and the arc is started on them.

Disposable run-off tabs should also be a material the same as or similar to the base metal. Copper tabs may be used, but they must be water cooled. The run-off tabs should be the same thickness as the base metal and securely attached to both plates at the end of the joint. The weld is completed in the cavity that they form, above the base plates.

Electrode Position

The position of the electrode will determine where the greatest amount of heat is generated. The electrode should normally be centered in the joint. However, if the electrode is cast toward one side, then the wire guide may be moved in the opposite direction to compensate for the cast. In the welding of corner and T-joints, or any joint where a fillet is used, the electrode may need to be offset to produce the required weld metal geometry.

Initiation and Termination of Welding

Unless molten slag is poured into the joint, the normal starting method is to strike an arc between the electrode and the starting plate. This may be done by two methods: (1) A steel wool ball is inserted between the electrode and the base metal, and the power is turned on; and (2) The electrode is advanced towards the starting plate with the power on. The latter method requires a chisel point on the end of the electrode. Once an arc is struck, flux is slowly added until the arc is extinguished. The process is now in the electroslag mode.

It is extremely important that the welding operation proceed without interruption. The equipment should be checked, and the electrode and flux supplies determined adequate before the weld is started. The weld must not be terminated to replenish the electrode supply. The welding equipment must be capable of operating continuously until the weld is finished.

When the weld has reached the run-off tabs, termination should follow a procedure which fills the crater, or crater cracking may occur. Usually the electrode feed is gradually reduced when the slag reaches the top of the run-off tabs and the crater fills. Welding amperage decreases simultaneously. When the electrode feed stops, the power source is turned off. The tabs and the weld metal are then cut off flush with the top and bottom edges of the weldment.

Slag Removal

A common chipping gun will be effective in slag removal, although a simple slag hammer or pick will also do the job. Slag will also adhere to the copper shoes and copper sump, if used. They must be cleaned before another weld is made. Eye protection should be worn during slag removal operations.

TYPES OF WELDS

Joint Designs

There is one basic type of joint, which is the square groove butt joint. The square groove weld joint preparation can be used to produce other types of joints such as the corner, T- and edge

joints. It is also possible to make transition joints, fillet welds, cross-shaped joints, overlays, and weld pads with the ESW process. Figure 7.8 shows typical ESW joint designs and the outlines of the final welds. Specially designed retaining shoes are needed for joints other than butt, corner, and T-joints.

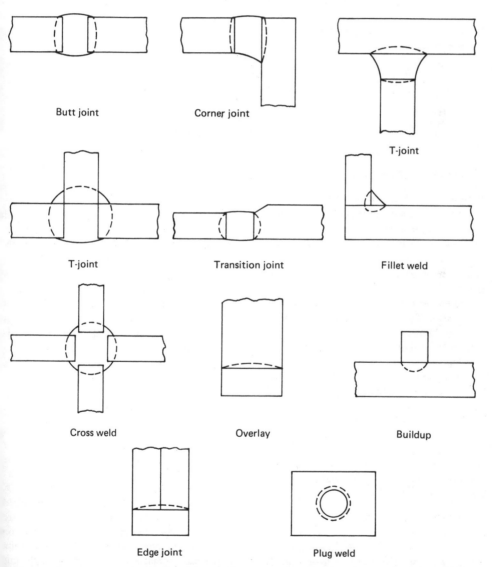

Butt joint

Corner joint

T-joint

T-joint

Transition joint

Fillet weld

Cross weld

Overlay

Buildup

Edge joint

Plug weld

Fig. 7. 8—Joint designs for electroslag welding—dashed line shows depth of fusion into the base metal

Circumferential Welds

A square groove butt joint is used to join the two round components end to end. The weld is made by rotating the parts and allowing the slag and molten weld metal to remain at the 3 o'clock position with clockwise rotation. The welding head is stationary until the finish of the weld. Special starting and run-off techniques are required to complete the seam. Small diameter weldments may not be economically feasible. However, as wall thickness and diameter increase, the cost savings may make the process attractive for the application.

WELD QUALITY

Welds made by the ESW process under proper operating conditions are high quality and free from harmful discontinuities. In any welding process, however, abnormal conditions that may occur during welding may cause discontinuities in the weld. Some of these discontinuities, their possible causes, and remedies are given in Table 7.3. The information is primarily applicable to electroslag welded joints in carbon and low alloy steels.

ELECTROGAS WELDING

FUNDAMENTALS OF THE PROCESS

Definition and General Description

Electrogas welding (EGW) is a variation of both the gas metal arc welding process (GMAW-EG) and the flux cored arc welding process (FCAW-EG), with retaining shoes (dams) to confine the molten weld metal for vertical position welding. Additional shielding may or may not be provided by an externally supplied gas or gas mixture. The mechanical aspects of the electrogas welding process are similar to those of the electroslag welding process from which it was developed.

A square groove weld joint is positioned so that the axis or length of the joint is vertical or nearly vertical. There is no repositioning of the joint once welding has started. Electrogas welding is a machine welding process, and like electroslag welding, once started it continues to completion. The welding action is quiet with a low level of spatter. Welding is usually done in one pass.

Principles of Operation

The consumable electrode, either solid or flux cored, is fed downward into a cavity formed by the groove faces of the plates to be welded and two water-cooled retaining shoes. An electric arc is initiated between the electrode and a starting plate at the bottom of the joint.

Heat from the electric arc melts the groove faces and the continuously fed electrode. Melted electrode and base metal continuously collect in a pool beneath the arc and solidify to form the weld metal. The electrode may be oscillated horizontally through the joint in thick sections for uniform distribution of heat and weld metal. As the cavity fills, one or both retaining shoes move upward with the welding head. Although the axis of the weld is vertical, the welding position is actually flat with vertical travel.

Process Variations

Solid Electrode. Figure 7.9 shows electrogas welding with a solid wire electrode (GMAW-EG). Normally, only one electrode is fed into the joint, although two are sometimes used with thick sections. The electrode is fed through a welding gun, which is also called an electrode guide or a contact tip holder. Gas shielding, normally carbon dioxide (CO_2) or an argon-CO_2 mixture is provided over the weld metal to shield it from air, similar to conventional gas metal arc welding (GMAW). There is no flux present on the top of the molten weld pool. Protection of the molten weld metal is provided solely by the gas shielding.

Table 7.3—Electroslag weld discontinuities, their causes and remedies

Location	Discontinuity	Causes	Remedies
Weld	1. Porosity	1. Insufficient slag depth 2. Moisture, oil, or rust 3. Contaminated or wet flux	1. Increase flux additions 2. Dry or clean workpiece 3. Dry or replace flux
	2. Cracking	1. Excessive welding speed 2. Poor form factor 3. Excessive center-to-center distance between electrodes or guide tubes	1. Slow electrode feed rate 2. Reduce current; raise voltage; decrease oscillation speed 3. Decrease spacing between electrodes or guide tubes
	3. Nonmetallic inclusions	1. Rough plate surface 2. Unfused nonmetallics from plate laminations	1. Grind plate surfaces 2. Use better quality plate
Fusion line	1. Lack of fusion	1. Low voltage 2. Excessive welding speed 3. Excessive slag depth 4. Misaligned electrodes or guide tubes 5. Inadequate dwell time 6. Excessive oscillation speed 7. Excessive electrode to shoe distance 8. Excessive center-to-center distance between electrodes	1. Increase voltage 2. Decrease electrode feed rate 3. Decrease flux additions; allow slag to overflow 4. Realign electrodes or guide tubes 5. Increase dwell time 6. Slow oscillation speed 7. Increase oscillation width or add another electrode 8. Decrease spacing between electrodes
	2. Undercut	1. Too slow welding speed 2. Excessive voltage 3. Excessive dwell time 4. Inadequate cooling of shoes 5. Poor shoe design 6. Poor shoe fit-up	1. Increase electrode feed rate 2. Decrease voltage 3. Decrease dwell time 4. Increase cooling water flow to shoes or use larger shoes 5. Redesign groove in shoe 6. Improve fit-up; seal gap with refractory cement or asbestos dam
Heat-affected zone	1. Cracking	1. High restraint 2. Crack-sensitive material 3. Excessive inclusions in plate	1. Modify fixturing 2. Determine cause of cracking 3. Use better quality plate

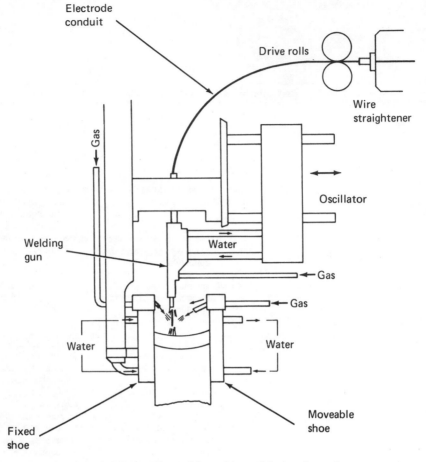

Fig. 7.9—Electrogas welding with a solid wire electrode

Electrogas welding with solid wire can be used to weld plates ranging in thickness from approximately 10 to 100 mm (3/8 to 4 in). However, thicknesses from 13 to 75 mm (1/2 to 3 in.) are most commonly welded. Electrode diameters most commonly used are 1.6 mm (1/16 in.), 2.0 mm (5/64 in.), and 2.4 mm (3/32 in.).

Flux Cored Electrode. Figure 7.10 shows electrogas welding with a flux cored wire (FCAW-EG). The principles of operation and character-istics are identical with the solid wire variation except a thin layer of slag forms on top of the molten weld pool. Gas shielding is used when the electrode type requires it.

Electrogas welding may be done with the the self shielding types of flux cored electrodes. Then no external gas shielding is required. The self shielded electrodes operate at higher amper-age levels, and deposition rates are higher than with the gas shielded types.[1]

---

1. Refer to Chapter 5 for further information on the oper ating characteristics of gas shielded and self shielde flux cored electrodes.

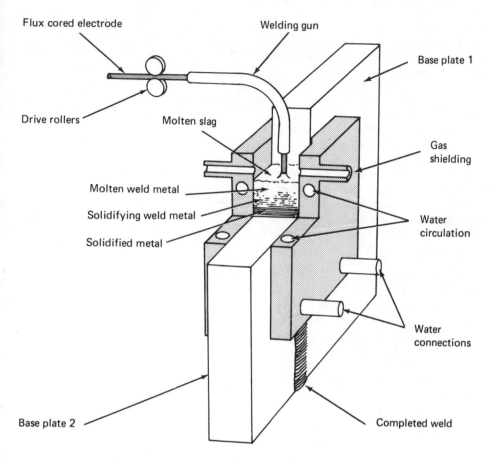

Flux cored electrode

Welding gun

Base plate 1

Drive rollers

Molten slag

Gas shielding

Molten weld metal

Solidifying weld metal

Solidified metal

Water circulation

Water connections

Base plate 2

Completed weld

Fig. 7.10—Electrogas welding with a flux cored electrode

EQUIPMENT

The basic mechanical equipment for electrogas welding is similar to that for conventional electroslag welding. The basic difference is the provision for introducing shielding gas coverage to the arc and the molten weld pool when shielding gas is required. Otherwise, there is no difference.

Essentially, the components of an electrogas welding system are a dc power supply, water-cooled shoes for retaining the molten weld metal, a welding gun, a device for feeding electrode wire, a mechanism for oscillating the welding gun, and the equipment needed for supplying shielding gas, when used. In a typical electrogas welding system, the essential components, with the exception of the power supply, are incorporated in an assembly (welding head) that moves vertically upward as welding progresses. Control devices for water flow, horizontal pressure on the retaining shoes, oscillation of the welding gun, wire feed, and vertical movement are similar to those used with electroslag welding. Metering control of shielding gases is required when electrogas welding with solid and some flux cored electrodes.[2]

2. See Chapters 3, 4, and 5 concerning shielding gases, their storage, and regulation equipment.

Power Sources

The power for electrogas welding is normally dc, reverse polarity (electrode positive). The power source may be either a constant voltage or a constant current type. When a constant voltage unit is used, vertical travel is controlled either manually or by a device, such as a photoelectric cell, that senses the height of the rising weld pool. With a constant current type of power source, vertical travel can be controlled by changes in arc voltage. When the arc voltage drops below a preset value, the travel mechanism is automatically actuated, and the unit moves up until the preset voltage is restored. Some manual control may be required at the beginning of welding and from time to time during the welding operation.

The power source must be rugged and capable of delivering the required current without interruption during the welding of a seam that may be several meters (feet) in length. Power sources used for electrogas welding are usually rated at 750 to 1000 A at 100 percent duty cycles. Direct current is usually supplied by transformer-rectifier power sources, although motor-generators are sometimes used.[3]

Electrode Wire Feeder

The electrode wire feeder, which is mounted as an integral part of the vertical-moving welding head, is the push type, such as used with automatic gas metal arc or self-shielded flux cored arc welding. It must be capable of supplying the electrode at a high rate of speed. Wire feeders are often equipped with start-weld speed controls, which automatically increase the electrode speed as the starting voltage changes to the preset welding voltage.

The electrode feed system may include a wire straightener located between the wire reel and the feed rolls. When properly set, the straightener eliminates any cast in the electrode. Since the electrode extension in electrogas welding is 38 mm (1-1/2 in.) or more, the projecting electrode must be straight.

Welding Guns

The welding gun for electrogas welding

performs the same functions as those for gas metal arc welding or flux cored arc welding. It guides the electrode into the desired position in the joint cavity; it transmits the welding current to the electrode; and, in some applications, it provides shielding gas coverage around the electrode and the arc. The gun may or may not be water cooled.

A main difference between an electrogas welding gun and a GMAW or FCAW welding gun is the restriction on its dimension parallel to the root opening between the plates. The gun nozzle must fit into this narrow opening. As the electrogas welding machine moves vertically up the seam, the welding gun (or at least part of it) rides inside the opening directly above the weld pool. There must be enough clearance to permit horizontal oscillation between the two shoes. For this reason, the width of a gun is often limited to 10 mm (3/8 in.) to fit in a minimum of 17 mm (11/16 in.) root opening. Larger welding guns may be used for welding thicker plate, where wider root openings are used. A larger gun is either water cooled or heavily insulated against the heat of the molten weld pool.

Welding Gun Oscillation

The arc heat must be applied uniformly through the weld joint. With plates 32 to 102 mm (1-1/4 to 4 in.) thick, the welding gun is oscillated horizontally over the weld pool to achieve uniform metal deposition and insure fusion to both root faces. The oscillator unit and controls provide a fixed traverse travel rate across the joint, with adjustable dwell times at each end. Horizontal oscillation is usually not needed for welding plates less than 32 mm (1-1/4 in.) thick. However, it is sometimes used with a thinner plate to control the depth of fusion into the base plates.

Retaining Shoes

As in electroslag welding, shoes or dams are used to retain the molten weld metal in the joint. Usually both shoes move upward as welding progresses. In some weldments, a stationary weld backing bar may replace one of the shoes.

To prevent molten metal from welding to

---

3. See Chapter 1 for additional information on arc welding power sources.

the copper shoes and to hasten solidification of the weld, the shoes are water cooled. The copper shoes can be given a concave contour to develop the desired weld reinforcement on each side of the plate.

In a typical setup, one of the shoes is pressed tightly against the joint by a rigid support from the welding head. The other shoe is movable normal to the plate surfaces to accommodate variations in the thickness of the plates. It is held against the joint by an air cylinder. In another type of arrangement, the two vertically movable shoes are held against the work by a spring loaded cantilever arm.

Gas Boxes

For electrogas welding, it is essential that a positive arc shield be provided. Rather than rely solely upon shielding gas from the nozzle of the welding gun, so-called "gas boxes" may be mounted on the retaining shoes to provide auxiliary gas shielding to the electrode, the welding arc, and the molten weld pool, as shown in Figs. 7.9 and 7.10. Gas from these compartments is directed out through ports, so placed and sized as to give uniform coverage of the arc and weld pool.

Controls

As in all automatic welding systems, various mechanical, electrical, and electronic control devices are required for optimum functioning. With the exception of the modes for vertical travel control and welding gun oscillation, these devices are primarily adaptations of the controls used with gas metal arc, flux cored arc, and electroslag welding.

Mechanical methods of travel, such as motor-driven hoists or track mounting, are often supplemented by electrical sensing of the weld surface level to increase or decrease travel speed, and to maintain a specific electrode extension.

Electrode Wire Supply

Most commercial electrogas welding machines mount the electrode wire supply at the rear of the machine. The wire supply provides a continuous supply of electrode. It should allow feeding of the electrode with minimal load on the wire feeder so that binding or wire stoppage

will not take place. The wire supply must be adequate to complete the entire weldment in a single pass.

Electrode diameters commonly used range from 1.6 to 3.2 mm (1/16 to 1/8 in.). The wire feeder must accommodate the required diameter and feed it at a speed necessary to meet the required deposition rate.

MATERIALS

Base Metals

Many low and medium carbon steels are electrogas welded. These include the following:

(1) Carbon steels, such as AISI 1018 and 1020

(2) Structural steels, such as ASTM-A36, A131, A441, and A573

(3) Pressure vessel steels, such as ASTM A285, A515, A516, and A537

Most applications of electrogas welding are for the fabrication of ship components, structural steel, and pressure vessel manufacture.

Consumables

Electrodes. There are generally two forms of electrodes used with the electrogas welding process, namely: (1) flux cored and (2) solid wires. Both types of wires are used commercially. AWS Specification A5.26 covers the requirements of these electrodes for welding carbon and high strength, low alloy steels (non heat-treatable types).

Flux cored electrodes are classified on the basis of whether or not a separate shielding gas is required, and also the chemical composition and as-welded mechanical properties of the deposited weld metal. Solid electrodes are classified on the basis of their chemical composition, as manufactured, and the as-welded mechanical properties of the deposited weld metal. There are six classifications based on the mechanical properties of deposited weld metal, separated into two levels. One level has a minimum tensile strength of 415 MPa (60 ksi), and for the other group, 485 MPa (70 ksi). Two classifications in each group have notch toughness requirements as determined by the Charpy V-notch impact test.

Flux cored electrogas electrodes contain a lower percentage of slagging type compounds in their flux fill than standard flux cored electrodes. The electrodes provide a thin slag layer between the retaining shoes or backing, which produces a smooth weld surface. Flux cored electrodes are available in standard sizes from 1.6 to 4.0 mm (1/16 to 5/32 in.) diameter.

Solid wires are generally identical to the wires used for gas metal arc welding. They are available in the same standard sizes as flux cored electrodes.

Both flux cored and solid wires are available in various chemical compositions. They are designed to introduce the necessary alloying elements into the weld metal to achieve the desired tensile and impact strength properties, or appropriate combinations of these properties. These properties are normally obtained in the as-welded condition, since this process is mostly used in field construction where postweld heat treatment is not normally done.

Shielding Gas. For welding steel with flux cored electrogas electrodes, carbon dioxide shielding gas is normally used. Recommended gas flow rates range from 14 to 66 L/min (30 to 140 ft³/h). A mixture of 80 percent argon—20 percent carbon dioxide is normally used for welding steel with solid electrodes, and it may be used with flux cored electrodes.

Some flux cored electrodes are the self shielded type. They generate a dense shielding vapor from the heat of the arc to protect the filler metal and the molten weld metal. This method of shielding is similar to the self shielded, flux cored arc welding process.

METALLURGICAL CONSIDERATIONS

The electrogas welding process is actually a consumable electrode, arc welding process. The welding energy density is significantly higher and more concentrated than in electroslag welding. However, the technique of welding is the same. Consequently, the macrostructure of welds made by the two processes is similar. Because the freezing pattern is faster and the molten weld pool is smaller, the grain sizes of the weld and heat-affected zones of electrogas welds tend to be smaller than those of electroslag welds. Consequently, mechanical properties of electrogas welds may be noticeably better.

As with electroslag welds, the ratio of width to the depth of the weld pool (form factor) is an important consideration. The effects of the form factor and the included angle between the principal dendrites in the weld are the same as with electroslag welding. A high form factor and an acute included angle are best.

Because of the similarity of the electrogas and electroslag welding processes and the base metals welded, the requirements for postweld heat treatments are the same. Normally, no preheat is used with either process.

MECHANICAL PROPERTIES

Typical weld-metal mechanical properties are listed in Table 7.4 for electrogas welds (FCAW-EG) in various steels.

One factor that should be considered in electrogas welding is the notch toughness properties of the weld metal and heat-affected zones in the as-welded condition. By using suitably alloyed electrodes and controlled welding conditions, the Charpy V-notch impact properties of the weld metal can be equal to or better than those of the base metal for most structural and pressure vessel steels.

Significant variations in electrogas weld metal notch impact valves may occur, depending upon the location of the test specimen. The weld center, or plane of neutral axis, usually has the lower values. The metal at the weld faces, where a failure due to impact loading would probably start, typically has considerably better notch toughness than the metal at the weld center.

GENERAL PROCESS APPLICATIONS

Electrogas welding is used for joining thick plates that must be welded in a vertical position or that can be positioned vertically for welding. Welding is usually done in one pass. The ecomomic practicality of the process depends on the thickness of the plate and the length of the joint. The process offers the advantages of flat position welding, but there must be a sufficient number of welds to justify the cost of the fixturing and the special welding equipment.

Large structures, such as ship hulls, bridges, caissons, storage tanks, offshore drilling rigs, and some elements for high-rise building con-

Table 7.4—Typical mechanical properties of weld metal from electrogas welds (FCAW-EG) in several steels

Base metal, ASTM	Thickness mm	in.	AWS electrode class.	Shielding gas[2]	Yield strength MPa	ksi	Tensile strength MPa	ksi	Elong., % in 51 mm (2 in.)	Reduction of area, %	Test temp. °C	°F	WMCL J	ft·lb	HAZ J	ft·lb
A36	19	3/4	EG70T1	None	390	56.6	550	79.8	24	–	−16	4	49	36	–	–
A131-C	38	1-1/2	EG70T1	None	415	60.2	542	78.6	27	31.6	−18	0	19-23	14-17	23-37	17-27
A441	51	2	EG70T1	None	442	64.1	563	81.7	24	–	−18	0	58	43	–	–
A441	51	2	EG72T4	CO₂	472	68.4	637	92.4	24	61	−18	0	30	22	–	–
A516-70	38	1-1/2	EG70T1	None	423	61.4	558	80.9	29	35	−18	0	61	45	39	29
A516-70	38	1-1/2	EG72T4	AG20	538	78	619	89.8	29	67	− 7	20	41	30	38-91	28-67
A537-A	25	1	EG72T4	CO₂	430	62.3	572	83	29	70	− 7	20	34	25	39	29
A537-A	29	1-1/8	EG72T4	AG20	510	74	690	100	26	68	− 8	18	46	34	–	–
A537-A	38	1-1/2	EG70T1	None	385	55.9	549	79.6	30	40	−18	0	23	17	23	17

Charpy V-notch impact strength[1] — the Test temp., WMCL, and HAZ columns above.

1. Test specimen notch locations: WMCL—weld metal center line, HAZ—heat-affected zone in the base metal.
2. AG20 is a mixture of 80% argon-20% CO₂.

struction may be profitably fabricated by the electrogas process. The longitudinal seams on large diameter pipes and circular pressure vessels are being butt welded with the electrogas process.

The process is used primarily for the welding of carbon and alloy steels, but it is also applicable to austenitic stainless steels and other metals and alloys that can be welded by GMAW. Plate material may range from 10 to 102 mm (3/8 to 4 in.) in thickness. Usually, when the plates are more than 76 mm (3 in.) thick, the electroslag process is preferred.

The longer the joint to be welded, the more efficient is the process. For field welding of vertical joints in large storage tanks, the process eliminates tedious and costly manual welding operations. The use of self shielded, flux cored electrodes is especially advantageous, since bulky shielding gas equipment is not needed. Electrogas welding is adaptable to two-pass welding of double-V joints. Specially shaped retaining shoes are required.

Figure 7.11 shows the field welding of a vertical seam in a storage tank. This electrogas

Fig. 7.11—Field electrogas welding of a storage tank

equipment uses a flux cored electrode without supplementary shielding gas.

The welding of a ship hull is shown in Fig. 7.12. Two electrogas machines are welding two joints simultaneously under the control of one operator.

PROCESS VARIABLES

The welding variables in electrogas welding are similar to those of electroslag welding.

Each variable, under the control of the welding operator, affects the weld form factor in the same manner as in electroslag welding. As with electroslag welding, the effect of each variable and weld joint geometry must be fully understood by the operator.

Weld penetration in conventional arc welding processes is in line with the axis of the electrode. Penetration is generally increased by increasing the welding current. However, in electrogas welding, the base metal to be melted

Fig. 7.12—Electrogas welding of a ship hull using two machines controlled by one operator

(joint faces) is perpendicular to the axis of the electrode wire.

Increasing the welding amperage or electrode feed speed in electrogas welding will decrease the weld width. Increasing the arc voltage will increase the depth of fusion and the width of the weld. Arc voltages will vary from 30 to 55 V, depending on process variation, electrode size, and section thickness.

As the welding amperage increases, so will the electrode feed speed, the deposition rate, and the rate of joint fill (travel speed). Excessive welding amperage and electrode feed speed for a given set of conditions may cause a severe reduction in weld width or depth of fusion into the workpieces.

In gas shielded electrogas welding, an electrode extension of approximately 40 mm (1-1/2 in.) is used. For self shielded electrogas welding, an electrode extension of 60 to 75 mm (2-1/2 to 3 in.) is used. As in conventional GMAW and FCAW, this long extension increases the electrode melting rate because the electrode is resistance heated before it is melted.

Oscillation of the electrode is generally required for welding sections thicker than 30 mm (1-1/4 in.). Horizontal oscillation is controlled so that the electrode guide stops at a distance of approximately 10 mm (3/8 in.) from each retaining shoe. Oscillation speed is commonly 7 to 8 mm/s (16 to 18 in./min). It is necessary to dwell for a time period at each end of oscillation travel to obtain complete fusion across the weld faces. To overcome the chilling effect of water-cooled retaining shoes, the dwell time should range from 1 to 3 seconds.

WELDING PROCEDURE

As with electroslag welding, most electrogas welding is done with a square groove joint. The plates to be joined are positioned vertically and assembled with a root opening of approximately 17 mm (11/16 in.). Strongbacks are welded across the joint on one side only. The plates should be as near vertical as practical.

Starting and run-off tabs are generally used when the plate thickness exceeds 25 mm (1 in.). Plates less than approximately 25 mm (1 in.) re-

quire only a starting plate and no run-off tabs. It is best to attach the work lead (ground) to the starting sump or plate. On repetitive joints, copper sumps and run-off tabs may be used.

Most electrogas welding is done with two water-cooled retaining shoes which are moved vertically with the welding head. In some applications, it may be advantageous to replace the shoe on the side of the joint opposite to the welding head with a fixed copper shoe or a steel backing strip (plate).

The welding arc is initiated in a manner similar to gas metal arc and flux cored arc welding. After the arc is initiated, appropriate control adjustments are made by the operator during welding to keep the process under control.

TYPES OF WELDS

Electrogas welding is used to produce the same types of joints as those made with electroslag welding. Normally, the joint designs applicable to the conventional method of electroslag welding are also applicable to electrogas welding, as shown in Fig. 7.8.

WELD QUALITY

Electrogas welding is basically a gas metal arc or flux cored arc welding process. All discontinuities found in conventional welds made by the two processes can be found in electrogas welds made by the same process. However, the cause of some discontinuities, such as lack of fusion, may be different with electrogas welding.

Welds made by the electrogas welding process under normal operating conditions result in high quality welds free of harmful discontinuities. However, abnormal welding conditions may result in a defective weld. Weld metal discontinuities which may be encountered are slag inclusions, porosity, and cracks.

Slag Inclusions

Slag inclusions may occur in electrogas welds. The process is usually a single pass process, and so slag removal between passes is not required. The weld metal solidification rate is relatively slow. Ample time is available for any molten slag to float to the surface of the molten weld metal.

However, when electrode oscillation is used, the weld may partially solidify near either shoe when the arc is near the other. When the arc returns, slag may be trapped if it is not remelted.

Porosity

Flux cored electrodes contain deoxidizing and denitriding ingredients in the core. A combination of shielding gas and slagging compounds in the flux core usually produces sound welds, free from porosity. However, if something interferes with the normal shielding gas coverage, porosity may result.

Other causes of porosity in electrogas welds may be excessive air drafts, water leaks in the retaining shoes, insufficient core material in FCAW electrodes, contaminated electrode or shielding gas, and air infusion at the beginning of welding.

When porosity is encountered in electrogas welds, it usually starts near the edges of the weld and runs toward the weld centerline, following the weld metal solidification path. Porosity in electrogas welds usually cannot be detected by visual inspection methods, except when the starting and run-off tabs are removed to expose the interior of the weld.

Cracks

Cracking in electrogas welds does not occur with normal welding conditions and a normal form factor. The relatively slow heating and cooling of the weld metal considerably reduces the risk of cold cracks developing in the weld. Also, the heat-affected zone has a high resistance to cold cracking.

If cracking does occur, it is usually the hot crack type. The cracks form at high temperatures in conjunction with, or immediately after, solidification. They are nearly always located in the center part of the weld.

Weld cracking is best overcome by modifying the solidification pattern of the weld. This can be accomplished by altering the weld pool shape through appropriate changes in the welding variables. The arc voltage should be increased and the amperage and travel speed decreased. Often, increasing the root opening between the plates will help, although this may be uneconomical. If cracks are caused by high carbon or high sulphur in the steel, penetration of the base metal should be kept low to minimize the dilution of base metal in the weld deposit. Further, an electrode with high manganese content should be used for welding high sulphur steel.

SUPPLEMENTARY READING LIST

Agic, T., and Hampton, J.A. Electroslag welding with consumable guide on the Bank of American World Headquarters Building. *Welding Journal* 47 (12), Dec. 1968, pp. 939-946.

Bentley, K.P. Toughness in electroslag welds. *British Welding Journal* 15 (8), Aug. 1968, pp. 408-410.

Brøsholen, A. Skaug, E, and Visser, J.J. Electroslag welding of large castings for ship construction. *Welding Journal* 56 (8), Aug. 1977, pp. 26-30.

Campbell, H.C. Electroslag, Electrogas, and Related Welding Processes. Bulletin No. 154. New York: Welding Research Council, 1970.

des Ramos, J.B., Pense, A.W., and Stout, R.D. Fracture toughness of electroslag welded A537G steel. *Welding Journal* (55) 12, Dec. 1976, pp. 389-s—399-s.

Dorschu, K.E., Norcross, J.E., and Gage, C.C. Unusual electroslag welding applications. *Welding Journal* 52 (11), Nov. 1973, pp. 710-716.

Hannahs, J.R. and Daniel, L. Where to consider electroslag welding. *Metal Progress* 98 (5), May 1970, pp. 62-64.

Harrison, J.D. Fatigue tests of electroslag welded joints. *Metal Construction and British Welding Journal* 1 (8), Aug. 1969, pp. 366-370.

Kenyon, N., Redfern, G.A., and Richardson, R.R. Electroslag welding of high nickel alloys. *Welding Journal* 54 (7), July 1977, pp. 235-s—239-s.

Lawrence, B.D. Electroslag welding curved and tapered cross-sections. *Welding Journal* 52 (4), Apr. 1973, pp. 240-246.

Norcross, J.E. Properties of electroslag and electrogas welds. *Welding Journal* 44 (3), Mar. 1965, pp. 135-s—140-s.

Normando, N.J., Wilcox, D.V., and Ashton, R.F. Electrogas vertical welding of aluminum. *Welding Journal* 52 (7). July 1973, pp. 440-448.

Okumura, M., et al. Electroslag welding of heavy section 2¼ Cr-1Mo steel. *Welding Journal* (55) 12, Dec. 1976, pp. 389-s—399-s.

Parrott, R.S., Ward, S.W., and Uttrachi, G.D. Electroslag welding speeds shipbuilding. *Welding Journal* 53 (4), Apr. 1974, pp. 218-222.

Patchett, B.M., and Milner, D.R. Slag-metal reactions in the electroslag process. *Welding Journal* 51 (10), Oct. 1972, pp. 491-s—505-s.

Paton, B.E. Electroslag welding of very thick material. *Welding Journal* 41 (12), Dec. 1962, pp. 1115-1123.

Paton, B.E., ed. *Electroslag Welding,* 2nd ed. Miami, Florida: American Welding Society, 1962.

Rote, R.S. Investigation of properties of electroslag welds in various steels. *Welding Journal* 43 (5), May 1964, pp. 421-426.

8

Stud Welding

Prepared by a committee consisting of

J.C. JENKINS, *Chairman*
  Nelson Division of TRW, Inc.
W. BODINE
  KSM Division, Omark Industries, Inc.

E. DASH
  Douglas Aircraft Company
A. HUBBARD
  Caterpillar Tractor Company

R. SHOLLE
Nelson Division of TRW, Inc.

8

Stud Welding

INTRODUCTION

Stud welding is a general term for joining a metal stud or similar part to a workpiece. Welding can be done by a number of welding processes including arc, resistance, friction, and percussion. Of these processes, the one that utilizes equipment and techniques unique to stud welding is arc welding. This process, known as *stud arc welding,* will be covered in this chapter. The other processes use conventionally designed equipment with special tooling for stud welding. Those processes are covered in other chapters of the Handbook.[1]

In stud arc welding, the base (end) of the stud is joined to the other work part by heating them with an arc drawn between the two. When the surfaces to be joined are properly heated, they are brought together under pressure. Stud welding guns are used to hold the studs and move them in proper sequence during welding. There are two basic power supplies used to create the arc for welding studs. One type uses dc power sources similar to those used for shielded metal arc welding. The other type uses a capacitor storage bank to supply the arc power.

The stud arc welding processes using these power sources are commonly known as *arc stud welding* and *capacitor discharge stud welding* respectively.

Arc stud welding, the more widely used of the two major stud welding processes, is similar in many respects to manual shielded metal arc welding. The heat necessary for welding of studs is developed by a dc arc between the stud (electrode) and the plate (work) to which the stud is to be welded. The welding current is supplied by either a dc motor-generator or a dc transformer-rectifier power source, similar to those used for shielded metal arc welding. Welding time and the plunging of the stud into the molten weld pool to complete the weld are controlled automatically. The stud, which is held in a stud welding gun, is positioned by the operator, who then actuates the unit by pressing a switch. The weld is completed quickly, usually in less than one second. This process generally uses a ceramic arc shield, called a *ferrule*. It surrounds the stud to contain the molten metal and shield the arc. A ferrule is not used with some special welding techniques, nor with some nonferrous metals.

Capacitor discharge stud welding derives its heat from an arc produced by the rapid discharge of electrical energy stored in a bank of

1. See Chapter 26, "Spot, Seam, and Projection Welding," and Chapter 27, "Flash, Upset and Percussion Welding," Section 2; Chapter 50, "Friction Welding," Section 3A, *Welding Handbook,* 6th ed. These chapters are scheduled for revision in Vol. 3, 7th ed.

capacitors. During or immediately following the electrical discharge, pressure is applied to the stud, plunging its base into the molten pool of the workpiece. The arc may be established either by rapid resistance heating and vaporization of a projection on the stud weld base or by drawing an arc as the stud is lifted away from the workpiece. In the first type, arc times are about three to six milliseconds; in the second type, they range from six to fifteen milliseconds. The capacitor discharge process does not require a shielding ceramic ferrule because of the short arc duration and small amount of molten metal expelled from the joint. It is suited for applications requiring small to medium sized studs.

For either process, the range of stud styles is wide. They include such types as threaded fasteners, plain or slotted pins, internally threaded fasteners, flat fasteners with rectangular cross section, and headed pins with various upsets. Studs may be used as holddowns, standoffs, heat transfer members, insulation supports, and in other fastening applications. Most stud styles can be rapidly applied with portable equipment.

PROCESS CAPABILITIES AND LIMITATIONS

Capabilities

Because stud arc welding time cycles are very short, heat input to the base metal is very small compared to conventional arc welding. Consequently, the weld metal and heat-affected zones are very narrow. Distortion of the base metal at stud locations is minimal. The local heat input may be harmful when studs are welded onto medium and high carbon steels. The unheated portion of the stud and base metal will cool the weld and heat-affected zones very rapidly, causing these areas to harden. The resulting lack of weld joint ductility may be detrimental under certain types of loading, such as cyclic loads. On the other hand, when stud welding precipitation hardened aluminum alloys, a short weld cycle minimizes overaging and softening of the adjacent base metal. Metallurgical compatibility between the stud material and the base metal must be considered.

Studs can be welded at the appropriate time during construction or fabrication without access to the back side of the base member. Drilling, tapping, or riveting for installation is not required. The absence of drilled holes and boss or pad weldments in a pressure vessel design improves its reliability against leakage.

Utilizing this process, designers do not have to specify thicker materials or provide heavy bosses and flanges to obtain required tap depths for threaded fasteners. With stud welded designs of lighter weight, not only can material be saved, but the amount of welding and machining needed to join parts can be reduced.

Small studs can be welded to thin sections by the capacitor discharge method. Studs have been welded to sheet as thin as 0.75 mm (0.03 in.) without melt-through. They have been joined to certain materials (stainless steel, for example) in thicknesses down to 0.25 mm (0.01 in.).

Capacitor discharge power permits the welding of dissimilar metals and alloys. Steel to stainless steel, brass to steel, copper to steel, brass to copper, and aluminum to die-cast zinc are a few of the acceptable combinations. Because the depth of melting is very shallow, capacitor discharge welds can be made without damage to a prefinished opposite side. No subsequent cleaning or finishing is required.

Limitations

Only one end of a stud can be welded to the workpiece. If a stud is required on both sides of a member, a second stud must be welded to the other side. Stud shape and size are limited because the stud design must permit chucking of the stud for welding. The stud base size is limited for thin base metal thicknesses.

Studs applied by arc stud welding usually require a disposable ceramic ferrule around the base. Flux in the stud base or a protective gas shield is needed to obtain a sound weld.

Most studs applied by capacitor discharge power require a close tolerance projection on the weld base to initiate the arc. Stud diameters that can be attached by this method generally range from 3.2 to 9.5 mm (1/8 to 3/8 in.). Above this size, arc stud welding is more economical.

A welding power source located convenient to the work area is required for stud welding. For arc stud welding, 230 or 460 volt (V) ac power is required to operate the dc welding power source. For most capacitor discharge welding, a single phase 110 V ac main supply will serve, but high production units require three phase ac, 230 or 460 V, for operation.

ARC STUD WELDING

PRINCIPLES OF OPERATION

The arc stud welding process involves the same basic principles as any of the other arc welding processes. Application of the process consists of two steps.

(1) Welding heat is developed with an arc between the stud and the plate (work).

(2) The two pieces are brought into intimate contact when the proper temperature is reached.

The equipment consists of the stud gun, a control unit (timing device), studs and ferrules, and an available source of dc welding current. A typical equipment connection is shown in Fig. 8.1. The mechanics of the process are illustrated in Fig. 8.2. The stud is loaded into the chuck, the ferrule (also known as an arc shield) is placed in position over the end of the stud, and the gun is properly positioned for welding (Fig. 8.2A). The trigger is then depressed, starting the automatic welding cycle.

A solenoid coil within the body of the gun is energized. This lifts the stud off the work and, at the same time, creates an arc (Fig. 8.2B). The end of the stud and the workpiece are melted by the arc. When the preset arc period is completed, the welding current is automatically shut off and the solenoid de-energized by the control unit. The mainspring of the gun plunges the stud into the molten pool on the work to complete the weld (Fig. 8.2C). The gun is then lifted from the stud and the ferrule is broken off (Fig. 8.2D).

The time required to complete a weld varies with the cross-sectional area of the stud. For example, weld time typically would be about 0.13 second for a 2.6 mm (10 gage) stud, and 0.92 second for a 22 mm (7/8 in.) diameter stud. Application rates vary with the size of the stud and other factors such as working conditions. An average rate is approximately six studs per minute, although a rate of fifteen studs per minute is common in many applications.

The equipment involved in stud welding compares with that of manual shielded metal arc welding with regard to portability and ease of operation. The initial cost of such equipment varies with the size of the studs to be welded.

The gun and the control unit are connected to a dc power source. The control unit connections shown in Fig. 8.1 are for power sources designed for secondary interruption, as is the case with motor-generator sets, battery units, and most rectifier type welding machines.

DESIGNING FOR ARC STUD WELDING

When a design calls for a stud type fastener or support, arc stud welding should be considered as a means for attaching them. Compared to threaded studs, the base (work) material thickness required to obtain full strength is less for arc stud welding. The use of arc welded studs may reduce the thickness of bosses at attachment points or eliminate them. Cover plate flanges may be thinner than those required for

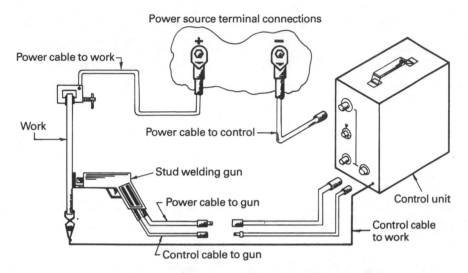

Fig. 8.1—Equipment setup for arc stud welding of steel

threaded fasteners. Thus, there is potential weight savings when the process is used.

The weld base diameters of steel studs range from 3.2 to 32 mm (1/8 to 1-1/4 in.). For aluminum, the range is 3.2 to 13 mm (1/8 to 1/2 in.); and for stainless steels, it is 3.2 to 25 mm (1/8 to 1 in.). For design purposes, the smallest cross-sectional area of the stud should be used for load determination, and adequate safety factors should be considered.

To develop full fastener strength, the plate (work) thickness should be a minimum of approximately one third the weld base diameter. A minimum plate thickness is required for each stud size to permit arc stud welding without melt-through or excessive distortion, as shown in Table 8.1. For steel, a 1:5 minimum ratio of plate thickness to stud weld base diameter is the general rule.

Fasteners can be stud welded with smaller edge distances than those required for threaded fasteners. However, loading and deflection requirements must be considered at stud locations.

STUDS

Stud Materials

The most common stud materials welded with the arc stud weld process are low carbon steel, stainless steel, and aluminum. Other materials are used for studs on a special application basis. Typical low carbon steel studs have

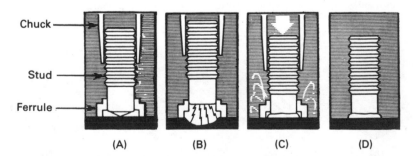

Fig. 8.2—Steps in arc stud welding: (A) gun is properly positioned; (B) trigger is depressed and stud is lifted, creating an arc; (C) arcing period is completed and stud is plunged into molten pool of metal on base metal; (D) gun is withdrawn from the welded stud and ferrule is removed

Table 8.1 — Recommended minimum steel and aluminium plate thicknesses for arc stud welding

Stud base diameter		Steel without backup		Aluminum			
				Without backup		With backup[a]	
mm	in.	mm	in.	mm	in.	mm	in.
4.8	3/16	0.9	0.04	3.2	0.13	3.2	0.13
6.4	1/4	1.2	0.05	3.2	0.13	3.2	0.13
7.9	5/16	1.5	0.06	4.7	0.19	3.2	0.13
9.5	3/8	1.9	0.08	4.7	0.19	4.7	0.19
11.1	7/16	2.3	0.09	6.4	0.25	4.7	0.19
12.7	1/2	3.0	0.12	6.4	0.25	6.4	0.25
15.9	5/8	3.8	0.15	–	–	–	–
19.1	3/4	4.7	0.19	–	–	–	–
22.2	7/8	6.4	0.25	–	–	–	–
25.4	1	9.5	0.38	–	–	–	–

a. A metal backup to prevent melt-through of the plate.

a chemical composition as follows (all values are maximum): 0.23 percent carbon, 0.90 percent manganese, 0.040 percent phosphorus, and 0.050 percent sulfur. They have a minimum tensile strength of 415 MPa (60 000 psi) and a minimum yield strength of 345 MPa (50 000 psi). The typical tensile strength for stainless steel studs is 585 MPa (85 000 psi), and for aluminum, it is 275 MPa (40 000 psi).

High strength studs, meeting the SAE steel fastener Grade 5 tensile strength of 825 MPa (120 000 psi) minimum, are also available. These studs are basically carbon steels that are heat treated to meet the tensile strength requirement.

Low carbon and stainless steel studs require a quantity of welding flux within or permanently affixed to the end of the stud. The main purposes of the flux are to deoxidize the weld metal and to stabilize the arc. Figure 8.3 shows

four methods for securing the flux to the base of the stud. The method most widely used is that of Fig. 8.3(C).

Aluminum studs do not use flux on the weld end. Argon or helium shielding is required to prevent oxidation of the weld metal and stabilize the arc. The studs usually have a small tip on the weld end to aid arc initiation.

Stud Designs

Most stud weld bases are round. However, there are many applications which utilize a square or rectangular shaped stud. With rectangular studs, the width-to-thickness ratio at the weld base should not exceed five to obtain satisfactory weld results. Figure 8.4 shows a wide variety of sizes, shapes, and types of stud weld fasteners. In addition to conventional straight threaded studs, they include eyebolts;

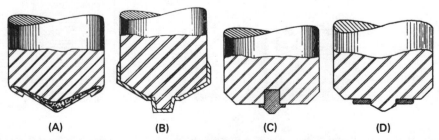

(A)   (B)   (C)   (D)

Fig. 8.3—Three methods of containing flux on the end of a welding stud: (A) granular flux, (B) flux coating, (C and D) solid flux

Fig. 8.4—*Studs and fastening devices commonly used for arc stud welding—stud stock may be round, square, or rectangular in cross section*

J-bolts; and punched, slotted, grooved, and pointed studs.

Stud designs are limited in that (1) welds can be made on only one end of a stud; (2) the shape must be such that a ferrule (arc shield) that fits the weld base can be produced; (3) the cross section of the stud weld base must be within the range that can be stud welded with available equipment; and (4) the stud must be of a size and shape that permit chucking or holding for welding. A number of standard stud designs are produced commercially. The stud manufacturers can provide information on both standard and special designs for various applications.

One important consideration in designing or selecting a stud is the material lost from the welding action. During welding, the stud and base metal melt. The molten metal is then expelled from the joint. The overall length is reduced by the approximate amounts for various stud diameters shown in Table 8.2. The stud length reductions shown in the table are typical. but they may vary to some degree depending upon the materials involved. The finished length

after welding is shorter than the original stud length by the amount of length reduction.

Part of the material from the length reduction appears as flash in the form of a fillet around the stud base. This fillet must not be confused with a conventional fillet weld because it is formed in a different manner. When properly formed and contained, the fillet indicates complete fusion over the full cross section of the stud base. It also suggests that the weld is free of contaminants and porosity. The stud weld

Table 8.2—Typical length reductions of studs in arc stud welding

Stud diameters		Length reductions	
mm	in.	mm	in.
4.8 thru 12.7	3/16 thru 1/2	3	1/8
15.9 thru 22.2	5/8 thru 7/8	5	3/16
25.4 and over	1 and over	5 to 6	3/16 to 1/4

Table 8.3—Weld fillet clearances for arc stud welds

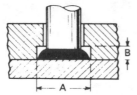 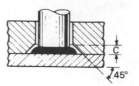

| Stud base diameter | | Counterbore | | | | 90 deg countersink | |
| | | A | | B | | C | |
mm	in.	mm	in.	mm	in.	mm	in.
6.4	1/4	11.1	0.437	3.2	0.125	3.2	0.125
7.9	5/16	12.7	0.500	3.2	0.125	3.2	0.125
9.5	3/8	15.1	0.593	3.2	0.125	3.2	0.125
11.1	7/16	16.7	0.656	4.7	0.187	3.2	0.125
12.7	1/2	19.1	0.750	4.7	0.187	4.7	0.187
15.9	5/8	22.2	0.875	5.5	0.218	4.7	0.187
19.1	3/4	28.6	1.125	7.9	0.312	4.7	0.187

fillet may not be fused along its vertical and horizontal legs. This lack of fusion is not considered detrimental to the stud weld joint quality.

The dimensions of the fillet are closely controlled by the design of the ferrule, where one is required. Since the diameter of the fillet is generally larger than the diameter of the stud, some consideration is required in the design of mating parts. Counterbore and countersink dimensions commonly used to provide clearance for the fillets of round studs are shown in Table 8.3. Fillet size and shape will vary with stud material and ferrule clearance. Therefore, test welds should be made and checked. Three other methods of accommodating fillets are shown in Fig. 8.5,

FERRULES

Ferrules are required for most arc stud welding applications. One of them is placed over the stud at the weld end where it is held in position by a grip or holder on the stud welding gun. The ferrule performs several important functions during welding, namely:

(1) Concentrates the heat of the arc in the weld area

(2) Restricts the flow of air into the weld area, which helps to control oxidation of the molten weld metal

(3) Confines the molten metal to the weld area

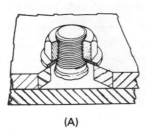 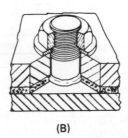 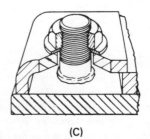

(A) (B) (C)

Fig. 8.5—Stud fillets may be accommodated by: (A) use of oversized clearance holes, (B) use of gasket material, and (C) use of a dog or hold down clip

(4) Prevents the charring of adjacent non-metallic materials

The ferrule also shields the operator from the arc. However, safety glasses with No. 3 filter lenses are recommended for eye protection.

There are two types of ferrules, expendable and semipermanent. Expendable ferrules, made of ceramic material, have the broadest applications. They are easily removed by breaking them. Because each one is used only once, its size is minimized for economy and its dimensions are optimized for the application. Stud configuration is not limited because the ferrule does not have to slip over the stud shank for removal.

Semipermanent ferrules are used primarily with automatic stud feeding systems for high production applications. They are usually graphite with heat resistant insulating rings between them and the mounts. The number of welds that can be obtained with a semipermanent ferrule depends on stud material and diameter, welding setup, and welding rate. The ferrule gradually wears away from heat and abrasion until weld quality becomes unacceptable. It must be replaced before this happens.

The standard ferrule, composed of a ceramic material, is cylindrical in shape and flat across the bottom for welding to flat surfaces. The base of the ferrule is serrated to vent gases expelled from the weld area. Its internal shape is designed to form the expelled molten metal into a fillet around the base of the stud. Special ferrule designs are used for unusual applications, such as welding at angles to the work and welding to contoured surfaces. Ferrules for such applications are designed so that their bottom faces match the required surface contours.

SPECIAL PROCESS TECHNIQUES

There are several special process techniques that employ the basic arc stud welding process, but each is limited to very specific types of applications.

One special process technique, referred to as "gas-arc," uses an inert gas for shielding the arc and molten metal from the atmosphere. A ferrule is not used. This technique is suitable for both steel and aluminum stud welding applications, but its primary use is with aluminum. It is usually limited to production type applications because a fixed setup must be maintained, and also the welding variables fall into a very narrow range. Without a ferrule, there is greater susceptibility to arc blow and poorer control of the fillet around the base.

Another special process technique, which also does not use a ferrule, is called "short cycle" welding. It uses a relatively high weld current for a very short time to minimize oxidation and nitrification of the molten metal. Short cycle welding is generally limited to small studs, 6.4 mm (1/4 in.) diameter and under, where the amount of metal melted is minimal. One application is the welding of studs to thin base materials where shallow penetration is required.

ARC STUD WELDING EQUIPMENT

The necessary equipment for stud welding consists of (1) a stud welding gun, (2) a control unit to control the time of the current flow, (3) a power source, and (4) proper connecting cables and accessories (Fig. 8.1).

Types of Guns

There are two types of stud welding guns, portable hand-held and fixed production type, shown in Fig. 8.6. The principle of operation is the same for both types.

The portable stud welding gun resembles a pistol. It is made of a tough plastic material and weighs between 2 and 4 kg (4-1/2 and 9 lb), depending upon the type of gun. A small gun is used for studs from 3.2 to 16 mm (1/8 to 5/8 in.) in diameter; a larger heavy duty gun is used for studs up through 32 mm (1-1/4 in.) diameter. The large gun can be used for the entire stud range. However, in applications where only small diameter studs are used, it is advantageous to use a small, lighter weight gun.

A gun consists basically of the body, a lifting mechanism, a chuck holder, an adjustable support for the ferrule holder, and the connecting weld and control cables (Fig. 8.6). The portable gun body is usually made of a high impact strength plastic. The stud lifting mechanism consists of a solenoid, a clutch, and a mainspring. The mechanism is actuated by the solenoid to provide positive control of the lift. The lift will be consistent over a range of 0.8 to 3.2 mm (1/32 to 1/8 in.), and it will be constant regardless of the length of stud protrusion (within limits of the gun). An added feature of some guns is a cushioning arrangement to control the plunging action of the stud to complete the weld. Controlled plunge eliminates the excessive spatter normally associated with the welding of large diameter studs with no control.

The adjustable support for the ferrule holder is designed so that the stud chucks can be easily changed. A welding cable conducts welding current from the control to the gun. The control cable carries the power for the gun solenoid and trigger circuit.

The fixed or production gun is mounted on an automatic positioning device, and it is usually air operated and electrically controlled. The workpiece is positioned under the gun with suitable locating fixtures. Tolerances of ± 0.13 mm (0.005 in.) on location and ± 0.26 mm (0.010 in.) in height may be obtained when a production gun is used. A production unit may contain a number of guns, depending upon the nature of the job and the production rate required.

Control Unit

The control unit consists fundamentally of a contactor suitable for conducting and interrupting the welding current, and a weld timing device with associated electrical controls. The adjustable weld timer is graduated in either cycles (60 Hz) or numbered settings. Once set, the control unit maintains the proper time interval for the size of stud being welded. The time interval may vary from 0.05 to 2 seconds depending upon the diameter of the stud.

In certain controls, the timer regulates the exact heat energy required for the weld regardless of line power fluctuations. The welding time is automatically increased or decreased in conjunction with an input power decrease or increase. The control unit has two connectors for the welding cables. One is for the cable from the terminal of the dc welding power source, and the other is for the cable to the stud welding gun. Most control units also have a ground control cable for connection to the workpiece. As with stud welding guns, control units are of two sizes. For welding studs up to 16 mm (5/8 in.) diameter, a small control unit can be used. A large control unit must be used for large diameter studs to accommodate a welding contactor of the appropriate size.

Power Sources

A dc type power source is used for arc stud welding. Alternating current is not suitable for stud welding processes. There are three basic types of dc power sources that can be used. They

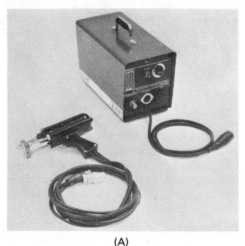

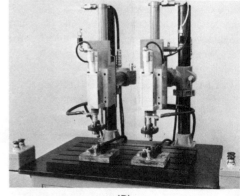

(A)   (B)

Fig. 8.6—Two types of arc stud welding guns: (A) portable hand operated type, (B) fixed production type

are (1) transformer-rectifier, (2) motor-generator (motor or engine driven), and (3) battery.

The general characteristics desired in a stud welding power source are

(1) High open-circuit voltage in the range of 70 to 100 V.

(2) A drooping output volt-ampere characteristic.

(3) A rapid output current rise to the set value.

(4) High current output for a relatively short time. The current requirements are higher, and the duty cycle is much lower for stud welding than for other types of arc welding.

There are many standard dc arc welding power sources available which meet these requirements. They are entirely satisfactory for stud welding. However, dc welding power sources with a constant voltage characteristic generally are not suitable for stud welding. With this type power source, weld current control can be difficult, and it may not be possible to obtain the proper weld current range for the application.

Standard dc power sources designed for manual shielded metal arc welding have excellent characteristics for arc stud welding. They have drooping volt-ampere characteristics and an open-circuit voltage of 70 or higher. Many stud welding current range requirements extend beyond those used for shielded metal arc welding. Therefore, standard dc power sources are generally used only for welding 13 mm (1/2 in.) diameter and smaller studs because of the maximum current output limitations. For larger studs, two standard dc type power sources must be used in parallel.

When the applications require high welding currents (sometimes over 2500 A) and short weld times (less than 1.5 s), special stud welding power sources are recommended. These special power sources yield higher efficiency not only from the standpoint of weld current output relative to their size and weight, but also from the fact that they cost less than two or more standard arc welding machines.

The basis for rating special stud welding power sources is different from that of conventional arc welding machines. Because stud welding requires a high current for a relatively short time, the current output requirements of a stud welding power source are higher, but the duty cycle is much lower than those for other types of arc welding. Also, the load voltage is normally higher for stud welding. Cable voltage drop is greater with stud welding than arc welding because of the higher current requirements.

The duration of a stud weld cycle is generally less than one second. Therefore, load ratings and duty cycle ratings are made on the basis of one second. The rated output of a machine is its average current output at 50 V for a period of one second. Thus, a rating of 1000 A at 50 V means that during a period of one second the current output will average 1000 A and the terminal voltage will average 50 V.

Oscillographic traces show that the current output of a motor-generator stud welding power source is higher at the start of welding than at the end. Thus, it is necessary to use the average current for rating purposes.

The duty cycle for stud arc welding machines is based on the formula

$$\text{Duty cycle, }\% = 1.7 \times \text{no. of 1 s loads per minute}$$

where the one second load is the rated output.

If a machine can be operated six times per minute at rated load without causing its components to exceed their maximum allowable temperatures, then the machine would have a 10 percent duty cycle rating.

Although the power sources that are available for stud welding include transformer-rectifier, motor-generator, and battery types, a transformer-rectifier power source is usually preferred. Transformer-rectifier power sources are of two basic types: (1) those that require a separate stud welding control unit, and (2) those that have the control and timing circuitry built in. The combination power-control machines utilize silicon controlled rectifiers (SCR's) for initiating and interrupting the weld current. Solid state components are also used for the gun control and timing circuitry.

Combination machines operate on either three phase or single phase incoming power. Three phase units are preferred for larger diameter studs, since they provide a balanced load on the incoming power line. Single phase units are primarily low cost, portable power sources

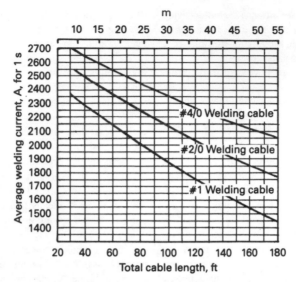

Fig. 8.7—Effect of cable size and length on available welding current from a 2000 A power source

for welding studs of 10 mm (3/8 in.) diameter and under. Although single phase units do have the capability of welding 13 mm (1/2 in.) diameter studs, three phase units are preferable for this size stud.

The choice between a rectifier type or a motor-generator type power source depends upon a number of conditions. Both types have advantages and disadvantages. Motor-generator power sources have a high current peak and excellent current stability at the start of the weld cycle. They are also relatively insensitive to small line voltage fluctuations. This is an advantage when welding large diameter studs, or where large line voltage fluctuations occur under heavy loads.

The transformer-rectifier machine is quiet, economical to operate, and easy to parallel. It operates efficiently alone or in parallel; its no-load power input is negligible.

Specially designed battery power sources provide a practical means for stud welding where the available power is inadequate or unsuitable for conventional power sources. Battery units have the advantages of mobility in that they can be operated for reasonable periods with no external power. They normally operate on 115 V, ac, which can be produced by a small engine powered generator.

Other major factors in connection with power sources for stud welding are the incoming power and the cable size and length (primary power and welding cables). Both motor-generator and rectifier power sources normally operate on 230 or 460 V, ac, three phase power. Because of the high currents required for stud welding, line voltage regulation sometimes becomes a problem. Satisfactory operation of either type of equipment can only be assured if the power line voltage regulation will remain within prescribed limits while a weld is in progress.

Welding Cable

The welding cable length, including both the gun and ground cables, and the cable size are very important in stud welding. Many times, there is significant power loss in the welding circuit caused by the use of either too small or too long welding cables. The current available for welding at a given machine setting may vary as much as 50 percent, depending upon the size and length of welding cables used.

Figure 8.7 illustrates the effect of cable size and cable length on welding current. The tests made to determine these curves were run with a 2000 A power supply at maximum setting. Only the cable length and cable size were changed. In this case, the maximum welding current was 2360 A with 9 m (30 ft) of AWG #1 cable. When the same size of cable was increased to 55 m (180 ft) long, the available current decreased 38 percent to 1450 A. On the other hand, when 55 m (180 ft) of #4/0 cable was used, the current was 2050 A, a decrease of only 13 percent. Thus, when the distance from the power source to the welding gun increases significantly, larger welding cable should be used.

Automatic Feed Systems

Stud welding systems with automatic stud feed are available for both portable and fixed welding guns. The studs are automatically oriented in a parts feeder, transferred to the gun (usually through a flexible feed tube), and loaded into the welding gun chuck. Generally, a ferrule is hand loaded for each weld. However, automatic ferrule feed is available with fixed gun production type systems. Also, a semipermanent ferrule may be used. Figure 8.8 shows an automatic stud feeding system for portable arc stud welding. Automated portable

equipment, that uses solid state controls and no ferrule, is available for welding 6 mm (1/4 in.) diameter studs and smaller.

STUD LOCATING TECHNIQUES

The method of locating studs depends upon the intended use of the studs and the accuracy of location required. For applications where extreme accuracy is required, special locating fixtures and fixed (production type) stud welding equipment is recommended. The extent of tooling will be a function of the required production rate as well as total production.

Several methods and procedures are used for positioning studs with a portable stud welding gun. The simplest and most common procedure is to lay out and center punch the locations either on the work or through a template. A stud is then located by placing the point of the stud in the punch mark. Although operator skill is always a factor in accuracy, location tolerances of ± 1.2 mm (3/64 in.) can be obtained. Cover plates that have been punched or drilled can be used as templates.

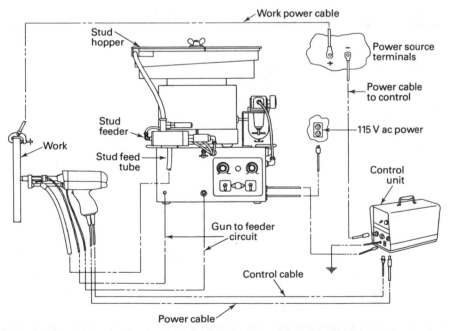

Fig. 8.8—Conventional portable arc stud welding equipment with an automatic stud feed system

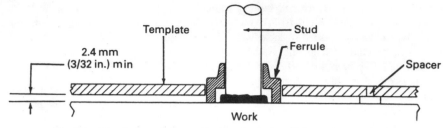

Fig. 8.9—Simple template used to locate studs within ±0.8 mm (1/32 in.)

When a number of pieces are to be stud welded, common practice is to weld directly through holes in a template without preliminary marking, as shown in Fig. 8.9. A simple template positions the stud by locating the ferrule. Because of manufacturing tolerances on ferrules, the tolerance on stud location with this method is usually ± 0.8 mm (1/32 in.).

When close stud location and alignment are required, a tube-type template is used. The stud is centered indirectly by inserting a tube adaptor on the gun in a locating bushing in the template. Figure 8.10 illustrates this type of template. The template uses a hardened and ground bushing with a closely machined tube adaptor. Because standard ferrule grips are used with this adaptor, standardization of templates is possible. It is only necessary to change ferrule grips to weld studs of different diameters. With this type of template, a tolerance of ± 0.4 mm (0.015 in.) can be held on stud location. This method also maintains perpendicular alignment of the stud.

WELDING CURRENT-TIME RELATIONSHIP

The current and time required for a proper arc stud weld are dependent on the cross-sectional area of the stud. The total energy input (joules or watt-seconds) is a function of welding current, arc voltage, and arc time. Arc voltage is determined by the lift distance set in the stud gun. Proper lift distance is usually recommended

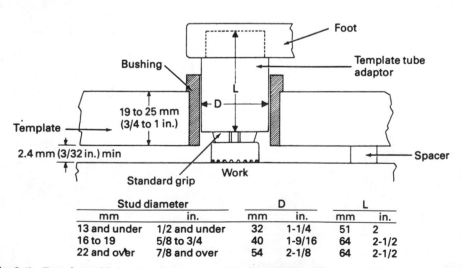

Stud diameter		D		L	
mm	in.	mm	in.	mm	in.
13 and under	1/2 and under	32	1-1/4	51	2
16 to 19	5/8 to 3/4	40	1-9/16	64	2-1/2
22 and over	7/8 and over	54	2-1/8	64	2-1/2

Fig. 8.10—Template with hardened and ground bushing and welding gun adaptor used to locate studs within ±0.4 mm (0.015 in.)

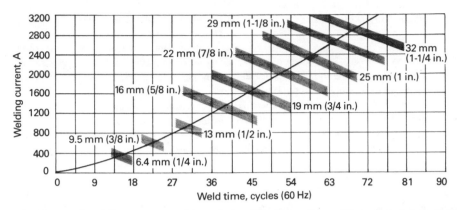

Fig. 8.11—Range of welding current vs time relationships for various mild steel stud sizes and the curve of average values

by the stud manufacturer. Therefore, arc energy is basically a function of the welding current and weld time settings, with a set lift distance.

The same energy input can be obtained by using a range of current and time settings. It is possible, within certain limitations, to compensate for low or high welding current by changing the weld time. Typical current and time relationships for various diameters of mild steel studs are shown in Fig. 8.11. There is a fairly broad range of combinations for each stud size. Under some conditions, such as welding studs to a vertical member or to thin gage material, the allowable range is much smaller.

Although energy input is a major criterion for satisfactory welds, it is not the only factor involved. Other factors such as arc blow, plate surface conditions (rust, scale, moisture, paint), and the operator technique can cause poor welds, even though the proper weld energy input was used.

METALLURGICAL CONSIDERATIONS

The metallurgical structures encountered in arc stud welds are generally the same as those found in any arc weld where the heat of an electric arc is used to melt both a portion of the base metal and the electrode (stud) in the course of welding. Acceptable mechanical properties are obtained when the stud and base material are metallurgically compatible. Properly executed stud welds are usually characterized by

the absence of inclusions, porosity, cracks, and other defects.

A typical stud weld macrosection, Fig. 8.12, shows that molten weld metal is pushed to the perimeter of the stud to form a fillet. A fairly large area of the unmelted stud is almost touching the unmelted base metal. The thickness of the weld metal (cast structure) in the joint is a minimum. Because of the short welding cycle, the heat-affected zones common to arc welding are present, but they are significantly smaller. Chapter 4, "Welding Metallurgy," *Welding Handbook,* Volume 1, 7th edition, contains significant information on welding metallurgy that is applicable to arc stud welding of various materials.

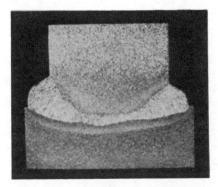

Fig. 8.12—Macrostructure of a typical arc stud weld

MATERIALS WELDED

Steels

Low Carbon Steel. Low carbon (mild) steels can be stud welded with no major metallurgical problems. The upper carbon limit for steel to be arc stud welded without preheat is 0.30 percent. If work sections are relatively thin for the stud diameters being welded (below those in Table 8.1), the carbon limit may be somewhat higher because of the decreased cooling effect of the work. If the section to which the stud is to be welded is heavy, stud welding of steel with more than 0.30 percent carbon, with normal techniques and without preheat, must be evaluated. The most important factor regarding work section thickness is that the material must be heavy enough to permit the welding of the studs without melt-through. Satisfactory welding currents and times for steel studs can be found in Fig. 8.11.

Medium and High Carbon Steel. If medium and high carbon steels are to be stud welded, it is imperative that preheat be used to prevent cracking in the heat-affected zones. In some instances, a combination of preheating and postheating after welding is recommended. In the case of tough alloy steels, either preheating or postheating may be used to obtain satisfactory results. In cases where the welded assemblies are to be heat-treated for hardening after the welding operation, the preheating or postheating operation may be eliminated if the parts are handled in a manner that prevents damage to the studs.

Low Alloy Steel. Generally, the high strength low alloy steels are satisfactorily stud welded when the carbon content is 0.15 percent or lower. This range generally fits the analyses of low alloy steels used in welding and forming operations. If the carbon content exceeds 0.15 percent, it may be necessary to use a low preheat temperature to obtain desired toughness in the weld area.

When the hardness of the heat-affected zones and fillet do not exceed 30 Rockwell C, studs can be expected to perform well under almost any type of severe service. Although good results have been obtained when the hardness ranges up to 35 Rockwell C, it is best to avoid extremely high working stresses and fatigue loading. In special cases where microstructures are important, the weld should be evaluated and qualified for the specific application considered. Since alloy steels vary in toughness and ductility at high hardness levels, weld hardness should not be used as the sole criterion for weld evaluation.

Heat Treated Structural Steel. Many structural steels used in shipbuilding and in other construction are heat-treated at the mill. Heat treated steels require that attention be given to the metallurgical characteristics of the heat-affected zone. Some of these steels are sufficiently hardenable that the heat-affected zones will be martensitic. This structure will be quite sensitive to underbead cracking, and it will have insufficient ductility to carry impact loads. Therefore, for maximum toughness in these steels, a preheat of 370° C (700° F) is recommended. Consideration of the application and end use of the stud will further influence the welding procedures to be followed.

Stainless Steels. Most classes of stainless steel may be arc stud welded. The exceptions are the free machining grades. However, only the austenitic stainless steels (3XX grades) are recommended for general application. The other types are subject to air hardening, and they tend to be brittle in the weld area unless annealed after welding. The weldable stainless steel grades include AISI Types 304, 305, 308, 309, 310, 316, 321, and 347. Types 304 and 305 are most commonly used for stud welding.

Stainless steel studs may be welded to stainless steel or to mild steel as the application may require. The welding setup used is the same as that recommended for low carbon steel except for an increase of approximately 10 percent in power requirement. Where stainless steel studs are to be welded to mild steel, it is essential that the carbon content of the base metal not exceed 0.20 percent. When welding stainless steel studs to mild steel with 0.20 to 0.28 percent carbon, or to low carbon hardenable steels, Type 308, 309, or 310 studs are recommended. Because of the composition of the weld metal when chromium-nickel alloy studs are welded to mild steel, the weld zone may be quite hard. The hardness will depend on the carbon content in the base metal. It is possible to overcome this by using studs with high alloy content such as Type 309 or 310. It is also suggested when welding stainless steel studs to mild steel that a fully annealed stainless steel stud be used.

Nonferrous Metals

Aluminum. The basic approach to aluminum stud welding is similar to that used for mild steel stud welding. The power sources, stud welding equipment, and controls are the same. The stud welding gun is modified slightly by the addition of a dampening device to control the plunging rate of the stud at the completion of the weld time. Also, a special gas adaptor foot ferrule holder is used to contain the high purity inert shielding gas during the weld cycle. Argon is generally used, but helium may be useful with large studs to take advantage of the higher arc energy. The equipment setup is shown in Fig. 8.13.

Reverse polarity is used with the stud (electrode) positive and the work negative. An aluminium stud differs from a steel stud in that no flux is used on the weld end. A cylindrical or cone shaped projection is used on the base of the stud. The projection dimensions on the welding end are designed for each size stud to give the best arc action. The projection serves to initiate the long arc used for aluminum stud welding.

Studs range in weld base diameters from 6.4 to 13 mm (1/4 to 1/2 in.). Their sizes and shapes are similar to steel studs.

Aluminum studs are commonly made of aluminum-magnesium alloys, including 5086 and 5356, that have a typical tensile strength of 275 MPa (40 000 psi). These alloys have high strength and good ductility; they are metallurgically compatable with the majority of aluminum alloys used in industry.

In general, all plate alloys of the 1100, 3000, and 5000 series are considered excellent for stud welding; alloys of the 4000 and 6000 series are considered fair; and the 2000 and 7000 series are considered poor. The minimum aluminum plate thickness, with and without backup, to which aluminum studs of 4.8 to 13 mm (3/16 to 1/2 in.) base diameter may be welded are given in Table 8.1.

Figure 8.14 illustrates a cross section of a typical aluminum alloy stud weld. Table 8.4 gives typical conditions for aluminum arc stud welding.

Magnesium. The gas shielded arc stud welding process used for aluminum produces high strength welds in magnesium alloys. A ceramic ferrule is not needed. Helium shielding

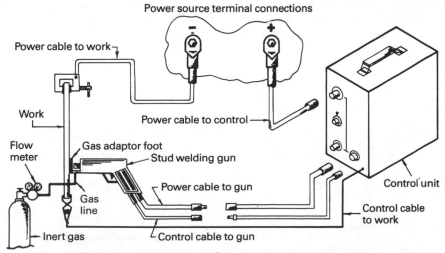

Fig. 8.13—Equipment setup for arc stud welding of aluminum

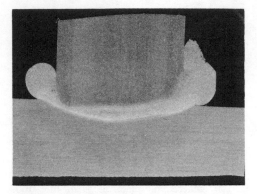

Fig. 8.14—Macrostructure of a 9.5 mm (3/8 in.) diameter Type 5356 aluminum alloy stud welded to a 6.4 mm (1/4 in.) Type 5053 aluminum alloy plate

gas and dc reverse polarity (electrode positive) should be used. A gun with plunge dampening will avoid spattering and base metal undercutting.

Breaking loads up to 6.7 kN (1500 lb) for 6.4 mm (1/4 in.) diameter studs and up to 20 kN (4500 lb) for 13 mm (1/2 in.) diameter studs have been obtained with AZ31B alloy studs welded to 6.4 mm (1/4 in.) thick AZ31B or ZE10A base metal.

Minimum base metal thicknesses to which 6.4 and 12.7 mm (1/4 and 1/2 in.) diameter studs may be attached, without melt-through or great loss in strength, are 3.2 and 6.4 mm (1/8 and 1/4 in.) respectively. If strength is not a consideration, 13 mm (1/2 in.) diameter studs can be welded to 4.8 mm (3/16 in.) thick plate without melt-through.

Other Materials. On a moderate scale, arc stud welding is being done in industry on various brass, bronze, nickel-copper, and nickel-chromium-iron alloys. The applications are usually very special ones requiring careful evaluation to determine suitability of design.

Nickel, nickel-copper, nickel-chromium-iron, and nickel-chromium-molybdenum alloys are best stud welded with dc using reverse polarity (electrode positive). Nickel, nickel-copper, and nickel-chromium-iron alloy stud welds tend to contain porosity and crevices. The mechanical strengths, however, are usually high enough to meet most requirements. The weld itself should not be exposed to corrosive media.

QUALITY CONTROL AND INSPECTION

Weld quality assurance requires the proper materials, equipment, setup, and operating procedures, and also a trained operator. Proper setup includes such things as gun retraction (lift), stud extension beyond the ferrule (plunge), and proper welding current and time.

Weld quality is maintained by close attention to the factors that may produce variations in the weld. To maintain weld quality and consistency, it is necessary to

(1) Have sufficient welding power for the size and type of stud being welded.

(2) Use dc straight polarity for steels and dc reverse polarity for aluminum and magnesium.

(3) Ensure a good ground connection to the work.

Table 8.4—Typical conditions for arc stud welding of aluminum alloys[a]					
Stud weld base diameter		Weld time,	Welding current,[b]	Shielding gas flow[c]	
mm	in.	cycles (60 Hz)	A	liter/min	ft³/h
6.4	1/4	20	250	7.1	15
7.9	5/16	30	325	7.1	15
9.5	3/8	40	400	9.4	20
11.1	7/16	50	430	9.4	20
12.7	1/2	55	475	9.4	20

a. Settings should be adjusted to suit job conditions.
b. The currents shown are actual welding current and do not necessarily correspond to power source dial settings.
c. Shielding gas—99.95% pure argon.

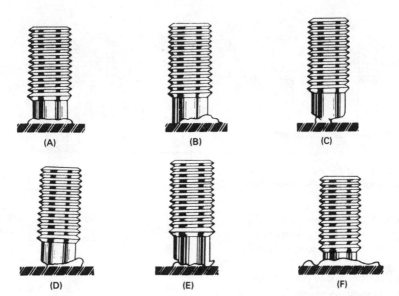

Fig. 8.15—Satisfactory and unsatisfactory arc stud welds: (A) satisfactory stud weld with a good fillet formation, (B) stud weld in which plunge is too short, (C) hang-up, (D) poor alignment, (E) stud weld made with low heat, and (F) stud weld made with high heat

(4) Have welding cables of sufficient size with good connections.

(5) Use correct accessories and ferrules.

(6) Adjust the gun so that the stud extends the recommended distance beyond the ferrule and also retracts it the proper distance for good arc characteristics. Stud extension will be about equal to the length reductions in Table 8.2.

(7) Hold the gun steady at the proper angle to the work. Generally it is perpendicular. Accidental movement of the gun during the weld cycle may cause a defective weld.

(8) Clean the work surface where the stud is to be welded.

(9) Keep stud welding equipment properly cleaned and maintained.

(10) Make test welds before starting and at selected intervals during the job.

Steel Studs

AWS D1.1-75, Structural Welding Code, contains provisions for the installation and inspection of steel studs welded to steel components. Quality control and inspection requirements for stud welding are also included. AWS C5.4, Recommended Practices for Stud Welding, briefly covers inspection and testing of both steel and aluminum stud welds.

Welded studs may be inspected visually for weld appearance and consistency, and also mechanically. Production studs can be proof tested by applying a specified load (force) on them. If they do not fail, the studs are considered acceptable. Production studs should not be bent or twisted for proof testing.

Visual Inspection. The weld fillet around the stud base is inspected for consistency and uniformity. Lack of a fillet may indicate a faulty weld. Figure 8.15(A) indicates a satisfactory stud weld with a good weld fillet formation. In contrast, Figure 8.15(B) shows a stud weld in which the plunge was too short. Prior to welding, the stud should always project the proper length beyond the bottom of the ferrule. (This type of defect may also be caused by arc blow.) Figure 8.15(C) illustrates "hang up." The stud did not plunge into the weld pool. This condition may be corrected by realigning the accessories to insure completely free movement of the stud during lift and plunge. Arc length may also require adjustment.

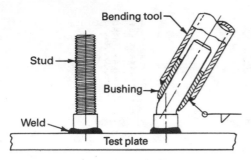

Fig. 8.16—Bend test for welded studs to determine acceptable welding procedures

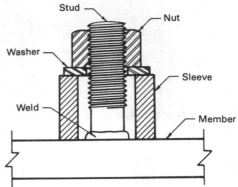

Fig. 8.17—Method of applying a tensile load to a welded stud using torque—a bolt can be used for an internal thread

Figure 8.15(D) shows poor alignment, which may be corrected by positioning the stud gun perpendicular to the work. Figure 8.15(E) shows the results of low weld power. To correct this problem, the ground and all connections should be checked. Also, the current setting, the time setting, or both, should be increased. It may also be necessary to adjust the arc length. The effect of too much weld power is shown in Fig. 8.15(F). Decreasing the current setting or the welding time, or both, will lower the weld power.

Mechanical Testing. Mechanical tests should be made before initiation of production welding to insure that the welding schedule is satisfactory. They may also be made during the production run or at the beginning of a shift to insure that welding conditions have not changed. Arc stud welds are tested by bending the stud or by applying a proof tensile load.

Bending may be done by striking the stud with a hammer or with a length of a tube or pipe, as shown in Fig. 8.16. The angle through which the stud will bend without weld failure will depend on the stud and base metal compositions and conditions (cold worked, heat treated) and stud design. Acceptable bending should be determined when the welding procedure specification is established or from the applicable welding code. Bend testing may damage the stud; therefore, it should be done on qualification samples only.

The method used to apply tensile load on an arc welded stud will depend on the stud design. Special tooling may be required to grip the stud properly without damage, and a special loading device may be needed. A simple method that can be used for straight threaded studs is shown in Fig. 8.17. A steel sleeve of appropriate size is placed over the stud. A nut of the same material as the stud is tightened against a washer bearing on the sleeve with a torque wrench. This applies a tensile load (and some shear) on the stud.

The relationship between nut torque, T, and tensile load, F, can be estimated using the equation

$$T = kFd$$

where

d is the nominal thread diameter

k is a constant related to such factors as thread angle, helix angle, thread diameters, and coefficients of friction between the nut and thread, and the nut and washer.

For mild steel, k is approximately 0.2 for all thread sizes and for both coarse and fine threads. However, the many factors that influence friction will influence the value of k. Several factors are the stud, nut, and washer materials and surface finishes, and also their lubrication. For other materials, k may have some other value because of the differences in friction between the parts.[2]

---

2. For a more detailed explanation, refer to J.E. Shigley, *Mechanical Engineering Design*, 2nd ed., McGraw-Hill, 1972, pp. 309-312.

Table 8.5—Typical nut torques causing failure of aluminum alloy studs

Thread size	Failure load	
	N•m	lbf•in.
1/4-20	7	60
5/16-18	13	115
3/8-16	22	195
7/16-14	33	290
1/2-13	49	435

Aluminum Studs

Visual inspection of aluminum stud welds for acceptance is limited because the appearance of the weld fillet does not necessarily indicate quality. Therefore, visual inspection of aluminum stud welds is recommended only to determine complete fusion and absence of undercut around the periphery of the weld.

Aluminum studs can be tested to establish acceptable welding procedures using the bend test shown in Fig. 8.16. If the stud bends about 15° or more from the original axis without breaking the stud or weld, the welding procedures should be considered satisfactory. Production studs should not be bent and then straightened because of possible damage to them. In this case, the torque test or separate qualification test plates may be substituted.

Torque testing of threaded aluminum studs is done in the same manner as that used for steel studs. Torque is applied to a predetermined value or until the stud fails. Typical torque tests gave the failure loads shown in Table 8.5. For a particular application, the acceptable proof load should be established by suitable laboratory tests relating applied torque to tensile loading.

APPLICATIONS

Arc stud welding has been widely accepted by all the metalworking industries. Specifically, stud welding is being used extensively in the following fields: automotive; boiler and building construction; farm, industrial, and domestic equipment manufacture; railroads; and shipbuilding. Defense industry applications include missile containers, armored vehicles, and tanks.

Some typical applications are attaching wood floors to steel decks or framework; fastening linings or insulation in tanks, boxcars, and other containers; securing inspection covers; and welding shear connectors and concrete anchors to structures.

CAPACITOR DISCHARGE STUD WELDING

DEFINITION AND GENERAL DESCRIPTION

Capacitor discharge stud welding is a stud arc welding process where dc arc power is produced by a rapid discharge of stored electrical energy with pressure applied during or immediately following the electrical discharge. The process uses an electrostatic storage system as a power source in which the weld energy is stored in capacitors of high capacitance. No ferrule or fluxing is required.

PRINCIPLES OF OPERATION

There are three different types of capacitor discharge stud welding: initial contact, initial gap, and drawn arc. They differ primarily in the manner of arc initiation. Initial contact and initial gap capacitor discharge stud welding utilize studs having a small, specially designed projection (tip) on the weld end of the stud. Drawn arc stud welding creates a pilot arc as the stud is lifted off the workpiece by the stud gun. It is similar to arc stud welding.

Initial Contact Method

In initial contact stud welding, the stud is first placed against the work as shown in Fig. 8.18(A). The stored energy is then discharged through the projection on the base of the stud. The small projection presents a high resistance to the stored energy, and it rapidly disintegrates from the high currency density. This creates an arc that melts the surfaces to be joined, Fig. 8.18(B). During arcing, Fig. 8.18(C), the pieces to be joined are being brought together by action of a spring, weight, or an air cylinder. When the two surfaces come in contact, fusion takes place, and a weld is produced between the stud and the workpiece, Fig. 8.18(D).

Initial Gap Method

The sequence of events in initial gap stud welding is shown in Fig. 8.19. Initially, the stud is positioned off the work, leaving a gap between it and the work, as shown in Fig. 8.19 (A). The stud is released and continuously moves toward the work under gravity or spring loading. At the same time, open-circuit voltage is applied between the stud and the work. When the stud contacts the work, Fig. 8.19(B), high current flashes off the tip and initiates an arc. The arc melts the surfaces of the stud and work as the stud continues to move forward, Fig. 8.19(C). Finally, the stud plunges into the work, and the weld is completed, Fig. 8.19(D).

With proper design of the electrical characteristics of the circuit and size of the projection, it is possible to produce a high current arc of such short duration (about 0.006 second) that its effect upon the stud and workpiece is purely superficial. A surface layer only a few hundredths of a millimeter (thousandths of an inch) in thickness on each surface reaches the molten state.

Drawn Arc Method

In the drawn arc method, arc initiation is accomplished in a manner similar to that of arc stud welding. The stud does not require a small tip on the weld face. An electronic control is used to sequence the operation. Weld time is controlled by an electronic circuit in the unit. The welding gun is similar to that used for arc stud welding.

The operating sequence is shown in Fig. 8.20. In sequence, the stud is positioned against the work as shown in Fig. 8.20(A). The trigger switch on the stud welding gun is actuated, energizing the welding circuit and a solenoid coil in the gun body. The coil motion lifts the stud from the work, Fig. 8.20(B), drawing a low amperage pilot arc between them. When the lifting coil is de-energized, the stud starts to return to the work. The welding capacitors are then discharged across the arc. The high amperage from the capacitors melts the end of the stud and the adjacent work surface. The spring action of the welding gun plunges the stud into the molten metal to complete the weld, Fig. 8.20(C). The completed weld is shown in Fig. 8.20(D).

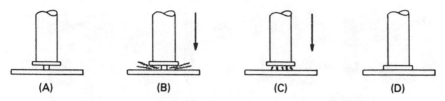

Fig. 8.18—Steps in initial contact capacitor discharge stud welding

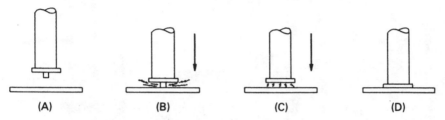

Fig. 8.19—Steps in initial gap capacitor discharge stud welding

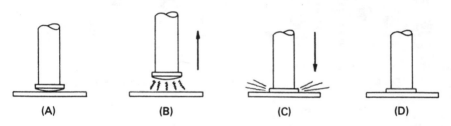

Fig. 8.20—Steps in drawn arc capacitor discharge stud welding

DESIGNING FOR CAPACITOR DISCHARGE STUD WELDING

The ability of the capacitor discharge method to weld studs to thin sections is an important design feature. Material as thin as 0.76 mm (0.030 in.) can be welded without melt-through. Studs have been successfully welded to some materials (stainless steel, for example) in thicknesses as low as 0.25 mm (0.010 in.).

Another design feature of this system of stud welding is its ability to weld studs to dissimilar metals. The penetration into the work from the arc is so shallow that there is very little mixing of the stud metal and the work metal. Steel to stainless steel, brass to steel, copper to steel, brass to copper, and aluminum to die cast zinc are a few of the combinations that may be used. Many unusual metal combinations, not normally considered weldable by fusion processes, are possible with this process.

Another feature is the elimination of post-weld cleaning or finishing operations on the showing (face) surface of the base metal. The process can be used on parts that have had the face surface painted, plated, polished, or coated with ceramic or plastic.

STUD MATERIALS AND DESIGNS

The materials that are commonly capacitor discharge stud welded are low carbon steel, stainless steel, aluminum, and brass. Low carbon steel and stainless steel studs are generally the same compositions as those used for arc stud welding. For aluminum, 1100 and 5000

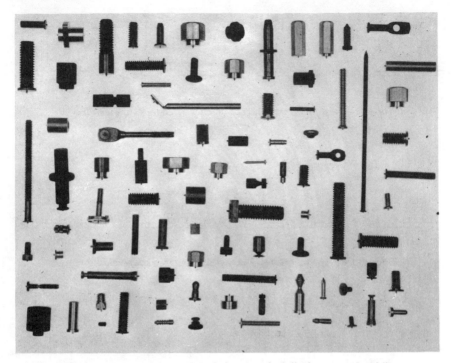

Fig. 8.21—Commonly used studs for capacitor discharge stud welding

series alloys are generally used. Copper alloy studs are mostly No. 260 and No. 268 compositions (brasses).

Stud designs for capacitor discharge stud welding range from standard shapes to complex forms for special applications. Usually, the weld base of the fastener is round. The shank may be almost any shape or configuration. These include threaded, plain, round, square, rectangular, tapered, grooved, and bent configurations, or flat stampings. The size range is 1.6 to 12.7 mm (1/16 to 1/2 in.) diameter, with the great bulk of attachments falling in the 3.2 to 9.5 mm (1/8 to 3/8 in.) diameter range. Figure 8.21 shows some common stud designs.

Initial contact and initial gap capacitor discharge studs are designed with a tip or projection on the weld end. The size and shape of this tip is important because it is one of the variables involved in the achievement of good quality welds. The standard tip is cylindrical in shape. For special applications, a conically shaped tip is used. The detailed weld base design is determined by the stud material, the base diameter, and sometimes by the particular application. The weld base is tapered slightly to facilitate the expulsion of the expanding gases that develop during the welding cycle. Usually the weld base diameter is larger than that of the stud shank. The weld area is larger than the stud cross section to provide a joint strength equal to or higher than that of the stud.

Drawn arc capacitor discharge studs are designed without a tip or projection on the weld end. However, the weld end is tapered or slightly spherical so that the arc will initiate at the center of the base. As with the other capacitor discharge methods, these studs are generally designed with a large base in the form of a flange.

Stud melt-off or reduction in length due to melting is almost negligible when compared to the arc stud welding method. Stud melt-off is generally in the range of 0.2 to 0.4 mm (0.008 to 0.015 in.).

(A)

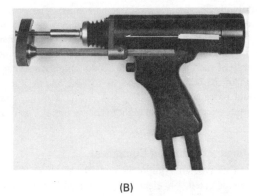
(B)

Fig. 8.22—Portable capacitor discharge stud welding equipment: (A) control-power source unit, (B) stud gun

WELDING EQUIPMENT

Capacitor discharge stud welding requires a stud gun and a combination power-control unit with associated interconnecting cables. Both portable and stationary production equipment is available.

Solid state control circuitry provides signals for automatic sequencing of several events during the welding cycle. The events include one or more of the following:

(1) Energize the gun solenoid or air cylinder for initial gap and drawn arc methods.

(2) Initiate the pilot arc in the drawn arc method.

(3) Discharge the welding current from the capacitor bank at the proper time in the welding sequence.

(4) De-energize the solenoid or air cylinder of the gun.

(5) Control the changing voltage of the capacitor bank.

Portable Units

The hand-held stud gun is usually made of high impact strength plastic. The gun holds and positions the stud for welding. A trigger initiates the welding cycle through a control cable to the power source-control unit. By changing the chuck that holds the stud, various diameters and shapes of studs can be accommodated.

The power source-control unit provides

the welding current, and it contains the necessary circuitry for charging the capacitors. Variable discharge currents are obtained by varying the charge voltage on the capacitors. Control of the charging and discharging currents is done

Fig. 8.23—Typical hand-fed, production-type, capacitor discharge stud welding machine

automatically by the welding machine. The units generally operate on 115 V, 60 Hz, power.

Typical portable capacitor discharge equipment is illustrated in Fig. 8.22. The stored energy of such a unit would be in the neighborhood of 70 000 μF charged to 170 V, and it would be capable of welding 6.4 mm (1/4 in.) diameter studs at a rate of eight to ten per minute.

Stationary Production Equipment

This type of equipment consists of either an air actuated, an electrically actuated, or a gravity drop stud gun (or guns) mounted above a work surface. The electrical controls for the air systems and for charging the capacitors are usually located under the work table.

The power-control units are generally designed for a specific application because automatic sequencing of clamping, indexing, and unloading devices may be incorporated. The capacitance of production units ranges from about 20 000 to 200 000 μF. The capacitor charge voltage does not exceed 200 V, and it is isolated from the stud chuck until welding is initiated. Power input is 230 or 460 V, single or three phase, for rapid charging rates.

With this equipment, high production rates can be obtained, depending upon the amount of automation in the fixturing and in the feeding of studs and parts to be welded. Up to 45 welds per minute have been made with a single gun. Figure 8.23 illustrates a typical single head, air actuated unit.

Fig. 8.24—Capacitor discharge production stud welding machine with automatic stud feed system

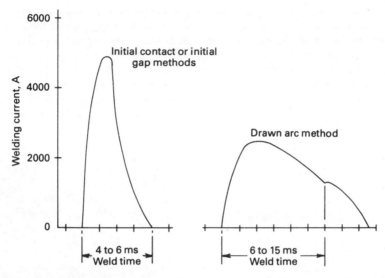

Fig.8.25—Typical current vs time curves for the three capacitor discharge stud welding methods

Automatic Stud Feed Systems

Capacitor discharge stud welding is well suited for high speed automatic stud feed applications because no ceramic ferrule is required. Portable drawn arc type capacitor discharge equipment with automatic stud feed is available for studs ranging from #6 through 6.4 mm (1/4 in.) diameter. Using this system, weld rates of approximately 42 studs per minute can be achieved.

Automatic stud feed is used with production units when high production rates are required. Such a unit is shown in Fig. 8.24.

STUD LOCATION

The method of locating studs depends on several factors: the accuracy and consistency of positioning required, the type of welding equipment to be used (portable or fixed), the required rate of production, and to some extent the physical proportions of the workpiece. In general, the fixed production type welding unit affords greater precision in stud location than does the portable hand-held unit.

Accuracy of location with a portable gun is usually dependent upon the care used in laying

Table 8.6—Typical combinations of base metal and stud metal for capacitor discharge stud welding

Base metal	Stud metal
Low carbon steel, 1006 to 1022	Low carbon steel, 1006 to 1010[a], stainless steel, series 300[b], copper alloy 260 and 268 (brass)
Stainless steel, series 300[b] and 400	Low carbon steel 1006 to 1010[a], stainless steel, series 300[b]
Aluminum alloys, 1100, 3000 series, 5000 series, 6061 and 6063	Aluminum alloy 1100, 5086, 6063
ETP copper, lead free brass, and rolled copper	Low carbon steel, 1006 to 1010[a], stainless steel, series 300[b], copper alloys 260 and 268 (brass)
Zinc alloys (die cast)	Aluminum alloys 1100 and 5086

a. 0.23% C max, 0.60% Mn max, 0.04% P max, and 0.05% S max.
b. Except for the free-machining Type 303 stainless steel.

out the location(s) on the workpiece. However, with the application of various types of spacers, bushings, and template, the accuracy range can be within a tolerance of ± 0.5 mm (0.020 in.).

Standard production type units will provide tolerance limits of ± 0.2 mm (0.008 in.). Special production units employing precision tooling for locating and work holding purposes can operate within tolerances of ± 0.08 mm (0.003 in.). Precision location requires not only fine welding equipment and tooling, but also exceptionally precise, high quality studs.

WELD ENERGY REQUIREMENTS

In capacitor discharge stud welding, arc power is obtained by discharging a capacitor bank through the stud to the work. Arc times are significantly shorter and welding currents are much higher than those used for arc stud welding. It is the very short weld time that accounts for the shallow weld penetration into the work and also the small stud melt-off length.

Depending upon stud size and type of equipment used, the peak welding current can vary from about 600 to 20 000 A. The total time to make a weld depends on the welding method used. For the drawn arc method, weld time is in the range of 6 to 15 milliseconds. Figure 8.25 illustrates typical current-time relationships for the three welding methods. Note that the arc current for the initial contact or initial gap method is much higher than for the drawn arc method.

MATERIAL WELDED

In general, the same metal combinations that can be joined by the arc stud welding method also can be joined by the capacitor discharge method. These include carbon steel, stainless steel, and aluminum alloys.

In addition, some dissimilar metal combinations that present metallurgical problems with arc stud welding can be successfully capacitor discharge stud welded. The reason for this is the small volume of metal melted in the very short capacitor discharge time. The small volume and its expulsion, when the stud

plunges into the plate, result in a very thin layer of weld metal in the joint. If the weld metal is sound and strong, the stud will carry its design load. Weld metal ductility is not a significant factor.

Weldable stud and base metal combinations of commonly used alloys are listed in Table 8.6. The applications are not limited to these mater-

(A)

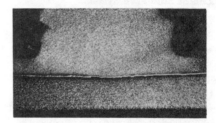

(B)

(C)

Fig. 8.26—Macrostructures of three capacitor discharge stud welds: (A) steel stud, 5 mm (3/16 in.) diameter, to 0.6 mm (0.024 in.) thick steel sheet; (B) brass stud, 6.4 mm (1/4 in.) diameter, to 1.6 mm (1/16 in.) thick mild steel; (C) 6061 T6 aluminum stud, 9.5 mm (3/8 in.) diameter, to 3.2 mm (1/8 in.) aluminum sheet of the same alloy

ials. The relative electrical conductivities or melting temperatures of the materials are not of great significance unless there are great differences between them.

Typical macrostructures of capacitor discharge stud welds are shown in Fig. 8.26. Note the very narrow weld line between the brass and steel sheet in Fig. 8.26(B).

Because of the very short weld times, flux or shielding is not normally required to prevent weld metal contamination from air. One exception is that argon shielding should be used when stud welding aluminum with the drawn arc method; arc time is long enough for harmful oxidation to occur. The stud gun should then be equipped with a gas adaptor foot. Welding grade argon (99.95 percent pure) should be used at the flow rate recommended by the manufacturer.

QUALITY CONTROL AND INSPECTION

Quality control of a capacitor discharge stud weld is more difficult than that of an arc stud weld because of the absence of a steady welding arc and a weld fillet. The operator does not hear and see the welding arc, nor can he use the characteristics of a weld fillet to evaluate weld quality. However, there should be some flash at the weld joint.

The best method of quality control for capacitor discharge stud welding is to destructively test studs that have been welded to base metal similar to that to be used in the actual production. The destructive test should be a bend, torque, or tensile test. Once a satisfactory welding schedule is established, the production

run can begin. It is best to check weld quality at regular intervals during production to ascertain that the welding conditions have not changed.

Some points to consider for producing and maintaining good capacitor discharge stud welds are

(1) Power source of sufficient size for stud size being welded.

(2) Properly maintained and operating equipment.

(3) Tight cable connections.

(4) Proper handling of studs and stud gun during the welding process.

(5) Welding surface cleanliness. The surface should be free from excessive oils, grease, and other lubricants and from rust, mill scale, and other oxides. These conditions contribute to high electrical resistance in areas of welding and grounding.

(6) Welding surface imperfections, such as extreme roughness, which can prevent complete fusion in the weld area.

(7) Perpendicularity of stud axis to the work surface. This is important for complete fusion.

(8) Proper weld end design on the stud. The tip size, face angle, and weld base diameter must be correct for the application.

Capacitor discharge stud welds may be inspected both visually and mechanically. The success of visual inspection methods depends on the interpretation of the appearance of the weld. Figure 8.27 illustrates good and bad capacitor discharge stud welds.

If a questionable weld is evident after the welds have been visually inspected, the weld should be mechanically tested.

(A)   (B)   (C)

Fig. 8.27—Examples of satisfactory and unsatisfactory capacitor discharge stud welds: (A) good weld, (B) weld power too high, and (C) weld power too low

MECHANICAL TESTING

Mechanical testing of capacitor discharge stud welds should be done by using the same methods described for arc stud welding. These include bend testing and proof tensile loading. The tests are used to establish welding conditions and also to qualify production studs.

Maximum nut torque values for proof testing studs by this method are given in Table 8.7 for various stud materials and sizes. The torque values listed will produce tensile stresses in those studs that are slightly below the material yield strengths.

The table also gives the tensile loads that will develop approximately the nominal tensile strength of the stud material in the different diameters. The maximum shear load that the studs can carry is also listed for information. For proof testing or stud selection, appropriate safety factors should be applied by the user.

APPLICATIONS

Some of the fields using capacitor discharge stud welding are aircraft and aerospace, appliances, building construction, maritime construction, metal furniture, stainless steel equipment, and transportation. They are used, where appropriate, as fasteners or supports.

Table 8.7—Torque, tensile, and shear loads for capacitor discharge welded studs of various materials and sizes

Stud material	Stud size	Maximum fastening torque[a]		Maximum tensile load[b]		Maximum shear load	
		N•m	lbf•in.	kn	lb	kN	lb
Low-carbon, copper-flashed steel	6-32	0.7	6	2.2	500	1.7	375
	8-32	1.4	12	3.4	765	2.6	575
	10-24	1.6	14	4.3	960	3.2	720
	1/4-20	4.9	43	7.8	1750	5.8	1300
	5/16-18	8.1	72	13	2900	9.8	2200
	3/8-16	12	106	19	4300	14	3250
Stainless steel 304 or 305	6-32	1.1	10	3.5	790	2.6	590
	8-32	2.3	20	5.6	1260	4.2	940
	10-24	2.6	23	6.8	1530	5.1	1150
	1/4-20	8.5	75	13	2880	9.6	2160
	5/16-18	14	126	17	3750	14	3100
	3/8-16	21	186	22	4850	20	4550
Aluminum alloy 1100	6-32	0.3	2.5	0.9	200	0.6	125
	8-32	0.6	5	1.3	295	0.8	185
	10-24	0.7	6.5	1.7	380	1.0	235
	1/4-20	2.4	21.5	3.0	670	1.9	415
	5/16-18	4.1	36	5.0	1125	3.1	695
	3/8-16	6.0	53	7.4	1660	4.4	1000
Aluminum alloy 5086	6-32	0.4	3.5	1.7	375	1.0	235
	8-32	0.8	7.5	2.6	585	1.6	365
	10-24	1.1	10	3.3	735	2.0	460
	1/4-20	3.7	32.5	6.1	1360	3.8	850
	5/16-18	6.2	54.5	10	2300	6.2	1400
	3/8-16	9.2	81	15	3400	9.4	2100
Copper alloy (brass) 260 and 268	6-32	0.9	8	2.7	600	1.7	390
	8-32	1.8	16	3.8	860	2.5	560
	10-24	2.1	18.5	4.6	1040	3.0	680
	1/4-20	6.4	61	8.7	1950	5.7	1275
	5/16-18	12	102	15	3280	9.5	2140
	3/8-16	16	150	21	4800	14	3160

a. These values should develop stud tensile stresses to slightly below the yield strengths of the materials.
b. These values should develop the nominal tensile strengths of the materials.

PROCESS SELECTION AND APPLICATIONS

PROCESS SELECTION CONSIDERATIONS

There are some types of applications for which the capabilities of the arc stud welding process and the capacitor discharge stud welding process overlap, but generally the selection between these two basic processes is well defined. A process selection chart is shown in Table 8.8. The area in which selection is usually more difficult is the method of capacitor discharge stud welding that should be used, i.e., contact, gap, or drawn arc.

The main criteria for selecting which basic type of stud welding process should be used are fastener size, base metal thickness, and base metal composition. Using these criteria, it is almost always possible to select the best method.

Fastener Size

For studs over 6.4 mm (1/4 in.) diameter, the arc stud welding process must be used for portable applications. The capacitor discharge stud welding process is limited to 6.4 mm (1/4 in.) diameter with hand-held guns and to 9.5 mm (3/8 in.) diameter studs with fixed or production type equipment. Applications suitable for the capacitor discharge type process with studs in the 7.9 to 9.5 mm (5/16 to 3/8 in.) diameter range generally involve thin base materials where avoidance of reverse side marking is the foremost requirement.

Base Metal Thickness

For base thicknesses under 1.6 mm (0.062 in.), the capacitor discharge stud welding process should be used. Using this process, the base metal can be as thin as 0.5 mm (0.020 in.) without melt-through occurring. On such thin material, the sheet will tear when the stud is loaded excessively. Reverse side marking is the principal effect involved in appearance.

Using the arc stud welding process, the base metal thickness should be at least 1/3 the weld base diameter of the stud to assure maxi-

mum stud strength. Where strength is not the foremost requirement, the base metal thickness may be a minimum of 1/5 the weld base diameter.

Base Metal Composition

For mild steel, austenitic stainless steel, and various aluminum alloys, either process can be used. For copper, brass, and galvanized steel sheet, the capacitor discharge process is best suited.

Capacitor Discharge Method

Using the above criteria, if the capacitor discharge process is chosen as the best one for the application, then the methods within this process must be evaluated. Since there is considerable overlap in the stud welding capabilities of the three methods, there are many applications where any one of them can be used. On the other hand, there are many instances where one method is best suited for the application. Setting up specific guidelines for selection of the best capacitor discharge method is rather difficult. However, usage of the three different methods is generally as follows:

Initial Contact Method. It is used only with portable equipment, principally for welding mild steel studs. Equipment simplicity makes it ideal for welding mild steel insulation pins to galvanized duct work.

Initial Gap Method. This method is used with portable and fixed equipment for welding mild steel, stainless steel, and aluminum. Generally, it is superior to both the drawn arc and initial contact methods for welding dissimilar metals and aluminum. Inert gas is not needed for aluminum welding.

Drawn Arc Method. The types of equipment and materials welded are the same as those of the initial gap method. The stud does not require a special tip. The method is ideally suited for high speed production applications involving automatic feed systems with either portable equipment or fixed production type equipment. Inert gas is required for aluminum welding.

Table 8.8—Stud welding process selection chart

Factors to be considered	Arc stud welding	Capacitor discharge stud welding	
		Initial gap and initial contact	Drawn arc
Stud shape			
Round	A	A	A
Square	A	A	A
Rectangular	A	A	A
Irregular	A	A	A
Stud diameter or area			
1.6 to 3.2 mm (1/16 to 1/8 in.) diam	D	A	A
3.2 to 6.4 mm (1/8 to 1/4 in.) diam	C	A	A
6.4 to 12.7 mm (1/4 to 1/2 in.) diam	A	B	B
12.7 to 25.4 mm (1/2 to 1 in.) diam	A	D	D
up to 32.3 mm² (0.05 in.²)	C	A	A
over 32.3 mm² (0.05 in.²)	A	D	D
Stud metal			
Carbon steel	A	A	A
Stainless steel	A	A	A
Alloy steel	B	C	C
Aluminum	B	A	B
Brass	C	A	A
Base metal			
Carbon steel	A	A	A
Stainless steel	A	A	A
Alloy steel	B	A	C
Aluminum	B	A	B
Brass	C	A	A
Base metal thickness			
under 0.4mm (0.015 in.)	D	A	B
0.4 to 1.6 mm (0.015 to 0.062 in.)	C	A	A
1.6 to 3.2 mm (0.062 to 0.125 in.)	B	A	A
over 3.2 mm (0.125 in.)	A	A	A
Strength criteria			
Heat effect on exposed surfaces	B	A	A
Weld fillet clearance	B	A	A
Strength of stud governs	A	A	A
Strength of base metal governs	A	A	A

Legend
A—Applicable without special procedures, equipment, etc.
B—Applicable with special techniques or on specific applications which justify preliminary trials or testing to develop welding procedure and technique.
C—Limited application.
D—Not recommended.

APPLICATION CONSIDERATIONS

Studs can be welded with the work in any position, i.e., flat, vertical, and overhead. The use of the gravity drop head principle, of course, is limited to the flat position. Also, welding studs to vertical work is presently limited to 19 mm (3/4 in.) diameter studs and smaller.

Studs can be welded to curved or angled surfaces. However, using the arc stud process which melts considerably more metal, the ceramic ferrule must be designed to fit the contour of the work surface.

The arc stud welding process is much more tolerant of work surface contaminants, such as light coatings of paint, scale, rust, or oil, than is the capacitor discharge stud welding process. The long arc duration with the arc stud welding process tends to burn the contaminants away. Also, the molten metal expulsion tends to wash away any residue out of the joint.

On the other hand, the percussive nature of the capacitor discharge arc tends to expel metallic coatings, such as those applied by electroplating and galvanizing, out of the joint. This makes the process suitable for welding small diameter studs to thin gage galvanized sheet metal. The arc stud welding process is suitable for welding through thick galvanized coatings using special welding procedures, provided the base material is thick enough to withstand the long arc time.

Arc stud welding and capacitor discharge stud welding have been widely accepted by all the metalworking industries.

SUPPLEMENTARY READING LIST

Automated system welds heat transfer studs. *Welding Journal* 53(1), Jan. 1974, pp. 29-30.

Baeslach, W.A., Fayer, G., Ream, S., and Jackson, C.E. Quality control in arc stud welding. *Welding Journal* 54(11), Nov. 1975, pp. 789-798.

Lockwood, L.F. Gas shielded steel welding of magnesium. *Welding Journal* 46(4), April 1967, pp. 168s to 174s.

Metals Handbook: Welding and Brazing, Vol. 6, 8th ed. Metals Park, Ohio: American Society for Metals, 1971, pp. 167-176.

Pease, C.C. Capability studies of capacitor discharge stud welding on aluminum alloy. *Welding Journal* 48 (6), June 1969, pp. 253-s —257-s.

Pease, C.C., and Preston, F.J. Stud welding through heavy galvanized decking. *Welding Journal* 51 (4), Apr. 1972, pp. 241-244.

Pease, C.C., Preston, F.J., and Taranto, J. Stud welding on 5083 aluminum and 9% nickel steel for cryogenic use. *Welding Journal* 52 (4), Apr. 1973, pp. 232-237.

Recommended Practices for Stud Welding, AWS C5.4. Miami: American Welding Society, 1974.

Shoup, T.E. Stud Welding. Bulletin 214. New York: Welding Research Council, April 1976.

Structural Welding Code, AWS D1.1. Miami: American Welding Society, 1975.

9

Plasma Arc Welding

Prepared by a committee consisting of

J.R. CONDRA, *Chairman*
Organic Chemicals Department
E.I. duPont de Nemours and Co., Inc.

R.D. MANN
Airco Inc.

J.W. McGREW
Industrial and Marine Division
Babcock and Wilcox

DANIEL O'HARA
Thermal Dynamics Corp.

M.J. TOMSIC
Hobart Brothers Company

E.J. WOODINGS
TRW, Inc.

9

Plasma Arc Welding

FUNDAMENTALS OF THE PROCESS

INTRODUCTION

The term "plasma arc" is used to describe a family of metal working processes that use a constricted electric arc. Constriction of the arc is usually accomplished by passing the arc through a water-cooled copper orifice. Its purpose is to control and increase the energy density of the arc stream. Plasma arc processes are employed for the welding, cutting, and surfacing of metals. Only welding will be considered in this chapter. Cutting is covered in Chapter 13. Surfacing is covered in Chapter 14. Thermal spraying, which utilizes a nontransferred arc, is covered in Chapter 29, Section 2, 6th edition.[1]

DEFINITIONS

Plasma arc welding (PAW) is an arc welding process where coalescence is produced by heating with a constricted arc between an electrode and the workpiece (transferred arc) or between the electrode and the constricting nozzle (nontransferred arc). Shielding is generally obtained from the hot, ionized gas issuing from the orifice of the constricting nozzle. An auxiliary source of shielding gas is often used when it is required to protect the weld from air. Shielding gas may be an inert gas or a mixture of gases. Filler metal may or may not be added.

Plasma arc welding is basically an extension of the gas tungsten arc welding (GTAW) processes. However, it has a much higher arc energy density and higher plasma gas velocity by virtue of the arc plasma being forced through a constricting nozzle, as shown in Fig. 9.1.

The orifice gas is the gas which is directed through the torch to surround the electrode. It becomes ionized in the arc to form the plasma and issues from the orifice in the torch nozzle as the plasma jet. For some operations, auxiliary shielding gas is provided through an outer gas cup, similar to gas tungsten arc welding. The purpose of the auxiliary shielding gas is to blanket the area of arc plasma impingement on the workpiece to avoid contamination of the weld pool.

The arc constricting nozzle through which the arc plasma passes has two main dimensions: orifice diameter and throat length. The orifice may be cylindrical or have a converging or diverging taper.

The distance that the electrode is recessed within the torch is the electrode setback. The dimension from the outer face of the torch nozzle to the workpiece is known as the torch standoff distance.

The plenum or plenum chamber is the space between the inside wall of the constricting nozzle and the electrode. The orifice gas is directed into this chamber and then through the orifice to the work. A tangential vector may be imparted to the gas flow to form a swirl through the orifice.

---

1. Chapter 29 is scheduled for revision in Vol. 3, 7th ed.

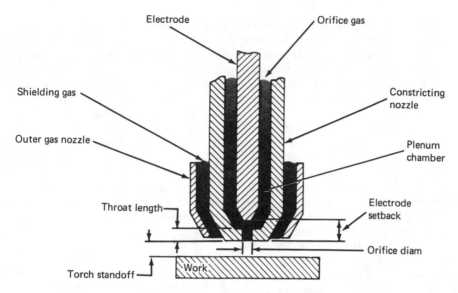

Fig. 9.1—Plasma arc torch terminology

PRINCIPLES OF OPERATION

Operating Principle

The plasma arc welding process is similar to the conventional gas tungsten arc welding (GTAW) process with the exception of a pilot arc starting circuit and a constricting orifice.[2] The basic arrangement of both systems is shown in Fig. 9.2. The electrode in the GTAW torch extends beyond the end of the shielding gas nozzle so that the arc is visible, and there is clearance between the nozzle and the work. The gas tungsten arc is not constricted and assumes an approximately conical shape. This produces a relatively wide heat pattern on the workpiece. The area of impingement of the cone-shaped arc on the workpiece varies with the electrode-to-work distance and arc amperage. Thus, a small change in arc length produces a relatively large change in heat input per unit area.

By contrast, the electrode in the plasma arc torch is recessed within the constricting nozzle. The arc is collimated and focused by the constricting nozzle on a relatively small area of the workpiece. Because the shape of the arc is essentially cylindrical, there is very little change in the area of contact with the workpiece as torch standoff varies. Thus, the PAW process is less sensitive to variations in torch-to-work distance than the GTAW process.

Since the electrode of the plasma arc torch is recessed inside the arc constricting nozzle, it is not possible to touch the electrode to the workpiece. This property greatly reduces the possibility of contaminating the weld with electrode metal.

As the orifice gas passes through the plenum chamber of the plasma arc torch, it is heated by the arc, expands, and exits through the constricting orifice at an accelerated rate. Since too powerful a gas jet can cause cutting or turbulence in the weld puddle, orifice gas flow rates are generally held to within 1.5 to 15 L/min (3 to 30 ft³/h). The orifice gas alone is not generally adequate to shield the weld pool from atmospheric contamination. Therefore, auxiliary shielding gas must be provided through an outer gas nozzle. Typical shielding gas flow rates are in the range of 10 to 30 L/min (20 to 60 ft³/h).

Purposes of Arc Constriction

Several improvements in performance over open arc operation (GTAW) can be obtained by passing the plasma arc through a small orifice. The most noticeable improvement is the direc-

---

2. The gas tungsten arc welding process is described in Ch. 3.

tional stability of the plasma jet. A conventional gas tungsten arc is attracted to the nearest work connection (ground), and it is deflected by low-strength magnetic fields. On the other hand, a plasma jet is comparatively stiff; it tends to go in the direction in which it is pointed; and it is less affected by magnetic fields.

High current densities and high energy concentration can be produced by arc constriction. The higher current densities result in higher temperatures in the constricted plasma arc.

The higher temperatures and electrical changes brought about by constricting an arc are shown in Fig. 9.3. The left side of this figure represents a normal nonconstricted tungsten arc operating at 200 A, dcsp, in argon at a flow rate of 19 L/min (40 ft³/h). The right side illustrates an arc, with the same current and gas flow, that is constricted by passing it through a 4.8 mm (3/16 in.) diameter orifice. Under these conditions, the constricted arc shows a 100 percent increase in arc power and a 30 percent increase in temperature over the open arc. The spectroscopic methods used to measure the temperatures of arcs are based on the analysis and interpretation of emission spectra.

The increased temperature of the constricted arc is not its chief advantage, since the temperature in the gas tungsten arc far exceeds the melting points of the metals generally welded by the process. The main advantages of the plasma arc are the directional stability and focusing effect brought about by arc constriction, and its relative insensitivity to variations in torch standoff distance.

The plasma arc offers better control over arc energy. The degree of arc collimation, arc force, energy density on the workpiece, and other characteristics are primarily functions of

(1) Plasma current
(2) Orifice diameter and shape
(3) Type of orifice gas
(4) Flow rate of orifice gas

The fundamental differences among the many plasma arc metal working processes lie in the relationship of these four factors. They can be adjusted to provide very high or very low thermal energies. For example, the high energy concentration and high jet velocity necessary for plasma arc cutting dictate a high arc current, a small diameter orifice, high orifice gas flow rate, and a gas having high thermal conductivity. For welding, on the other hand, a low plasma jet velocity is necessary to prevent weld metal expulsion from the workpiece. This calls for larger orifices, considerably lower gas flow rates, and lower transferred arc currents.

The constricted arc is much more effective than an open arc for heating the gas to be used for a particular operation. When the gas passes directly through a constricted arc, it is exposed to higher energy concentrations than when it passes around a conventional gas tungsten arc, as shown in Fig. 9.2.

Arc Modes

Two arc modes are used in plasma arc welding: transferred arc and nontransferred arc. With a transferred arc, the main arc is established between the electrode and the workpiece. With a nontransferred arc, the arc is established between the electrode and the constricting orifice inside the torch. The arc plasma is then forced through the orifice by the plasma gas. The workpiece is not in the arc circuit. Transferred arcs have the advantage of greater energy transfer to the work, but they require an electrically conductive workpiece. Nontransferred arcs are useful for cutting and joining nonconductive workpieces or for applications where very low energy concentration is desirable. Figure 9.4 illustrates the two modes of arc transfer.

In the transferred arc mode, the workpiece is part of the electrical circuit and the arc transfers from the electrode to the work. The transferred arc produces heat both from the anode spot on the work and the plasma stream. This mode is the one generally used for welding.

In the nontransferred arc mode, the terminals of the arc are contained within the torch. The arc is maintained between the electrode and the constricting nozzle. Useful heat is obtained from the hot plasma gas issuing from the orifice.

For a given power input into the system, a specific nozzle contour, and a set rate of gas flow through the orifice, a transferred arc will provide more thermal energy than the nontransferred arc.

Basic Equipment

The basic equipment for PAW is the same

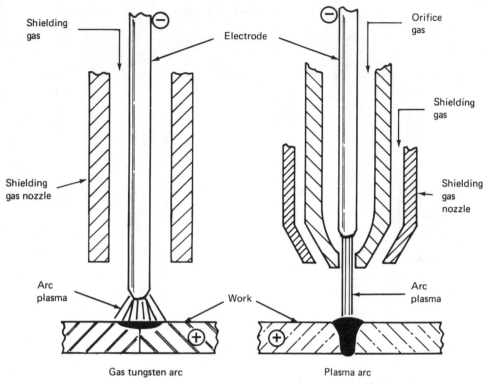

Shielding gas

Electrode

Orifice gas

Shielding gas

Shielding gas nozzle

Shielding gas nozzle

Arc plasma

Work

Arc plasma

Gas tungsten arc

Plasma arc

Fig. 9.2—Comparison of gas tungsten arc and plasma arc welding processes

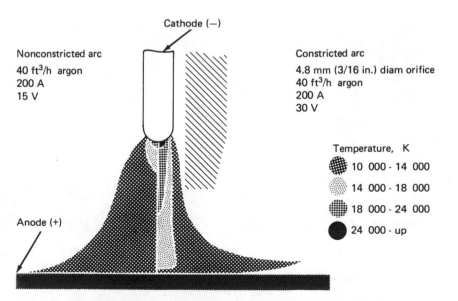

Cathode (−)

Nonconstricted arc
40 ft³/h argon
200 A
15 V

Constricted arc
4.8 mm (3/16 in.) diam orifice
40 ft³/h argon
200 A
30 V

Temperature, K
10 000 - 14 000
14 000 - 18 000
18 000 - 24 000
24 000 - up

Anode (+)

Fig. 9.3—Effect of arc constriction on temperature and voltage

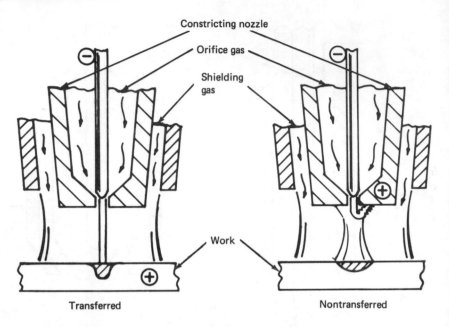

Fig. 9.4—Transferred and nontransferred plasma arc modes

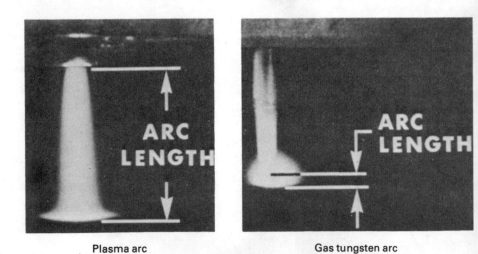

Fig. 9.5—Comparison of plasma arc and gas tungsten arc lengths commonly used for welding very thin metal sections at 10A

as for GTAW. It consists of a welding torch, a welding power source, a means for initiating the arc, and gas and cooling water supplies. The welding torches are similar in appearance to GTAW welding torches. The welding power sources for PAW are usually direct current output with the characteristic drooping volt-ampere output of a constant current type.[3]

When filler metal addition is required, it can be added manually in rod form by the welder or from a wire spool with a filler wire feeder for machine or automatic welding. The techniques for filler metal addition are the same as with gas tungsten arc welding.

Welding Current

Direct current, straight polarity (electrode negative) power is used with a tungsten electrode and a transferred arc for most applications. Direct current, reverse polarity (electrode positive) is used to a limited extent with a tungsten electrode or a water-cooled copper electrode for welding aluminum. Reverse polarity is also used with specially designed torches for joining titanium and zirconium sponge compacts when freedom from electrode contamination is a prime consideration. The current range for dc welding is from approximately 0.1 to 500 A.

Alternating current, with continuous high frequency stabilization, can be used to weld aluminum in the current range of approximately 10 to 100 A. With special power sources, amperages up to 200 A can be used.

Arc Length

The columnar nature of the constricted arc makes the plasma arc process less sensitive to variations in arc length than the gas tungsten arc process. Since the unconstricted gas tungsten arc has a conical shape the area of heat input to the workpiece varies as the square of the arc length. A small change in arc length, therefore, causes a relatively large change in the unit area heat transfer rate. With the essentially cylindrical plasma jet, however, as the arc length is varied within normal limits, the area of heat input and the intensity of the arc are virtually constant.

3. Arc welding power sources are discussed in Ch. 1.

The collimated plasma jet, obtained by passing an arc through a small diameter orifice, is chiefly responsible for the lack of sensitivity of the process to changes in arc length. This characteristic also permits the use of a much longer torch-to-work distance (torch standoff) than is possible with the GTAW process and greatly reduces the operator skill required to manipulate the torch. Typical arc lengths used to weld thin gage material at approximately 10 amperes are shown in Fig. 9.5. The plasma arc is approximately 6.4 mm (0.25 in.) long as compared to the 0.6 mm (0.025 in.) long gas tungsten arc.

Process Variations

Melt-in Technique. Conventional fusion welding, such as that done with the gas tungsten arc process, can also be done with the plasma arc at low amperages and orifice gas flow rates. This "melt-in" technique is the only one usually done with manual welding. Filler metal in rod form can be added to the molten weld pool.

Keyhole Welding. In plasma arc welding of certain ranges of metal thicknesses, special combinations of plasma gas flow, arc current, and weld travel speed will produce a relatively small weld pool with a hole penetrating completely through the base metal. It is called a "keyhole" and is illustrated in Fig. 9.6. The plasma arc process is the only gas shielded welding process with this unusual characteristic.

In a stable keyhole operation, molten metal is displaced to the top bead surface by the plasma stream (in penetrating the weld joint) to form the characteristic keyhole. As the plasma arc torch is mechanically moved along the weld joint, metal melted by the arc is forced to flow around the plasma stream and to the rear where the weld pool is formed and solidified. This motion of molten metal and the complete penetration of the metal thickness allows the impurities to flow to the surface and the gases to be expelled more readily before solidification. This action is similar to "magnetic stirring" developed for gas tungsten arc welding. The maximum weld pool volume and the resultant root surface profile are largely determined by the effects of a force balance between the molten weld metal surface tension and

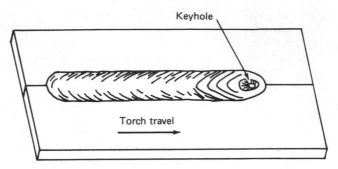

Fig. 9.6—Pictorial representation of the keyhole in plasma arc welding

the plasma stream velocity characteristics. The bead appearance for a butt weld made in the keyhole mode is pictured in Fig. 9.7.

The high current keyhole technique of welding operates just below conditions that would actually cut rather than weld. For cutting, slightly higher orifice gas velocity blows the molten metal away. In welding, the gas velocity is just low enough that surface tension of the molten metal holds it in the joint. Consequently, orifice gas flow rates for welding are critical and must be closely controlled. Variation of no more than 0.12 L/min (1/4 ft³/h) in flow rate is recommended. This value is quite small.

When the melt-in welding technique is used, orifice gas flow rates are lower and less critical than for the keyhole technique. However, the melt-in technique is the same as with gas tungsten arc welding. Therefore, plasma arc welding must have other desirable features for the application to justify its use in place of gas tungsten arc welding.

EQUIPMENT

The basic equipment for plasma arc welding is shown in Fig. 9.8. Plasma arc welding is done with both manual and mechanized equipment.

A complete system for manual plasma arc welding consists of a torch, control console, power source, orifice and shielding gas supplies, source of torch coolant, and accessories such as an on-off switch, gas flow timers, and remote current control. Equipment is presently available for operation in the current range of 0.1 to 225 A, dc, straight polarity (electrode negative).

Mechanized equipment must be used to achieve the fast welding speeds and the penetration advantages associated with high current plasma arc welding. A typical mechanized installation consists of a power source, control unit, machine welding torch, torch stand or travel carriage, coolant source, high frequency power generator, and supplies of shielding gases. Accessory units such as an arc voltage control and filler wire feed system may be used as required. Machine welding torches are available for welding with currents up to approximately 500 A, dc, straight polarity, but 400 A is approximately the highest current used.

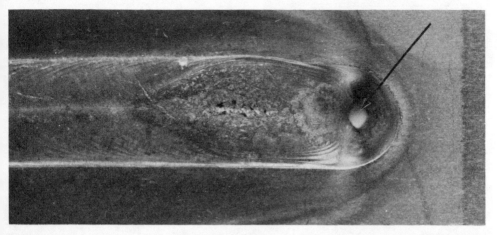

Face of bead

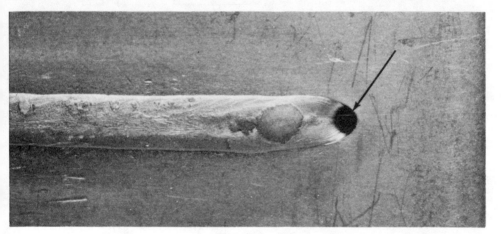

Root of bead

Fig. 9.7—Keyhole at the end of a weld bead

ARC INITIATION

The plasma arc cannot be started with the normal techniques used with gas tungsten arc welding. Since the electrode is recessed in the constricting nozzle, it cannot be touch-started against the workpiece. It is first necessary to ignite a low-current pilot arc between the electrode and the constricting nozzle. Pilot arc power may be provided by a separate power source or by the welding power source. The pilot arc can be started in two ways. With a low amperage torch, the electrode can be advanced until it touches the nozzle and then retracted to draw an arc. With a high amperage torch, either high frequency, ac power (high amperage, low amperage), or one or more high voltage, low power pulses are superimposed on the welding circuit. The high voltage power ionizes the orifice gas so that it will conduct the pilot arc current.

The basic circuitry with a high frequency generator is shown in Fig. 9.9. The constricting

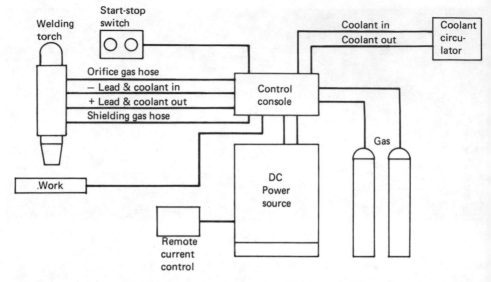

Fig. 9.8—Typical equipment for plasma arc welding

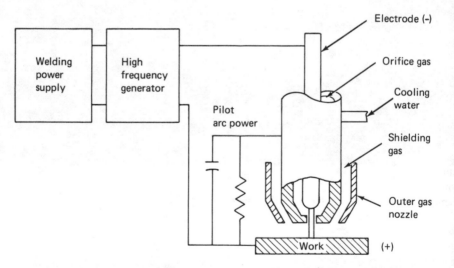

Fig. 9.9—Plasma arc welding system with high frequency pilot arc initiation

nozzle is connected to the power source positive terminal through a current limiting resistor. A low current pilot arc is initiated between the electrode and the nozzle by the high frequency generator. The electrical circuit is completed through the resistor. The ionized gas from the pilot arc forms a low resistance path between the electrode and the nozzle or between the electrode and work. When the power source is energized, the main arc is initiated between the electrode and the work or the nozzle, depending on the welding circuit arrangement. The pilot arc is used only to assist in starting the main arc. After the main arc starts, the pilot arc may be extinguished.

WELDING TORCHES

Torches for plasma arc welding are more complex than those for gas tungsten arc welding. Separate passages are required for orifice gas and shielding gas, and the constricting nozzle must be water-cooled.

A water-cooled power cable is used to bring power and cooling water into the torch; other hoses are provided for orifice gas input, shielding gas input, and water return. The electrode holder in a plasma arc welding torch is designed to center the electrode very accurately with respect to the central part in the nozzle. Misalignment of the electrode with the central part tends to cause melting of the copper nozzle near the orifice and to shorten its life.

The relatively low orifice gas flows used for welding do not provide adequate protection for the molten weld pool. Accordingly, auxiliary shielding gas is supplied through a shielding gas nozzle on the torch. For some applications, additional trailing gas shields may be required.

A variety of constricting nozzle designs exists for different welding applications. The diameter of the nozzle orifice used in a particular application depends on the welding current to be used. Higher currents require larger diameter orifices.

For each size, a torch is further rated for a given orifice diameter. For a given orifice size and a particular torch, there is a maximum current rating for the unit. This maximum current rating also requires a given plasma gas type and flow rate. For example, a 2.3 mm (0.089 in.) diameter orifice might be rated at 100 A with 0.7 L/min (1-1/2 ft³/h) flow of argon gas. If the flow rate of the orifice gas falls below 0.7 L/min (1-1/2 ft³/h), the maximum amperage rating of the orifice would have to be decreased.

Manual Torches

A typical torch configuration for manual plasma arc welding is shown in Fig. 9.10. The construction of a torch head is shown in Fig. 9.11. This torch has a handle for holding, a means for securing the tungsten electrode in position and conducting current to it, separate passages for orifice and shielding gases, a water-cooled constricting nozzle (copper), and a shielding gas nozzle (usually of ceramic material).

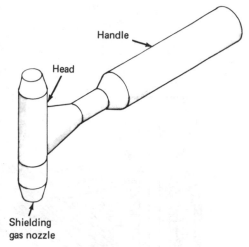

Fig. 9.10—Manual torch configuration

Low-current plasma arc torches are lightweight and designed primarily for manual welding. A supply of cooling water is necessary to dissipate heat generated in the constricting nozzle by the pilot and main welding arcs. The tungsten electrode is automatically centered in relation to the constricting orifice. Auxiliary shielding gas is supplied through a separate gas system. Manual plasma arc welding torches are available for operation on dc, straight polarity (dcsp) at currents up to 225 A. Torch holders are available to mount the torch for mechanized use.

Machine Torches

Torches for mechanized plasma arc welding are similar to manual torches, except some are straight-line, offset torches, as shown in Fig. 9.12.

Mechanized plasma arc welding torches are available commercially for operation on either straight or reverse polarity. Straight polarity power is used with a tungsten electrode for most welding applications. Reverse polarity is used to a limited extent with tungsten or water-cooled copper electrodes for welding aluminum. Reverse polarity is also used with specially designed torches and copper electrodes for joining titanium and zirconium sponge compacts where freedom from tungsten or copper contamination is a prime consideration. Machine torches are available commercially for operation on either dcsp or dcrp at currents up to 500 A.

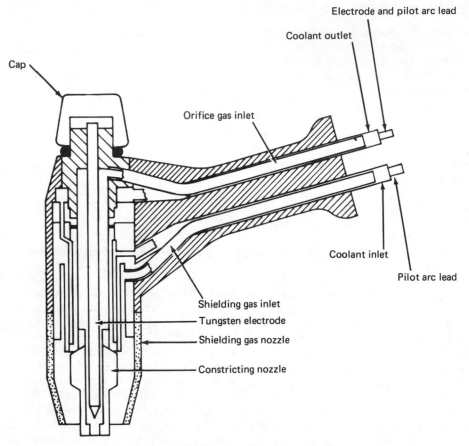

Fig. 9.11 — Typical manual torch head design

Arc-Constricting Nozzles

A wide variety of nozzles has been made and evaluated. These include the single port nozzles and multiple port nozzles with holes arranged in circles, rows, and other geometric patterns. Single port nozzles are most widely used. Among the multiple port nozzles, the most widely used design has the center orifice bracketed by two smaller ports, with a common centerline for all three openings. These two common nozzle types are shown in Fig. 9.13.

The electrode in the plasma arc torch is recessed in the arc-constricting nozzle. As the arc passes through the nozzle, it is collimated and focused so that the arc heat is concentrated on a relatively small area of the workpiece. This increased heat concentration, coupled with the characteristically more forceful plasma stream, can produce a narrower weld fusion zone.

With the single port nozzle, the arc and all of the orifice gas pass through the single orifice. With the multiple port nozzle, the arc and some of the orifice gas pass through the larger center orifice. The remainder of the orifice gas is discharged through two smaller ports that bracket the center orifice. A single port nozzle is generally used for dc reverse polarity work, since multiple port nozzles do not provide the advantages realized with straight polarity welding.

With appropriately designed weld joints, the multiple port nozzle, shown in Fig. 9.13, can be used to advantage. When the multiple port nozzle is aligned to place the common centerline of the side ports perpendicular to the weld groove

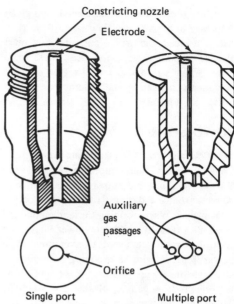

Fig. 9.13—*Single and multiple port constricting nozzles*

and electrical insulation to protect the inside surface of the nozzle. As a result of this, water-cooled copper (the most common nozzle material, with a melting point of 1080° C), can be used to constrict an arc which may contain plasma temperatures in excess of 24 000° C. If this protective gas layer is disturbed by an occurrence, such as insufficient orifice gas flow or excessive arc current for a given nozzle geometry, the nozzle may be damaged by double arcing. In double arcing, the metallic torch nozzle forms part of the current path to the work, and, in essence, two arcs are formed. The first arc is from the electrode to the nozzle and the second is from the nozzle to the work. Heat generated at the cathode and anode spots, formed where the two arcs attach to the nozzle, invariably causes damage to this part.

Influence of Gas Flow

The orifice gas flow influences the penetrating power of the plasma arc torch. The higher the orifice gas flow, the more penetrating power is available. However, as the gas flow is increased, undercutting will be greater at the edges of the weld. Generally, it is recommended that the lowest plasma gas flow that will do the job be

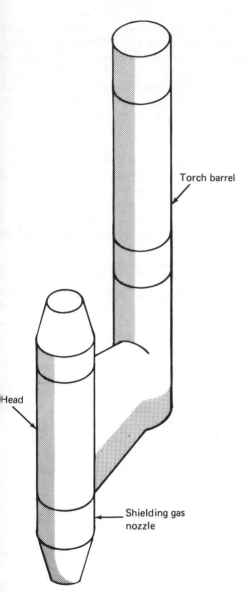

Fig. 9.12—*Machine torch configuration*

axis, the arc is elongated in line with the joint. For the keyhole technique, this allows an increase in welding speed of from 30 to 50 percent over that obtained with single port nozzles.

During normal operation, the arc column within the torch nozzle is surrounded by an annular layer of cooler gas with a very steep thermal gradient. The layer of relatively cool, nonconductive gas at the nozzle wall provides thermal

used. The orifice size is usually rated for a given amperage at a given gas flow rate. If the flow rate is reduced, the orifice current rating must also be proportionately decreased. Thus, for a given orifice size, the practical region of torch operation is defined by the welding current and orifice gas flow rate.

POWER SOURCES

Power sources for plasma arc welding are available with various amperage outputs from 0.1 to several hundred amperes. A combination dc power source and control of 0.1 to 10 A output is available. When welding current output exceeds 10 A, usually a separate power source and control console are used. Conventional type power sources with a drooping volt-ampere characteristic are preferred for dc straight polarity (electrode negative) plasma arc welding. These units are typically the same types of power sources that are used for gas tungsten arc welding, and they are available with current ratings from 100 to 400 A, with 60 or 100 percent duty cycle.

Rectifier type power sources are preferred over motor-generator types because of their electrical output characteristics. A rectifier power source with an open-circuit voltage in the range of 65 to 80 V is satisfactory for plasma arc welding with pure argon or with argon-hydrogen gas mixtures containing up to 7 percent hydrogen for orifice gas. However, if pure helium or an argon-hydrogen gas mixture containing more than 7 percent hydrogen is used, higher open circuit voltage is required for reliable arc ignition. This may be obtained by connecting two power sources in series. If erratic arc ignition is experienced, the arc may be started in pure argon and then automatically switched to the argon-hydrogen mixture or helium for welding. Constant current power sources are available with several programmed options such as upslope of weld current, taper of weld current, and downslope of weld current. These special features of the power source are used for various applications, primarily automatic welding.

Pulsed direct current can also be used for plasma arc welding. Pulsed current power sources are similar to those used with gas tungsten arc welding. A conventional drooping volt-ampere characteristic power source with the capability of pulsing between a low level current, referred to as a "background current," and a high level current called "peak current" is employed. Pulsed current power sources used for plasma arc welding have variable pulse frequencies and sometimes variable pulse width ratios.

For conventional steady current power sources, there are available add-on packages which will provide pulsed current within a limited range of pulse frequencies. The same features of upslope of weld current, taper of weld current, and downslope of weld current can be obtained on pulse current power sources similar to steady current power sources.

The electrical circuit generally used for low-current (0.1 to 10 A) plasma arc welding is shown in Fig. 9.14. The system utilizes two separate power sources, one for the pilot arc and one for the transferred arc. The pilot arc power source is preset to deliver approximately 5 A continuously.

The pilot arc may be initiated by touching the electrode to the nozzle and then retracting it, or with a high frequency discharge in a manner similar to high current equipment. The pilot arc circuit is always energized when the unit is in use. The main transferred arc current is energized by closing a contactor in the work lead (ground connected to the workpiece).

CONTROLS

A manual welding control contains the pilot arc power supply, gas flowmeters, and solenoid valves for controlling orifice and shielding gas and water flow rates. Indicating current meters, control switches, and connections for remote current control are generally available. For low current applications, the control and main power supply are combined in one console.

A typical control system for mechanized high current plasma arc welding consists of a main console control, a junction box for gas and water hoses, an operator's pendant control box, and a high-frequency generator. The main control unit sequences the power source; high-frequency generator; orifice, shielding, and backing gases; torch travel; auxiliary wire feeding; and cooling

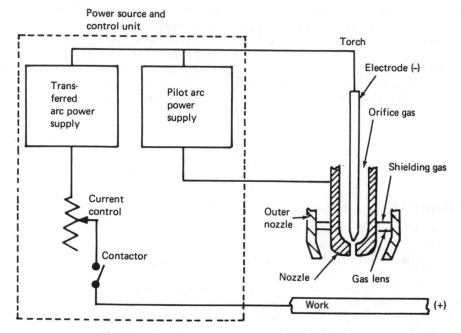

Fig. 9.14—Low-current plasma arc welding equipment

water for the torch. Travel speed and wire feed rates are controlled by electronic governors.

Flowmeters are provided for metering the orifice, shielding, and backing gases. In addition, timers and a gas flow tapering device are provided for up and down sloping of the orifice gas. An adequate supply of cooling water for the torch is provided by a circulation system. A flow switch and interlocking circuitry assure that cooling water is flowing to the torch before the arc can be ignited.

ACCESSORY EQUIPMENT

Wire Feeders

Conventional filler wire feed and hot wire feed systems can be used with the plasma arc process.[4] Filler metal is added to the leading edge of the weld pool or keyhole by the use of conventional wire feeders in the same manner as with the GTAW process. Hot wire systems feed into the trailing edge of the weld pool. Initiation and termination of wire feed may be programmed with automatic welding equipment.

---

4. Hot wire feed systems are described in Ch. 3.

Torch Coolant Systems

Because of particular electrolysis problems associated with plasma arc torch design and the small passages, good cooling practices are mandatory. The water-cooling unit for plasma arc welding can be a tank type water circulator or a water circulator-radiator type. The primary function of the water-cooling unit is to remove a specific quantity of heat per hour to properly cool the torch. The water-cooling unit capacity should be defined by the welding equipment manufacturer. The unit must filter out all particles that can possibly block the small torch passages. The torch manufacturer should specify the type of water or coolant recommended.

Welding Programmers

Various types of programmers are available for automatic welding. The important variables for automatic welding are upslope of the welding current, feedback control of the welding current to maintain a constant amperage, taper of the welding current, and downslope of the welding current. Other variables that are usually programmed are the sequence of the shielding gas, the torch cooling water, upslope of orifice gas flow from a low level to a high level for key-hole welding, and downslope of orifice gas flow for closing the keyhole at the end of the weld.

Arc Voltage Control

Arc voltage control equipment is not necessary for most applications because of the relative insensitivity of the plasma arc process to arc length variations. However, arc voltage control can be used with plasma welding to follow contoured shapes. This control must be de-energized when current or gas slope control is energized because changing these variables also causes changes in the arc voltage.

MATERIALS

BASE METALS

The plasma arc welding process is used to join most of the metals commonly welded by the gas tungsten arc welding process. These metals include carbon and low alloy steels, stainless steels, copper alloys, nickel and cobalt base alloys, and titanium alloys. Thicknesses ranging from 0.03 to 6.4 mm (0.001 to 0.25 in.) can be welded in one pass. They are welded with a transferred arc using dc straight polarity (electrode negative).

Plasma arc welding of aluminum alloys can be done with dc reverse polarity (electrode positive) and alternating current. The problem with reverse polarity current is that it produces excessive heating of the tungsten electrode. As a result, the current rating of the torch is usually decreased to approximately one-half that of its straight polarity current rating. However, there are special ac power sources available that control the straight polarity current and time independent of the reverse polarity current and time. Thus, the reverse polarity time can be kept to a minimum and yet have reverse polarity current for adequate cleaning of the aluminum. By using this special power source, the maximum ac rating of a given plasma torch can be higher than with a standard ac power source. Aluminum alloys up to 11 mm (7/16 in.) thick can be welded in one pass using the keyhole technique with a special power source.

The metallurgical effects that the process has on metals are the same as with other welding processes, and appropriate procedures must be followed to preserve metallurgical integrity. Preheat, postheat, gas shielding, and gas selection are similar to those with the gas tungsten arc welding process. Metals to be joined will determine the gases to be used, particularly hydrogen additions to orifice and shielding gas. Hydrogen will severely degrade the quality of the welds in titanium and low alloy steels.

CONSUMABLES

Filler Metals

Filler metals used for welding various base metals are the same as those used with gas tungsten arc and gas metal arc welding processes. They are added in wire form as rods for manual welding or as continuous lengths for mechanized welding. Filler metals are covered by the AWS specifications listed in Table 9.1. All of them except zirconium filler metals are discussed in Chapter 94, "Filler Metals," *Welding Handbook*, Section 5, 6th edition.[5]

Electrodes

The tungsten electrodes used with a plasma arc torch are the same as those used for gas tungsten arc welding. They are basically tungsten rods, either pure or modified with additions to improve arc operation. The electrodes are manufactured to AWS Specification A5.12, Specification for Tungsten Arc-Welding Electrodes.

There are basically three types of tungsten electrodes: (1) pure tungsten, (2) tungsten with 1 or 2 percent thoria, and (3) tungsten with 0.25 percent zirconia. Pure tungsten electrodes have lower current carrying capacity than the others.

---

5. Chapter 94 is scheduled for revision in Vol. 4, 7th ed.

Table 9.1—AWS specifications for filler metals used for plasma arc welding	
AWS specification	Filler metals
A5.7	Copper and copper alloy welding rods
A5.9	Corrosion resistant chromium and chromium nickel steel bare electrodes and welding rods
A5.10	Aluminum and aluminum alloy welding rods and bare electrodes
A5.14	Nickel and nickel alloy bare welding rods and electrodes
A5.16	Titanium and titanium alloy bare welding rods and electrodes
A5.18	Mild steel electrodes for gas metal arc welding
A5.19	Magnesium alloy welding rods and bare electrodes
A5.24	Zirconium and zirconium alloy bare welding rods and electrodes

They are generally used for ac welding. Tungsten-thoria electrodes have higher electron emission characteristics than the other two types and, thus, higher welding current capacity. They are intended for use with dc, straight polarity (electrode negative). Tungsten-zirconia electrodes are generally used with ac, and they can be used at higher welding currents than pure tungsten electrodes.

Tungsten electrodes exhibit reduced current capacity when connected for dc, reverse polarity operation. Therefore, a larger diameter tungsten electrode is used for reverse polarity welding. The directional nature of the plasma jet produces an arc that is inherently stable, compared to the arc encountered in reverse polarity gas tungsten arc operation. Further information on tungsten electrodes is found in Chapter 3.

Influence of Electrode Shape

The arcing end of the electrode used for plasma arc welding is ground to an included angle of between 20 and 60 degrees, and the tip is either sharp or flattened slightly. The diameter of the flat tip is not critical, but generally it is approximately 0.8 mm (1/32 in.) for 3.2 mm (1/8 in.) and 4 mm (5/32 in.) diameter electrodes, and proportionately less for smaller diameters. The shape of the tungsten electrode must be absolutely round and concentric. The concentricity may be provided either by a precision grinding jig or by one of the commercial chemical pointing agents.

If double arcing occurs from the tungsten electrode to the orifice and then to the workpiece, a tungsten electrode tip with a smaller included angle helps to avoid this problem. Also, larger electrode setback usually decreases the double arcing tendency.

Gases

Gases are used for both the orifice gas and the shielding gas. The orifice gas must be inert with respect to the tungsten electrode to avoid rapid deterioration of the electrode. Orifice and shielding gases need not be inert as long as they do not adversely affect the weld joint properties.

The choice of gases to be used for plasma arc welding depends on the metal to be welded. For high current welding, the shielding gas is usually the same as the orifice gas because variations in the consistency of the arc would be inevitable if two different gases were used. Typical gases used to weld various metals with high current are shown in Table 9.2.

Argon is the preferred orifice gas for low current plasma arc welding because its low ionization potential assures reliable arc starting and a dependable pilot arc. Since the pilot arc is used only to maintain ionization in the plenum chamber, pilot arc current is not critical; it can be fixed for a wide variety of operating conditions. The recommended orifice gas flow rates are usually less than 2 L/min (1 ft³/h) and the pilot arc current is generally fixed at five amperes.

Table 9.2—Gas selection guide for high current plasma arc welding[a]

Metal		Thickness		Welding technique	
		mm	in.	Keyhole	Melt-in
Carbon steel	under	3.2	1/8	Ar	Ar
(aluminum killed)	over	3.2	1/8	Ar	75% He-25% Ar
Low alloy steel	under	3.2	1/8	Ar	Ar
	over	3.2	1/8	Ar	75% He-25% Ar
Stainless steel	under	3.2	1/8	Ar, 92.5% Ar-7.5% H_2	Ar
	over	3.2	1/8	Ar, 95% Ar-5% H_2	75% He-25% Ar
Copper	under	2.4	3/32	Ar	75% He-25% Ar, He
	over	2.4	3/32	Not recommended[b]	He
Nickel alloys	under	3.2	1/8	Ar, 92.5% Ar-7.5% H_2	Ar
	over	3.2	1/8	Ar, 95% Ar-5% H_2	75% He-25% Ar
Reactive metals	under	6.4	1/4	Ar	Ar
	over	6.4	1/4	Ar-He (50 to 75% He).	75% He-25% Ar

a. Gas selections are for both orifice and shielding gases.
b. The underbead will not form correctly. The technique can be used for copper-zinc alloys only.

Typical shielding gases for low current welding are given in Table 9.3. Argon is used for welding carbon steel, high-strength steels, and reactive metals such as titanium, tantalum, and zirconium alloys. Even minute quantities of hydrogen in the gas used to weld these metals may result in porosity, cracking, or reduced mechanical properties.

Although argon is suitable as the orifice and shielding gas for welding all metals, it does not necessarily produce optimum welding results. As in gas tungsten arc welding, additions of hydrogen to argon produce a hotter arc and efficient heat transfer to the workpiece. In this way, higher welding speeds are obtained with a given arc current. The amount of hydrogen that can be used in the mixture is limited because excessive hydrogen additions tend to cause porosity or cracking in the weld bead. With the plasma arc keyhole technique, a given metal thickness can be welded with higher percentages of hydrogen than are possible in the gas tungsten arc welding process. The ability to use higher percentages of hydrogen without inducing porosity may be associated with the keyhole effect and the different solidification pattern it produces.

Argon-hydrogen mixtures are used as the orifice and shielding gases for making keyhole welds in stainless steel, nickel base, and copper-nickel alloys. Permissible hydrogen percentages vary from the 5 percent used on 6.4 mm (1/4 in.) thick stainless steel to the 15 percent used for the highest welding speeds on 3.8 mm (0.150 in.) and thinner wall stainless tubing in tube mills. In general, the thinner the workpiece, the higher the permissible percentage of hydrogen in the gas mixture, up to 15 percent maximum. However, when argon-hydrogen mixtures are used as an orifice gas, the rating of the orifice diameter for a given welding current is usually reduced because of the higher arc temperature.

Helium additions to argon produce a hotter arc for a given arc current. A mixture must contain at least 40 percent helium before a significant change in heat can be detected; mixtures containing over 75 percent helium behave about the same as pure helium. Argon-helium mixtures containing between 50 and 75 percent helium are generally used for making keyhole welds in heavier titanium sections and for filler passes on all metals when the additional heat and wider heat pattern are desirable.

The shielding gas provided through the gas nozzle can be argon, an argon-hydrogen mixture, or an argon-helium mixture, depending

Table 9.3—Shielding gas selection guide for low current plasma arc welding[a]

Metal		Thickness		Welding technique	
		mm	in.	Keyhole	Melt-in
Aluminum	under	1.6	1/16	Not recommended	Ar, He
	over	1.6	1/16	He	He
Carbon steel	under	1.6	1/16	Not recommended	Ar, 25% He-75% Ar
(aluminum killed)	over	1.6	1/16	Ar, 75% He-25% Ar	Ar, 75% He-25% Ar
Low alloy steel	under	1.6	1/16	Not recommended	Ar, He, Ar-H_2 (1-5% H_2)
	over	1.6	1/16	75% He-25% Ar, Ar-H_2 (1-5% H_2)	Ar, He, Ar-H_2 (1-5% H_2)
Stainless steel		All		Ar, 75% He-25% Ar, Ar-H_2 (1-5% H_2)	Ar, He, Ar-H_2 (1-5% H_2)
Copper	under	1.6	1/16	Not recommended	25% He-75% Ar, 75% He-25% Ar, He
	over	1.6	1/16	75% He-25% Ar, He	He
Nickel alloys		All		Ar, 75% He-25% Ar, Ar-H_2 (1-5% H_2)	Ar, He, Ar-H_2 (1-5% H_2)
Reactive metals	under	1.6	1/16	Ar, 75% He-25% Ar, He	Ar
	over	1.6	1/16	Ar, 75% He-25% Ar, He	Ar, 75% He-25% Ar

a. Gas selections are for shielding gas only. Argon is the orifice gas in all cases.

on the welding application. Shielding gas flow rates are usually in the range of 10 to 15 L/min (20 to 30 ft³/h) for low current applications; for high current welding, flow rates of 15 to 30 L/min (30 to 60 ft³/h) are used.

Use of helium as an orifice gas increases the heat load on the torch nozzle and reduces its service life and current capacity. Because of the lower mass of helium, it is difficult, at reasonable flow rates, to obtain a keyhole condition with this gas. Therefore, helium is used only for making melt-in welds.

Since the shielding gas does not contact the tungsten electrode, gases such as CO_2 can be used. Flow rates for CO_2 are in the range of 10 to 15 L/min (20 to 30 ft³/h).

The orifice and shielding gas flow rates for plasma arc welding are controlled with flow meters. One- or two-stage pressure regulators are used to set the inlet pressure to the flow meters.

GENERAL PROCESS APPLICATIONS

The plasma arc welding process is accepted in aerospace, nuclear, electronic, shipbuilding, and many other commercial industries. It offers process fabrication latitude and economy while maintaining high quality and reliability. Most metals weldable with the gas tungsten arc welding process can be satisfactorily welded with the plasma arc welding process. Accordingly, established gas tungsten arc welding acceptance specifications for weldments can be extended to plasma arc welding.

Manual welding is usually done with the melt-in technique, and, therefore, it is similar to gas tungsten arc welding. The welding of very thin metal sections can be accomplished with greater ease and reliability with the melt-in technique. There are applications, such as pipe fabrication, where skilled welding operators use the low current keyhole technique to achieve complete and uniform weld penetration. In this regard, the plasma arc welding process offers good versatility and is suited for welding complicated structures made of thin metal sections.

Several low-current plasma arc welding applications are

(1) Thin wire mesh screen filters
(2) Small wire butt welds
(3) Relay case fabrication
(4) Bellows assemblies
(5) Thermal shields
(6) Thin-wall pressure vessels
(7) Vacuum tube components
(8) Thermocouple junctions

Mechanized high current welding is commercially important for several applications because of the keyhole characteristics, the high welding speeds, and the high quality joints produced. Typical applications are

(1) Stainless steel and titanium tubing (longitudinal welds)
(2) Girth joints in pipe fabrication
(3) Missile tankage
(4) Turbine engine components
(5) T-joints for structural members

For plate thickness in the 6.4 mm (1/4 in.) range, square-groove butt joints are welded without filler metal addition at travel speeds up to twice those of the gas tungsten arc welding process with the keyhole technique. At thicknesses of 0.05 to 1.6 mm (0.002 to 1/16 in.), the low current melt-in technique of plasma arc welding is applicable. Speed is no longer an advantage for these thicknesses, but the other advantages associated with the process remain. Keyholing can be accomplished on sections down to approximately 1.6 mm (1/16 in.) thick, but welding speeds are generally lower than normal GTAW speeds.

ADVANTAGES AND LIMITATIONS

ADVANTAGES

The collimated shape of the plasma jet is chiefly responsible for the lack of process sensitivity to changes in arc length. Arc cross-sectional area and total arc current establish the average current density at any location in the arc. Current density establishes the rate at which heat is transferred to a unit area of weld pool or work surface when the work is placed at a given torch standoff distance. A well collimated plasma arc can tolerate relatively large variations in arc length before its melting capability is affected seriously. This usually obviates the need for sensing and maintaining a constant arc length. The longer permissible torch-to-work (standoff) distance affords the welder better visibility.

Because the electrode in the plasma arc torch is recessed within the arc constricting nozzle, it is not possible to touch the electrode to the workpiece. This feature greatly reduces the possibility of tungsten inclusions in the weld, and it substantially extends the period between electrode dressings. The life of the electrode is also improved by a constant flow of inert orifice gas which minimizes erosion of the electrode.

One of the chief advantages of the plasma arc welding process is the keyhole effect. It can be obtained with the plasma arc when welding square-groove butt joints in certain thicknesses of most metals. The keyhole is a positive indication of complete penetration and weld uniformity.

Many metallurgical advantages are possible with the keyhole technique. Lower heat input preserves the strength of the joint in heat-treated metals, and limits grain growth for better ductility. The higher welding speeds result in less time for embrittlement of stainless steels and superalloys by carbides and complex intermetallic compounds.

The greater penetrating power of the plasma jet, as compared to conventional gas tungsten arc welding, can be used to produce higher depth-to-width ratios in the weld. This is illustrated in Fig. 9.15 which shows a cross sec-

tion of a single-pass plasma arc weld in 5 mm (0.21 in.) thick Type 304 stainless steel. This weld has a depth-to-width ratio of approximately 1:1, compared to the 1:3 ratio generally obtained with gas tungsten arc welding. This high ratio tends to equalize shrinkage stresses through the weld cross section and produce less distortion in the joint. The surface condition of filler wire is a major cause of porosity. The plasma arc welding process requires less filler wire for keyhole welding, and therefore porosity may be low.

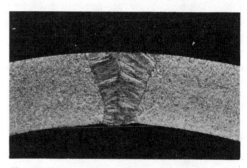

Fig. 9.15—Typical plasma arc weld in 5 mm (0.21 in.) thick Type 304 stainless steel without filler metal

Various fabrication advantages are possible with the plasma arc welding process. They include few weld passes, low cost for filler metal, and minimum possibility of human error due to less manipulation in general. Another advantage is that requirements for interpass cleaning, back gouging, and temperature maintenance, if needed, are low compared to GTAW. Tooling can be simpler and less costly because distortion tendencies are reduced. For keyholing applications, a square groove can be substituted for the V- or U-groove that might have been necessary with other processes. Thus, machining is reduced.

LIMITATIONS

The plasma arc welding process is limited to metal thicknesses of 25 mm (1 in.) and lower for butt welds. Further development is needed to extend its application to thicker sections.

Mechanized plasma arc welding is generally restricted to the flat and horizontal positions. (Manual plasma arc welding can be used in all positions.)

In general, plasma arc welding requires greater knowledge on the part of the welder than does gas tungsten arc welding. The welding torch is more complicated. It requires proper electrode tip configuration and positioning, selection of correct orifice size for the application, and setting of both orifice and shielding gas flow rates.

JOINT DESIGNS

The common types of welds made with the plasma arc welding process are square-groove, single- and double-U-groove; and single- and double-V-groove. These are generally used for welding butt joints from one or both sides of the joint with single or multiple pass welding. Fillet welds can be made using a technique similar to gas tunsten arc welding. T-joints can be welded with the plasma arc where its penetration characteristics are an advantage.

Square-groove butt welds are usually made in sections approximately 1.6 to 6.4 mm (1/16 to 1/4 in.) with the keyhole welding technique in one pass. Metal thicknesses over 6.4 to 25 mm (1/4 to 1 in.) require a U- or V-groove preparation for butt welding when welding from one side. Wide root faces up to 6.4 mm (1/4 in.) can be used. The first pass is welded with the keyhole technique and the fill passes with the melt-in technique. A square-groove weld can be made in sections up to 16 mm (5/8 in.) by welding from both sides (2 passes). Figure 9.16 shows a comparison of V-groove geometries for plasma arc and gas tungsten arc weld joints in 9.5 mm (3/8 in.) thick steel.

In metal thickness of 0.05 to 0.25 mm

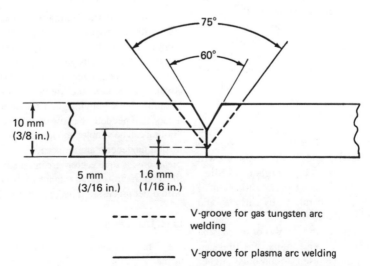

- - - - - - V-groove for gas tungsten arc welding

——————— V-groove for plasma arc welding

Fig. 9.16—Comparison of typical V-groove designs for gas tungsten arc and plasma arc welds on 10 mm (3/8 in.) thick steel

(0.002 to 0.010 in.), edge-flange welds can be made with the plasma arc welding process by the melt-in technique. Typical flange heights are shown in Table 9.4.

In 0.25 to 1.6 mm (0.010 to 0.060 in.) thick metal sections, butt joints with square-groove design are commonly welded. In this thickness range, the melt-in technique of plasma arc welding is used. T, Edge, and corner joints are easily welded with or without a filler metal addition.

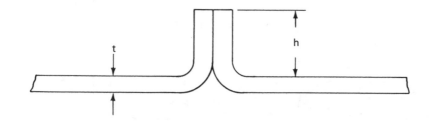

Table 9.4—Flange heights for edge flange welds

Metal thickness, t			Flange height, h	
mm	in.		mm	in.
0.05	0.002		0.25 to 0.51	0.010 to 0.020
0.13	0.005		0.51 to 0.64	0.020 to 0.025
0.25	0.010		0.76 to 1.0	0.030 to 0.040

WELDING PROCEDURES

FIT-UP AND FIXTURING

Although a low current plasma arc is easier to manage than a gas tungsten arc as a heat source, the melting behavior is the same for both processes. Therefore, joint fixturing requirements are much the same for both processes. For example, joint edges must be in contact or sufficiently close to assure bridging across the joint by the weld metal. In general, the root opening between adjacent edges should not be more than 10 percent of the metal thickness. Where maintenance of this tolerance becomes too difficult, filler metal must be added. Alternatively, butt joints made with the edges turned up (flanged welds) can be substituted for preplaced filler metal. Generally, this is done with foil less than 0.13 mm (0.005 in.) thickness.

Good fixturing is required to assure proper joint fit-up before and during welding. Heat buildup must not cause warping of joint edges because gaps will then appear that the weld metal cannot bridge. Copper chill bars in fixtures can prevent heat buildup and help to equalize the melting rates of dissimilar metal thicknesses. The best practice is to provide equal joint edge thicknesses by machining the thicker edge.

In general, tooling requirements for plasma arc welding are simpler and less expensive than for gas tungsten arc or gas metal arc welding. The keyhole technique essentially equalizes contraction stresses at the top and bottom of the weld. Therefore, tooling has to counteract a smaller distortion tendency when using the keyhole technique of welding.

Dimensional tolerances are comparable to gas tungsten arc welding. Sheared edges up to 6.4 mm (1/4 in.) thick may be satisfactory, but

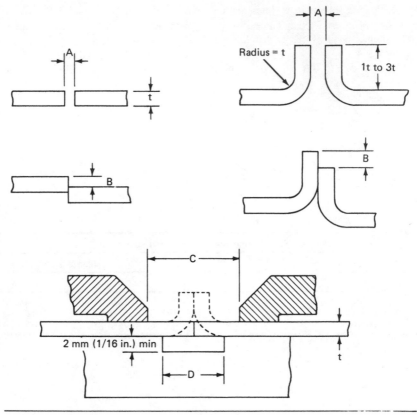

Type of weld	Gap A max	Mismatch B max	Clamp spacing C		Backup groove[a] D	
			min	max	min	max
Square-groove	0.2t	0.4t	10t	20t	4t	16t
flanged[b]	0.6t	1t	15t	30t	4t	16t

a. Gas underbead shielding, either argon or helium, is required.
b. Edge flange-weld is recommended for butt joints in thicknesses below 0.25 mm (0.010 in.).

Fig. 9.17—Tolerances for butt joints in foil and sheet up to 0.76 mm (0.030 in.) thick

machined joints are preferable. Metal-to-metal fit is best, although a gap up to 0.5 mm (0.020 in.) is permissible on 6.4 mm (1/4 in.) thick or heavier metal sections. On thinner thicknesses, a proportionately smaller gap is permitted. Mismatch up to 1.6 mm (1/16 in.) is permissible on thicknesses 6.4 mm (1/4 in.) or greater. Proportionately less mismatch to approximately 25 percent of thickness is permissible on thinner sections.

For metal thickness of 0.76 mm (0.030 in.) and less, illustrations of tolerances for joint fit-up and fixturing are shown in Figs. 9.17 and 9.18. Figure 9.17 shows the tolerances for joint gap mismatch, and fixture dimensions for hold-down clamps and backup grooves. The tolerances are listed in relation to the metal thickness. Thus, for a square-groove weld in a butt joint, a maximum joint gap of $0.2t$ is permitted. This is a very small gap. The fixturing for such weldments must be tooled with precision.

The basic requirement in welding thin sections is to make sure, by whatever means possible, that both joint edges are in continuous contact and that both edges melt simultaneously to form a single weld pool. Separation between the joint edges before or during welding will cause the edges to melt back and remain separate.

Increased latitude in butt joint fixturing tolerances can be obtained by flanging the edges. The turned-up edges act as preplaced filler metal to fill the gap and assure fusion across the joint.

They also stiffen the joint edges to minimize warpage from heat buildup during welding.

Figure 9.18 shows the fit-up and fixturing tolerances for edge joints. The permissible tolerances are much greater than those for butt joints. Because of this wide tolerance, an edge joint is the easiest and most reliable joint for welding foil thicknesses. Successful welding of foil thickness assemblies is greatly assured by using edge or flange joints wherever possible.

WELD BACKING

When welding with the melt-in technique, the weld backing is generally the same as that with tungsten arc welding. A grooved copper bar is used to support the molten weld pool and to remove heat from the base metal. The groove geometry should be designed to produce the desired root reinforcement. Shielding of the back side of the weld is recommended to minimize contamination from air. It is a requirement when welding reactive and refractory metals, such as titanium and tantalum.

Since the molten pool of a keyhole plasma arc weld is supported by surface tension, it is not necessary to employ close fitting backing bars as with the melt-in technique. A backing bar with a simple relief groove is used to support the weldment, contain the underbead shielding gas, and provide a vent space for the plasma jet. Groove dimensions are generally 13 mm (1/2 in.) wide

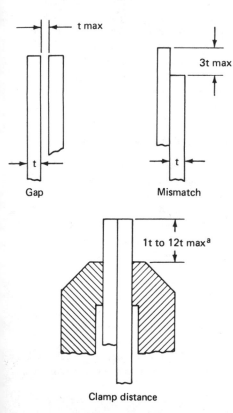

a. Wrinkling of joint edges begins at 8t.

Fig. 9.18 — Tolerances for edge joints in foil and sheet up to 0.76 mm (0.030 in.) thick

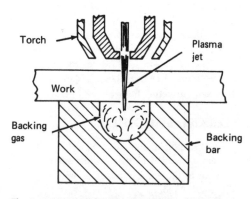

Fig. 9.19 — Typical backing bar for keyhole plasma arc welding

by 19 mm (3/4 in.) deep. A typical backing bar for keyhole welding is shown in Fig. 9.19.

GROUNDING THE WORK

The type and location of the work lead connection are governed by the same requirements as for other arc welding processes. The requirements are discussed in Chapters 2 and 6.

WELDING POSITIONS

Like gas tungsten arc welding, plasma arc welding can be done manually in all welding positions. Mechanized plasma arc welding is usually done in the flat and horizontal positions. Automatic pipe welding can be done in the flat position while rolling the pipe with the pipe axis horizontal (1G position) and in the horizontal position with the pipe axis vertical (2G position).

TORCH POSITION

For manual welding, the torch head is positioned with the plasma jet at a travel angle of 25 to 35 degrees from the vertical and pointing in the direction of welding (forehand technique). The torch (and filler rod, if used) is manipulated in the same manner as a gas tungsten arc torch to control weld bead shape, size, and penetration.

For mechanized welding, the travel angle is 10 to 15 degrees with the plasma jet pointed in the direction of travel. For keyhole welding of butt joints, the torch is placed perpendicular to the adjacent work surfaces in a plane transverse to the joint. For 1G pipe welding, the plasma torch is usually placed in the 11:00 o'clock position with the pipe rotating clockwise.

The torch standoff distance is normally about 5 mm (3/16 in.). However, the distance may vary from 3 to 6 mm (1/8 to 1/4 in.) without significantly affecting the welding operation.

FILLER METAL ADDITION

For the melt-in technique of welding, filler metal can be added to the leading edge of the weld pool in the same manner as with the gas tungsten arc welding process. Hot wire systems feed the filler wire into the trailing edge of the weld pool. Wire height adjustments are not so critical with plasma arc welding because the wire can be lifted off the plate and melted into the plasma stream without contaminating the electrode.

For the keyhole technique, filler metal can be added to the leading edge of the pool formed by the keyhole. The molten weld metal will flow around the keyhole to form a reinforced weld bead. Depending on fit-up and bead contour requirements, this technique may be used to make single pass butt welds with a square-groove joint preparation in metals up to approximately 6 mm (1/4 in.) thick. On heavier sections, a joint preparation is selected that will allow the plasma jet to melt the maximum amount of base metal supportable by surface tension. For this reason, filler metal is generally not added in making the root pass of a multiple pass weld.

Keyhole circumferential welds in butt joints with square-groove preparation require close control of timing and slope rates for arc current and orifice gas flow during keyhole initiation and withdrawal. The addition of filler metal in making such welds may complicate the keyhole withdrawal operation and may be undesirable for such applications.

AUXILIARY WELD SHIELDING

It is sometimes useful to supply auxiliary weld shielding when welding titanium and other metals that react with air and suffer degradation of mechanical and metallurgical properties as a result. Also, air can substantially alter the fluidity of the molten weld metal in several base metals. This can change the keyholing characteristics. For such applications, an auxiliary gas shield is beneficial in reducing variations in weldability. It is also necessary when welding more common, but less reactive, metals at high travel speeds. An important part of an auxiliary shield is the insulator between the torch and the shield. It should be a good electrical insulator and also possess good temperature stability and mechanical strength.

Where a trailing shield is necessary, shield length must be 1-1/2 to 2 times that of shields used with gas tungsten arc welding because of the higher travel speeds of plasma arc welding. Openings in the shield to accommodate filler wire or attachment to the torch should be as small as possible. Such openings tend to let air which may contaminate the weld metal into the shield.

MANUAL WELDING

Manual plasma arc welding is generally used for applications of up to 100 A, and where contour welding is necessary, with apparatus designed to be handheld. The apparatus uses a pilot arc system and a foot-controlled contactor to transfer welding current through the plasma stream. The pilot plasma arc, visible to the welder wearing protective lenses, facilitates accurate positioning of the torch for starting the weld. Initiation of welding current is positive and instantaneous. It is not subject to the difficulties inherent in the gas tungsten arc welding process when starting with low welding currents.

In one system, the pilot arc is started by moving the electrode forward until it touches the nozzle and then retracting it. The pilot arc power is always energized when the unit is in use. The transferred arc power is then energized by closing the welding contractor. Another system utilizes high frequency to start the pilot arc and a single power source.

Like manual gas tungsten arc welding, manual plasma arc welding is better adapted to melt-in welding. High-current plasma arc welding is usually mechanized to obtain the benefits of the keyhole technique. However, low-current plasma arc welding can also be mechanized. Manual plasma arc welding can be used in all positions common to gas tungsten arc welding.

The low-current volt-ampere characteristics of a gas tungsten arc and a plasma arc operating in argon are compared in Fig. 9.20. The slope of the 6.4 mm (0.25 in.) long plasma arc volt-ampere curve is very nearly flat with very little change in arc voltage. By contrast, when the current of a 1.3 mm (0.05 in.) long gas tungsten arc is reduced from 10 to 0.1 A, the voltage increases from 12 to 22 V, or 83 percent. The increase in arc voltage as the current decreases causes operational difficulties with gas tungsten arc welding at low currents. The relatively stable arc voltage of the plasma arc permits good control when welding at low currents.

The molten weld pool in thin metal sections

and foils behaves quite differently from one in thicker sections. This is due to surface tension, which is the dominant force. For example, as thinner sections are used, say below 0.76 mm (0.030 in.), the effect of weight or gravity diminishes until at 0.38 mm (0.015 in.) it disappears. The strong effect of surface tension then determines the shape of the molten metal regardless of the welding position.

Good gas shielding and fixturing are especially important. Weld defects leading to nonrepairable damage and part rejection are largely related to shielding and fixturing because of their effects on the surface tension of melted joint edges. Common oversights in foil welding procedures include

(1) Excessive joint gaps, which cannot be bridged by the melted edges

(2) Oxidation of the weld or base metal, which prevents good wetting by the molten weld metal

(3) Unbalanced part geometries, which allow weld metal contraction in only one direction

(4) Inadequate clamping, which allows joint warpage during welding

The basic requirement in welding thin sections is to make sure that both joint edges are in continuous contact, and that both edges melt simultaneously to form a single weld pool. Separation between the joint edges before or

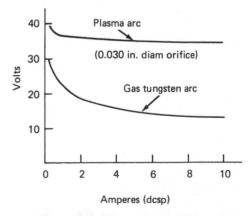

Fig. 9.20—Comparison of gas tungsten arc and plasma arc volt-ampere characteristic curves below 10A in argon

during welding will cause the edges to melt separately and remain separated.

MECHANIZED WELDING

Mechanized welding is usually required for high-current plasma arc applications, such as making keyhole welds or high-current filler passes. Mechanization is required because of the high travel speeds, the need for accurate joint alignment, the narrow plasma arc weld fusion zones, and the types of apparatus available. For repetitive operations, mechanized welding is applicable just as it is for other welding processes. Mechanized welding is also applied to low current applications where advantageous.

A mechanized plasma arc torch utilizes high frequency to initiate a pilot arc between the electrode and the constricting nozzle. The ionized gas generated by the low current pilot arc flows through the constricting orifice to the workpiece and completes the circuit for the main welding arc. In most systems, the high frequency generator and pilot arc are turned off after the main arc is initiated.

A high current, mechanized plasma arc welding system is capable of utilizing the keyhole technique in approximate metal thicknesses of 1.6 to 7 mm (1/16 to 1/4 in.). Low current, mechanized plasma arc welding can utilize the keyhole technique down to an approximate metal thickness of 0.8 mm (1/32 in.). The velocity of the plasma jet is quite critical for welding thickness ranges under 1.6 mm (1/16 in.). To be able to repeat identical conditions on successive welds in the thinner sections, the plasma jet velocity must be measured either directly or indirectly. Since the nozzle is a diaphragm plate with an orifice, the flow rate and velocity of the gas flowing through the torch orifice can be calculated from its size and the pressure drop across it. The orifice gas exhausts into the atmosphere. Therefore, by measuring the gas pressure in the orifice gas line between the torch and pressure regulator, the pressure drop is known and the gas velocity and flow rate can be calculated. This technique provides greater accuracy than is practical with a flowmeter. Some commercially available plasma arc welding units are equipped with such a pressure gage.

Metal thicknesses above 7 mm (1/4 in.) may require a V- or U-groove butt joint design, but a root face up to 7 mm (1/4 in.) wide can be used. An addition of filler metal to fill the prepared joint, using multiple pass welding procedures, is commonly done in plasma arc welding. The practical thickness limit for plasma arc welding is approximately 25 mm (1 in.). Above this thickness, the plasma torch nozzle prevents accessibility to the root of the joint.

When welding metal thicknesses under 3 mm (1/8 in.), keyhole welds in straight seams and circumferential welds can be started at full operating current, travel speed, and orifice gas flow rate. In this thickness range, the keyhole is developed with little disturbance in the weld pool, and the weld surface and underbead are reasonably smooth.

It is important to differentiate between straight seam welds, where run-on and run-off tabs can be utilized to isolate the keyhole initiation and withdrawal areas, and circumferential or girth type welds where the keyhole initiation and withdrawal must be included within the actual weld joint. The operating currents and orifice gas velocities required for keyhole welds in thicknesses greater than approximately 3 mm (1/8 in.) usually produce a plasma stream that tends to gorge or tunnel underneath the molten metal during keyhole initiation. Because the gouging action may cause plasma gas entrapment, voids, and severe surface irregularities, run-on and run-off tabs are generally used to make straight seam welds with the keyhole technique in these thicknesses. When circumferential joints are welded, run-off tabs cannot be used. A suitable keyhole initiation zone can be achieved by the use of a programmed increase in welding current and orifice gas flow rate with travel speed set at the welding rate. Any gas entrapped in the keyhole initiation area is removed by the overlapping keyhole and is not generally a quality problem. These important sloping functions for programming the rise in welding current and orifice-gas flow rate to start the run, and their gradual decay to close the keyhole are normally produced automatically by commercial welding equipment and accompanying controls. Figure 9.21 is a schematic representation of slope control cycles for the current and gas flow.

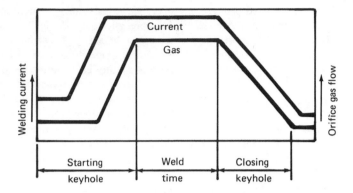

**Fig. 9.21—Welding current and orifice gas slope control cycle for
starting and terminating a keyhole weld**

As the current and gas flow increases, the penetration into the joint increases until a hole is formed. The proper orifice gas flow is just below that which would maintain a cutting action.

If the keyholing plasma arc is abruptly interrupted at the end of a circumferential weld, a variety of weld defects may occur. Although abrupt keyhole termination may be desirable in straight seam welds, it is necessary to employ very different keyhole termination procedures for circumferential welds in the various metals and thicknesses. The principal problem area in keyhole circumferential welds is the keyhole withdrawal zone.

Proper termination of the keyhole in a circumferential weld requires downsloping the arc current and orifice gas flow rate, as shown in Fig. 9.21. The net effect of this procedure is to reduce gradually the keyholing force of the plasma arc while retaining the arc heat necessary to maintain weld metal fluidity as the keyhole fills in. Travel speed is generally maintained at the welding rate. Specific metals generally require variations in downslope time phasing and duration using this procedure, and in some cases it may be beneficial to simultaneously reduce the travel speed to help in maintaining weld metal fluidity during the time the keyhole is being filled.

Plasma welding of T-joints can be done by penetrating the top plate into the upright member to provide equally distributed fillets on each side of the upright member. A uniform bead is generated on the welded side, either flush or slightly convex, depending on the amount of filler wire added to the molten pool. Simple copper backup bars, with a radius on the inner top edge, are clamped on both sides of the upright member to assist in forming and controlling the size and contour of the fillets.

Plug or slot welding requires tooling and fixturing to provide adequate interface contact at the weld area and also positive clamping adjacent to the weld. The upper sheet is thus maintained in position so it does not lift from thermal expansion. Spot timers and current decay control can be used to eliminate crater cracking. Wire feed control timers used in conjunction with current decay and post gas flow controls can also be used to provide still greater assurance of consistent weld quality.

MULTIPLE PASS WELDING

A multiple pass plasma arc weld involves a keyhole root pass and one or more melt-in filler passes, with or without filler metal addition. The melt-in plasma arc uses a plasma jet with lower force than that used for keyhole welding. Total orifice gas flow rate is reduced; torch standoff is increased. Varying percentages of helium or hydrogen may be mixed with argon in both the orifice and shielding gas circuits to dissipate the arc heat over a larger surface area of the weld joint. Helium can be used, and is favored, for some applications because it provides a wider heat input pattern and produces a flatter weld bead. The very wide range of weld penetration characteristics attainable with these high current plasma arc techniques provides substantially greater process flexibility than is available with other gas shielded arc welding processes.

REVERSE POLARITY WELDING

Tungsten electrodes exhibit reduced current capacity when connected for reverse polarity operation, whether they are used for plasma arc or gas tungsten arc welding. Therefore, a water-cooled electrode is preferable for reverse polarity welding. Compared to the arcs encountered in reverse polarity gas tungsten arc operation, the plasma jet produces an inherently stable arc. A single port nozzle is generally used with reverse polarity because multiple port nozzles do not provide the advantages realized with straight polarity welding.

WELDING CONDITIONS

Metals to be joined, joint configuration, and thickness are the major considerations for plasma arc welding. The requirements for metal cleanliness, joint alignment, fit-up, and the tolerance for mismatch for plasma arc welding are commensurate with the quality required and the general welding results. Preheat and post-weld thermal treatments and heating methods are similar to other welding processes and are dependent upon metallurgical considerations for the base metals.

Tables 9.5 through 9.10 give typical welding conditions for plasma arc welding various metals and thicknesses. The plasma arc welding conditions listed are nominal for the thicknesses and applications. They are a summation of conditions used in production and process development for both low current manual and high current mechanized applications.

For the high current keyhole welding of metal thicknesses of 1.6 mm (1/16 in.) and above, welding conditions are critical. It is recommended that the following tolerances be applied:

(1) Weld current: ±5 A
(2) Travel speed: ±0.05 mm/s (1/8 in./min)

(3) Orifice gas flow rate: ±0.12 L/min (0.25 ft³/h)
(4) Shielding-gas flow rate: ±2.4 L/min (5 ft³/h)
(5) Torch standoff: ±1.6 mm (1/16 in.)

For the filler passes, where filler metal is added, the control of the welding conditions is not quite so critical as for the root pass using the keyhole technique. The following tolerances for filler passes are recommended:

(1) Weld current: ±10 A
(2) Travel speed: ±0.2 mm/s (1/2 in./min)
(3) Orifice gas flow rate: ±0.24 L/min (0.5 ft³/h)
(4) Shielding gas flow rate: ±2.4 L/min (5 ft³/h)
(5) Torch standoff: 3.2 to 9.5 mm (1/8 to 3/8 in.)

Pure argon is generally used for the keyhole root pass. Either helium or mixtures of helium and argon can be used for fill passes. Where pure helium is specified, high flow rates are required, particularly for shielding. The use of a trailing shield, in addition to the normal torch shielding, is desirable for making multiple pass welds on some metals.

Table 9.5—Typical plasma arc welding conditions for butt joints in stainless steel

Thickness		Travel speed		Current (dcsp)	Arc voltage	Nozzle type[a]	Gas flow[b]				Remarks[d]
							Orifice[c]		Shield[c]		
mm	in.	mm/s	in./min	A	V		L/min	ft³/h	L/min	ft³/h	
2.4	0.092	10	24	115	30	111M	3	6	17	35	Keyhole, square-groove weld
3.2	0.125	13	30	145	32	111M	5	10	17	35	Keyhole, square-groove weld
4.8	0.187	7	16	165	36	136M	6	13	21	45	Keyhole, square-groove weld
6.4	0.250	6	14	240	38	136M	8	18	24	50	Keyhole, square-groove weld

a. Nozzle type: number designates orifice diameter in thousandths of an inch; "M" designates design.
b. Gas underbead shielding is required for all welds.
c. Gas used: 95% Ar-5% H.
d. Torch standoff: 4.8 mm (3/16 in.).

Table 9.6—Typical plasma arc welding conditions for butt joints in carbon and low alloy steels

Metal	Thickness		Travel speed		Current (dcsp)	Arc voltage	Nozzle type[a]	Gas flow[b]				Remarks[d]
								Orifice[c]		Shield[c]		
	mm	in.	mm/s	in./min	A	V		L/min	ft³/h	L/min	ft³/h	
Mild steel	3.2	0.125	5	12	185	28	111M	6	13	28	60	Keyhole, square-groove weld
4130 steel	4.3	0.170	4	10	200	29	136M	6	12	28	60	Keyhole, square-groove weld, 1.2 mm (3/64 in.) diam filler wire added at 13 mm/s (30 in./min)
D6ac steel	6.4	0.250	6	14	275	33	136M	7	15	28	60	Keyhole, square-groove weld, 315° C (600° F) preheat

a. Nozzle type: number designates orifice diameter in thousandths of an inch; "M" designates design.
b. Gas underbead shielding is required for all welds.
c. Gas used: argon.
d. Torch standoff: 1.2 mm (3/64 in.) for all welds.

Table 9.7—Typical plasma arc welding conditions for butt joints in titanium

Thickness		Travel speed		Current (dcsp)	Arc voltage	Nozzle type[a]	Gas flow[b]				Remarks[c]
mm	in.	mm/s	in./min	A	V		Orifice		Shield		
							L/min	ft³/h	L/min	ft³/h	
3.2	0.125	8.5	20	185	21	111M	3.8	8[d]	28	60[d]	Keyhole, square-groove weld
4.8	0.187	5.5	13	175	25	136M	9	18[d]	28	60[d]	Keyhole, square-groove weld
9.9	0.390	4.2	10	225	38	136M	15	32[e]	28	60[e]	Keyhole, square-groove weld
12.7	0.500	4.2	10	270	36	136M	13	27[f]	28	60[f]	Keyhole, square-groove weld

a. Nozzle type: number designates orifice diameter in thousandths of an inch; "M" designates design.
b. Gas underbead shielding is required for all welds.
c. Torch standoff: 4.8 mm (3/16 in.).

d. Gas used: argon.
e. Gas used: 75% He-25% Ar.
f. Gas used: 50% He-50% Ar.

Table 9.8—Typical plasma arc welding conditions for welding stainless steels—low amperage

Thickness		Type of weld	Travel speed		Current (dcsp)	Orifice diam		Gas flow orifice[a,b,c]		Torch standoff		Electrode diam		Remarks
mm	in.		mm/s	in./min	A	mm	in.	L/min	ft³/h	mm	in.	mm	in.	
0.76	0.030	Square-groove weld, butt joint	2	5.0	11	0.76	0.030	0.3	0.6	6.4	0.250	1.0	0.040	Mechanized
1.5	0.060	Square-groove weld, butt joint	2	5.5	28	1.2	0.047	0.4	0.8	6.4	0.250	1.5	0.060	Mechanized
0.76	0.030	Fillet weld, tee joint	—	—	8	0.76	0.030	0.3	0.6	6.4	0.250	1.0	0.040	Manual, filler metal[d]
1.5	0.060	Fillet weld, tee joint	—	—	22	1.2	0.047	0.4	0.8	6.4	0.250	1.5	0.060	Manual, filler metal[d]
0.76	0.030	Fillet weld, lap joint	—	—	9	0.76	0.030	0.3	0.6	9.5	0.375	1.0	0.040	Manual, filler metal[d]
1.5	0.060	Fillet weld, lap joint	—	—	22	1.2	0.047	0.4	0.8	9.5	0.375	1.5	0.060	Manual, filler metal[e]

a. Orifice gas: argon.
b. Shielding gas: 95% Ar-5% H at 10 L/min (20 ft³/h).
c. Gas underbead shielding: argon at 5 L/min (10 ft³/h).

d. Filler wire: 1.1 mm (0.045 in.) diameter 310 stainless steel.
e. Filler wire: 1.4 mm (0.055 in.) diameter 310 stainless steel.

Table 9.9—Typical plasma arc welding conditions for butt joints in thin gage metals[a]

Metal	Thickness mm	Thickness in.	Current (dcsp) A	Shielding gas[b]	Travel speed mm/s	Travel speed in./min
Stainless steel	0.03[c]	0.001	0.3	99% Ar-1% H_2	2	5
	0.08[c]	0.003	1.6	99% Ar-1% H_2	2.5	6
	0.13	0.005	2.4	99% Ar-1% H_2	2	5
	0.25	0.010	6.0	99% Ar-1% H_2	3.4	8
	0.80	0.030	10.0	99% Ar-1% H_2	2	5
Titanium	0.08[c]	0.003	3.0	50% Ar-50% He	2.5	6
	0.20	0.008	5.0	Ar	2	5
	0.38	0.015	5.8	Ar	2	5
	0.56	0.022	10.0	75% He-25% Ar	3	7
Ni-19% Cr-19% Fe	0.30	0.012	6.0	99% Ar-1% H_2	6	15
Ni-21% Cr-19% Fe	0.13	0.005	4.8	99% Ar-1% H_2	4	10
	0.25	0.010	5.8	99% Ar-1% H_2	3.4	8
	0.51	0.020	10.0	99% Ar-1% H_2	4	10
Copper	0.08[c]	0.003	10.0	75% He-25% Ar	2.5	6

a. Orifice gas: 0.24 L/min. (0.5 ft^3/h) argon through a 0.80 mm (0.030 in.) diameter nozzle.
b. Shielding gas flow: 10 L/min. (20 ft^3/h).
c. Flanged butt joint.

Table 9.10—Typical plasma arc welding conditions for edge joints in thin gage metals[a]

Metal	Thickness mm	Thickness in.	Current (dcsp) A	Shielding gas[b]	Travel speed mm/s	Travel speed in./min
Stainless steel	0.03	0.001	0.3	99% Ar-1% H_2	2	5
	0.13	0.005	1.6	99% Ar-1% H_2	6	15
	0.25	0.010	4.0	99% Ar-1% H_2	2	5
Titanium	0.08	0.003	1.6	Ar	2	5
	0.20	0.008	3.0	Ar	2	5
Ni-21% Cr-19% Fe	0.13	0.005	1.5	99% Ar-1% H_2	4	10
	0.25	0.010	3.0	Ar	1.3	3
	0.51	0.020	6.5	Ar	3	7
Fe-28% Ni-18% Co	0.26	0.011	9.0	95% Ar-1% H_2	8.5	20

a. Orifice gas flow: 0.24 L/min. (0.5 ft^3/h) argon through a 0.80 mm (0.030 in.) diameter nozzle.
b. Shielding gas flow: 10 L/min. (20 ft^3/h).

WELD QUALITY

The quality of plasma arc welded joints is equal to the quality of those made by gas tungsten arc welding. Usual considerations must be given to the cleanliness of parts and consumables, the proper gas shielding of the face and root of the weld, and the metallurgical characteristics of the metal being joined.

Plasma arc weld discontinuities include both surface and subsurface types, as listed in Table 9.11. Surface discontinuities, such as reinforcement, underfill, undercut, and mismatch, are associated with weld bead contour and joint alignment. They are easily detected visually or dimensionally. Lack of penetration is also detected visually through the absence of a root surface or a concave root surface. Welding cracks which are open to the surface are usually detected by the use of liquid penetrant inspection. Finally, surface contamination, which results from insufficient shielding gas coverage, is usually denoted by the severe discoloration of the weld bead and adjacent heat-affected zone.

Subsurface weld discontinuities are generally more prevalent in manual than in mechanized welding. In either case, subsurface discontinuities are detected primarily by the use of radiography and ultrasonic methods. Porosity is by far the most common discontinuity encountered.

Table 9.11—Plasma arc weld discontinuities

Surface discontinuities	Internal discontinuities
Reinforcement	Porosity
Underfill	Tunneling (voids)
Undercut	Lack of fusion
Mismatch	Contamination
Lack of penetration	Cracks
Cracks	
Contamination	

Detection is possible by the use of either radiography or ultrasonics. Sensitivity is greater if both the crown and root beads are machined flush with the base metal. Ultrasonic techniques can be used to detect porosity, provided the joint can be machined flush and the joint thickness exceeds approximately 1.3 mm (0.050 in.).

Tunneling is a severe void which runs along the interface between the weld metal and the base metal. This discontinuity seems to result from a combination of torch alignment and welding conditions (particularly travel speed). Tunneling is easily detected by the use of radiography.

Lack of fusion discontinuities are most prevalent in repair welds, either single or multiple pass. The discontinuity results from heat input insufficient to obtain complete fusion of the particular weld pass to the base metal. Lack of fusion can be detected with radiography or ultrasonics. Depending on the orientation of the discontinuity, one process may have an advantage over the other, so optimum inspection would employ both methods.

Regardless of their cause, subsurface weld cracks are easily detected with radiography and ultrasonics, particularly when employed together. With radiography, the crack must be aligned with the radiation for successful detection.

Subsurface contamination results when copper from the torch nozzle is expelled into the weld. This condition usually occurs in manual welding when the torch nozzle gets too close to the weld, particularly in a groove. The resulting contamination, which may be detrimental, is undetectable after welding by conventional nondestructive testing procedures. The only way of detecting copper contamination is observation by the welder. The contaminated metal must be machined out.

SAFETY RECOMMENDATIONS

For detailed safety information, refer to the manufacturer's instructions and the latest editions of ANSI Z49.1, *Safety in Welding and Cutting,* and AWS A6.1, *Recommended Safe Practices for Gas-Shielded Arc Welding.* For mandatory federal safety regulations established by the U.S. Labor Department's Occupational Safety and Health Administration, refer to the latest edition of OSHA Standards, Code of Federal Regulations, Title 29 Part 1910, available from the Superintendent of Documents, U.S. Printing Office, Washington, DC 20402.

When welding with transferred arc currents up to 5 A, spectacles with side shields or other types of eye protection with a No. 6 filter lens are recommended. Although face protection is not normally required for this current range, its use depends on personal preference. When welding with transferred arc currents between 5 and 15 A, a full face plastic shield is recommended in addition to eye protection with a No. 6 filter lens. At current levels over 15 A, a standard welder's helmet with the proper shade of filter plate for the current being used is required.

When a pilot arc is operated continuously, normal precautions should be used for protection against arc flash and heat burns. Suitable clothing must be worn to protect exposed skin from arc radiation. Welding power should be turned off before electrodes are adjusted or replaced. Adequate eye protection should be used when observation of a high frequency discharge is required to center the electrode.

Accessory equipment, such as wire feeders, arc voltage heads, and oscillators should be properly grounded. If they are not grounded, insulation breakdown might cause these units to become electrically "hot" with respect to ground.

Adequate ventilation should be used, particularly when welding metals with high copper, lead, zinc, or beryllium contents.

SUPPLEMENTARY READING LIST

Ashauer, R.C., and Goodman, S. Automatic plasma arc welding of square butt pipe joints. *Welding Journal* 46 (5), May 1967, pp. 405-415.

Cooper, G., Palermo, J., and Browning, J.A. Recent developments in plasma welding. *Welding Journal* 44 (4), Apr. 1965, pp. 268-276.

Filipski, S.P. Plasma arc welding. *Welding Journal* 43 (11), Nov. 1964, pp. 937-943.

Garrabrant, E.C., and Zuchowski, R.S. Plasma arc-hot wire surfacing—A new high deposition process. *Welding Journal* 48 (5), May 1969, pp. 385-395.

Gorman, E.F. New developments and applications in manual plasma arc welding. *Welding Journal* 48 (7), July 1969, pp. 547-556.

Gorman, E.F., Skinner, G.M., and Tenni, D.M. Plasma needle arc for very low current work. *Welding Journal* 45 (11), Nov. 1966, pp. 899-908.

Hackman, R.L. Plasma Arc Techniques. ASTME Technical Paper No. SP60-135.

Langford, G.J. Plasma arc welding of structural titanium joints. *Welding Journal* 47 (2), Feb. 1968, pp. 102-113.

MacAbee, P.T., Dyar, J.R., and Bratkovich, N.F. Plasma Arc Welding of Thin Materials. Technical Report AFML-TR-66-177, June 1967.

Metcalfe, J.C., and Quigley, M.B.C. Heat transfer in plasma-arc welding. *Welding Journal* 54 (3), Mar. 1975, pp. 99-103.

Metcalfe, J.C., and Quigley, M.B.C. Keyhole stability in plasma arc welding. *Welding Journal* 54 (11), Nov. 1975, pp. 401-404.

Miller, H.R., and Filipski, S.P. Automated plasma arc welding for aerospace and cryogenic fabrications. *Welding Journal* 45 (6), June 1966, pp. 493-501.

O'Brien, R.L. Applications of the Plasma Arc. ASTME Technical Paper No. SP63-56.

O'Brien, R.L. Arc Plasmas for Joining, Cutting, and Surfacing. Bulletin No. 131. New York: Welding Research Council, July, 1968.

Recommended Practices for Plasma-Arc Welding, C5.1. Miami: American Welding Society, 1973.

Ruprecht, W.J., and Lundin, C.D. Pulsed current plasma arc welding. *Welding Journal* 53 (1), Jan. 1974, pp. 11-19.

Steffans, H.D., and Kayser, H. Automatic control for plasma arc welding. *Welding Journal* 51 (6), June 1972, pp. 408-418.

10

Oxyfuel Gas Welding

Prepared by

A. W. PENSE
Lehigh University

Reviewed by

R. D. GREEN
Mapp Products
J. E. McQUILLEN
Air Products and Chemicals, Inc.

10
Oxyfuel Gas Welding

FUNDAMENTALS OF THE PROCESS

GENERAL DESCRIPTION[1]

Oxyfuel gas welding (OFW) includes any welding operation that makes use of a fuel gas combined with oxygen as a heating medium. The process involves the melting of the base metal and a filler metal, if used, by means of the flame produced at the tip of a welding torch. Fuel gas and oxygen are mixed in the proper proportions in a mixing chamber which may be part of the welding tip assembly. Molten metal from the plate edges and filler metal, if used, intermix in a common molten pool and, upon cooling, coalesce to form a continuous piece.

An advantage of this welding process is the control a welder can exercise over the rate of heat input, the temperature of the weld zone, and the oxidizing of reducing potential of the welding atmosphere. Weld bead size and shape and weld puddle viscosity are also controlled in the welding process because the filler metal is added independently of the welding heat source. OFW is ideally suited to the welding of thin sheet, tubes, and small diameter pipe, and also for repair welding. Thick section welds, except for repair work, are not economical.

The equipment used in oxyfuel gas welding is low in cost, usually portable, and versatile enough to be used for a variety of related operations, such as bending and straightening, preheating, postheating, surfacing, braze welding,

and torch brazing. With relatively simple changes in equipment, manual and mechanized oxygen cutting operations can be performed. Metals normally welded include steels, especially low alloy steels, and most nonferrous metals but generally not refractory or reactive metals.

Commercial fuel gases have one common property: they all require oxygen to support combustion. To be suitable for welding operations, a fuel gas, when burned with oxygen, must have

(1) High flame temperature
(2) High rate of flame propagation
(3) Adequate heat content
(4) Minimum chemical reaction of the flame with base and filler metals

Among the commercially available fuel gases, acetylene most closely meets all these requirements. Other gases, fuel such as methylacetylene-propadiene products, propylene, propane, natural gas, and proprietary gases based on these, have sufficiently high flame temperatures but exhibit low flame propagation rates. These gas flames are excessively oxidizing at oxygen-to-fuel gas ratios high enough to produce usable heat transfer rates. Flame holding devices, such as counterbores on the tips, are necessary for stable operation and good heat transfer, even at the higher ratios. These gases, however, are used for oxygen cutting. They are also used for torch brazing, soldering, and many other operations where the demands upon the flame characteristics and heat transfer rates are not the same as those for welding.

---

1. A brief introduction to the process and its areas of application are presented in Vol. 1, 7th ed., pp. 17-18.

CHARACTERISTICS OF FUEL GASES

GENERAL CHARACTERISTICS

Table 10.1 lists some of the pertinent characteristics of commercial gases. In order to appreciate the significance of the information in this table, it is necessary to understand some of the terms and concepts involved in the burning of fuel gases.

Specific Gravity

The specific gravity of a fuel gas with respect to air is an indication of how the gas may accumulate in the event of a leak. For example, gases having a specific gravity less than one tend to rise and dissipate; those with a specific gravity greater than one tend to accumulate in low, still areas.

Volume-to-Weight Ratio

A specific quantity of gas at a standard temperature and pressure can be described by volume or weight. The values shown in Table 10.1 give the volume per unit weight at 15.6° C (60° F) and atmospheric pressure. Multiplying these figures by the known weight will give the volume. If the volume is known, multiplying the reciprocal of the figures shown by the volume will give the weight.

Combustion Ratio

Table 10.1 indicates the volume of oxygen theoretically required for complete combustion of each of the fuel gases shown. These oxygen-to-fuel gas ratios (called stoichiometric mixtures) are obtained from the balanced chemical equations given in Table 10.2. The values shown for complete combustion are useful in calculations. They do not represent the oxygen-to-fuel gas ratios actually delivered by an operating torch because, as will be explained later, part of the complete combustion depends on oxygen supplied by the surrounding air.

Heat of Combustion

As shown in Table 10.1, the total heat of combustion (heat value) of a hydrocarbon fuel gas is the sum of the heat generated in the primary

and secondary reactions that take place in the overall flame. The combustion of hydrogen takes place in a single reaction. The theoretical basis for these chemical reactions and their heat effects is discussed in Volume 1.[2]

Typically, the heat content of the primary reaction is generated in an inner, or primary, flame where combustion is supported by oxygen supplied by the torch. The secondary reaction takes place in an outer, or secondary, flame envelope where combustion of the primary reaction products is supported by oxygen from the air.

Although the heat of the secondary flame is quite important in most applications, the more concentrated heat of the primary flame is a major factor of the welding capability of an oxyfuel gas system. The primary flame is said to be neutral when the chemical equation for the primary reaction is exactly balanced, yielding only carbon monoxide and hydrogen. Under these conditions the primary flame atmosphere is neither carburizing nor oxidizing, but it is reducing.

Since the secondary reaction is necessarily dependent upon the primary reaction products, the term "neutral" serves as a convenient reference point for describing combustion ratios and for comparing the various heat characteristics of different fuel gases.

Flame Temperature

The flame temperature is a physical property of a fuel gas. This temperature will vary depending on the oxygen-to-fuel ratio. Although the flame temperature gives an indication of the heating ability of the fuel gas, it is only one of the many physical properties that must be considered in evaluating a fuel gas. Flame temperatures are usually calculated since there is, at present, no simple method of physically measuring these values.

The flame temperatures listed in Table 10.1 are for the so-called neutral flame, i.e., the primary flame that is neither oxidizing nor carbur-

---

2. See Vol. 1, 7th ed., *Welding Handbook,* pp. 41-43.

Table 10.1—Characteristics of the common fuel gases

Fuel gas	Formula	Specific gravity 15.6°C (60°F) Air=1	Volume to weight ratio (15.6°C)		Oxygen-to-fuel gas combustion ratio[a]	Flame temperature for oxygen[b]		Heat of combustion					
								Primary		Secondary		Total	
			m³/kg	ft³/lb		°C	°F	MJ/m³	Btu/ft³	MJ/m³	Btu/ft³	MJ/m³	Btu/ft³
Acetylene	C_2H_2	0.906	0.91	14.6	2.5	3087	5589	19	507	36	963	55	1470
Propane	C_3H_8	1.52	0.54	8.7	5.0	2526	4579	10	255	94	2243	104	2498
Methylacetylene-propadiene (MPS)[c]	C_3H_4	1.48	0.55	8.9	4.0	2927	5301	21	571	70	1889	91	2460
Propylene	C_3H_6	1.48	0.55	8.9	4.5	2900	5250	16	438	73	1962	89	2400
Natural gas (methane)	CH_4	0.62	1.44	23.6	2.0	2538	4600	0.4	11	37	989	37	1000
Hydrogen	H_2	0.07	11.77	188.7	0.5		4820					12	325

a. The volume units of oxygen required to completely burn a unit volume of fuel gas according to the formulae shown in Table 10.2. A portion of the oxygen is obtained from the atmosphere.
b. The temperature of the neutral flame.
c. May contain significant amounts of saturated hydrocarbons.

izing in character. Flame temperatures higher than this may be achieved, but, in every case, the flame will be oxidizing, an undesirable condition in the welding of many metals.

Combustion Velocity

A characteristic property of a fuel gas, combustion velocity (flame propagation rate) is an important factor in the heat output of the oxyfuel gas flame. This is the velocity at which a flame front travels normal to its surface through the adjacent unburned gas. It is a factor which influences the size and temperature of the primary flame. It also affects the velocity at which gases may flow from the torch tip without causing a flame standoff or backfire.

As shown in Fig. 10.1, the combustion velocity of a fuel gas varies in a characteristic manner according to the proportions of oxygen and fuel in the mixture.

Combustion Intensity

Flame temperatures and heating values of fuels have been used almost exclusively as the criteria for evaluating fuel gases. These two factors alone, however, do not provide sufficient information to allow complete appraisal of fuel gases for heating purposes. A concept known as combustion intensity or specific flame output is used to evaluate different oxygen-fuel gas combinations. Combustion intensity takes into account the burning velocity of the flame, the heating value of the mixture of oxygen and fuel gas, and the area of the flame cone issuing from the tip.

Combustion intensity may be expressed as
$$C_i = C_v \times C_h$$
where

C_i = combustion intensity in $J/m^2 \cdot s$ ($Btu/ft^2 \cdot s$)
C_v = normal combustion velocity of flame in m/s (ft/s)
C_h = heating value of the gas mixture under consideration in J/m^3 (Btu/ft^3)

Combustion intensity (C_i), therefore, is maximum when the product of the normal burning velocity of the flame (C_v) and the heating value of the gas mixture (C_h) is maximum.

Like the heat of combustion, the combustion intensity of a gas can be expressed as the sum of the combustion intensities of the primary and secondary reactions. However, the combustion intensity of the primary flame, located near the torch tip where it can be concentrated on the workpiece, is of major importance in welding. The secondary combustion intensity influences the thermal gradient in the vicinity of the weld.

Figures 10.2 and 10.3 show the typical rise and fall of the primary and secondary combustion intensities of several fuels with varying proportions of oxygen and fuel gas. Figure 10.4 gives the total combustion intensities for the same gases. These curves show that, for the gases plotted, acetylene produces the highest combustion intensities.

ACETYLENE

A hydrocarbon compound, C_2H_2, acetylene contains a larger percentage of carbon by weight than any of the other hydrocarbon fuel gases. Colorless and lighter than air, it has a distinctive odor resembling that of garlic. Acetylene contained in cylinders is dissolved in acetone and therefore has a slightly different odor from that of pure generated acetylene.

At temperatures above 780° C (1435° F) or at pressures above 207 kPa (30 psig), gaseous acetylene is unstable and an explosive decomposition may result even without the presence of oxygen. This characteristic has been taken into consideration in the preparation of a code of safe

Table 10.2—Chemical equations for the complete combustion of the common fuel gases

Fuel gas	Reaction with oxygen
Acetylene	$C_2H_2 + 2.5O_2 \rightarrow 2CO_2 + H_2O$
Methylacetylene-propadiene (MPS)	$C_3H_4 + 4O_2 \rightarrow 3CO_2 + 2H_2O$
Propylene	$C_3H_6 + 4.5O_2 \rightarrow 3CO_2 + 3H_2O$
Propane	$C_3H_8 + 5O_2 \rightarrow 3CO_2 + 4H_2O$
Natural gas (methane)	$CH_4 + 2O_2 \rightarrow CO_2 + 2H_2O$
Hydrogen	$H_2 + 0.5O_2 \rightarrow H_2O$

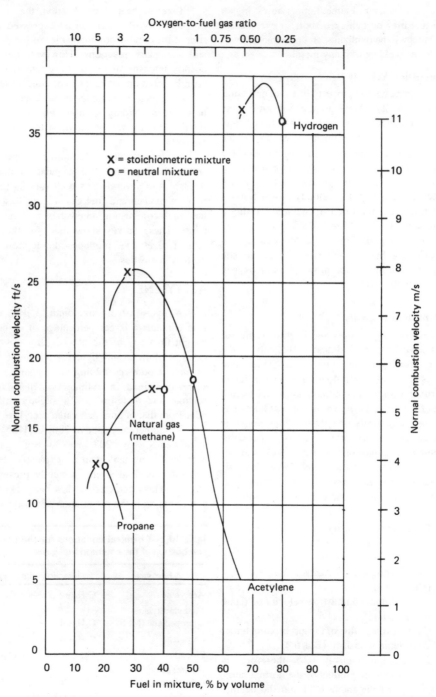

Fig. 10.1—Normal combustion velocity (flame propagation rate) of various fuel gas-oxygen mixtures

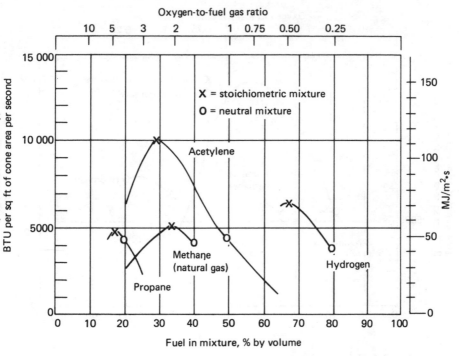

Fig. 10.2—*Primary combustion intensity of various fuel gas-oxygen mixtures*

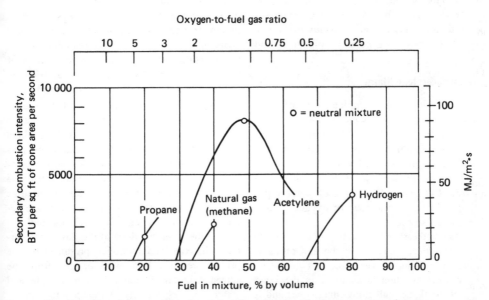

Fig. 10.3—*Secondary combustion intensity of various fuel gas-oxygen mixtures*

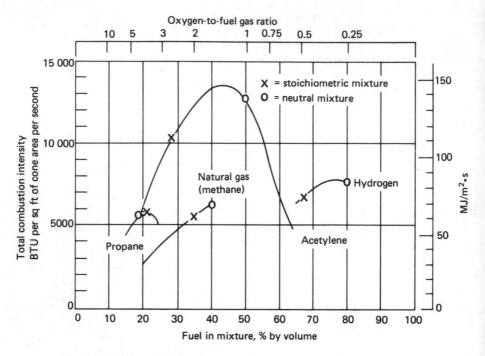

Fig. 10.4—Total combustion intensity of various fuel gas-oxygen mixtures

practices for the generation, distribution, and use of acetylene gas. The accepted safe practice is never to use acetylene at pressures exceeding 103 kPa (15 psig) in generators, pipelines, and hoses.

The Oxyacetylene Flame

Theoretically, the complete combustion of acetylene is represented by the chemical equation

$$C_2H_2 + 2.5\,O_2 \rightarrow 2CO_2 + H_2O \qquad (1)$$

This equation indicates that one volume of acetylene (C_2H_2) and 2.5 volumes of oxygen (O_2) react to produce two volumes of carbon dioxide (CO_2) and one volume of water vapor (H_2O). The volumetric ratio of oxygen to acetylene is 2.5 to one.

As noted earlier, the reaction of equation (1) does not proceed directly to the end products shown. Combustion takes place in two stages. The primary reaction takes place in the inner zone of the flame (called the inner cone) and is represented by the chemical equation

$$C_2H_2 + O_2 \rightarrow 2CO + H_2 \qquad (2)$$

where one volume of acetylene and one volume of oxygen react to form two volumes of carbon monoxide and one volume of hydrogen. The heat content and high temperature (Table 10.1) of this reaction result from the decomposition of the acetylene as well as from the partial oxidation of the carbon resulting from the decomposition.

When the gases issuing from the torch tip are in the one to one ratio indicated in equation (2), the reaction produces the typically brilliant blue inner cone. This relatively small flame produces the combustion intensity needed for welding steel. The flame is termed neutral because there is no excess carbon or oxygen to carburize or oxidize the metal. The end products are actually reduced, a benefit when welding steel.

In the outer envelope of the flame, the carbon monoxide and hydrogen produced by the primary reaction burn with oxygen from the surrounding air, forming carbon dioxide and water vapor respectively, as shown in the following secondary reaction:

$$2CO + H_2 + 1.5\,O_2 \rightarrow 2CO_2 + H_2O$$

Although the heat of combustion of this outer flame is greater than that of the inner, its combustion intensity and temperature are lower because of the large cross-sectional area. The final end products are produced in the outer flame because they cannot exist in the high temperature of the inner cone.

The oxyacetylene flame is easily controlled by valves on the welding torch. By a slight change in the proportions of oxygen and acetylene flowing through the torch, the chemical characteristics in the inner zone of the flame and the resulting action of the inner cone on the molten metal can be varied over a wide range. Thus, by adjusting the torch valves, it is possible to produce a neutral, oxidizing, or carburizing flame.

Production

Acetylene is the product of a chemical reaction between calcium carbide (CaC_2) and water. In the reaction, the carbon in the calcium carbide combines with the hydrogen of the water, forming gaseous acetylene, while the calcium combines with oxygen and hydrogen to form a calcium hydroxide residue. The chemical equation is

$$CaC_2 + 2H_2O \rightarrow C_2H_2 + Ca(OH)_2$$

The carbide used in this process is produced by the smelting of lime and coke in an electric furnace. When removed from the furnace and cooled, it is crushed, screened, and packed in air-tight containers, the most common of which holds 45 kg (100 lb) of the hard, grayish solid. One kg (2.2 lb) of carbide will generate approximately 0.28 m³ (10 ft³) of acetylene.

Acetylene Generators

The two principal methods currently employed in the generation of acetylene are (1) carbide-to-water and (2) water-to-carbide. In the United States, the carbide-to-water method is used almost exclusively. The construction of the generator used for this method allows small lumps of carbide to be discharged from a hopper into a relatively large body of water. This type of generator is shown schematically in Fig. 10.5. The details of its construction vary with different manufacturers. All carbide-to-water generators may generally be classified as low-pressure or medium-pressure types. The former operates at

about 1 psig or less, while the latter produces acetylene from 1 to 15 psig.

The water-to-carbide type of acetylene generator is rarely used in this country but has a greater popularity in Europe. Fundamentally, the operating principal is the same as that of the carbide-to-water type, but the method differs. Water from a tank is allowed to drip onto a bed of carbide, and the gas evolved is piped from the generator. The carbide used in this type of generator is usually in the form of bricks or cakes which limit the surface area exposed to the water.

The generation of acetylene evolves a considerable amount of heat which must be dissipated because of the unstable characteristics of acetylene at elevated temperatures. The relatively large volume of water employed in the carbide-to-water generator makes this type highly suitable for the dissipation of the heat. The water-to-carbide type, on the other hand, uses the minimum amount of water, and heat dissipation is slow.

Acetylene generators are available in both stationary and portable units in a large range of sizes and gas production rates. Generating capacities of these units range from 0.34 m³/h (12 ft³/h) for small portable units to about 170 m³/h (6000 ft³/h) for the large stationary units at industrial installations. Most generators currently in use

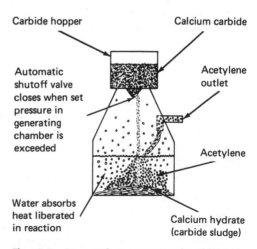

Fig. 10.5—An acetylene generator in which calcium carbide is added to water

operate automatically after an initial setting of operating pressure.

Acetylene Cylinders

Because free acetylene, under certain pressure and temperature conditions, may dissociate explosively into its constituents, cylinders to be filled with acetylene are initially packed with a porous filler. Acetone, a solvent capable of absorbing 25 times its own volume of acetylene per atmosphere of pressure, is added to the filler. By so dissolving the acetylene and dividing the interior of the cylinder into small, partly separated cells, a safe acetylene-filled container is produced.

Acetylene cylinders are available in sizes containing from 0.28 to 11 m^3 (10 to 380 ft^3) of the gas. The cylinders are equipped with fusible safety plugs made of a metal that melts at about 99°C (210°F) and allows the gas to escape if the cylinders should be subjected to excessive heat.

MPS FUEL GAS TYPES

Several commercially prepared fuel gas mixtures are available for welding, but they are not generally used for this purpose. They are more extensively used for cutting, torch brazing, and other heating operations. One group of mixed fuel gases has compositions approximating methylacetylene-propadiene (MPS) and containing mixtures of propadiene, propane, butane, butadiene, and methylacetylene. One of the characteristics of these mixed fuel gases is that the heat distribution within the flame is more even than with acetylene, thus requiring less manipulation of the gas torch for controlling the heat input. The flame temperatures of these gases are lower than that of acetylene with neutral gas mixtures. Temperatures can be increased by making the flames oxidizing. The gases are popular because they are more stable than acetylene and have narrower flammability limits.

PROPYLENE

A single component fuel gas, propylene (C_3H_6) is an oil refinery product having performance characteristics similar to the MPS type gases. Although not suitable for welding or hardfacing, propylene is used for oxygen cutting, torch brazing, flame spraying, and flame hardening. The gas makes use of equipment similar in design to that of the MPS type gases.

PROPANE

Propane (C_3H_8) is used primarily for preheating in oxygen cutting or for heating operations. The main source of this gas is the crude oil gas mixtures obtained from active oil and natural gas wells. Propane is also produced in certain oil refining processes and from the recycling of natural gas. It is sold and transported in steel cylinders containing up to 45 kg (100 lb) of the liquified gas. Large consumers are supplied by tank car and bulk delivery. Small self-contained propane torch sets are available for home workshop use as well as for incidental heating operations.

NATURAL GAS (METHANE)

Natural gas is obtained from wells and distributed by pipelines. Its chemical composition varies widely, depending upon the locality from which it is obtained. The principal constituents of most natural gases are methane (CH_4) and ethane (C_2H_6). The volumetric requirement of natural gas is, as a rule, about 1-1/2 times that of acetylene to provide an equivalent amount of heat. Natural gas finds its principal use in the welding industry as a fuel gas for oxygen cutting and heating operations.

HYDROGEN

The relatively low heat content of the oxyhydrogen flame restricts the use of hydrogen to certain torch brazing operations and to the welding of aluminum, magnesium, lead, and similar metals. Other welding processes, however, are largely supplanting oxyfuel gas welding for many of these materials. Hydrogen is also used with oxygen in preheating operations for underwater cutting.

Hydrogen is available in seamless, drawn steel cylinders charged to a pressure of about

2000 psig at a temperature of 21° C (70° F). It may also be supplied as a liquid either in individual cylinders or in bulk. At the point of use, the liquid hydrogen is vaporized into gas.

OXYGEN

Oxygen in the gaseous state is colorless, orderless, and tasteless. It occurs abundantly in nature. One of its chief sources is the atmosphere, which contains approximately 21 percent oxygen by volume. Although there is sufficient oxygen in the air to support fuel gas combustion, the use of pure oxygen speeds up burning reactions and increases flame temperatures.

Most oxygen used in the welding industry is extracted from the atmosphere by liquefaction techniques. In the extraction process, air may be compressed to approximately 20 MPa (3000 psig), although some types of equipment operate at much lower pressure. The carbon dioxide and any impurities are removed; the air passes through coils and is allowed to expand to a rather low pressure. The air is substantially cooled during the expansion, and then it is passed over the coils, further cooling the incoming air, until liquefaction occurs. Liquid air is sprayed on a series of evaporating trays or plates in a rectifying tower. Nitrogen and other gases boil at lower temperatures than the oxygen and, as these gases escape from the top of the tower, high purity liquid oxygen remains in a receiving chamber at the base. Some plants are designed to produce liquid oxygen; in other plants, gaseous oxygen is withdrawn for compression into cylinders.

OXYFUEL GAS WELDING EQUIPMENT

STORAGE AND DISTRIBUTION

Gas is distributed to various sections of an industrial plant in single cylinders, from portable manifolds, from stationary piped manifolds, from generators, and by pipelines fed from bulk supply systems.

Single Cylinders

Individual cylinders of gaseous oxygen and acetylene provide an adequate supply of gas for welding and cutting torches consuming limited quantities of gas. Cylinder trucks are used extensively to provide a convenient, safe support for a cylinder of oxygen and a cylinder of acetylene. Gas can be transported readily by such means.

Oxygen may be transported to the user in individual cylinders as a compressed gas or as a liquid; there are also several bulk distribution methods. Gaseous oxygen in cylinders is usually under a pressure of approximately 2200 psig. Cylinders of various capacities are used, holding approximately 2, 2.3, 3.5, 6.9, and 8.5 m³ (70, 80, 122, 244, and 300 ft³) of oxygen. Liquid oxygen cylinders contain the equivalent of approximately 85 m³ (3000 ft³) of gaseous oxygen. These cylinders are used for applications which do not warrant a bulk oxygen supply system but which are too large to be supplied conveniently by gaseous oxygen in cylinders. The liquid oxygen cylinders are equipped with liquid-to-gas converters.

A single cylinder of acetylene cannot be used if the volumetric demand is high, since acetone may be drawn from the cylinder with the acetylene. It has, therefore, become standard practice to limit the withdrawal of acetylene from a single cylinder to an hourly rate not exceeding one seventh of the cylinder's volumetric contents.

Cylinders for liquefied fuel gas contain no filler material. These welded steel or aluminum cylinders hold the liquefied fuel gas under pressure. The pressure in the cylinder is a function of the temperature. These liquefied fuel gas cylinders have relief valves which are set for 375 psig, a pressure reached at approximately 93° C (200° F). Should these temperatures be reached, the rapid discharge of fuel gas through the relief valve causes the cylinder to cool down and the relief valve to close. In a fire, the cylinder relief valve opens and shuts intermittently until all the fuel in the cylinder has been discharged or the source of the extreme heat has been removed.

The withdrawal rate of liquefied fuel gases from cylinders is a function of

(1) The temperature

(2) The amount of fuel in the cylinder

(3) The desired operating pressure

Pertinent information should be obtained from the gas supplier.

Manifolded Cylinders

Individual cylinders cannot supply high rates of gas flow, particularly for continuous operation over long periods of time. Manifolding of cylinders is one answer to this problem. A reasonably large volume of fuel gas is provided by this means, which can be discharged at a moderately rapid rate.

Manifolds are of two types, portable and stationary. Portable manifolds may be installed with a minimum of effort and are useful where moderate volumes of gas are required for jobs of a nonrepetitive nature, either in the shop or the field.

Stationary manifolds are installed in shops where even larger volumes of gas are required. Such a manifold feeds a pipeline system distributing the gas at various stations throughout the plant. This arrangement enables many operators to work from a common pipeline system without interruption. Alternatively, it may supply large automatic torch brazing or oxygen cutting operations.

Bulk Systems

To satisfy the large consumption of some industries, gaseous oxygen may be transported from the producing plant to the user in multiple cylinder portable banks or in long, high-pressure tubes mounted on truck trailers. The trailers may hold as many as 850 to 1417 m³ (30 000 to 50 000 ft³) in the larger units and 283 m³ (10 000 ft³) in the smaller units.

Bulk oxygen may also be distributed as a liquid in large insulated containers mounted on truck trailers or railroad cars. The liquid oxygen is transferred to an insulated storage holder on the consumer's property. The oxygen is withdrawn, converted to gas, and passed into the distribution pipelines, as needed, by means of automatic regulating equipment. For users having moderately large consumption, the liquid oxygen may be converted to gas by equipment on the truck, pumped under high pressure into storage tubes, and permanently connected into the user's distribution piping system.

Liquefied fuels may be distributed from on-site bulk tanks of 3785 to 7570 liters (1000 to 2000 gal) capacity. The tanks are filled periodically by truck delivery.

Other Systems

When stationary manifolds, truck trailers, or bulk supply systems are used as sources of oxygen and large acetylene generators are used to produce acetylene in a consumer's plant, the gases are distributed by pipelines to the points of use. The pipelines should be designed properly to handle and distribute the gases in sufficient volume without undue pressure drop. They should also incorporate all necessary safety devices, such as relief values, bursting disks, and hydraulic arrestors, on acetylene systems.

Connections to the pipeline at points of use are made with suitable valves, regulators, relief valves, filters, and hydraulic back pressure valves. Hydraulic back pressure valves and relief valves used on an acetylene or fuel gas station drop should be vented to the outdoors.

SAFE PRACTICES

Oxygen by itself does not burn or explode, but it does support combustion. Oxygen under high pressure may react violently with oil or grease. Cylinders, fittings, and all equipment to be used with oxygen should be kept away from oil and grease at all times. Oxygen cylinders should never be stored near highly combustible materials. Oxygen should never be used to operate pneumatic tools, to start internal combustion engines, to blow out pipelines, to dust clothing, or to create a head pressure in tanks.

Acetylene is a fuel gas and will burn readily. It must, therefore, be kept away from open fire. Acetylene cylinder and manifold pressures must always be reduced through gas pressure-reducing regulators. Cylinders should always be protected against excessive temperature rises and should be stored in well ventilated, clean, and dry locations, free from other combustibles. They must be stored and used with the valve end up. Loose carbide should not be scattered about or allowed to remain on floors because it will absorb moisture from the air and generate acetylene.

Acetylene in contact with copper, mercury, or silver may form acetylides, especially if impurities are present. These compounds are violently explosive and can be detonated by a slight shock or by the application of heat. Alloys containing more than 67 percent copper should not be used in any acetylene system, unless such alloys have been found to be safe in the specific application by experiment or test.

Cylinders of other fuel gases are also pressurized and should be handled with care. These cylinders should also be stored in clean, dry locations.

Liquefied gas cylinders are of double wall construction with a vacuum between the inner and outer shell. They should be handled with extreme care to prevent damage to the internal piping and the loss of vacuum. Such cylinders should always be transported and used in an upright position.

Cylinders can become a hazard if tipped over. The greatest care should be exercised to avoid this possibility. Acetylene cylinders, in particular, should always be used in an upright position. When gas is taken from an acetylene cylinder that is lying on its side, acetone is readily withdrawn, contaminating the flame and resulting in welds of inferior quality. It is standard practice to fasten the cylinders on a cylinder truck or to secure them against a rigid support. All hoses connecting the equipment should be kept as short as possible and free from sharp twists.

BASIC WELDING EQUIPMENT

The minimum basic equipment needed to perform oxyfuel gas welding is shown schematically in Fig. 10.6. This equipment setup is completely self-sufficient and relatively low in cost. It consists of fuel gas and oxygen cylinders, each with gas regulators for reducing cylinder pressure, hoses for conveying the gases to the torch, and a torch and tip combination for adjusting the gas mixtures and producing the desired flame.

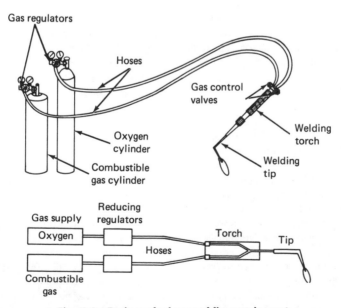

Fig. 10.6—Basic oxyfuel gas welding equipment

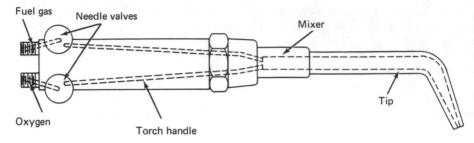

Fig. 10.7—Basic elements of an oxyfuel gas welding torch

Each of these units plays an essential part in the control and utilization of the heat necessary for welding. The same basic equipment is used for torch brazing and for many heating operations. By a simple substitution of the proper torch and tip combination, the equipment is readily available for manual and carriage-controlled oxygen cutting. Since the use of this equipment is controlled by the operator, he must be thoroughly familiar with the capabilities and limitations of the equipment and the rules of safe operation.

A variety of equipment is obtainable for most welding operations. Some of this equipment is designed for general use and some is produced for specific operations. The most suitable equipment should be selected for each particular operation. It is very important in oxyacetylene welding that the correct mixture of gases and tip size be used. Manufacturer's instructions should be carefully followed.

Welding Torches

The welding torch is a device used for mixing and controlling the flow of gases to the tip. It also provides a means of holding and directing the tip. Figure 10.7 is a simplified schematic drawing of the basic elements of a welding torch. There are two throttling or control valves. The gases, after passing the valves and handle, are conducted to the gas mixture at the front end. The tip is shown as a simple tube, narrowed down at the front end to produce a suitable welding cone. Sealing rings or surfaces are provided in the torch head or on the mixer seats to facilitate leak-tight assembly.

Welding torches are manufactured in a variety of styles and sizes from the small torch for extremely light (low gas flow) work to the extra heavy (high gas flow) torches generally used for localized heating operations. Several different styles and sizes of torch bodies, mixers, and tips are shown in Fig. 10.8.

A typical small welding torch used for sheet metal welding will pass acetylene at volumetric rates ranging from about 0.007 to 1 m^3/h (0.25 to 35 ft^3/h). The medium sized torch is designed to provide for acetylene flows from about 0.028 to 2.8 m^3/h (1 to 100 ft^3/h). Heavy duty heating torches may provide for acetylene flows as high as 11 m^3/h (400 ft^3/h). Fuel gases other than acetylene may use even larger torches with fuel gas flow rates as high as 17 m^3/h (600 ft^3/h).

Types of Torches. The two general types of torches are the positive pressure (also called equal or medium pressure) type and the injector or low pressure type.

The positive pressure type torch requires that the gases be delivered to the torch at pressures generally above 1 psig. In the case of acetylene, the pressure should be between 1 and 15 psig. Oxygen generally is supplied at approximately the same pressure. There is, however, no restrictive limit on the oxygen pressure. It can, and sometimes does, range up to 25 psig on the larger sizes of tips when used in positive pressure torches.

The purpose of the injector type torch is to increase the effective use of fuel gases supplied at pressures of 7 kPa (1 psig) or lower. In this torch, oxygen is supplied at pressures ranging from 70 to 275 kPa (10 to 40 psig), the pressure increasing with the tip size. The relatively high velocity of the oxygen flow is used to aspirate or draw in more fuel gas than would normally

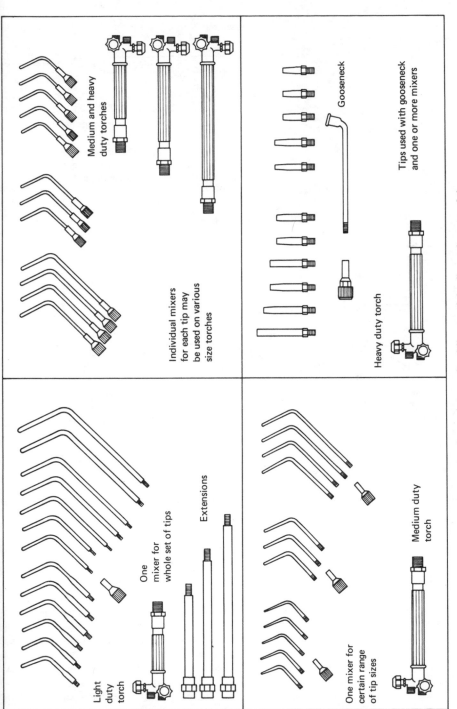

Medium and heavy duty torches

Individual mixers for each tip may be used on various size torches

Gooseneck

Tips used with gooseneck and one or more mixers

Heavy duty torch

Light duty torch

One mixer for whole set of tips

Extensions

Medium duty torch

One mixer for certain range of tip sizes

Fig. 10.8—Several styles and sizes of welding torch bodies, mixers, and tips

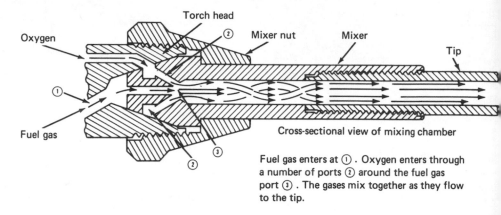

Fuel gas enters at ① . Oxygen enters through a number of ports ② around the fuel gas port ③ . The gases mix together as they flow to the tip.

(A) Positive pressure type

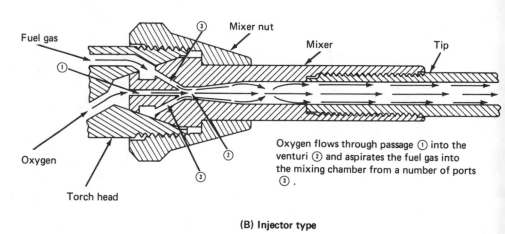

Oxygen flows through passage ① into the venturi ② and aspirates the fuel gas into the mixing chamber from a number of ports ③ .

(B) Injector type

Fig. 10.9—Typical design details of gas mixers for positive pressure and injector type welding torches

flow at the low supply pressures. This action is achieved by specially designed gas mixtures.

Gas Mixers. As indicated in Fig. 10.8, gas mixers come in various styles and sizes according to the manufacturer's design. The chief function of these units is to mix the fuel gas and oxygen thoroughly to assure smooth combustion. Because of their construction, mixers also serve as a heat sink to help prevent the possibility of the flame flashing back into the torch or hose.

A typical mixer for a positive pressure torch is shown in Fig. 10.9(A). The fuel gas enters through a central duct, and the oxygen enters through several angled ducts to effect the mixing. Mixing turbulence decreases to a laminar flow as the gas passes through the tip.

Gas mixers designed for injector type torches employ the principle of the venturi tube to increase the fuel gas flow. In this case (Fig. 10.9B), the high-pressure oxygen passes through the small central duct creating a high velocity jet. The oxygen jet crosses the openings of the angled fuel gas ducts at a point where the venturi tube is restricted. This action produces a pressure drop at the fuel gas openings, causing the low-pressure fuel gas flow to increase as the mixing gases pass into the enlarged portion of the venturi.

Care of Torches. Each welding torch either has a gas mixer unit as an integral part or provision is made to attach a mixer by screwing it onto the front end of the torch body (Fig. 10.8). The welding tip is held in place by threads. This means that the front portion of a welding torch should not be mishandled, or the gas seals may be damaged. When attaching mixers to the front end of torches, particular care should be exercised to prevent the seating surfaces from being scored. Slight damage to these seating surfaces will cause gas leaks, indicated by frequent popping sounds. Also, escaping fuel gas may burn at the leak.

Welding torches are built to withstand rough usage, but the following rules should be observed when handling them:

(1) Keep torches away from all oil and grease.

(2) If the throttle value does not shut off when hand-tightened in the normal manner, do not use a wrench to tighten or seat the valve stem. If foreign matter cannot be blown off the seat, remove the stem assembly and wipe the seat clean before reassembling.

(3) Never clamp the torch handle tightly in a vise, because the handle may collapse and damage the interior of the gas tubes.

(4) Keep the mixer seat free of dust and other foreign matter at all times.

(5) Before using a torch for the first time, check the packing nuts on the throttle valves to insure they are tight.

Welding Tips

The welding tip is that portion of the torch through which the gases pass just prior to their ignition and burning. The tip enables the welder to guide the flame and direct it to the work with maximum ease and efficiency.

Tips are generally made of a nonferrous metal, such as a copper alloy, with high thermal conductivity which reduces the danger of overheating. Tips are generally manufactured by drilling bar stock to the proper orifice size or by swaging tubing to the proper diameter over a

mandrel. The bore in both types must be smooth in order to produce the required flame cone. The front end of the tip should also be shaped to permit easy use and provide a clear view of the welding operation being performed.

Welding tips are available in a great variety of sizes, shapes, and constructions (Fig. 10.8). Two general methods of using tip and mixer combinations are employed. One class uses a tip and mixer unit that provides the proper mixer for each size of tip, and the other general class employs one or more mixers for the entire range of tip sizes. In the latter class, the tip unscrews from its mixer, and each size of mixer has a particular thread size to prevent improper grouping of a tip and a mixer.

A single mixer is used for some classes of welding. It has a "gooseneck" into which the various sizes of tips may be threaded.

Since tips generally are made of a relatively soft copper alloy, care must be taken to guard against damage to them. The following precautions should be observed:

(1) Tips should be cleaned using tip cleaners designed for this purpose.

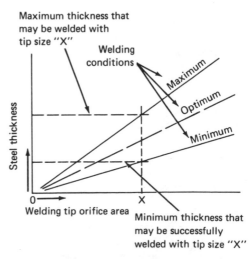

Fig. 10.10—Graphic representation of the linear relationship between steel thickness and the area of a welding tip orifice

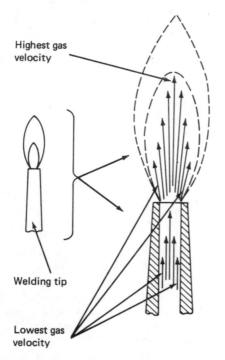

Highest gas velocity

Welding tip

Lowest gas velocity

Fig. 10.11—Vector representation of laminar flow velocity in a welding tip and in the formation of a uniform flame cone

(2) Tips should never be used for moving or holding the work.

(3) Tip seat or threads, or both, always should be absolutely clean and free from foreign matter in order to prevent scoring when tightening the tip nut.

When performing a welding operation, care should be taken to obtain the correct flame adjustment with the proper size of torch, mixer, and tip. The proper methods for obtaining the desired flame characteristics are given elsewhere in this chapter.

The relationship between the thickness of a steel piece and the size of tip that is best for welding is indicated in Fig. 10.10. When a series of welding tips is selected for a variety of thicknesses of metal, the metal thickness range covered by one tip should slightly overlap that covered by the next tip. Since there is no single standard for tip size designations, the manufacturer's recommendations should be followed.

Volumetric Rate of Flow. The important factor in determining the usefulness of a torch tip is the action of the flame on the metal. If it is too violent, it may blow the metal out of the molten pool. Under such conditions, the volumetric flow rates of acetylene and oxygen have to be reduced to a point at which the metal can be welded. This point represents the maximum volumetric flow rate that can be handled by a given size of welding tip. As a general rule, the larger the volumetric rate of gas that can be handled by a specific size of tip, the greater the heat. The flame may also be too soft for easy welding. When the flame is too soft, the gas volumetric flow rates must be increased.

Tips having a hooded or cup-shaped end for some of the slower burning gases, such as propane, are available from most equipment manufacturers. These tips are usually used for heating, brazing, or soldering.

Flame Cones. The purpose of a welding flame is to raise the temperature of metal to the point where it can be welded. This can best be accomplished when the welding flame (or cone) permits the heat to be directed easily. Consequently, the cone characteristics become important. Laminar of streamlined gas flow becomes of paramount importance throughout the length of the tip, especially during passage through the front portion.

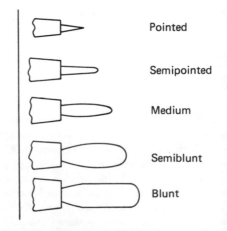

Pointed

Semipointed

Medium

Semiblunt

Blunt

Fig. 10-12—Representative flame cone shapes produced by welding tips

A high velocity flame cone presents striking proof of the velocity gradient extending across a circular orifice when the existing flow is laminar (Fig. 10.11). Since the greatest velocity exists at the center of the stream, the flame at the center is the longest. Similarly, since the velocity of the gas stream is lowest at the wall of the tip (bore) where the flowing friction is greatest, that portion of the flame bordering the wall is the shortest.

From the analysis of the principles that underlie the formation of a flame cone, it is possible to understand the flow conditions that exist along the last portion of the gas passageway in a tip. The shape of the flame cone will depend upon a number of factors, such as the smoothness of the bore, the ratio of lead-in to final run diameter, and the sharpness of neck-down.

Generally speaking, the cone produced by a small tip will vary from a pointed to a semipointed shape. Cones from medium-sized tips will vary from a semipointed to a medium shape, and cones from a large-sized tip will vary from a semiblunt to a blunt shape (Fig. 10.12).

Hoses

Hoses used in oxyfuel gas welding and allied operations are manufactured specifically to meet the utility and safety requirements for this service. For mobility and ease of manipulation in welding, the hoses must be flexible. They must also be capable of withstanding high line pressures at moderate temperatures. To avoid serious accidents which may occur if fuel gas and oxygen hoses should be mistakenly interchanged, special safety devices are incorporated into their manufacture.

For rapid identification, all fuel gas hoses are colored red. As a further precaution, the swivel nuts used for making fuel gas hose connections are identified by a groove cut into the outside of the nut. The nuts also have left-hand threads to match the fuel gas regulator outlet and the fuel gas torch inlet.

Oxygen hoses are colored green, and the connections each have a plain nut with right-hand threads matching the oxygen regulator outlet and the oxygen torch inlet.

The standard means of specifying hose is by inside diameter and application. Nominal inside diameters most commonly used are 3.2, 4.8, 6.4, 7.9, 9.5, and 12.7 mm (1/8, 3/16, 1/4, 5/16, 3/8, and 1/2 in.), although larger sizes are available. Standard industrial welding hose and fittings have a maximum working pressure of 200 psig. Although higher pressures were formerly permitted, standard hose and fittings should not be used at higher pressures.

Wherever possible, hoses should be supported in an elevated position to avoid damage by falling objects, truck wheels, or hot metal. Damaged hose should be replaced or repaired with the proper fittings intended for this purpose.

Although a pressure drop occurs along the hose length, the pressures supplied by cylinders and manifolds are too high for direct use by oxyfuel gas torches. Nevertheless, the gas pressure delivered to the torch must be sufficient to achieve and maintain the flame adjustment best suited to the application involved. For these reasons, hose working pressures must be regulated.

Regulators

A regulator can be described as a mechanical device for maintaining the delivery of a gas at some substantially constant reduced pressure even though the pressure at the source may change. Regulators used in oxyfuel gas welding and allied applications are adjustable pressure reducers designed to operate automatically after an initial setting. Except for minor modifications, all of these regulators operate on the same basic principle. They fall into different application categories, however, according to their designed capabilities for handling specific gases, different pressure ranges, and different volumetric flow rates.

Despite their variety, regulators are generally classed as single stage or two stage depending on whether the pressure is reduced in one or two steps. As to performance, the output pressure of the single stage types exhibits a characteristic drift. When these regulators are attached to high-pressure cylinders, periodic readjustment is necessary to maintain steady output as cylinder pressure drops with gas withdrawal. Two stage regulators maintain the initial pressure setting within close limits throughout the

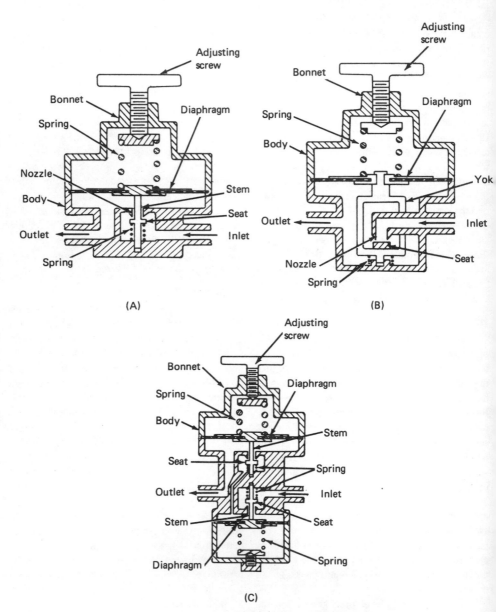

Fig. 10.13—Cross sections of three typical regulators showing the major components of (A) a single stage stem type, (B) a single stage nozzle type, and (C) a two stage type

usuable pressure range of the cylinder. They are essentially two single stage regulators operating in series within one housing.

Operating Principle. The components of a pressure reducing regulator are shown schematically in Fig. 10.13(A). The principal operating elements are

(1) An adjusting screw that controls the thrust of a bonnet spring

(2) A bonnet spring that transmits this thrust to a diaphragm

(3) A diaphragm connected with a movable valve seat

(4) A valve consisting of a nozzle and the movable valve seat

(5) A small spring located under the movable valve seat

The bonnet spring force tends to hold the valve open while the forces on the underside of the diaphragm tend to cause the valve to close. When gas is withdrawn at the outlet, the pressure under the diaphragm is reduced, thus further opening the valve and admitting more gas until the forces on either side of the diaphragm are equal.

A given set of conditions, such as constant inlet pressure, constant volumetric rate of flow, and constant outlet pressure, will produce a balanced condition where the nozzle and its mating seat member will maintain a fixed relationship. As noted earlier, the inlet pressure from cylinders drops as gas is used, causing a gradual drift in regulator outlet pressure. The factors affecting the extent of this drift depend on the type of single stage regulator.

Single Stage Types. There are two basic types of pressure reducing regulators:

(1) The stem type or inlet pressure closing (sometimes referred to as the inverse or negative type), illustrated in Fig. 10.13(A)

(2) The nozzle type or inlet pressure opening (sometimes referred to as the direct acting or positive type), illustrated in Fig. 10.13(B)

In the stem type regulator, the inlet pressure tends to close the seat member (pressure closing) against the nozzle. The outlet pressure of this type of regulator has a tendency to increase somewhat as the inlet pressure decreases. This increase is caused by a decrease of the force produced by

the inlet gas pressure against the seating area as the inlet pressure decreases.

The gas outlet pressure for any particular setting of the adjusting screw is regulated by a balance of forces between the bonnet spring thrust and the opposing forces created by

(1) The gas pressure against the underside of the diaphragm

(2) The force created by the inlet pressure against the valve seat

(3) The force of the small spring located under the valve seat

When the inlet pressure decreases, its force against the seat member decreases, allowing the bonnet spring force to move the seat member away from the nozzle. Thus, more gas pressure is allowed to build up against the diaphragm to reestablish the balanced condition.

In the nozzle type regulator, the inlet pressure tends to move the seat member away (pressure opening) from the nozzle, thus opening the valve. The outlet pressure of this type of regulator decreases somewhat as the inlet pressure decreases because the force tending to move the seat member away from the nozzle is reduced as the inlet pressure decreases. A small outlet pressure on the underside of the diaphragm is then required to close the seat member against the nozzle.

Two Stage Regulators. The two stage regulator provides more precise regulation over a wide range of varying inlet pressures. A two stage regulator, as illustrated in Fig. 10.13(C), is actually two single stage regulators in a series, incorporated as one unit.

The outlet pressure from the first stage is usually preset to deliver a specified inlet pressure to the second stage. In this way, a practically constant delivery pressure may be obtained from the outlet of the regulator as the supply pressure decreases.

The combinations employed to make a two stage regulator are as follows:

(1) Nozzle type first stage and stem type second stage

(2) Stem type first stage and nozzle type second stage

(3) Two stem types, as illustrated in Fig. 10.13(C)

(4) Two nozzle types

Regardless of the combination used, the increase or decrease in outlet pressure is usually so slight (and apparent only at very low inlet supply pressures) that, for all practical purposes, the variation in delivery pressure is disregarded in welding or cutting operations. Two stage regulators are suggested for precise work, such as continuous machine cutting, in order to maintain a constant working pressure and a controlled volumetric flow rate at the welding or cutting torch.

Applications of Regulators. Regulators are produced with different capacities for pressure and volumetric flow, depending on the application and the source of supply. They should, therefore, be used only for the purpose intended. In oxyfuel gas welding, the requirements for cylinder regulators are considerably different from those of station regulators.

In the commonly used one-torch setup shown in Fig. 10.6, oxygen and acetylene are supplied from single cylinders; each is connected to a cylinder regulator which may be of either the single stage or the two stage type. Each regulator is equipped with two pressure gages, one indicating the inlet or cylinder pressure, the other indicating the outlet or torch working pressure. Cylinder regulators and cylinder pressure gages are built to withstand high pressures with a safe overload margin. Working pressure gages are built and graduated to accommodate the intended service application.

Pipeline pressures for oxygen seldom exceed 200 psig; for acetylene, the pressure should not exceed 15 psig. Station regulators are, therefore, built for low-pressure operation, although they may have volumetric flow capacity. Station regulator requirements are adequately met by single stage types equipped only with a working pressure gage. Due to their capacity limitations, station regulators should never be substituted for cylinder regulators because of the possibility of a serious accident.

Regulator Inlet and Outlet Connections. Cylinder outlet connections are of different sizes and shapes to preclude the possibility of connecting a regulator to the wrong cylinder. Regulators must, therefore, be made with different inlet

connections to fit the various gas cylinders. The Compressed Gas Association (CGA) has standardized noninterchangeable cylinder valve outlet connections. These specifications are published by the American National Standards Institute as ANSI B57.1, *Compressed Gas Cylinder Valve Outlet and Inlet Connections,* latest edition, which should be consulted for information.

Regulator outlet fittings also differ in size and thread, depending upon the gas and regulator capacity. Oxygen outlet fittings have right-hand threads and fuel outlet fittings have left-hand threads with grooved nuts.

Recommended Safe Practices. The following safety precautions should be observed when using gas regulators:

(1) Clean the cylinder valve outlet with a clean, lint-free, dry cloth, and blow any dust from the outlet by opening the valve momentarily before connecting the regulator to it.

(2) Always use the correct size wrench to connect the regulator to the cylinder outlet, and never force the connection.

(3) Before opening the cylinder valve, the regulator valve must be closed. The regulator adjusting screw is turned to the left (counterclockwise) until it runs free.

(4) Always open the cylinder valves very slowly so that the high-pressure gas does not surge into the regulator. When doing this, stand to one side of the regulator rather than directly in front or in back of it.

(5) The adjusting screws should be turned in (clockwise) slowly.

(6) Check the gages periodically to insure their correct operation.

(7) Never use oil or grease on a regulator. The only lubricant that should be used is that specified by the manufacturer, and it should be used only when specified.

(8) If a leak is suspected, use a grease-free, soapy water solution to locate it by bubble indications.

(9) Regulators should only be repaired by qualified, trained mechanics.

Manifolds

Portable Manifolds. Two types of portable manifolds are in general use. In one type, shown

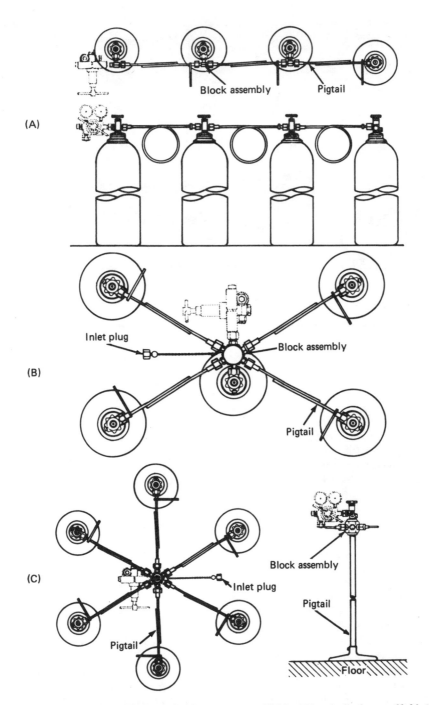

Fig. 10.14—Typical arrangements for portable oxygen manifolds (A) four cylinder manifold, (B) five cylinder manifold, and (C) six cylinder manifold

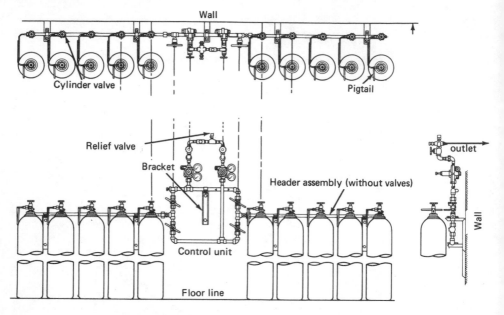

Fig. 10.15—Typical arrangement for a stationary gas manifold

in Fig. 10.14(A), tees are connected to the individual cylinder valves and leads, or pigtails, successively join the tees together. The gas from each cylinder passes through the tee, into the main gas stream, and, finally, to a single regulator that serves the entire group of cylinders.

In the second type of portable manifold (Figs. 10.14B and C), the cylinders are connected by means of individual cylinder leads (pigtails) to a common coupler block, which is connected to a pressure reducing regulator.

These systems are extremely flexible and permit the manifolded cylinders to be located near the point of use. Such manifolds can be assembled rapidly when needed.

Stationary Manifolds. A stationary manifold may be installed where very large volumes of gas are regularly required or where a centralized gas supply is desired. This type of manifold consists of an adequately supported high-pressure header to which a number of cylinders are connected by means of pigtails. One or more permanently mounted regulators reduce and regulate the pressure of a gas flowing from the manifold into the plant piping system.

An example of a stationary manifold is illustrated in Fig. 10.15. In this type of stationary manifold, two banks of cylinders are used. One bank of cylinders supplies the system, while the other is held in reserve. When the bank in use becomes exhausted, the other is automatically switched to the line. In this way, uninterrupted service is assured.

Pipelines of various diameters are used to deliver the high volumetric flow supplied by stationary manifolds. As noted earlier, the relatively low (compared to those used in cylinders) pressure gases are delivered to station drops through station regulators.

An important protective device for the system is the hydraulic seal or hydraulic flashback arrestor. This device will stop a flashback originating at a station from passing further into the system. It consists of a small pressure vessel partly filled with water through which the acetylene supply flows. The gas continues through the space above the water level and through the vessel heat to the station regulator. A flashback or high-pressure backup will set off a relief valve in the vessel head which will vent the pressure to the atmosphere outside. A check valve prevents the water from backing up into the line.

Regulations and Safe Practices for Manifolds and Pipelines. The rules and regulations set forth in the current issue of the *Standard for the Installation and Operation of Oxy-fuel Gas Systems for Welding and Cutting,* as Recommended by the National Fire Protection Association (NFPA No. 51)[3] govern the installation of oxygen and acetylene manifolds and pipelines. Local regulations or ordinances should be consulted also to be sure of compliance with them. Manifolds should, in all cases, be obtained from reliable manufacturers and installed by personnel familiar with proper construction and installation of oxygen and acetylene manifolds and pipelines.

Accessories

In addition to the equipment and materials described above, an almost limitless variety of auxiliary equipment may be used in the process of gas welding. Only a brief description of such items is included here.

One of the most universally required articles is the friction lighter, which should always be used to ignite the gas.

Other accessories, such as tip cleaners, cylinder trucks, clamps, holding jigs and fixtures, and gas savers for automatically shutting off the torch flames are also important auxiliary aids for gas welding.

Welders should at all times use goggles or eyeshields as a protection against the excessive glare and radiated heat of the flame and molten metal. Suitable gloves, leather aprons, sleeves, and leggings should also be worn.

---

3. Obtainable from the National Fire Protection Association, 470 Atlantic Ave., Boston, Mass. 02110.

APPLICATION OF OXYFUEL GAS WELDING

METALS WELDED

Oxyacetylene welding can be used on a wide range of commercial ferrous and nonferrous metals and alloys. As in any welding process, however, physical dimensions and chemical composition may limit the weldability of certain materials and pieces.

During welding, the metal is taken through a temperature range almost the same as that of the original casting procedure. The base metal in the weld area loses those properties that were given to it by prior heat treatment or cold working. The ability to weld such materials as high carbon and high alloy steels is limited by the equipment available for heat treating after welding. These metals are successfully welded when the size or nature of the piece permits postheat treating operations.

The welding procedure for plain carbon steels is straightforward and offers little difficulty to the welder. Sound welds are produced in other materials by variation in technique, heat treating, preheating, and fluxing.

The oxyacetylene welding process can be used for repair welding metal of considerable thickness and for the usual assemblies encountered in maintenance and repair. Very thick cast iron machinery frames have been repaired by braze welding or by welding with a cast iron filler rod.

Steels and Cast Irons

Low carbon, low alloy, and cast steels are the materials most easily welded by the oxyacetylene process. Fluxes are not required when welding these materials.

In oxyacetylene welding, plain carbon steels having more than 0.35 percent carbon are considered high carbon steels and require special care to maintain their particular properties. Alloy steels of the air-hardening type require extra precautions to maintain their properties, even though the carbon content may be 0.35 percent or less. The joint area usually is preheated to retard the cooling of the weld by conduction of heat into surrounding base metal. Slow cooling prevents the hardness and brittleness associated with rapid cooling. However, a full furnace anneal may be required immediately after welding air-hardening steels.

The welder should use a neutral or slightly carburizing flame for welding and should be careful not to overheat and decarburize the base metal. The preheating temperature required depends upon the composition of the steel to be welded. Temperatures ranging from 150° to 540° C (300° to 1000° F) have been used.

In addition to finding the proper preheat temperature, it is important that it be maintained uniformly during welding. A uniform temperature is maintained by protecting the part with an asbestos blanket (Fig. 10.16). Other means of shielding can also be used to retain the temperature in the part. Generally, the interpass temperature should be maintained within 65° C (150° F) of the preheat temperature. Lower interpass temperatures cause excessive shrinking forces that can result in either distortion or cracking at the weld or other sections. In the welding of circular structures made of brittle metals, such as cast iron, this type of cracking frequently occurs. Sections with excessive internal stresses, such as the water boxes shown in Fig. 10.16, are typical examples.

Modifications in procedures are required for stainless and similar steels. Because of their high chromium or chromium-nickel content, these steels have relatively low thermal conductivity. A smaller flame than that used for equal thicknesses of plain carbon steel is recommended. Because chromium oxidizes easily, a neutral flame is employed to minimize oxidation. A flux is used to dissolve oxides and protect the weld metal. Filler metal of high chromium or nickel-chromium steel is used. Table 10.3 summarizes the basic information for welding ferrous metals.

Cast iron, malleable iron, and galvanized iron all present particular problems in welding by any method. The gray cast iron structure in cast iron can be maintained through the weld area by the use of preheat, a flux, and an appropriate cast iron welding rod.

Fig. 10.16 *— Salvaging a large cast iron water box by oxyfuel gas welding — preheat temperature is maintained with an asbestos blanket*

Nodular iron requires materials in the welding filler metal that will assist in promoting agglomeration of the free graphite to maintain ductility and shock resistance in the heat-affected area. The filler metal manufacturer should be consulted to obtain information on preheat and interpass temperature control for the filler metal being used.

Although there are instances in which cast irons are welded without preheat, particularly in salvage work, preheat from 200° to 320° C (400° to 600° F), with control of interpass temperature and provision for slow cooling, will assure more consistent results. Protection such as the asbestos covering shown in Fig. 10.16 can be used to assure slow and uniform cooling. Care should be taken that localized cooling is not allowed to occur. It is also important to stress that in the salvage of cast iron, removal of all foundry sand and slag is necessary for consistent repair results.

Nonferrous Metals

The particular properties of each nonferrous alloy should be considered when selecting the most suitable welding technique. When the necessary precautions are taken, little difficulty due to the nature of the metal should be encountered.

Aluminum, for instance, gives no warning by changing color prior to melting, but it appears to collapse suddenly at the melting point. Consequently, practice in welding is required to learn to control the rate of heat input. Aluminum and its alloys suffer from hot shortness, and welds should be supported adequately in all the areas during welding. Finally, any exposed aluminum surface is always covered with a layer of oxide that, when combined with the flux, forms a fusible slag which floats on top of the molten metal.

When copper is welded, allowances are necessary for the chilling of the welds because of the very high thermal conductivity of the metal.

Preheating is often required. Considerable distortion can be expected in copper because the thermal expansion is higher than in other commercial metals. These characteristics obviously pose difficulties that must be surmounted for satisfactory welding.

Parts to be welded should be fixtured or tack welded securely in place. The section least subject to distortion should be welded first so that it forms a rigid structure for the balance of the welding.

When the design of the structure permits welding from both sides, distortion can be minimized by welding alternately on each side of the joint. Strongbacks or braces can be applied to sections most likely to distort. The welds may be peened to reduce distortion. This method, if performed properly, can overcome severe warpage. A backstep sequence of welding may be used to control distortion. This method consists primarily of making short weld increments in the direction opposite to the progress of welding the joint. A weldment should be so designed that distortion during welding will be minimized.

Metallurgical Effects

The temperature of the base metal varies during welding from that of the molten weld puddle to room temperature in the areas most remote from the weld. When steels are involved, the weld and the adjacent heat-affected zones are heated considerably above the transformation temperature of the steel. This results in a coarse grain structure in the weld and adjacent base metal. The coarse grain structure can be refined by a normalizing heat treatment (heating to the austenitizing temperature range and cooling in air) after welding.

Hardening of the weld and adjacent base metal heated to above the transformation temperature of the steel can occur if the steel contains sufficient carbon and the cooling rate is high

Table 10.3—General conditions for oxyacetylene welding of various ferrous metals

Metal	Flame adjustment	Flux	Welding rod
Steel, cast	Neutral	No	Steel
Steel pipe	Neutral	No	Steel
Steel plate	Neutral	No	Steel
Steel sheet	Neutral	No	Steel
	Slightly oxidizing	Yes	Bronze
High carbon steel	Reducing	No	Steel
Wrought iron	Neutral	No	Steel
Galvanized iron	Neutral	No	Steel
	Slightly oxidizing	Yes	Bronze
Cast iron, gray	Neutral	Yes	Cast iron
	Slightly oxidizing	Yes	Bronze
Cast iron, malleable	Slightly oxidizing	Yes	Bronze
Cast iron pipe, gray	Neutral	Yes	Cast iron
	Slightly oxidizing	Yes	Bronze
Cast iron pipe	Neutral	Yes	Cast iron or base metal composition
Chromium-nickel steel castings	Neutral	Yes	Base metal composition or 25-12 chromium-nickel steel
Chromium-nickel steel (18-8 and 25-12)	Neutral	Yes	Columbium stainless steel or base metal composition
Chromium steel	Neutral	Yes	Columbium stainless steel or base metal composition
Chromium iron	Neutral	Yes	Columbium stainless steel or base metal composition

enough. Hardening can be avoided in most hardenable steels by using the torch to keep heat on the weld for a short time after the weld is completed. If air-hardening steels are welded, the best heat treatment is a full furnace anneal of the weldment.

The oxyacetylene flame allows a degree of control to be maintained over the carbon content of the deposited metal and over the portion of the base metal that is heated to its melting temperature. When an oxidizing flame is used, a rapid reaction results between the oxygen and the carbon in the metal. Some of the carbon is lost in the form of carbon monoxide, and the steel and the other constituents are also oxidized. When the torch is used with an excess-acetylene flame, carbon is introduced into the molten weld metal.

When heated to a temperature range between 430° and 870° C (800° and 1600° F), carbide precipitation occurs in unstabilized austenitic stainless steel. Chromium carbides gather at the grain boundaries and lower the corrosion resistance of the heat-affected zone. If this occurs, a heat treatment after welding is required, unless the steel is stabilized by the addition of columbium or titanium and welded with the aid of a columbium-bearing stainless steel welding rod. The columbium combines with the carbon and minimizes the formation of chromium carbide. All the chromium is left dissolved in the austenitic matrix, the form in which it can best resist corrosion.

Another factor to be considered in welding is the possible tendency toward hot shortness of the metal (a marked loss in strength at high temperatures). Some of the copper base alloys have this tendency to a high degree. If the base metal has this tendency, it should be welded with care to prevent hot cracking in the weld zone. Allowances should be made in the welding technique used with these metals, and jigging or clamping should be done with caution. Proper welding sequence and multiple layer welding with narrow stringer beads help to reduce hot cracking.

Oxidation and Reduction

Certain metals have such a high affinity for oxygen that oxides form on the surface almost as rapidly as they are removed. In oxyacetylene operations, these oxides are usually removed by means of fluxes. This affinity for oxygen can be a very useful characteristic in certain welding operations. Manganese and silicon, for example, are elements common to plain carbon steel. They are important in oxyacetylene welding because they react with oxygen when the metal is molten.

The reaction produces a very thin slag covering that prevents porosity and tends to prevent any oxygen from contacting the weld metal. When the viscosity of the slag covering is properly controlled, the molten metal may be kept in position even against the pull of gravity. The action of these elements in oxyacetylene welding is the same as in steelmaking. The correct manganese and silicon content of steel welding rods is, therefore, important.

The type of flame used in welding various metals plays an important part in securing the most desirable weld metal deposit. The proper type of flame with the correct welding technique can be used as a shielding medium which will reduce the effect of oxygen and nitrogen (in the atmosphere) on the molten metal. Such a flame also has the effect of stabilizing the molten weld metal and preventing the loss of carbon, manganese, and other alloying elements.

The proper type of flame for any application is determined by the type of base and filler metals involved and the thickness of the base metal. For most metals a neutral flame is used. An exception is the welding of aluminum, where a slightly reducing flame is used.

WELDING FILLER METALS

The properties of the weld metal should closely match those of the base metal. Because of this requirement, welding rods of various chemical compositions are available for the welding of many ferrous and nonferrous materials. Obviously, it is important that the correct welding filler metal be selected.

The welding process itself influences the filler metal composition, since certain elements are lost during welding. Filler metals are available for joining almost all common base materials. The standard diameters of rods vary from 1.6 to 10 mm (1/16 to 3/8 in.), and the standard lengths for rods are 610 and 914 mm (24 and 36 in.).

The chemical composition of a filler metal must be within the limits specified for that particular material. There are many proprietary filler metals on the market recommended for specific applications. Filler metal should be free from porosity, pipes, nonmetallic inclusions, and any other foreign matter. The metal should deposit smoothly.

Allowances for changes taking place during welding are made in the chemical composition of good welding rods so that the deposited metal will be of the correct composition. Deposits should be made with free flowing filler metal that unites readily with the base metal to produce sound, clean welds.

In maintenance and repair work, it is not always necessary that the composition of the welding rod match that of the base metal. A steel welding rod of nominal strength can be used to repair parts made of alloy steels broken by overloading or accident. Every effort should be made, however, to match the filler metal and base metal. Where it is necessary to heat-treat a steel part after welding, carbon can be added to a deposit of mild steel by the judicious use of the carburizing flame. It is preferable, however, to use a welding rod of low alloy steel.

The AWS Committee on Filler Metal has prepared a number of specifications. Many of the oxyfuel gas welding filler metals meet these specifications. Information on the welding and selection of filler materials will be found in the chapters on the various metals and alloys in the *Welding Handbook,* Section 4, 6th edition.

Welding rods for steel are listed in AWS A5.2, *Specification for Iron and Steel Gas Welding Rods.* The rods are classified on the basis of strength. The most commonly used is RG60 which has properties compatible with most low carbon steels.

For the oxyacetylene and braze welding of cast iron, both cast iron and copper base welding rods are used. See AWS A5.15, *Specification for Welding Rods and Covered Electrodes for Welding Cast Iron.* These filler metals are classified on the basis of chemical composition.

FLUXES

One of the most important factors in weld quality is the removal of oxides from the surface of the metal to be welded. Unless the oxides are removed, fusion may be difficult, the joint may lack strength, and inclusions may be present. The oxides will not flow from the weld zone but will remain to become entrapped in the solidifying metal, interfering with the addition of filler metal. These conditions may occur when the oxides have a higher melting point than the base metal, and a means must be found to remove those oxides. Fluxes are applied for this purpose.

Steel and its oxides and slags which form during welding do not fall into the above category, and no flux is needed. Aluminum, however, forms an oxide with a very high melting point, and the oxide must be removed from the welding zone before satisfactory results can be obtained. Certain substances will react chemically with the oxides of most metals, forming fusible slags at welding temperature. These substances, either singly or in combination, make efficient fluxes.

A good flux should assist in removing the oxides during welding by forming fusible slags that will float to the top of the molten puddle and and not interfere with the deposition and fusion of filler metal. The flux should protect the molten puddle from the atmosphere and also prevent the puddle from absorbing, or reacting with, gases in the flame. This must be done without obscuring the welder's vision or hampering his manipulation of the molten puddle.

During the preheating and welding periods, the flux should clean and protect the surfaces of the base metal and, in some cases, the welding rod. Flux should not be used as a substitute for base metal cleaning during joint preparation. Flux is an excellent metal cleaner, but if its activity is used for cleaning dirty metal, it will be useless for its primary functions.

Flux may be prepared as a dry powder, a paste or thick solution, or a preplaced coating on the welding rod. Some fluxes operate much more favorably if they are used dry. Braze welding fluxes and fluxes for use with cast iron are usually in this class. These fluxes are applied by heating the end of the welding rod and dipping it into the powdered flux. Enough will adhere to the rod for adequate fluxing. Dipping the hot rod into the flux again will coat another portion. Dropping some of the dry powder on the base metal ahead of the welding zone will sometimes help, especially in the repair of dirty and oil-soaked castings.

Fluxes in paste form are usually painted on the base metal with a brush, and the welding rod can either be painted or dipped. Commercial precoated rods can be used without further preparation, and when required, additional flux can be placed on the base metal. Sometimes a precoated rod will have to be dipped in powdered flux if the flux melts off too far from the end of the rod during welding.

The common metals and welding rods requiring fluxes are bronze, cast iron, brass, silicon bronze, stainless steel, and aluminum.

TYPICAL APPLICATIONS

The wide field of possible applications, as well as the convenience and the economy of oxyacetylene welding, are recognized in most metalworking industries. The process is used for fabricating sheet metal, tubing, pipe, and other metal shapes in industries such as industrial piping and automotive. It is also used in welding pipelines of small and medium diameter up to 51 mm (2 in.).

The process is almost universally used and accepted in the field of maintenance and repair, where its flexibility and mobility result in great savings of time and labor. The typical self-contained unit, consisting of a welding torch, an oxygen cylinder, and an acetylene cylinder on a two-wheeled cart, can be readily moved about in a plant. It can be carried easily into the field on a small truck wherever a breakdown may have occurred. Oxyacetylene welding is also well suited for use in machine and automobile repair shops. It is equally useful in shops devoted entirely to welding, where the repair of large and small industrial, agricultural, and household equipment may be the main business.

Many surfacing operations are conducted with the oxyacetylene process, some of which are not possible with the arc welding processes. For instance, the application of materials high in zinc content, such as admiralty metal, can be accomplished with the gas welding torch. Automatic procedures are used for this type of surfacing application, such as on tube sheets or heat exchangers. Surfacing applications are discussed in Chapter 14 of this volume.

The same gases and equipment can be used in cutting, brazing, soldering, heat treating, surfacing, and welding. This makes the oxyacetylene process particularly attractive from the viewpoint of initial investment.

WELDING PROCEDURES

OPERATING PRINCIPLES

The oxyfuel gas welding torch serves as the medium for mixing the combustible and combustion-supporting gases and provides the means for applying the flame at the desired location. A range of tip sizes is provided for obtaining the required volume or size of welding flame which may vary from a short, small diameter needle flame to a flame 4.8 mm (3/16 in.) or more in diameter and 50 mm (2 in.) or more in length.

The inner cone or vivid blue flame of the burning mixture of gases issuing from the tip is called the working flame. The closer the end of the inner cone is to the surface of the metal being heated or welded, the more effective is the heat transfer from flame to metal. The flame can be made soft or harsh by varying the gas flow. Too low a gas flow for a given tip size will result in a soft, ineffective flame sensitive to backfiring. Too high a gas flow will result in a harsh, high velocity flame that is hard to handle and will blow the molten metal from the puddle.

The chemical action of the flame on a molten pool of metal can be altered by changing the ratio of the volume of oxygen to acetylene issuing from the tip. Most oxyacetylene welding is done with a neutral flame having approximately a 1:1 gas ratio. An oxidizing action can be obtained by increasing the oxygen flow, and a reducing action will result from increasing the acetylene flow. Both adjustments are valuable aids in welding.

Flame Adjustment

Torches should be lighted with a friction lighter or a pilot flame. The instructions of the equipment manufacturer should be observed when adjusting operating pressures at the gas regulators and torch valves before the gases issuing from the tip are ignited. Three types of oxyacetylene flame adjustment are shown in Fig. 10.17.

The neutral flame is obtained most easily by adjustment from an excess-acetylene flame, which is recognized by the feather extension of the inner cone. The feather will diminish as the flow of acetylene is decreased or the flow of oxygen is increased. The flame is neutral just at the point of disappearance of the "feather" extension of the inner cone. This flame is actually reducing in nature but is neither carburizing nor oxidizing.

A practical method of determining the amount of excess acetylene in a reducing flame is to compare the length of the feather with the length of the inner cone, measuring both from the torch tip. A $2\times$ excess-acetylene flame has an acetylene feather that is twice the length of the inner cone. Starting with a neutral flame adjustment, the welder can produce the desired acetylene feather by increasing the acetylene flow (or by decreasing the oxygen flow). This flame also has a carburizing effect on steel.

The oxidizing flame adjustment is sometimes given as the amount by which the length of a neutral inner cone should be reduced, for example, one tenth. Starting with the neutral flame, the welder can increase the oxygen (or decrease the acetylene) until the length of the inner cone is decreased the desired amount.

Forehand and Backhand Welding

Oxyacetylene welding may be performed with the torch tip pointed forward in the direction the weld progresses, as shown in Fig. 10.18; this method is called the forehand technique. Welding can also be done in the opposite direction where the torch points toward the completed weld; this method is known as backhand welding. Each method has its advantages, depending upon the application, and each method imposes some variations in deposition technique.

In general, the forehand method is recommended for welding material up to 3.2 mm (1/8 in.) thick, because it provides better control of the small weld puddle, resulting in a smoother weld at both top and bottom. The puddle of molten metal is small and easily controlled. A great deal of pipe welding is done using the forehand technique, even in 9.5 mm (3/8 in.) wall thicknesses.

Increased speeds and better control of the puddle are possible with the backhand technique when metal 3.2 mm (1/8 in.) and thicker is weld-

(A) Carburizing

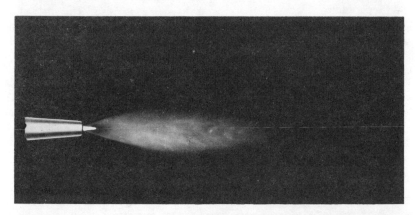

(B) Oxidizing

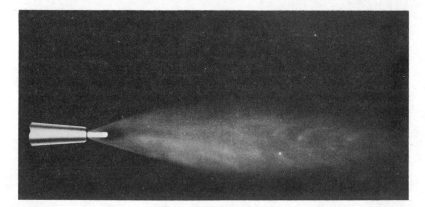

(C) Neutral

Fig. 10.17—Three types of oxyacetylene flame adjustment: (A) carburizing or reducing, (B) oxidizing, and (C) neutral

ed. This recommendation is based on the careful study of the speeds normally achieved with this technique and on the greater ease of obtaining fusion at the root of the weld. Backhand welding may be used with a slightly reducing flame (slight acetylene feather) when it is desirable to melt a minimum amount of steel in making a joint. The increased carbon content obtained from this flame lowers the melting point of a thin layer of steel and increases welding speed. This technique greatly increases the speed of making pipe joints where the wall thickness is 6.4 to 7.9 mm (1/4 to 5/16 in.) and the groove angle is less than normal. Backhand welding is sometimes used in surfacing operations.

Base Metal Preparation

Cleanliness along the joint and on the sides of the base metal is of the utmost importance. Dirt, oil, and oxides can cause incomplete fusion, slag inclusions, and porosity in the weld.

The spacing between the parts to be joined should be considered carefully. The root opening for a given thickness of metal should permit the gap to be bridged without difficulty, yet it should be large enough to permit full penetration. Speci-

fications for root openings should be followed exactly.

The thickness of the base metal at the joint determines the type of edge preparation for welding. Thin sheet metal is easily melted completely by the flame. Thus, edges with square faces can be butted together and welded. This type of joint is limited to material under 4.8 mm (3/16 in.) in thickness. For thicknesses of 4.8 to 6.4 mm (3/16 to 1/4 in.), a slight root opening or groove is necessary for complete penetration, but filler metal must be added to compensate for the opening.

Joint edges 6.4 mm (1/4 in.) and greater in thickness should be beveled. Beveled edges at the joint provide a groove for better penetration and fusion at the sides. The angle of bevel for oxyacetylene welding varies from 35 to 45 degrees which is equivalent to a variation in the included angle of the joint from 70 to 90 degrees, depending upon the application. A root face 1.6 mm (1/16 in.) wide is normal, but feather edges are sometimes used. Plate thicknesses 19 mm (3/4 in.) and above are double beveled when welding can be done from both sides. The root face can vary from 0 to 3.2 mm (1/8 in.). Beveling both

Fig. 10.18—Demonstration of oxyacetylene welding of plate showing the forehand technique

sides reduces by approximately one-half the amount of filler metal required. Gas consumption per unit length of weld is also reduced.

A square groove edge preparation is the easiest to obtain. This edge can be machined, chipped, ground, or oxygen cut. The thin oxide coating on an oxygen-cut surface does not have to be removed, because it is not detrimental to the welding operation or to the quality of the joint. A bevel angle can be oxygen cut.

MULTIPLE LAYER WELDING

Multiple layer welding is used when maximum ductility of a steel weld in the as-welded or stress-relieved condition is desired or when several layers are required in welding thick metal. Multiple layer welding is accomplished by depositing filler metal in successive passes along the joint until it is filled. Since the area covered with each pass is small, the weld puddle is reduced in size. This procedure enables the welder to obtain complete joint penetration without excessive penetration and overheating while the first few passes are being deposited. The smaller puddle is more easily controlled, and the welder can thus avoid oxides, slag inclusions, and incomplete fusion with the base metal.

An increase in ductility in the deposited steel results from grain refinement in the underlying passes as they are reheated. The final layer will not possess this refinement unless an extra pass is added and removed or unless the torch is passed over the joint to bring the last deposit up to normalizing temperature.

Oxyacetylene welding is not recommended for high strength, heat treatable steels, especially when they are being fabricated in the heat treated condition. In welding heat treated steels, the slow rate of heat input with oxyacetylene welding can cause metallurgical changes in the heat-affected area and destroy the heat treated base metal properties. This type of metal should be welded with one of the arc welding processes. Interpass welding temperature on the final passes is generally held between 150° and 230° C (300° and 450° F), and in addition, the heat energy must be maintained within the limits recommended by the manufacturer of the material being welded.

WELD QUALITY

The appearance of a weld does not necessarily indicate its quality. If discontinuities exist in a weld, they can be grouped into two broad classifications: those that are apparent to visual inspection and those that are not. Visual examination of the underside of a weld will determine whether there is complete penetration or whether there are excessive globules of metal. Inadequate joint penetration may be due to insufficient beveling of the edges, too wide a root face, too great a welding speed, or poor torch and welding rod manipulation.

Oversized and undersized welds can be observed readily. Weld gages are available to determine whether a weld has excessive or insufficient reinforcement. Undercut or overlap at the sides of the welds can usually be detected by visual inspection.

Although other discontinuities, such as incomplete fusion, porosity, and cracking, may or may not be externally apparent, excessive grain growth or the presence of hard spots cannot be determined visually. Incomplete fusion may be caused by insufficient heating of the base metal, too rapid travel, or gas or dirt inclusions. Porosity is a result of entrapped gases, usually carbon monoxide, which may be avoided by more careful flame manipulation and adequate fluxing where needed. Hard spots and cracking are a result of metallurgical characteristics of the weldment. For further details, see *Fundamentals of Welding,* the *Welding Handbook,* Volume 1, 7th edition.

Inspection

The term inspection usually implies the formal inspection, prescribed by a code or by the requirements of a purchaser, that is given to welds and welded structures. As far as welding is concerned, the minimum requirements of the codes are inflexible and must be met. Codes vary somewhat in their requirements with respect to the testing and inspection of the finished structure.

WELDING WITH OTHER FUEL GASES

PRINCIPLES OF OPERATION

Hydrocarbon gases, such as propane, butane, city gas, and natural gas, are not suitable for welding ferrous materials due to their oxidizing characteristics. In some instances, many nonferrous and ferrous metals can be braze welded with care taken in the adjustment of flame and the use of flux. It is important to use tips designed for the fuel gas being employed. These gases are extensively used for brazing and soldering operations, utilizing both mechanized and manual methods.

These fuel gases have relatively low flame propagation rates, with the exception of some manufactured city gases containing considerable amounts of hydrogen. When standard welding tips are used, the maximum flame velocity is so low that it interferes seriously with heat transfer from the flame to the work. The highest flame temperatures of the gases are obtained at high oxygen-to-fuel gas ratios. These ratios produce highly oxidizing flames which prevents the satisfactory welding of most metals.

Tips should be used having flame-holding devices, such as skirts, counterbores, and holder flames, to permit higher gas velocities before they leave the tip. This makes it possible to use these fuel gases for many heating applications with excellent heat transfer efficiency.

Air contains approximately 80 percent nitrogen by volume which does not support combustion. Fuel gases burned with air, therefore, produce lower flame temperatures than those burned with oxygen. The total heat content is also lower. The air-fuel gas flame is suitable only for welding light sections of lead and for light brazing and soldering operations.

EQUIPMENT

Standard oxyacetylene equipment, with the exception of torch tips and regulators, can be used to distribute and burn these gases. Special regulators may be obtained, and heating and cutting tips are available. City gas and natural gas are supplied by pipelines; propane and butane are stored in cylinders or delivered in liquid form to storage tanks on the user's property. Delivery is made through the pipeline equipment to the points of use.

The torches for use with air-fuel gas generally are designed to aspirate the proper quantity of air from the atmosphere to provide combustion. The fuel gas flows through the torch at a supply pressure of 2 to 40 psig and serves to aspirate the air. For light work, the fuel gas usually is supplied from a small cylinder that is easily transportable.

The plumbing, refrigeration, and electrical trades use propane in small cylinders for many heating and soldering applications. The propane flows through the torch at a supply pressure from 3 to 60 psig and serves to aspirate the air. The torches are used for soldering electrical connections, the joints in copper pipelines, and light brazing jobs.

The usual safety precautions should be observed when using these fuel gases. Storage and distribution systems should be installed according to the applicable national, state, or local codes.

APPLICATIONS

Air-fuel gas is used for welding lead to approximately 6.4 mm (1/4 in.) in thickness. Perhaps the greatest field of application, however, is in the plumbing and electrical industry. The process is used extensively for soldering copper tubing.

SUPPLEMENTARY READING LIST

Fay, R.H. Heat transfer from fuel gas flames. *Welding Journal* 46 (8), Aug. 1967, pp. 380-s—383-s.

Gas Systems for Welding and Cutting. NFPA No. 51. Boston, Mass.: National Fire Protection Association.

Oxyacetylene Welding and Its Applications. New York: International Acetylene Association, 1958.) Obtained from the Compressed Gas Association.)

Koziarski, J. Hydrogen vs acetylene vs inert gas in welding aluminum alloys. *Welding Journal* 36 (2), Feb. 1957, pp. 141-148.

Kugler, A.N. *Oxyacetylene Welding and Oxygen Cutting Instruction Course*. New York: Airco, Inc., revised 1966.

Lewis, B., and Von Elbe, G. *Combustion Flames and Explosions of Gases*. New York: Academic Press, Inc., 1961.

Moen, W.B., and Campbell, J. Evaluation of fuels and oxidants for welding and associated processes. *Welding Journal* 34 (9), Sept. 1955, pp. 870-876.

Postman, B.F. Safety in installation and use of welding equipment. *Welding Journal* 35 (10), Oct. 1956, pp. 1021-1025.

Potter, M.H. *Oxyacetylene Welding*. American Technical Society, 1956.

Roper, E.H. An improved method of oxyfuel combustion. *Welding Journal* 34 (4), Apr. 1955, pp. 337-344.

The Oxyacetylene Handbook, 2nd ed. New York: York: Union Carbide Corp., Linde Div., 1960.

11
Brazing

Prepared by a committee consisting of

M.M. SCHWARTZ, *Chairman*
Rohr Industries, Inc.

S. CORICA
General Motors Corporation

G.D. CREMER
International Harvester Co.

W.H. KEARNS
American Welding Society

R.L. PEASLEE
Wall Colmonoy Corporation

M.N. RUOFF
General Electric Company

A.M. SETAPEN
Handy and Harman

G.M. SLAUGHTER
Oak Ridge National Laboratory

11
Brazing

FUNDAMENTALS

DEFINITION AND GENERAL DESCRIPTION

Brazing is a group of welding processes which produces coalescence of materials by heating to a suitable temperature and by using a filler metal having a liquidus above 450°C (840°F) and below the solidus of the base metals. The filler metal is distributed between the closely fitted surfaces of the joint by capillary attraction. Brazing is distinguished from soldering in that soldering employs a filler metal having a liquidus below 450°C (840°F).

Brazing with silver alloy filler metals is sometimes called "silver soldering," a nonpreferred term. Silver brazing filler metals have liquidus temperatures above 450° C (840° F).

Brazing does not include the process known as braze welding. Braze welding is a method of welding which uses a brazing filler metal. However, the filler metal used in braze welding is melted and deposited in grooves and fillets exactly at the points where it is to be used. Capillary action is not a factor in the distribution of the brazing filler metal. Limited base metal fusion may also occur in braze welding.

Brazing must meet each of three criteria:

(1) The parts must be joined without melting the base metals.

(2) The filler metal must have a liquidus temperature above 450°C (840°F).

(3) The filler metal must wet the base metal surfaces and be drawn onto or held in the joint by capillary attraction.

To achieve a good joint using any of the various brazing processes described in this chapter, the parts must be properly cleaned and must be protected by either the flux or atmosphere during the heating process to prevent excessive oxidation. The parts must be designed to afford a capillary for the filler metal when properly aligned, and a heating process must be selected that will provide the proper brazing temperature and heat distribution.

PRINCIPLES OF OPERATION

Capillary flow is the dominant physical principle that assures good brazements providing that both faying surfaces to be joined are wet by the molten filler metal. The joint must also be properly spaced to permit efficient capillary action and resulting coalescence. More specifically, capillarity is a result of surface tension between base metal(s), filler metal, flux or atmosphere, and the contact angle between base metal and filler metal. In actual practice, brazing filler metal flow characteristics are also influenced by dynamic considerations involving fluidity, viscosity, vapor pressure, gravity, and especially by the effects of any metallurgical reactions between the filler metal and the base metal.

The brazed joint, in general, is one of a relatively large area and very small thickness. In the simplest application of the process, the surfaces to be joined are cleaned to remove contaminants and oxide. Next, they are coated with a flux. A flux is a material which is capable of dissolving solid metal oxides still present and also preventing new oxidation. The joint area is then heated until the flux melts and cleans the base metals, which are protected against further oxidation by a layer of liquid flux.

Brazing filler metal is then melted at some point on the surface of the joint area. Capillary attraction between the base metal and the filler metal is much higher than that between the base metal and the flux. Therefore, the flux is displaced by the filler metal. The joint, upon cooling to room temperature, will be filled with solid filler metal, and the solid flux will be found on the joint periphery.

Joints to be brazed are usually made with relatively small clearances of 0.025 to 0.25 mm (0.001 to 0.010 in.). The fluidity of the filler metal, therefore, is an important factor. High fluidity is a desirable characteristic of brazing filler metal since capillary attraction may be insufficient to cause a viscous filler metal to flow into close fitting joints.

Brazing is sometimes done with an active gas, such as hydrogen, or in an inert gas or vacuum. Atmosphere brazing eliminates the necessity for post cleaning and insures absence of corrosive mineral flux residue. Carbon steels, stainless steels, and super alloy components are widely processed in atmospheres of reacted gases, dry hydrogen, dissociated ammonia, argon, and vacuum. Large vacuum furnaces are used to braze zirconium, titanium, stainless steels, and the refractory metals. With good processing procedures, aluminum alloys can also be vacuum furnace brazed with excellent results.

Brazing is an economically attractive process for the production of high strength metallurgical bonds while preserving desired base metal properties.

BRAZING PROCESSES

Brazing processes customarily are designated according to the sources or methods of heating. Those methods currently of industrial significance are

(1) Torch brazing
(2) Furnace brazing
(3) Induction brazing
(4) Resistance brazing
(5) Dip brazing
(6) Infrared brazing

Whatever the process used, the filler metal has a melting point above 450°C (840°F) but below that of the base metal and is distributed in the joint by means of capillary attraction.

TORCH BRAZING

Torch brazing is accomplished by heating with one or more gas torches.[1] Depending upon the temperature and the amount of heat required, the fuel gas (acetylene, propane, city gas, etc.) may be burned with air, compressed air, or oxygen. Manual torch brazing is shown in Fig. 11.1.

Brazing filler metal may be preplaced at the joint in the forms of rings, washers, strips, slugs, powder, etc., or it may be fed from hand-held filler metal, usually in the form of wire or rod. In any case, proper cleaning and fluxing are essential.

For manual torch brazing the torch may be equipped with a single tip, either single or multiple flame. Manual torch brazing is particularly useful on assemblies involving sections of unequal mass. Machine operations can be set up, where the rate of production warrants, using one or more torches equipped with single or multiple flame tips. The machine may be designed to move either the work or the torches, or both. For premixed city gas-air flames, a refractory type burner is used.

---

1. Chapter 10 contains information on gas torches used for welding and brazing.

Fig. 11.1—Manual torch brazing

FURNACE BRAZING

Furnace brazing, as illustrated in Fig. 11.2, is used extensively where the parts to be brazed can be assembled with the filler metal preplaced near or in the joint. This process is particularly applicable for high production brazing. The preplaced brazing filler metal may be in the form of wire, foil, filings, slugs, powder, paste, tape, etc. Fluxing is employed except when an atmosphere is specifically introduced in the furnace to perform the same function. Most of the high production brazing is done in a reducing gas

atmosphere, such as hydrogen and either exothermic or endothermic combusted gas. Pure inert gases, such as argon or helium, are used to obtain special atmospheric properties.

A large volume of furnace brazing is performed in a vacuum, which prevents oxidation and often eliminates the need for flux. Vacuum brazing is widely used in the aerospace and nuclear fields, where reactive metals are joined or where entrapped fluxes would be intolerable. If the vacuum is maintained by continuous pumping, it will remove volatile constituents liberated during brazing. There are several base metals and filler metals that should not be brazed in a vacuum because low boiling point or high vapor pressure constituents may be lost. The types of furnaces generally used are either batch or continuous. These furnaces are usually heated by electrical resistance elements, gas, or oil, and should have automatic time and temperature controls. Cooling is sometimes accomplished by cooling chambers, which either are placed over the hot retort or are an integral part of the furnace design. Forced atmosphere injection is another method of cooling. Parts may be placed in the furnace singly, in batches, or on a continuous conveyor.

Vacuum is a relatively economical method of providing an accurately controlled brazing atmosphere. Vacuum provides the surface cleanliness needed for good wetting and flow of filler metals without the use of fluxes. Base metals containing chromium and silicon can be easily

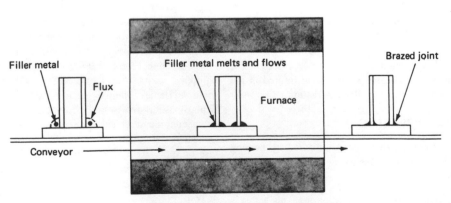

Fig. 11.2—Illustration of furnace brazing operation

vacuum brazed where a very pure, low dew point atmosphere gas would otherwise be required.

INDUCTION BRAZING

The heat necessary for brazing with this process is obtained from an electric current induced in the parts to be brazed, hence the name induction brazing. For induction brazing, the parts are placed in or near a water-cooled coil carrying alternating current. They do not form a part of the electrical circuit. A laboratory setup for induction brazing in vacuum is shown in Fig. 11.3. The brazing filler metal is usually preplaced. Careful design of the joint and the coil setup are necessary to assure that the surfaces of all members of the joint reach the brazing temperature at the same time. Typical coil designs are shown in Fig. 11.4. Flux is employed except when an atmosphere is specifically introduced to perform the same function. The three common sources of high frequency electric current used for induction brazing are the motor-generator, resonant spark gap, and vacuum tube oscillator.

RESISTANCE BRAZING

The heat necessary for resistance brazing is obtained from the resistance to the flow of an electric current through the electrodes and the joint to be brazed. The parts comprising the joint form a part of the electric circuit. The brazing filler metal, in some convenient form, is preplaced or face fed. Fluxing is done with due attention to the conductivity of the fluxes. (Most fluxes are insulators when dry.) Flux is employed except when an atmosphere is specifically introduced to perform the same function. The parts to be brazed are held between two electrodes, and proper pressure and current are applied. The pressure should be maintained until the joint has solidified. In some cases, both electrodes may be located on the same side of the joint with a suitable backing to maintain the required pressure.

Fig. 11.3—Induction brazing in vacuum

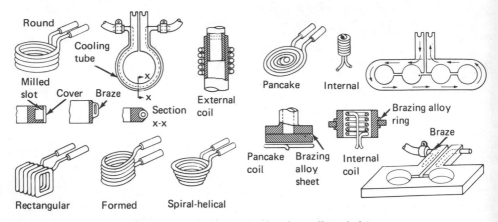

Fig. 11.4—Typical induction brazing coils and plates

The equipment consists of tongs or clamps with the electrodes attached at the end of each arm. The tongs should preferably be water-cooled to avoid overheating. The arms are current carrying conductors attached by leads to a transformer. Direct current may be used but is comparatively expensive. Resistance welding machines are also used. The electrodes may be carbon, graphite, refractory metals, or copper alloys according to the required conductivity.

DIP BRAZING

There are two methods of dip brazing: chemical bath dip brazing and molten metal bath dip brazing. In chemical bath dip brazing the brazing filler metal, in suitable form, is pre-placed and the assembly is immersed in a bath of molten salt, as shown in Fig. 11.5. The salt bath furnishes the heat necessary for brazing and usually provides the necessary protection from oxidation; if not, a suitable flux should be used. The salt bath is contained in a metal or other suitable pot, also called the furnace, which is heated (1) from the outside through the wall of the pot, (2) by means of electrical resistance units placed in the bath, or (3) by the I^2R loss in the bath itself.

In molten metal bath dip brazing the parts are immersed in a bath of molten brazing filler metal contained in a suitable pot. The parts must be cleaned and fluxed if necessary. A cover of flux should be maintained over the molten bath to protect it from oxidation. This method is largely confined to brazing small parts, such as wires or narrow strips of metal. The ends of the wires or parts must be held firmly together when they are removed from the bath until the brazing filler metal has fully solidified.

INFRARED BRAZING

Infrared heat is radiant heat obtained below the red rays in the spectrum. While with every "black" source there is some visible light, the principal heating is done by the invisible radiation. Heat sources (lamps) capable of delivering up to 5000 watts of radiant energy are commercially available. The lamps do not necessarily need to follow the contour of the part to be heated even though the heat input varies inversely as the square of the distance from the source. Reflectors are used to concentrate the heat.

Assemblies to be brazed are supported in a position that enables the energy to impinge on the part. In some applications only the assembly itself is enclosed. There are, however, applications where the assembly and the lamps are placed in a bell jar or retort that can be evacuated, or in which an inert gas atmosphere can be maintained. The assembly is then heated to a controlled temperature, as indicated by thermocouples. Figure 11.6 shows an infrared brazing arrangement. The part is moved to the cooling platens after brazing.

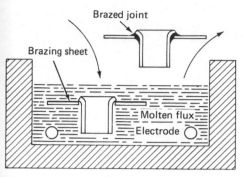

Fig. 11.5—Illustration of chemical bath dip brazing

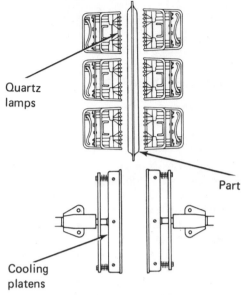

Fig. 11.6—Infrared brazing apparatus

SPECIAL PROCESSES

Blanket brazing is another of the processes used for brazing. A blanket is resistance heated, and most of the heat is transferred to the parts by two methods, conduction and radiation, the latter being responsible for the majority of the heat transfer.

Exothermic brazing is another special process by which the heat required to melt and flow a commercial filler metal is generated by a solid state exothermic chemical reaction. An exothermic chemical reaction is defined as any reaction between two or more reactants in which heat is given off due to the free energy of the system. Nature has provided us with countless numbers of these reactions; however, only the solid state or nearly solid state metal-metal oxide reactions are suitable for use in exothermic brazing units.

Exothermic brazing utilizes simplified tooling and equipment. The process employs the reaction heat in bringing adjoining or nearby metal interfaces to a temperature where preplaced brazing filler metal will melt and wet the metal interface surfaces. The brazing filler metal can be a commercially available one having suitable melting and flow temperatures. The only limitations may be the thickness of the metal that must be heated through and the effects of this heat, or any previous heat treatment, on the metal properties.

BASE METALS

SELECTION OF BASE METAL

In addition to the normal mechanical requirements of the base metal in the brazement, the effect of the brazing cycle on the base metal and the final joint strength must be considered. Cold-work strengthened base metals will be annealed when the brazing process temperature and time are in the annealing range of the base metal being processed. "Hot-cold worked" heat resistant base metals can also be brazed; however, only the annealed physical properties will be available in the brazement. The brazing cycle by its very nature will usually anneal the cold worked base metal unless the brazing temperature is very low and the time at heat is very short. It is not practical to cold work the base metal after the brazing operation.

When a brazement is required to have strength above the annealed properties of the base metal after the brazing operation, a heat treatable base metal should be selected. The base metal can be an oil quench type, an air quench type that can be brazed and hardened in the same or separate operation, or a precipitation hardening type in which the brazing cycle and solution treatment cycle may be combined. Hardened parts may be brazed with a low temperature filler metal using short times at temperature to maintain the mechanical properties.

The strength of the base metal has a profound effect on the strength of the brazed joint. This must be clearly kept in mind when designing the joint for specific properties. Some base metals are also easier to braze than others, particularly by specific brazing processes. For example, a nickel base metal containing high titanium or aluminum additions will present special problems in furnace brazing. Nickel plating is sometimes used as a barrier coating to prevent the oxidation of the titanium or aluminum, and it presents a readily wettable surface to the brazing filler metal.

ALUMINUM AND ALUMINUM ALLOYS

The nonheat treatable wrought aluminum alloys that are brazed most successfully are the ASTM 1XXX and 3XXX series, and low magnesium alloys of the ASTM 5XXX series. The alloys containing a higher magnesium content are more difficult to braze by the usual flux methods because of poor wetting by the filler metal and excessive penetration into the base metal. There are available filler metals that melt below the solidus temperatures of all commercial wrought, nonheat treatable alloys. Of the heat treatable wrought alloys, those most commonly brazed are the ASTM 6XXX series. The ASTM 2XXX and 7XXX series of aluminum alloys are low melting and, therefore, not normally brazeable, with the exception of 7072 and 7005 alloys.

Alloys that have a solidus above 590°C (1100°F) are easily brazed with commercial binary aluminum-silicon filler metals. Stronger, lower melting alloys can be brazed with proper attention to filler metal selection and temperature control, but the brazing cycle must be short to minimize penetration by the molten filler metal. Aluminum sand and permanent mold casting alloys with a high solidus temperature are brazeable; the most commonly brazed are ASTM 443.0, 356.0, and 712.0 alloys. Aluminum die castings are generally not brazed because of blistering from their high gas content.

Table 11.1 lists the common aluminum base metals that can be brazed. The alloys that are generally brazed are 1100, 3003, 6061, 6063, 6951, and 7072. Other aluminum base metals can be brazed, but careful consideration must be given to the specific application and processing requirement.

All commercial filler metals for brazing aluminum alloys are aluminum base. These filler metals are available as wire or shim stock. A convenient method of preplacing filler metal is

Table 11.1 – Nominal composition and melting range of common brazeable aluminum alloys

Commercial designation	ASTM alloy	Brazeability rating[b]	Nominal composition[a]						Approximate melting range	
			Copper	Silicon	Manganese	Magnesium	Zinc	Chromium	°C	°F
EC	EC	A		Al 99.45% min					646-657	1195-1215
1100	1100	A		Al 99% min					643-657	1190-1215
3003	3003	A	—	—	1.2	—	—	—	643-654	1190-1210
3004	3004	B	—	—	1.2	1.0	—	—	629-651	1165-1205
3005	3005	A	0.3	0.6	1.2	0.4	0.25	0.1	638-657	1180-1215
5005	5005	B	—	—	—	0.8	—	—	632-654	1170-1210
5050	5050	B	—	—	—	1.2	—	—	588-649	1090-1200
5052	5052	C	—	—	—	2.5	—	—	593-649	1100-1200
6151	6151	C	—	1.0	—	0.6	—	0.25	643-649	1190-1200
6951	6951	A	0.25	0.35	—	0.65	—	—	615-654	1140-1210
6053	6053	A	—	0.7	—	1.3	—	—	596-651	1105-1205
6061	6061	A	0.25	0.6	—	1.0	—	0.25	593-651	1100-1205
6063	6063	A	—	0.4	—	0.7	—	—	615-651	1140-1205
7005	7005	B	0.1	0.35	0.45	1.4	4.5	0.13	607-646	1125-1195
7072	7072	A	—	—	—	—	1.0	—	607-646	1125-1195
Cast 43	Cast 443.0	A	—	5.0	—	—	—	—	629-632	1065-1170
Cast 356	Cast 356.0	C	—	7.0	—	0.3	—	—	557-613	1035-1135
Cast 406	Cast 406	A		Al 99% min					643-657	1190-1215
Cast A612	Cast A712.0	B	—	—	—	0.7	6.5	←	596-646	1105-1195
Cast C612	Cast C712.0	A	—	—	—	0.35	6.5	—	604-643	1120-1190

a. Percent of alloying elements; aluminum and normal impurities constitute remainder.
b. Brazeability ratings: A = Alloys readily brazed by all commercial methods and procedures.
 B = Alloys that can be brazed by all techniques with a little care.
 C = Alloys that require special care to braze.

to use brazing sheet, which is an aluminum alloy base metal that is coated with a brazing filler metal. Brazing sheet is coated on one or both sides. Core alloys, 3003 and 6951 (a heat treatable alloy), are generally used. A third method of applying brazing filler metal is to use a paste mixture of flux and filler metal powder. Common aluminum brazing metals contain silicon as the melting point depressant with or without additions of zinc, copper, and magnesium.

Aluminum can be brazed by most of the standard practices. Most aluminum brazing is done by the torch, dip, or furnace processes. Furnace brazing may be done in air or controlled atmosphere, including vacuum. Other methods including induction, infrared, and resistance brazing are used for specific applications, usually of the readily brazeable aluminum alloys. Regardless of the process, the temperature must be closely controlled for successful brazing.

Vacuum furnace brazing is being used for aluminum fabrication. Since it is done without flux, the joints are free from corrosion problems commonly associated with residual or entrapped flux. Moreover, brazed assemblies containing inaccessible recesses can be fabricated efficiently. Furnaces operating in 0.0013 Pa (10^{-5} torr) range are used. The success of the operation depends on the use of magnesium vapor as a getter of oxygen on the aluminum surface. New aluminum brazing sheets with magnesium alloyed in the aluminum-silicon coating (BA1Si-6, BA1Si-7, and BA1Si-8) are available, as well as new brazing filler metals. Furnace brazing in vacuum is being used for production of automotive and aircraft heat exchangers.

Additional information on the brazing of aluminum and aluminum alloys is contained in Chapter 69, Section 4 of the *Welding Handbook,* 6th edition,[2] and Chapter 12, *Brazing Manual,* 3rd edition.

MAGNESIUM AND MAGNESIUM ALLOYS

Brazing techniques similar to those used for aluminum are used for magnesium alloys. Furnace, torch, and dip brazing can be employed, although the latter process is the most widely used.

Magnesium alloys that are considered brazeable and recommended filler metals are given in Table 11.2. Furnace and torch brazing experience is limited to M1A alloy. Dip brazing can be used for AZ10A, AZ31B, AZ61A, K1A, M1A, ZE10A, ZK21A, and ZK60A alloys.

The filler metals used for brazing magnesium are also summarized in Table 11.2. BMg-1 and BMg-2a brazing filler metals are suitable for the torch, dip, or furnace brazing processes. The BMg-2a alloy is usually preferred in most brazing applications because of its lower melting range. A zinc base filler metal known as GA432 is an even lower melting composition suitable only for dip brazing use. Additional information on brazing magnesium, including filler metals, cleaning, fluxes, and brazing procedures, is given in Chapter 70, Section 4 of the *Welding Handbook,* 6th edition, and Chapter 13, *Brazing Manual,* 3rd edition.

BERYLLIUM[3]

Brazing is the preferred method for metallurgically joining beryllium. Because of its chemical and metallurgical reactivity, brazing techniques must be highly specialized. Suitable brazing filler metals are few. Brazing alloy

---

2. Section 4 is scheduled for revision in the 7th ed. of the *Welding Handbook.*

3. Beryllium and its compounds are toxic. Proper handling and identification of beryllium metal is required by some state laws and by federal regulations.

systems and brazing temperature ranges include
(1) Zinc: 427-454°C (800-850°F)
(2) Aluminum-silicon: 566-677°C (1050-1250°F)
(3) Silver-copper: 649-904°C (1200-1660°F)
(4) Silver: 882-954°C (1620-1750°F)

Strictly speaking, zinc does not quite meet the AWS definition for a brazing filler metal. Nevertheless, it is generally accepted as the lowest melting filler metal for brazing of beryllium.

Aluminum-silicon filler metals can be used in high strength, wrought beryllium assemblies because brazing is performed well below the base metal recrystallization temperature. BA1Si-4 type filler metal is considered highly satisfactory with fluxes. Fluxless brazing requires stringent processing, but it is being used effectively. A significant advantage of aluminum base filler metals over silver base ones is that metallurgical interaction with the base metal is minimal. This is of prime concern when thin beryllium sections or foils are to be joined.

Silver and silver base brazing alloys find use in structures exposed to elevated temperatures. An added advantage with these alloy systems is that atmosphere brazing is straightforward and may be performed in purified atmospheres or vacuum.

COPPER AND COPPER ALLOYS

Copper and copper alloys are commonly brazed with copper and silver base brazing alloys. The copper alloy base metals include copper-zinc alloys (brass), copper-silicon alloys (silicon bronze), copper-aluminum alloys (aluminum bronze), copper-tin alloys (phosphor bronze), copper-nickel alloys, and several others. The brazing of copper and copper alloys and appropriate filler metals are discussed in detail in Chapter 68, Section 4, *Welding Handbook,* 6th edition, and Chapter 14, *Brazing Manual,* 3rd edition.

LOW CARBON AND LOW ALLOY STEELS

Low carbon and low alloy steels can usually be brazed without difficulty. The low carbon

Table 11.2 – Brazeable magnesium alloys and filler metals

AWS A5.8 classification	ASTM alloy designation	Avail. forms	Solidus °C	Solidus °F	Liquidus °C	Liquidus °F	Brazing range °C	Brazing range °F	Suitable filler BMg-1	Suitable filler BMg-2a
			Base metal							
—	AZ10A	E	632	1170	643	1190	582-616	1080-1140	X	X
—	AZ31B	E, S	566	1050	627	1160	582-593	1080-1100		X
—	K1A	C	649	1200	650	1202	582-616	1080-1140	X	X
—	M1A	E, S	648	1198	650	1202	582-616	1080-1140	X	X
—	ZE10A	S	593	1100	646	1195	582-593	1080-1100		X
—	ZK21A	E	626	1159	642	1187	582-616	1080-1140	X	X
			Filler metal							
BMG-1	AZ92A	W, R	443	830	599	1110	604-616	1120-1140	—	—
BMg-2a	AZ125A	W, R, ST, P	410	770	566	1050	582-610	1080-1130	—	—

E = Extruded shapes and structural sections
S = Sheet and plate
C = Castings
W = Wire
R = Rod
ST = Strip
P = Powder

steels can be brazed very economically with the commonly used brazing processes. These steels are frequently brazed at temperatures in excess of 1080°C (1980°F) with copper filler metal in a controlled atmosphere, or at lower temperatures with silver base filler metals. When brazed with copper, the steels will have lower tensile and yield strength properties and increased ductility as the brazing time or temperature, or both, are increased.

These changes in properties are a result of either decarburization of the steels in some types of atmospheres or changes in their grain size, or both. Original grain size can be restored by subsequent heat treatment below the remelt temperature of the copper braze. Loss of carbon through decarburization is generally unimportant in low carbon steels. However, surface hardness of some low alloy steels may be substantially lowered.

The filler metal should have a solidus well above the heat treating temperature of alloy steels to avoid damage to joints which must be heat treated after brazing. When brazing temperatures exceed the critical temperatures of some low alloy steels, hardening may result if the rate of cooling is rapid enough. The critical cooling rates of these alloys should not be exceeded when annealed properties are required; however, in some cases, air hardening types can be brazed and then hardened by quenching from the brazing temperature. A filler metal with a brazing temperature that is lower than the critical temperature of the steel can be used when no change in the metallurgical properties of the base metal is wanted.

Corrosion resistant brazing filler metals, such as the nickel base series, are sometimes used where the assembly is to be ceramic coated, chromized, or aluminized for corrosion protection, or when the brazement may be subject to corrosive media that would attack copper and silver base filler metals.

HIGH-CARBON AND HIGH-SPEED TOOL STEELS

High carbon steels are those containing more than 0.45 percent carbon. High-carbon tool steels usually contain 0.60-1.40 percent carbon.

The brazing of high carbon steels is best accomplished prior to or at the same time as the hardening operation. The hardening temperatures for carbon steels range from 760-820°C (1400 to 1500°F). Filler metals having brazing temperatures above 820°C (1500°F) should be used. When brazing and hardening are done in one operation, filler metals having a solidus at or below the austenitizing temperature should be used. Localized heating for brazing may decrease the hardness of heat-treated steels when the brazing temperature is above the tempering temperature of the steel.

Alloy tool steels have a wide range of chemical compositions and, therefore, wide differences in behavior when subjected to heat treatment and brazing temperatures. Each of these alloy steels should be studied carefully to determine the proper heat treating cycle, the quench rate and media necessary, the best brazing alloy, and the proper procedures for combining heat treatment with the brazing operation.

The tempering and brazing operations can be combined when brazing some high-speed tool steels and some high-carbon, high-chromium alloy tool steels which have tempering temperatures in the range of 540°C-650°C (1000-1200°F). Filler metals with brazing temperatures in the 590°-650°C (1100-1200°F) range must be used. The part is removed from the tempering furnace, brazed by localized heating methods, and then returned to the furnace for completion of the tempering cycle.

CAST IRONS

The brazing of cast irons generally requires special consideration. The types of cast iron include white, gray, malleable, and ductile. The white cast iron is seldom brazed.

Prior to brazing, the faying surfaces are generally treated by electrochemical surface cleaning, seared with an oxidizing flame, grit blasted, or chemically cleaned. Cast iron is not readily wet by high-temperature brazing filler metals, such as copper, without pretreatment because of the formation of silicon oxides and the presence of graphitic carbon on the surfaces. Where the silicon and graphitic carbon contents are relatively low, the brazing alloys will wet without difficulty. Where the contents are high, wetting is difficult. When low-melting silver brazing alloys are used, the oxidation effect is lower and wetting by the brazing alloy is easier. The metallurgical structure of ductile and malleable cast irons may be damaged if they are heated above 760°C (1400°F). Brazing should be done below this temperature.

When brazing temperatures exceed the critical temperature (transformation to austenite) of the cast iron, the cooling rate should be consistent with normal heat-treating cooling rates to avoid undesirable martensitic or fine pearlitic-cementite structures.

Cast iron with high carbon content has a relatively low melting point. When it is brazed with copper, the brazing temperature should be as low as possible to avoid melting of localized areas of the cast iron, particularly in light sections.

STAINLESS STEELS

Stainless steels include a wide variety of iron base alloys containing chromium which are used primarily for applications requiring heat or corrosion resistance. Success in the fabrication of stainless steel components by brazing depends on (1) knowledge of the characteristics of the various types of stainless steels, and (2) rigid adherence to certain items of process control required by these characteristics.

All of the stainless steel alloys are difficult to wet because of their high chromium content. Brazing of these alloys is best accomplished in a purified (dry) hydrogen atmosphere or in a vacuum. Dew points of −51°C (−60°F) or lower must be maintained because problems with wetting may arise following the formation of chromium oxide. When torch brazing these base metals, fluxes are required to reduce any chromium oxides present.

Most of the silver alloy, copper, and copper-zinc filler metals are used for brazing stainless steels. The nickel-containing silver alloys are generally best for corrosion resistance. Silver brazing filler metals containing phosphorus should not be used on highly stressed parts because brittle nickel and iron phosphides may be formed at the joint interface.

The nickel base filler materials should be used for all applications above 425°C (800°F) to obtain maximum corrosion resistance. The selection of the particular nickel alloy filler metal will depend on the service application and the base metal used. The boron-containing filler metals are generally best for base metals containing titanium or aluminum, or both, because boron has a mild fluxing action which aids in wetting these base metals.

Chromium-Nickel Stainless Steels

These steels are readily brazed using the proper brazing alloys, fluxes, brazing methods, and procedures. Certain precautions must be taken with respect to brazing temperature to avoid subsequent corrosion problems in service. Also, the components must be properly supported during brazing with some filler metals to avoid stress cracking and intergranular attack of the base metal. Brazing of the austenitic chromium-nickel stainless steels is discussed further in Chapter 65, Section 4, *Welding Handbook,* 6th edition, and Chapter 18, *Brazing Manual,* 3rd edition.

Chromium Irons and Steels

The martensitic stainless steels (403, 410, 414, 416, 420, and 431) air harden upon cooling from above their austenitizing temperature range. Therefore, they must be heat treated after brazing or during the brazing operation. These steels are also subject to stress cracking with certain brazing alloys.

The ferritic stainless steels (405, 406, and 430) cannot be hardened and their grain structure cannot be refined by heat treatment. These alloys degrade in properties when held at temperatures above 980°C (1800°F) because of excessive grain growth. Long heating times at temperatures between 340 and 600°C (650 and 1100°F) cause these steels to lose ductility.

Precipitation-Hardening Stainless Steels

These steels are basically stainless steels with additions of one or more of the elements copper, molybdenum, aluminum, and titanium. Such alloying additions make it possible to strengthen the alloys by precipitation hardening heat treatments. When alloys of this type are brazed, the brazing cycle and temperature must match the heat treatment cycle of the alloy. Since heat treatments differ among alloys and some alloys have more than one standard heat treatment, no specific rules can be formulated. Most manufacturers of these special alloys have developed recommended brazing procedures for their particular alloys.

NICKEL AND HIGH NICKEL ALLOYS

Nickel and the high nickel alloys may be divided into commercially pure nickel, nickel-copper alloys, and some nickel-chromium alloys, all of which contain over 50 percent nickel.

The principal precautions to be observed in the brazing of nickel and its alloys relate to their embrittlement by sulfur and low melting metals. such as zinc, lead, bismuth, and antimony. Nickel and nickel alloy parts should be thoroughly cleaned prior to brazing to assure the absence of substances that may contain any of these elements. Sulfur and sulfur compounds must also be excluded from the brazing atmospheres. Nickel and its alloys are subject to stress cracking in the presence of molten brazing filler metals. Parts should, therefore, be annealed prior to brazing to remove residual stresses or carefully stress relieved during the braze cycle.

Nickel and its alloys are commonly joined with silver brazing filler metals. For corrosive environments, alloys containing high silver are preferred. Cadmium-free brazing alloys are preferred where stress corrosion cracking is a problem. The brazing of nickel with copper is similar to the brazing of carbon and low alloy steel except that the copper brazing filler metal will characteristically alloy to a greater extent with nickel than with iron. Alloying during brazing makes capillary flow difficult. Therefore, the assembly should be heated as rapidly as practicable to the brazing temperature

Nickel base brazing filler metals offer the greatest corrosion and oxidation resistance and elevated temperature strength. Because of their similar compositions, the brazing alloys are very compatible with nickel and its alloys.

Dispersion-strengthened nickel and nickel alloys are useful materials for high temperature applications. Submicron size thorium oxide or yttrium oxide particles provide inherent metallurgical stability up to 1300°C (2400°F). The desirable properties of nickel, such as high thermal and electrical conductivity and ease of fabrication, are preserved while the remarkable strength is due to the dispersed phase. Brazing is one of the preferred methods for joining dispersion-strengthened nickel alloys that must function at elevated temperatures. High strength brazements have been made with special nickel base brazing alloys and then tested up to 1300°C (2400°F).

HEAT RESISTANT ALLOYS

Heat resistant alloys are suitable for use under moderate to high loading in the temperature range of 540-1100°C (1000-2000°F). These metals are complex austenitic alloys based on nickel or cobalt, or both. They have often been termed "super alloys." Their greatest use is in the construction of gas turbine engines and hot airframe components.

Heat resistant alloys are generally brazed in hydrogen atmosphere or high vacuum furnaces with nickel base or special filler metals. Since the brazing temperatures are high, the effect of the brazing thermal cycle on the base metals should be taken into account. The nonheat treatable alloys will suffer moderate strength losses due to grain growth during brazing. Cold worked alloys should not be brazed unless the severe loss in strength from annealing during brazing is considered in the design.

The cobalt base alloys are the easiest of the super alloys to braze because most of them do not contain titanium or aluminum. Alloys that are high in titanium or aluminum, or both, are difficult to braze in dry hydrogen because titanium and aluminum oxides are not reduced at brazing temperatures. Parts made of such alloys are usually nickel plated prior to hydrogen furnace brazing, or else they are brazed in a vacuum furnace. It may be desirable to heat treat the brazement after brazing to attain optimum base metal properties.

TITANIUM AND ZIRCONIUM

The common characteristic of these metals is their reactivity. Titanium and zirconium combine very readily with oxygen and react to form brittle intermetallic compounds with many other metals. The reactivity with oxygen requires that these metals be very carefully cleaned before brazing and brazed immediately after cleaning. The reactivity with other metals imposes rather stringent limitations on brazing filler metals. Still another problem arises from the reactivity of these metals with hydrogen and nitrogen. Titanium and zirconium are embrittled by absorption of these gases.

Although considerable use was made of silver and silver base alloys in early brazing of titanium, brittle intermetallic formation and crevice corrosion attack are problems with these brazing alloys. Type 3003 aluminum foil may be used to join thin, lightweight structures, such as complex honeycomb sandwich panels. Service temperature limitation for such brazements is about 260° C (500° F). Another approach is the electroplating of various elements on the base metal faying surfaces which react in situ with the titanium during brazing to form a titanium alloy eutectic. Excellent flow and filleting occur as a transient liquid phase is formed which subsequently solidifies due to interdiffusion.

Other brazing filler metals with high service capability and corrosion resistance include Ti-Zr-Ni-Be, Ti-Zr-Ni-Cu, and Ti-Ni-Cu alloys. These special filler metals are formulated to melt and flow between the titanium alloy beta transus temperatures. The best braze processing is obtained in high vacuum furnaces using closely controlled temperatures in the range of 900-955° C (1650-1750° F).

The development of brazing filler metals for zirconium fabrication has been directed toward alloys that will produce sound, corrosion resistant joints in tubing exposed to pressurized water at elevated temperatures for nuclear reactors. Brazing alloys that are produced by 4 to 5 percent beryllium additions to the zirconium base metal composition are currently under investigation for this application. Other promising brazing alloys are Ni-7% P and Ni-20% Pd-3% In compositions.

Additional information on the brazing of titanium and zirconium is contained in Chapter 73, Section 4 of the *Welding Handbook*, 6th edition, and Chapter 20, *Brazing Manual*, 3rd edition.

CARBIDES AND CERMETS

Carbides of the refractory metals tungsten, titanium, and tantalum that are bonded with cobalt are used for cutting tools and dies. These carbides are joined to metal support structures for cutting tools. Closely related materials called cermets may be used for applications involving high temperature, corrosion, and wear resistance. The cermets are ceramic particles bonded with various metals. Their high-temperature strength is intermediate between the ceramic materials and the binder metals employed. Their greatest disadvantage is their brittleness.

The brazing of carbides and cermets is generally more difficult than is the brazing of metals. The materials that are high in tungsten and tantalum carbide content are relatively easy to braze, but titanium carbide is more difficult to braze because of the stability of titanium oxide formed during heating. Torch, induction, or furnace brazing is used for joining these materials. The brazing of carbides and cermets must be carefully done because of their brittleness and low thermal expansion. When they are brazed to hard metal parts, a sandwich brazing technique is often used. A layer of weak, ductile metal (often pure nickel or pure copper) is interposed between the hard metal and the carbide or cermet. The cooling stresses cause the soft metal to deform instead of cracking the carbide or cermet.

The silver base brazing alloys as well as copper-zinc alloys and copper are often used on carbide tools. Silver alloys containing nickel are preferred because of their improved wettability. The 85%Ag-15%Mn and 85%Cu-15%Mn alloys also work well, as does manganese-nickel alloy. The nickel base alloys containing boron and a 60% Pd-40%Ni alloy may be satisfactory for brazing nickel- and cobalt-bonded cermets of tungsten carbide, titanium carbide, and columbium carbide.

CERAMICS

The increasing use of ceramics in industrial and developmental applications results from their good insulating and high temperature properties. The joining of ceramics to metals is frequently required. Alumina, zirconia, magnesia, forsterite (Mg_2SiO_4), beryllia, and thoria are ceramic materials which can be joined by brazing.

Ceramic materials are inherently difficult to wet with conventional brazing filler metals. Most filler metals merely ball up at the joint with little or no wetting. When bonding does occur, it can be attributed to interlocking particles or penetration into surface pores and voids, while a chemical bond derives strength from material transfer between the filler metal and the base material.

Another basic problem in brazing ceramics is the difference in thermal expansion between the base material and the brazing filler metal itself, or between the ceramic and the metal to which it is joined. In addition, ceramics are poor conductors of heat, and therefore it takes them longer to reach equilibrium temperature than it does the metals. Both of these factors may result in cracking when trying to make such a joint. Since ceramics generally have low tensile and shear strength, crack propagation occurs at relatively low stresses. In addition, the low ductility permits very little distribution of any stresses set up by stress concentration.

In brazing applications where the ceramic is premetallized to facilitate wetting, filler metals such as copper, silver-copper, and gold-nickel are frequently used. Utilizing the affinity of titanium and zirconium for ceramics, highly active titanium or zirconium can be made available at the ceramic-metal interface by hydride decomposition of a powder slurry on the ceramic surface. The action of the titanium or zirconium with the ceramic, or with additional metals placed at the interface, forms an intimate bond. In some cases, the titanium or zirconium is merely painted on the ceramic surface and then placed in contact with a suitable filler metal and base metal to which the joint is to be made.

This process has an advantage over some others in that only one firing operation is required.

Commercial filler metals which are used for brazing nonmetallized ceramics are silver-copper clad or nickel clad titanium wires. Some titanium and zirconium alloys which may prove useful are Ti-Zr-Be, Ti-V-Zr, Zr-V-Cb, Ti-V-Be, and Ti-V-Cr. Most are not commercially available.

PRECIOUS METALS

The precious metals silver, gold, platinum, and palladium are often used industrially for electrical contact materials and for various articles of jewelry or tableware. The brazing of these metals and their alloys presents very few difficulties. The precious metals form only very thin oxide films which are readily removed by fluxes or reducing atmospheres. Some brazing problems do arise in certain electrical contact materials when nonmetallics, such as carbon or refractory oxides, are mixed with the precious metals to prevent contact sticking.

Resistance or furnace brazing is most commonly used for electrical contacts. Carbon or tungsten electrodes are preferred for resistance brazing because of the generally high electrical conductivity of the contacts.

The silver (BAg) and precious metal (BAu) filler metals are most commonly used for brazing of contacts to the holders. They are generally used as brazing preforms; however, contacts are often supplied preclad with the brazing filler metal for convenience in assembly. A filler metal containing phosphorus will wet contact materials containing molybdenum and the silver-cadmium oxide type.

REFRACTORY METALS

Tungsten, molybdenum, tantalum, and columbium all melt above 2000°C (4000°F) and considered to be refractory metals. The technology of brazing the refractory metals is still in the developmental stages with new techniques and brazing materials being reported continually.

Tungsten

Tungsten is one of the several materials

being employed advantageously for components designed to operate at high temperatures. It is produced in pressed and sintered and arc cast forms that are warm worked to wrought product forms. It has the highest melting temperature of any of the known metals. A combination of high melting point and high-temperature strength makes tungsten advantageous for elevated temperature applications. An unusually high ductile-to-brittle transition temperature approximately 260-370°C (500-700°F) prevents the working of tungsten at room temperature. The recrystallized metal exhibits reduced strength and ductility at elevated temperatures. Tungsten can be brazed to itself and to other metals and nonmetals when proper procedures and processes are used. Tungsten alloys are not generally used commercially. Because of its extremely high melting point, low vapor pressure, and its reduced tendency for material transfer under arcing and sparking conditions, tungsten is utilized in electrical contact applications.

Care must be taken when handling tungsten parts during assembly operations because of their inherent brittleness. An excess of nickel base filler metals should be avoided since the interaction between tungsten and nickel results in a recrystallized base metal. Contact with graphite should be avoided to prevent the formation of brittle tungsten carbides. Stresses in tungsten parts, caused by previous forming or welding, should be relieved by heat treatment prior to brazing. In most cases, filler metals and brazing cycles should be selected to provide a minimum interaction between the filler metal and the tungsten to maintain the integrity of the base metal.

Molybdenum

Molybdenum and its alloys are available in pressed and sintered, arc cast, and wrought forms. The former is used for electrical contacts, the other two have application where high strength at elevated temperatures is required. Molybdenum has a high modulus of elasticity and is about as strong as steel. It is half as dense as tungsten and has good thermal conductivity. Like tungsten, it has excellent high temperature properties; however, poor oxidation resistance requires coating protection at high operating temperatures. The presence of minute quantities of oxygen, nitrogen, and carbon lower the ductility of molybdenum. Recrystallized molybdenum has reduced strength and ductility and should be avoided if possible. Molybdenum is brittle at room temperature, but it can be formed at moderately elevated temperatures 150-260°C (300-500°F). The addition of an alloying element raises the recrystallization temperature, allowing greater use of different brazing filler metals.

Molybdenum has application in the electric and electronics industries (e.g., electronic tubes, electrical contacts, supports for transducers and transistors), the aerospace industry (nozzles, leading edges, radiation shields), the nuclear industry (heat exchangers, support grids), and in the chemical, glass, and metallizing industries.

The portion of a brazing cycle above 1090°C (2000°F) should be short to avoid the recrystallization of molybdenum. Another consideration is the formation of intermetallic compounds between molybdenum and filler metals. Recrystallization occurs either by exceeding the recrystallization temperature of molybdenum or lowering the recrystallization temperature by the diffusion of elements, such as nickel, into the molybdenum from the filler metal. Palladium base filler metals and molybdenum base metals with high recrystallization temperatures (Mo-0.5Ti) are used to minimize these conditions. Chromium plate, as a barrier layer, may be used to prevent the formation of intermetallic compounds. Most high-temperature brazing filler metals are suitable for oxidation resistant service. Filler metals with very high melting points are most suitable for coating applications.

Tantalum and Columbium

Tantalum is available in most forms, including sheet, tubing, and rod. It is midway between tungsten and molybdenum in density and melting point. It can be worked easily at room temperature. The thermal conductivity of tantalum is one-fourth that of molybdenum, and its coefficient of expansion is one-third greater. Its elevated temperature strength is low compared to tungsten and molybdenum. Its corrosion resistance is unusually good in most commercial combination of acids, except in hot sulfuric acid solutions. Pure tantalum recrystallizes at approximately 1200°C (2200°F), depending upon the amount of cold work. The addition of alloying elements, such as tungsten (Ta-30%Cb-10%W), raises the recrystallization temperature to approximately 1650°C (3000°F). Other tantalum alloys (Ta-20%Ti-5%Al) have a lower recrystallization temperature of 1010°C (1850°F) attributed to the titanium additions. A few tantalum alloys are available commercially.

Columbium is very similar to tantalum, and several alloys are available in the arc cast and wrought forms. Columbium has the lowest melting point, modulus of elasticity, and thermal conductivity, as well as the highest thermal expansion of the refractory metals. It also has the lowest strength and lowest density. Its high melting point warrants its use at temperatures above the maximum service temperatures of the iron, nickel, and cobalt base metals. It has excellent ductility and fabricability. Pure columbium has a recrystallization temperature range of 980-1090°C (1800-2000°F). However, this temperature increases with the addition of alloying elements. Strengthening mechanisms, such as solid solution and dispersion hardening, are employed in columbium alloys.

Special techniques must be used to satisfactorily braze tantalum and columbium. All gases that have any reactivity must be removed from the brazing atmosphere. These include not only oxygen, but carbon monoxide, ammonia, hydrogen, nitrogen, and carbon dioxide. Tantalum forms oxides, carbides, nitrides, and hydrides very readily with these gases, which contribute to a loss of ductility. At high temperatures, tantalum and columbium must be protected from oxidation. One method is to electroplate the surfaces with copper or nickel. It is therefore necessary for the fillet metal to be compatible with any plating used.

DISSIMILAR METAL COMBINATIONS

Many dissimilar metal combinations may be brazed. In fact, brazing can often be used where metallurgical incompatibility precludes the use of welding processes.

One of the most important criteria when brazing dissimilar metals is a consideration of the difference in thermal expansion between them. If a metal with a high thermal expansion surrounds a low expansion metal, clearances at room temperature which are satisfactory for promotion of capillary flow will be too great at brazing temperature. Conversely, if the low expansion metal surrounds the high expansion metal, no clearance may exist at brazing temperature. For example, when brazing a molybdenum plug in a copper block, the parts must be press fit at room temperature. But, if a copper plug is to be brazed in a molybdenum block, a properly centered loose fit at room temperature is required.

In brazing tube and socket type joints between dissimilar base metals having different coefficients of thermal expansion, optimum joint design minimizes the possibility that residual stresses will cause failure during or after brazing or during service. The tube should be the low expansion metal and the socket the high expansion metal. At brazing temperature, the clearance will be maximum and the capillary will fill with brazing alloy. When the joint cools to room temperature, the brazed joint and the tube will be in compression. In this case, there is a residual tensile circumferential stress in the socket. In other words, the outer tube is compressing the braze metal layer and the tube. Another design is to use a tongue-in-groove joint with the groove in the low expansion material. The fit at room temperature is designed to give capillary joint clearances on both sides of the tongue at brazing temperature. In addition to the assumption that no plastic deformation occurs during heating, these joint designs ignore the longitudinal shear stresses in the braze metal that dictate the use of small overlap distances.

A technique often used when brazing materials with different coefficients of thermal expansion is "sandwich brazing." A common application of this technique is used in the manufacture of carbide-tipped metal cutting tools. A relatively ductile metal is coated on both sides with the brazing filler metal, and the composite is used in the joint. This places a third material in that joint ·which will deform during cooling and reduce the stresses caused by differential contraction of the two parts brazed together.

There are other criteria which must also be considered for successful brazing of dissimilar metals. The filler metal must be compatible with both base metals. Wide differences in base metal softening temperatures must be considered when choosing the filler metal. Where corrosion or oxidation resistance is needed, the filler metal should have properties at least equal to the poorest of the two metals being brazed. In addition, the conditions of the application should avoid the formation of galvanic couples which may promote crevice corrosion in the braze area. Brazing filler metals which form low melting phases with the base metals require more attention to brazing cycle, filler metal place-

ment, quantity of filler metal, and joint design. However, many of the filler metal base metal combinations do possess mutual solubility, and, in special cases, a technique is used to form the final filler metal in situ, for example, a silver plated copper joint brazed at 816°C (1500°F).

The metallurgical reactions which occur during brazing or subsequent thermal treatments between the brazing filler metal and the dissimilar base metals are important. In some cases, they provide beneficial properties while in other cases they may be objectionable. One example is in the brazing of aluminum to copper. The copper reacts with the aluminum to form a low melting brittle compound. Such problems can be overcome by coating one of the base metals with a metal which is compatible with the brazing alloy. In the case of aluminum to copper, the copper can be coated with silver, or a high silver alloy, and then the joint made with a standard aluminum brazing filler metal.

BRAZING FILLER METALS

CHARACTERISTICS

For satisfactory use in brazing applications, brazing filler metals must possess the following properties:

(1) The ability to form brazed joints possessing suitable mechanical and physical properties for the intended service application.

(2) A melting point or melting range compatible with the base metals being joined and sufficient fluidity at brazing temperature to flow and distribute into properly prepared joints by capillary action.

(3) A composition of sufficient homogeneity and stability to minimize separation of constituents (liquation) under the brazing conditions to be encountered.

(4) The ability to wet the surfaces of the base metals being joined and form a strong, sound bond.

(5) Depending on the requirements, ability to produce or avoid base metal-filler metal interactions.

MELTING AND FLUIDITY

Pure metals melt at a constant temperature and are generally very fluid. Binary compositions (two metals) have different characteristics, depending upon the relative contents of the two metals. Figure 11.7 is the equilibrium diagram for the copper-silver binary system. The solidus line, ADCEB, represents the start of the melting of the alloys, while the liquidus line, ACB, represents the temperatures where the alloys are completely liquid. At point C the two lines meet (72 percent silver-28 percent copper) indicating that a particular alloy melts at constant temperature (the eutectic temperature). This alloy is the eutectic composition. This compo-

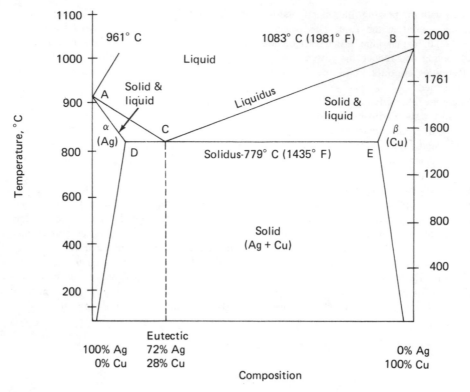

Fig. 11.7 — Copper-silver equilibrium diagram

tion. This composition is essentially as fluid as a pure metal, while the other alloy combinations are mushy between the solidus and liquidus temperatures. The wider the temperature spread, the more sluggish are the alloys with respect to flow in a capillary joint.

The α region is a solid solution of copper in silver, the β region is a solid solution of silver in copper. The central solid zone consists of an intimate mixture of α and β solid solutions. Above the liquidus line the copper and silver atoms are thoroughly interdispersed as a liquid solution.

For versatility, most brazing filler metals contain more than two elements and their phase relations are thus more complex than the above simple binary system. The interaction of elements results in a series of alloys, most of which have a substantial melting range. Rarely is a true

eutectic formed, although some of the binary and higher systems may exhibit more than one eutectic.

There are exceptions to the binary systems discussed above (for example, gold-nickel and gold-copper), where the solid and liquid compositions are very nearly the same; however, they do have separate solidus and liquidus lines and a eutectic type lowest melting temperature and composition. Some systems, such as silver-gold, form only one type of solid solution. The greatest spread between the liquidus and solidus lines generally occurs near the center of the diagram, where the content of the two metals is approximately balanced. They do not have a eutectic alloy composition.

A brazing filler metal will have a liquidus below the melting point of the highest melting metal in the alloy; thus, the alloy is suitable for brazing that metal. There are many possible alloy combinations that have melting points or ranges compatible with the metal or alloy to be joined. Above the liquidus line, all alloy compositions can be as fluid as pure metals, and they will flow into joints by capillary action, where clearance is close and uniform.

LIQUATION

Because the alloy phases of the solid and liquid brazing filler metal generally differ, the composition will undergo gradual change as the temperature increases from the solidus to the liquidus. If the portion that melts first is allowed to flow out, the remaining solid has a higher melting point than the original composition. It may never melt and remain behind as a residue or "skull." Filler metals with narrow melting ranges do not tend to separate, but they flow quite freely in joints with extremely narrow clearance. Rapid heating of filler metals with wide melting ranges, or their application to the joint after the base metal reaches brazing temperature, will minimize separation or liquation. However, liquation cannot be entirely eliminated. Therefore, wide melting range filler metals, which tend to have a more sluggish flow, require wider joint clearances and form larger fillets at the joint extremities.

Some brazing filler metals become sufficiently fluid below the actual liquidus, and satisfactory joints are achieved with them even though the liquidus temperature is not reached. When sluggish behavior is needed, such as filling large joint gaps, brazing is occasionally accomplished within the melting range of the filler metal. Normally, the brazing temperature is usually 10-90°C (50-200°F) above the liquidus of the filler metal. The actual temperature required is influenced by factors such as heating rate, brazing environment (atmosphere or flux), thickness of parts, thermal conductivity of the base metals, and type of joint to be made.

WETTING AND BONDING

To be effective, a brazing filler metal must alloy with the surface of the base metal without undesirable diffusion into the base metal, dilution with the base metal, base metal erosion, or formation of brittle compounds. These effects are dependent upon (1) the mutual solubility between the brazing filler metal and the base metal, (2) the amount of brazing alloy present, and (3) the temperature and the time of the brazing cycle. Some filler metals will diffuse excessively, leading to changes in base metal properties. Diffusion must be controlled by filler metal selection, by a minimum application of filler metal, and by the use of the appropriate brazing cycle. If the filler metal wets the base metal, capillary flow is enhanced. If the filler metal must flow into long capillaries between the metal parts, mutual solubility can change the filler metal composition by alloying. This can raise its liquidus temperature and cause it to solidify before completely filling the joint. Base metal erosion occurs if the base metal and the brazing filler metal are mutually soluble. Sometimes the alloying produces brittle intermetallic compounds that reduce the joint ductility.

Compositions of brazing filler metals are different to allow for the above factors, as well as provide desirable characteristics, such as corrosion resistance to specific media, specific brazing temperatures, or material economies. Thus, the limited alloying ability (wettability) of silver-copper alloys with iron and steel can be enhanced by the addition of zinc or cadmium, or both, to lower the liquidus and solidus temperatures. Tin is added in place of zinc or cadmium when high vapor pressure constituents are objectionable. Similarly, silicon is used in lowering the liquidus and solidus temperatures of aluminum or nickel base brazing alloys. Some brazing filler metals contain elements, such as lithium, phosphorus, or boron, which reduce the surface oxides on the base metal and form compounds with melting temperatures below the brazing temperature. These molten oxides flow out of the joint, leaving a clean metal surface for brazing. These filler metals are essentially self-fluxing. The effectiveness of this action depends on the type and amount of the oxides present, the brazing atmosphere, and the brazing cycle.

FILLER METAL SELECTION

The following factors should be considered when selecting a brazing filler metal:

(1) Compatibility with base metal and joint design.

(2) Service requirements for the brazed assembly. Compositions should be selected to suit operating requirements, such as service temperature (high or cryogenic), thermal cycling, life expectancy, stress loading, corrosive conditions, radiation stability, and vacuum operation.

(3) Brazing temperature required. Low brazing temperatures are usually preferred to economize on heat energy; minimize heat effects on base metal (annealing, grain growth, warpage, etc.); minimize base metal-filler metal interaction; and increase the life of fixtures and other tools. High brazing temperatures are preferred in order to take advantage of a higher melting, but more economical, brazing filler metal; to combine annealing, stress relief, or heat treatment of the base metal with brazing; to permit subsequent processing at elevated temperatures; to promote base metal-filler metal interactions to increase the joint remelt temperature; or to promote removal of certain refractory oxides by vacuum or an atmosphere.

(4) Method of heating. Filler metals with narrow melting ranges—less than 28°C (50°F) between solidus and liquidus—can be used with any heating method, and the brazing filler metal may be preplaced in the joint area in the form of rings, washers, formed wires, shims, powder, or paste. Alternatively, such alloys may be manually or automatically face fed into the joint after the base metal is heated. Filler metals that tend to liquate should be used with heating methods that bring the joint to brazing temperature quickly, or allow the introduction of the brazing filler metal after the base metal reaches the brazing temperature.

To simplify filler metal selection, AWS A5.8, Specification for Brazing Filler Metal, divides filler metals into eight categories and various classifications within each category.[4]

4. For the chemical compositions of the various classifications and their liquidus, solidus, and brazing temperature ranges, refer to the specification.

metals are the commonly used, commercially available ones. Suggested base metal-filler metal combinations are given in Table 11.3. Other brazing filler metals not currently covered by the specification are available for special applications.

ALUMINUM-SILICON FILLER METALS

This group is used for joining the following grades of aluminum and aluminum alloys: 1060, 1100, 1350, 3003, 3004, 5005, 5050, 6053, 6061, 6062, 6063, 6951, and cast alloys A712.0 and C712.0. Brazed assemblies can generally be subjected to continuous service temperatures up to 150°C (300°F). Short time operating temperatures of 200°C (400°F) may be permissible, depending on the operating environment. All types are suited for furnace and dip brazing, while some types are also suited for torch brazing, using lap joints rather than butt joints. Flux should be used in all cases and removed after brazing, except when vacuum brazing. In joints with less than 6.4 mm (1/4 in.) overlap, joint clearances of 0.15 to 0.25 mm (0.006 to 0.010 in.) are commonly used, while clearances up to 0.64 mm (0.025 in.) are used for greater overlaps.

Three of the aluminum filler metals are suitable for vacuum brazing of aluminum. Their magnesium content ranges from 1 to 3 percent. During the vacuum brazing process, magnesium is depleted from these filler metals, causing their solidus temperature to increase and approach the solidus temperature of similar nonvacuum filler metals that are low in magnesium.

A convenient method of supplying aluminum filler metal is to use brazing sheet or tubing that consists of a core of aluminum alloy and a coating of lower melting filler metal. The coatings are aluminum-silicon alloys and may be applied to one or both sides of sheet. Brazing sheet or tubing is frequently used as one member of an assembly with the mating piece made of an unclad brazeable alloy. The coating on the brazing sheet or tubing melts at brazing temperature and flows by capillary attraction and gravity to fill the joints.

MAGNESIUM FILLER METALS

Because of its higher melting range, one magnesium filler metal (BMg-1) is used for joining AZ10A, K1A, and M1A magnesium alloys, while the other alloy (BMg-2a), with a lower melting range, is used for the AZ31B and ZE10A compositions. Both filler metals are suited for torch, dip, or furnace brazing processes. Heating must be closely controlled with both filler metals to prevent melting of the base metal. Joint clearances of 0.10 to 0.25 mm (0.004 to 0.010 in.) are best for most applications. Corrosion resistance is good if the flux is completely removed after brazing. Brazed assemblies are generally suited for service up to 120°C (250°F) continuous service or 150°C (300°F) intermittent service, subject to the usual limitations of the actual operating environment.

COPPER AND COPPER-ZINC FILLER METALS

These brazing filler metals are used for joining various ferrous metals and nonferrous metals. They are commonly used for lap and butt joints with various brazing processes. However, the corrosion resistance of the copper-zinc alloy filler metals is generally inadequate for joining copper, silicon bronze, copper-nickel alloys, or stainless steel.

The essentially pure copper brazing filler metals are used for joining ferrous metals, nickel base, and copper-nickel alloys. They are very free flowing and are often used in furnace brazing with a combusted gas, hydrogen, or dissociated ammonia atmosphere without flux. However, with metals that have constituents with difficult-to-reduce oxides (chromium, manganese, silicon, titanium, vanadium, and aluminum), a higher quality atmosphere or mineral flux may be required. Joint clearances from press-fit to 0.05 mm (0.002 in.) maximum should be used. Operating temperatures range from 200°C (400°F) for continuous service to 480°C (900°F) for short time service, modified to account for any operating environment effects. Copper filler metals are available in wrought and powder forms.

One copper filler metal is supplied as a copper oxide suspension in an inorganic vehicle, although the nominal composition does not include the requirements for the vehicle. The oxide is reduced to copper during brazing. The flow of this filler metal is somewhat more sluggish than the other BCu types, although the applications are similar.

Copper-zinc alloy filler metals are used on most common base metals. A mineral flux is commonly used with the filler metals and accurate temperature control is important to prevent overheating and the formation of voids in the joint by entrapped zinc vapors. Joint clearances from 0.05 to 0.13 mm (0.002 to 0.005 in.) should be used. Service temperatures range up to 200°C (400°F) for continuous service and 316°C (600°F) for intermittent service, depending upon the operating environment.

Copper-zinc filler metals are used on steel, copper, copper alloys, nickel and nickel base alloys, and stainless steel where corrosion resistance is not a requirement. They are used with the torch, furnace, and induction brazing processes. Fluxing is required, and a borax-boric acid flux is commonly used.

COPPER-PHOSPHORUS FILLER METALS

These filler metals are primarily used for joining copper and copper alloys and have some limited use for joining silver, tungsten, and molybdenum. They should not be used on ferrous or nickel base alloys, or on copper-nickel alloys with more than 10 percent nickel. Corrosion resistance is satisfactory except where the joint is exposed to sulfurous atmospheres at elevated temperatures. Brazed assemblies can generally be subjected to continuous service temperatures up to 150° C (300° F). Short times at 200° C (400° F) may be permissible, depending on the operating environment. These filler metals are suited for all brazing processes and have self fluxing properties when used on copper. However, flux is recommended with all other metals, including copper alloys. Lap joints are recommended, but butt joints may be used if requirements are less stringent. These filler metals have a tendency to liquate if heated slowly. However, with proper heating they can produce satisfactory joints at

Table 11.3—Base metal-filler metal combinations

	Al & Al alloys	Mg & Mg alloys	Cu & Cu alloys	Carbon & low alloy steels	Cast iron	Stainless steels	Ni & Ni alloys	Ti & Ti alloys	Be, Zr, & alloys (reactive metals)	W, Mo, Ta, Cb & alloys (refractory metals)	Tool steels
Al & Al alloys	BA1Si										
Mg & Mg alloys	X	BMg									
Cu & Cu alloys	X	X	BAg, BAu, BCuP, RBCuZn								
Carbon & low alloy steels	BA1Si	X	BAg, BAu, RBCuZn	BAg, BAu, BCu, RBCuZn, BNi							
Cast iron	X		BAg, BAu, RBCuZn	BAg, RBCuZn	BAg, RBCuZn, BNi						
Stainless steel	BA1Si	X	BAg, BAu	BAg, BAu, BCu, BNi	BAg, BAu, BCu, BNi	BAg, BAu, BCu, BNi					
Ni & Ni alloys	X	X	BAg, BAu, RBCuZn	BAg, BAu, BCu, RBCuZn, BNi	BAg, BCu, RBCuZn	BAg, BAu, BCu, BNi	BAg, BAu, BCu, BNi				

Table 11.3(continued) – Base metal-filler metal combinations

	Al & Al alloys	Mg & Mg alloys	Cu & Cu alloys	Carbon & low alloy steels	Cast iron	Stainless steels	Ni & Ni alloys	Ti & Ti alloys	Be, Zr, & alloys (reactive metals)	W, Mo, Ta, Cb & alloys (refractory metals)	Tool steels
Ti & Ti alloys	BA1Si	X	BAg	BAg	BAg	BAg	BAg	Y			
Be, Zr, & alloys (reactive metals)	X BA1Si(Be)	X	BAg	BAg, BNi*	BAg, BNi*	BAg, BNi*	BAg, BNi*	Y	Y		
W, Mo, Ta, Cb & alloys (refractory metals)	X	X	BAg	BAg, BCu, BNi*	BAg, BCu, BNi*	BAg, BCu, BNi*	BAg, BCu, BNi*	Y	Y	Y	
Tool steels	X	X	BAg, BAu, RBCuZn, BNi	BAg, BAu, BCu, RBCuZn, BNi	BAg, BAu, RBCuZn, BNi	BAg, BAu, BCu, BNi	BAg, BAu, BCu, RBCuZn, BNi	X	X	X	BAg, BAu, BCu, RBCuZn, BNi

Note: Refer to AWS Specification 5.8 for information on the specific compositions within each classification.
X–Not recommended; however, special techniques may be practicable for certain dissimilar metal combinations.
Y–Generalizations on these combinations cannot be made. Refer to chapters 20 and 21 for usable filler metals.
* –Special brazing filler metals are available and are used successfully for specific metal combinations.

Filler metals:
BA1Si– Aluminum silicon   BCuP–Copper phosphorus
BAg–Silver base   RBCuZn–Copper zinc
BAu–Gold base   BMg–Magnesium base
BCu–Copper   BNi–Nickel base

temperatures below their liquidus. Joint clearances should be 0.03 to 0.13 mm (0.001 to 0.005 in.). The range of clearance depends on the fluidity of the particular alloy.

SILVER FILLER METALS

These filler metals are used for joining most ferrous and nonferrous metals, except aluminum and magnesium, with all methods of heating. They may be preplaced in the joint or fed into the joint area after heating. Lap joints are generally used with joint clearances of 0.05 to 0.13 mm (0.002 to 0.005 in.) when mineral type fluxes are used, and up to 0.05 mm (0.002 in.) when gas phase fluxes (atmospheres) are used. However, butt joints may be used if the service requirements are less stringent. Fluxes are generally required, but fluxless brazing with filler metals free of cadmium and zinc can be done on most metals in an inert or reducing atmosphere (such as dry hydrogen, dry argon, vacuum, combusted fuel gas, etc.).

Copper forms alloys with iron, cobalt, and nickel much more readily than silver does. Also, copper wets many of these metals and their alloys satisfactorily, where silver does not. Consequently the wettability of silver-copper alloys decreases as the silver content increases when brazing steel, stainless steel, nickel-chromium alloys, and other metals. Thus, a high silver-containing filler metal does not wet steel well when brazing is done in air with a flux. When brazing in certain protective atmospheres without flux, silver-copper alloys will wet and flow freely on most steels.

The addition of cadmium to the silver-copper-zinc alloy system dramatically lowers the melting and flow temperatures of the filler metal. Cadmium also increases the fluidity and wetting action of the filler metal on a variety of base metals. Cadmium bearing filler metals should be used with caution. If they are improperly used and subjected to overheating, cadmium oxide fumes can be generated. Cadmium oxide fumes are a health hazard, and excessive inhalation of these fumes must be avoided. Since cadmium bearing filler metals are not intended for fluxless brazing, an appropriate flux should always be used with these filler metals when brazing either in air or furnace atmospheres.

Of the elements that are commonly used to lower the melting and flow temperatures of copper-silver alloys, zinc is by far the most helpful wetting agent when joining alloys based on iron, cobalt, or nickel. Alone, or in combination with cadmium or tin, zinc produces alloys that wet the iron group metals but do not alloy with them to any appreciable depth.

Tin has a low vapor pressure at normal brazing temperatures. It is used in silver brazing filler metals in place of zinc or cadmium when volatile constituents are objectionable, such as when brazing is done without flux in atmosphere or vacuum furnaces, or when the brazed assemblies will be used in high vacuum at elevated temperatures. Tin additions to silver-copper alloys produce filler metals with wide melting ranges. Alloys containing zinc wet ferrous metals more effectively than those containing tin, and where zinc is tolerable, it is preferred to tin.

Generally, as the combined zinc and cadmium contents are increased beyond 40 percent, the ductility of the alloys decreases. This fact puts a practical limit on how much the flow temperatures of silver brazing alloys can be lowered.

Stellites, cemented carbides, and other molybdenum and tungsten rich refractory alloys are difficult to wet with the alloys previously mentioned. Manganese, nickel, and infrequently, cobalt are often added as wetting agents in brazing filler metals for joining these materials. An important characteristic of silver brazing filler metals containing small additions of nickel is improved resistance to corrosion under certain conditions. They are particularly recommended where joints in stainless steel are to be exposed to salt water corrosion.

When stainless steels and other alloys that form refractory oxides are to be brazed in reducing or inert atmospheres without flux, silver brazing filler metals containing lithium as the wetting agent are quite effective. The heat of formation of Li_2O is very high; consequently, lithium is capable of reducing the adherent oxides on the base metal. The resultant lithium oxide is readily displaced by the brazing alloy. Lithium bearing alloys are advantageously used in very pure dry hydrogen or inert atmospheres.

Continuous service temperatures for silver base filler metals range up to 200°C (400°F), with intermittent service up to 315° C (600° F), adjusted for the actual operating environment.

GOLD FILLER METALS

These filler metals are used for (1) joining parts in electron tube assemblies where volatile components are undesirable, and (2) the brazing of iron, nickel, and cobalt base metals where resistance to oxidation or corrosion is required. Because of their low rate of interaction with the base metal, they are commonly used on thin sections, usually with induction, furnace, or resistance heating in a reducing atmosphere or in vacuum without flux. For certain applications, a borax-boric acid flux may be used. These filler metals are generally suited for continuous service at 430° C (800° F) and intermittent service at 540° C (1000° F), depending upon the operating environment. Joint clearances of 0.03 to 0.10 mm (0.001 to 0.004 in.) are used.

NICKEL FILLER METALS

These brazing filler metals are generally used for their corrosion resistance and heat resistant properties up to 980°C (1800°F) continuous service, and 1200°C (2200°F) short time service, depending on the specific filler metals and operating environment. They are generally used on 300 and 400 series stainless steels and nickel and cobalt base alloys. Other base metals such as carbon steel, low alloy steels, and copper are also brazed when specific properties are desired. The filler metals also exhibit satisfactory room temperature and cryogenic temperature properties down to the liquid point of helium. The filler metals are normally applied as powders, pastes, or in the form of sheet or rod with plastic binders.

These filler metals are particularly suited for vacuum systems and vacuum tube applications for operation at elevated temperatures because of their very low vapor pressure. Chromium is the limiting element in vacuum applications for those filler metals in which it is used. When phosphorus is combined with some elements, these compounds have very low vapor pressures and can braze readily in a vacuum of 0.13 Pa (1×10^{-3} torr) at 1066°C (1950°F) without loss of the phosphorus.

The phosphorus containing filler metals exhibit the lowest ductility because of the presence of nickel phosphides. The boron containing filler metals should not be used for brazing thin sections because of their erosive action. The quantity of filler metal and time at brazing temperatures should be controlled because of the high solubility of some base metals in these filler metals.

Best joint quality can be obtained by brazing in an atmosphere which is reducing to both the base metal and the brazing filler metal. A vacuum below 0.67 Pa (5×10^{-3} torr), a pure, dry hydrogen atmosphere with a dew point of $-51°$ C ($-60°$F), or a pure, dry argon atmosphere with a dew point of $-62°$C ($-80°$F) are commonly used.

COBALT FILLER METAL

This filler metal is generally used for its high temperature properties and its compatibility with cobalt base metals. For optimum results, brazing should be performed in a high quality atmosphere. Special high temperature fluxes are available. The brazing technique requires a degree of skill. By using diffusion brazing procedures, the filler metal can be used for service temperature up to 1040°C (1900°F) with excursions to 1150°C (2100°F) and above.

FILLER METALS FOR REFRACTORY METALS

Brazing is an attractive means for fabricating many assemblies of refractory metals, in particular those involving thin sections. The use of brazing to join these materials is somewhat restricted by the lack of filler metals specifically designed for brazing them. Although several references to brazing are present in the literature, the reported filler metals that are suitable for applications involving both high temperature and high corrosion are very limited.

Some of the filler metals and pure metals which may be used to braze refractory metals are given in Table 11.4. The liquidus temperatures of the brazing filler metals range from 650° to 2095°C (1200° to 3800°F). Low melting filler metals, such as silver-copper-zinc, copper-phosphorus, and copper, are used to join tungsten for electrical contact applications. These filler metals are limited in their applications, however, because they cannot operate at very high temperatures. The use of higher melting metals, such as tantalum and columbium, is warranted in those

cases. Nickel base and precious-metal base filler metals may be used for joining tungsten.

A wide variety of brazing filler metals may be used to join molybdenum. The brazing temperature range is the same as that for tungsten. Each filler metal should be evaluated for its particular applicability. The service temperature requirement in many cases dictates the brazing filler metal selection. However, consideration must be given to the effect of brazing temperature on the base metal properties, specifically recrystallization. When brazing above the recrystallization temperature, time should be kept as short as possible. When high temperature service is not required, copper and silver base filler metals may be used. For electronic parts and other non-

Table 11.4—Brazing filler metals for refractory metals[a]

Brazing filler metal	Liquidus temperature		Brazing filler metal	Liquidus temperature	
	°C	°F		°C	°F
Cb	2416	4380	Mn-Ni-Co	1021	1870
Ta	2997	5425			
Ag	960	1760	Co-Cr-Si-Ni	1899	3450
Cu	1082	1980	Co-Cr-W-Ni	1427	2600
Ni	1454	2650	Mo-Ru	1899	3450
Ti	1816	3300	Mo-B	1899	3450
Pd-Mo	1571	2860	Cu-Mn	871	1600
Pt-Mo	1774	3225	Cb-Ni	1190	2175
Pt-30W	2299	4170			
Pt-50Rh	2049	3720	Pd-Ag-Mo	1306	2400
			Pd-Al	1177	2150
Ag-Cu-Zn-Cd-Mo	619-701	1145-1295	Pd-Ni	1205	2200
Ag-Cu-Zn-Mo	718-788	1325-1450	Pd-Cu	1205	2200
Ag-Cu-Mo	780	1435	Pd-Ag	1306	2400
Ag-Mn	971	1780	Pd-Fe	1306	2400
			Au-Cu	885	1625
Ni-Cr-B	1066	1950	Au-Ni	949	1740
Ni-Cr-Fe-Si-C	1066	1950	Au-Ni-Cr	1038	1900
Ni-Cr-Mo-Mn-Si	1149	2100	Ta-Ti-Zr	2094	3800
Ni-Ti	1288	2350			
Ni-Cr-Mo-Fe-W	1305	2380	Ti-V-Cr-Al	1649	3000
Ni-Cu	1349	2460	Ti-Cr	1481	2700
Ni-Cr-Fe	1427	2600	Ti-Si	1427	2600
Ni-Cr-Si	1121	2050	Ti-Zr-Be[b]	999	1830
			Zr-Cb-Be[b]	1049	1920
			Ti-V-Be[b]	1249	2280
			Ta-V-Cb[b]	1816-1927	3300-3500
			Ta-V-Ti[b]	1760-1843	3200-3350

a. Not all the filler metals listed are commercially available.
b. Depends on the specific composition.

structural applications requiring higher temperatures, gold-copper, gold-nickel, and copper-nickel filler metals can be used. Higher melting metals and alloys may be used as brazing filler metals at still higher temperatures.

A number of refractory or reactive-metal base brazing filler metals have been specifically developed for joining columbium and tantalum and their alloys. The metal systems Ti-Zr-Be and Zr-Cb-Be are typical. Platinum, palladium, platinum-iridium, platinum-rhodium, and titanium have also been used. Nickel base filler metals (such as nickel-chromium-silicon alloys) have been used to braze tantalum and columbium. These filler metals are satisfactory for service temperatures below 980°C (1800°F). Copper-gold alloys containing less than 40 percent gold can also be used as filler metals, but gold in amounts between 46 and 90 percent tends to form age hardening compounds which are brittle. Although silver base filler metals have been used to join tantalum and columbium, they are not recommended because of a tendency to embrittle the base metals.

FLUXES AND ATMOSPHERES

Metals and alloys tend to react with various constituents of the atmosphere to which they are exposed. This tendency increases as the temperature is raised. The most common reaction is oxidation, but nitrides and carbides are sometimes formed. These reactions result in conditions which hinder the production of consistently sound brazed joints.

Fluxes, gas atmospheres, and vacuum are used to exclude reactants and thus prevent undesirable reactions during brazing. Under some conditions, fluxes and atmospheres may also reduce oxides that are present. Caution must be observed in the use of atmospheres because some materials are embrittled by various gases. Notable among these are titanium, zirconium, columbium (niobium), and tantalum which become permanently embrittled when brazed in any atmosphere containing hydrogen, oxygen, or nitrogen. Also the hydrogen embrittlement of copper that has not been thoroughly deoxidized, must be avoided.

The use of any flux or atmosphere does not eliminate the need for thorough cleaning of parts prior to brazing. Recommended cleaning procedures for various metals are contained in Chapter 7 of the AWS *Brazing Manual,* 3rd edition, 1976.

FLUXES

To effectively protect the surfaces to be brazed, the flux must completely cover and protect them until the brazing temperature is reached. It must remain active throughout the brazing cycle. Since the molten filler metal should displace the flux from the joint at the brazing temperature, the viscosity and surface tension of the flux and the interfacial energy between the flux and the surfaces of parts are important. Therefore, recommended fluxes should be used in their proper temperature ranges and on the materials for which they are designed.

Certain brazing filler metals contain alloy additions of deoxidizers, such as phosphorus, lithium, and other elements which have strong affinities for oxygen. In some instances, these additions make the filler metals self-fluxing without the application of prepared fluxes or controlled atmospheres. In other cases, they are used in conjunction with protective atmospheres or fluxes to increase wetting tendencies.

Constituents of Fluxes

In general, the ingredients of brazing fluxes are chlorides, fluorides, fluoborates, borax, borates, boric acid, wetting agents, and water. Most brazing fluxes are proprietary mixtures of several of the above ingredients. They are mixed and reacted in ways that give satisfactory results for specific purposes. For a discussion of the functions of the individual flux ingredients, see Chapter 4, AWS *Brazing Manual,* 3rd edition.

Grouping and Selection of Fluxes

There is no single flux which is best for all brazing applications. Fluxes are classified into six groups according to their performance on certain groups of base metals in rather specific temperature ranges. These groupings are listed in Table 11.5, which also lists the appropriate base materials and temperature ranges for each flux.

Table 11.5 is not a substitute for a thorough evaluation for an optimum flux for a specific high production joint. For successful use, a flux must be chemically compatible with all the base metals and filler metals involved in the brazement. It must be active throughout the brazing temperature range and the time at brazing temperature. If the brazing cycle is long, a less active and longer-lived flux should be used. Conversely, if the cycle is short, a more active flux with a shorter effective life may be used. Where more than one flux is suitable for the application, other considerations, such as safety and cost, should be evaluated.

Within a particular AWS type, there are several criteria for choosing a specific flux for maximum efficiency:

(1) For dip brazing, water (including water of hydration) must be avoided.

(2) For resistance brazing, the flux must permit the passage of current. This usually requires a wet, dilute flux.

(3) Ease of flux residue removal should be considered.

(4) Corrosive action on the base and filler metals should be minimal.

Application of Fluxes

Fluxes for brazing are generally available in the form of powder, paste, or liquid. The form selected depends upon the individual work re-quirements, the brazing process, and the brazing procedure. Fluxes are most commonly applied in paste form because of the ease of application to small parts and their adherence in any position. The particle size of paste or dry flux should be uniform and small for the most effective application. It is frequently helpful to heat the paste slightly before application.

Powdered flux may be applied to the joint as follows: (1) dry, (2) mixed with water and alcohol as a paste, or (3) in torch brazing, by dipping the heated filler rod into the flux as needed. Liquid fluxes, where the fluxing ingredients are completely in solution, may be sprayed on the joint or entrained in the fuel gas. Liquid fluxes are sometimes used in torch brazing. Mixtures of powdered filler metal and flux are sometimes used where it is desirable to have both preplaced.

When flux is employed in the brazing of a joint, the following points should be observed:

(1) After degreasing and prior to flux application, the parts should be handled as little as possible.

(2) Powdered flux should be kept dry prior to application.

(3) When a portion of flux is removed from a container, the remainder of the flux should be kept in an uncontaminated condition. If there is any question regarding the cleanliness of the flux, it is more economical to discard it and open a new container.

Special considerations are involved where fluxes surround and blanket the entire part during dip brazing. Safety in dip brazing is obtained by using fluxes from which all moisture (including water of crystallization) has been removed. During the brazing operation, the flux should have the proper viscosity for dipping. It must be checked continuously to ascertain that the proper analysis is maintained during the brazing cycle.

Flux Removal

Residual fluxes should be removed completely and quickly after brazing is completed. Fluxes may cause damaging corrosion in the joint if they are not completely removed. The

Table 11.5—Fluxes for brazing

AWS brazing flux type no.	Recommended base metals	Recommended filler metals[c]	Recommended useful temp. range	Ingredients	Forms supplied
1	All brazeable aluminum alloys	BAlSi	700-1190F 371- 643C	Chlorides Fluorides	Powder
2	All brazeable magnesium alloys	BMg	900-1200F 482- 649C	Chlorides Fluorides	Powder
3A	All except those listed under 1, 2, and 4[a]	BCuP,BAg	1050-1600F 566- 871C	Boric acid Borates Fluorides Fluoborates Wetting agent	Powder Paste Liquid
3B	All except those listed under 1, 2, and 4	BCu, BCuP, BAg, BAu, RBCuZn, BNi	1350-2100F 732-1149C	Boric acid Borates Fluorides Fluoborates Wetting agent	Powder Paste Liquid
4	Aluminum bronze, aluminum brass, and iron or nickel base alloys containing minor amounts of Al or Ti, or both[b]	BAg (all) BCuP (Copper base alloys only)	1050-1600F 566- 871C	Chlorides Fluorides Borates Wetting agent	Powder Paste
5	All except those listed under 1, 2, and 4	Same as 3B (excluding BAg-1 through -7)	1400-2200F 760-1204C	Borax Boric acid Borates Wetting agent	Powder Paste Liquid

Note: This table provides a guide for classification of most of the proprietary fluxes available commercially. When used alone, the information given here is generally not adequate for a specific application.

a. Some Type 3A fluxes are specifically recommended for base metals listed under Type 4.
b. In some cases, Type 1 flux may be used on base metals listed under Type 4.
c. See Table 11.3 for filler metals designated by symbols listed.

joint design and brazing procedure should permit all of the flux inside the joint to be displaced by the molten filler metal, leaving only the exposed outer surfaces of the joint to be freed from residual flux after brazing. The residue of flux after brazing is usually in a brittle, glassy condition. Brazed parts can usually be freed from residual flux by washing in hot water and then air drying.

If a moderate thermal shock will not impair the properties of the brazed joint, a rapid immersion of the hot parts in cold water is very effective in removing residual flux. Thermal shock will tend to crack off the flux residue.

Tenaciously adhering flux particles may be removed either by dipping the parts in one of several proprietary chemical dips or by mechanical means such as fiber brushing, wire brushing, shot blasting, or chipping. The method used should be compatible with the required properties of the base metals. If, for example, stainless steel parts are cleaned with a wire brush, a stainless steel one should be used. Soft metals, such as aluminum and copper, should be cleaned by methods which will not damage or roughen their surfaces. Cleaning with nitric acid is not recommended where brazing filler metals containing silver or copper are used. Parts should be thoroughly washed and dried after a chemical dip.

CONTROLLED ATMOSPHERES

Controlled atmospheres and vacuum operations are employed to prevent the formation of oxides during brazing and, in many instances, to remove the oxide films present on metals so that the filler metal can wet and flow. The reactions resulting from the use of gas atmospheres and vacuum atmospheres are diverse. Certain conditions, however, apply to both.

Controlled atmospheres are most commonly used in furnace or retort brazing operations. However, they may also be used with induction and resistance brazing. Inert atmospheres are useful in induction brazing, especially where titanium, zirconium, and refractory metals are concerned.

The general techniques of atmosphere brazing can involve

(1) Gaseous atmospheres alone
(2) Gaseous atmosphere together with solid or liquid fluxes preplaced at the interfaces
(3) High vacuums
(4) Combinations of vacuum and gas atmospheres

Controlled atmospheres, where additional solid or liquid fluxes are not required, have the following advantages:

(1) The entire part is maintained in a clean, unoxidized condition throughout the brazing cycle. Parts may, therefore, frequently be machined to finished size prior to brazing.

(2) Usually no postbraze cleaning operation is necessary.

(3) Intricate sealed parts from which fluxes cannot be removed, such as electronic tubes, can be brazed satisfactorily.

(4) Large surface areas can be brazed integrally and continuously without danger of brittle, entrapped flux pockets at the interfaces.

Gas Atmospheres

The principle followed in the use of controlled gas atmospheres involves the preparation of a special protective gas and its introduction into the furnace or brazing retort at pressures above atmospheric. As the gas is continuously supplied to the furnace and circulated through it, the furnace becomes purged of air. The protective gas atmosphere is maintained at slight pressure, which prevents air from seeping into the brazing retort or furnace. In some operations, work is placed in a cold retort or furnace prior to purging, and the retort or furnace is not opened until the brazing cycle is completed. Where parts must be fed continuously or periodically into a furnace which is at brazing temperature, gas curtains or intermediate chambers are provided to avoid contamination of the furnace atmosphere.

The ability to control the composition and therefore the effectiveness of a furnace atmosphere depends not only on the condition and proper operation of the atmosphere-producing equipment but also on the proper setup and operation of the furnace being used.

When certain types of controlled atmospheres, such as those containing hydrogen, are employed, extreme care must be taken to prevent the formation of explosive mixtures of gas. Mixtures of hydrogen with air, ranging from 4 to 75 percent hydrogen, are explosive.

As a safety precaution when potentially explosive gas atmospheres are used, the furnace or retort should be thoroughly purged with the gas to ensure the removal of all air before heat is applied. Waste gases from the furnace can be either continuously burned or directed into the open air outside the building.

Some atmospheres, such as those containing carbon monoxide, are toxic. Proper burning off or disposal of the waste gases from these atmospheres is especially important for safety. When brazing toxic metals, such as beryllium, waste gases should be carefully filtered or piped to an outside area. (See American National Standard Z49.1.)

Compositions. The compositions of controlled atmospheres recommended for brazing cover a wide range, some of which are tabulated in Table 11.6. These data are not intended as a comprehensive tabulation of atmosphere-metal combinations but rather as a general outline of some of the more widely used combinations. Competent metallurgical advice should be sought when unusual metals or more complicated dissimilar-metal combinations are to be brazed because of the varied effects of atmospheres and heat treatments on the mechanical and metallurgical characteristics of the metals.

Dew Point Control. The combustion of gas mixtures results in a controlled atmosphere containing entrained moisture which is largely undesirable in brazing. The moisture can sometimes be removed by condensation. The use of certain brazing metals, however, requires cooling in conjunction with absorption type dryers to reduce the dew point to satisfactory levels. Accurate dew point control is especially important when dry hydrogen atmospheres are required because of the sensitivity to moisture of the metals usually brazed in this type of atmosphere. Dissociated ammonia atmospheres do not always require such accurate control.

Figure 11.8 is a graphical presentation of 21 metal-metal oxide equilibria in pure hydrogen atmosphere. For any combination of dew point and temperature lying above or left of the curve, the oxide tends to form. Below or right of the curve, the oxide tends to be reduced to the metallic state by the hydrogen. Thus, more water vapor may be tolerated at higher temperatures in hydrogen. When the metal is alloyed, the element having the most stable oxide is the governing curve. For example, chromium is the governing element for chromium stainless steels. Accurate control of dew point cannot be overemphasized in gas atmosphere brazing if sound and completely bonded joints are to be produced in metals which form oxides of high stability.

Oxides of aluminum, titanium, beryllium, and magnesium cannot be reduced by hydrogen at ordinary brazing temperatures. If these elements are present in small amounts, satisfactory brazing can be done in gas atmospheres. When these elements are present in quantities exceeding 1 or 2 percent, the metal surface should be plated with a pure metal that is easily cleaned by hydrogen, or a flux should be used in addition to the hydrogen.

The components of brazing atmospheres shown in Table 11.6 have individual characteristics which affect their suitability for brazing various metals and alloys. These components and the precautions to be taken in their use are discussed individually.

Carbon Dioxide (CO_2). Carbon dioxide is an inert constituent of some brazing atmospheres, except when it is decomposed to carbon monoxide or carbon and oxygen, all of which are reactive with metals. It may be used in certain applications to dilute an atmosphere provided the proper ratio of CO_2 to carbon monoxide (CO) is maintained. However, in applications such as the brazing of carbon steels it must be removed from the atmosphere to avoid oxidation and decarburization of the metal surfaces.

Carbon Monoxide (CO). Carbon monoxide is an active reducing agent, but it is not as active as hydrogen. It is useful for reducing oxides of copper, nickel, iron, and cobalt at elevated temperatures; and it will carburize carbon and alloy steels under certain conditions. When decomposed, it may release oxygen, which is undesirable in these controlled atmospheres. Carbon monoxide is toxic and unless the waste gases are burned or otherwise disposed of, adequate ventilation should be provided in the area where it is used.

Nitrogen (N_2). Nitrogen serves as a diluent in some brazing atmospheres and may be used alone as a protective atmosphere in some applications. In the nascent condition (N), it tends to form nitrides on certain metal surfaces that inhibit the wetting action of the filler metal on the base metal surfaces, resulting in weak and incomplete bonds.

Hydrogen (H_2). As described previously, pure hydrogen is one of the most active agents for reducing a wide range of metal oxides at elevated temperatures. Dew point control is especially necessary in dry hydrogen brazing and should be carefully controlled throughout the brazing cycle.

Caution should be exercised in considering hydrogen as a protective brazing atmosphere, because it may cause embrittlement in such metals as copper, titanium, zirconium, columbium, and tantalum. Also, it is explosive in concentrations of 4 percent or more in air. Waste gas should be burned or exhausted out-of-doors.

Table 11.6—Atmospheres for brazing

AWS brazing atmosphere type number	Source	Maximum dew point incoming gas	Approximate composition, %				Filler metals[e]	Application		Remarks
			H_2	N_2	CO	CO_2		Base metals		
1	Combusted fuel gas (low hydrogen)	Room temp.	5-1	87	5-1	11-12	BAg,[a] BCuP, RBCuZn[a]	Copper, brass[a]		
2	Combusted fuel gas (decarburizing)	Room temp.	14-15	70-71	9-10	5-6	BCu, BAg,[a] RBCuZn,[a] BCuP	Copper,[b] brass,[a] low-carbon steel, nickel, monel, medium carbon steel[c]		Decarburizes
3	Combusted fuel gas, dried	−40° C (−40° F)	15-16	73-75	10-11		Same as 2	Same as 2 plus medium and high-carbon steels, monel, nickel alloys		
4	Combusted fuel gas, dried (carburizing)	−40° C (−40° F)	38-40	41-45	17-19		Same as 2	Same as 2 plus medium and high-carbon steels		Carburizes
5	Dissociated ammonia	−54° C (−65° F)	75	25			BAg,[a] BCuP, RBCuZn,[a] BCu, BNi	Same as for 1, 2, 3, 4 plus alloys containing chromium[d]		
6	Cylinder hydrogen	Room temp.	97-100				Same as 2	Same as 2		Decarburizes
7	Deoxygenated and dried hydrogen	−59° C (−75° F)	100				Same as 5	Same as 5 plus cobalt, chromium, tungsten alloys and carbides[d]		
8	Heated volatile materials	Inorganic vapors (i.e., zinc, cadmium, lithium, volatile fluorides)					BAg	Brasses		Special purpose. May be used in conjunction with 1 thru 7 to avoid use of flux
9	Purified inert gas	Inert gas (e.g., helium, argon, etc.)					Same as 5	Same as 5 plus titanium, zirconium, hafnium		Special purpose. Parts must be very clean and atmosphere must be pure

Table 11.6 (continued)—Atmospheres for brazing

AWS brazing atmo- sphere type number	Source	Pressure	Approximate composition, %				Filler metals[e]	Application	
			H₂	N₂	CO	CO₂		Base metals	Remarks
10	Vacuum	Vacuum above 2 torr[f]					BCuP, BAg	Cu	
10A	Vacuum	0.5 to 2 torr					BCu, BAg	Low carbon steel, Cu	
10B	Vacuum	0.001 to 0.5 torr					BCu, BAg	Carbon and low alloy steels, Cu	
10C	Vacuum	1 × 10⁻³ torr and lower					BNi, BAu, BAlSi, Ti alloys	Ht. and corr. resisting steels, Al, Ti, Zr, refractory metals	

Note: AWS Types 6, 7, and 9 include reduced pressures down to 2 torr.

a. Flux is required in addition to atmosphere when alloys containing volatile components are used.
b. Copper should be fully deoxidized or oxygen free.
c. Heating time should be kept to a minimum to avoid objectionable decarburization.
c. Flux must be used in addition if appreciable quantities of aluminum, titanium, silicon, or beryllium are present.
e. See Table 11.3 for explanation of filler metals.
f. 1 Torr=133 Pa.

Partial pressure of water vapor
(May be read as vacuum furnace pressure)

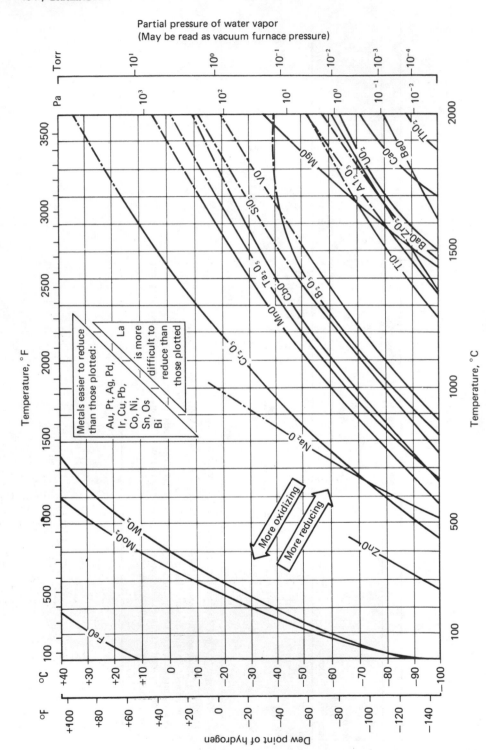

Fig. 11.8—Metal-metal oxide equilibria in pure hydrogen atmosphere

Inert Gases. Argon and helium are inert to metals. These gases may be used in applications where the base metals or the filler metals contain constituents which will volatilize in vacuum at brazing temperatures.

Methane. Methane may come from generated atmosphere gases or organic materials left on the metal surfaces from improper cleaning. Methane is sometimes added to certain atmospheres to balance decarburizing gases present.

Vacuum

Brazing in a furnace or chamber evacuated of air is an economical method of joining many similar and dissimilar base metal combinations. Flux is not used.

Vacuum conditions are especially suited for brazing very large, continuous areas where (1) solid or liquid fluxes cannot be removed adequately from the interfaces during brazing, and (2) gaseous atmospheres are not completely efficient because of their inability to purge occluded gases evolved at close-fitting brazing interfaces. Vacuum is also suitable for brazing many similar and dissimilar base metals, including titanium, zirconium, columbium, molybdenum, and tantalum. The characteristics of these metals are such that even very small quantities of atmospheric gases may result in embrittlement and sometimes disintegration at brazing temperatures. These metals and their alloys may also be brazed in inert-gas atmospheres if the gases are of sufficiently high purity to avoid contamination and the resultant loss in properties of the metals. It is interesting to note, however, that a vacuum system evacuated to 0.0013 Pa (10^{-5} torr) contains only 0.000001 percent residual gases. A typical commercial vacuum furnace is illustrated in Fig. 11.9.

Vacuum brazing has the following advantages and disadvantages compared with other high purity brazing atmospheres:

(1) Vacuum removes essentially all gases from the brazing area, thereby eliminating the necessity of purifying a supplied atmosphere. Commercial vacuum brazing is generally done at pressure varying from 0.065 to 65 Pa (0.5 to 500 millitorr) and above. The actual pressures used depend upon the materials brazed, the filler metals being used, the area of the brazing interfaces, and the degree to which gases are expelled from the base metals during the brazing cycle.

(2) Certain oxides of base metals will dissociate in vacuum at brazing temperatures. Vacuum is used widely to braze stainless steel, superalloys, aluminum alloys, and refractory materials with special techniques.

(3) Difficulties sometimes experienced with contamination of brazing interfaces due to base metal expulsion of gases are negligible in vacuum brazing. Occluded gases are removed from the interfaces immediately upon evolution from the metals.

(4) The low pressure existing around the base and filler metals at elevated temperature removes volatile impurities and gases from the metals. Frequently the properties of the base metals themselves are improved. This characteristic is also a disadvantage where the filler, base metals, or elements of them volatilize at brazing temperatures because of the low surrounding pressures. This tendency may, however, be corrected by proper vacuum brazing techniques.

There are two general types of vacuum brazing: (1) brazing in a high vacuum, and (2) brazing in a partial vacuum. High vacuum is particularly suited for brazing base metals containing hard-to-dissociate oxides.

Partial vacuums are used where the base metal or filler metals, or both, volatilize at their brazing temperatures under high-vacuum conditions. The lowest pressure at which the metals will remain in the solid or liquid phases at the brazing temperatures is determined by calculations or experimentation. The brazing chamber is evacuated to high-vacuum conditions. The heating cycle proceeds under high vacuum until just below the temperature where vaporization would begin. High-purity argon, helium, or in some instances, hydrogen is gradually introduced in sufficient amounts to overcome the vapor pressure of the volatile metals at brazing temperature. This technique appre-

Fig. 11.9—A typical top-loading high vacuum brazing and heat treating furnace

ciably widens the range of materials for which vacuum brazing is effective.

Vacuum purging prior to high-purity dry-hydrogen brazing is frequently employed where extra precautions must be taken to ensure optimum freedom of the atmosphere from even small amounts of foreign or contaminating gases. Similarly, dry-hydrogen or inert gas purging prior to evacuation is sometimes helpful in obtaining improved brazing results in a high vacuum atmosphere.

Zirconium, titanium, and other elements with high affinities for oxygen and other gases are sometimes strategically placed close to, but not in contact with, the part being brazed in a high vacuum atmosphere. These so-called "getters" rapidly absorb very small quantities of oxygen, nitrogen, and other occluded gases which may be evolved from the metals being brazed and thus improve the quality of the brazing atmosphere.

JOINT DESIGN

Many variables are to be considered in manufacturing a reliable mechanical design of the assembly. From the mechanical standpoint, the design of a brazement is no different from the design of any other part. The rules applying to concentrated loads, stress risers, stress concentration, static loading, dynamic loading, etc., that apply to machined or other fabricated parts also apply to the brazed components.

The design of a brazed joint does have specific requirements that must be met when adequate operating characteristics are to be achieved. Some of the more important factors are

(1) Composition of the base metal and filler metal (members to be brazed may be of similar or dissimilar materials)

(2) Type and design of joint

(3) Service requirements: mechanical performance, electrical conductivity, pressure tightness, corrosion resistance, and service temperature

Reference should be made to the appropriate sections in this chapter for information on filler metals and base metals.

TYPES OF JOINTS

Several factors influence the selection of the type of joint to be used. These factors

include the brazing process to be used, fabrication techniques prior to brazing, the number of items to be brazed, the method of applying filler metal, and the ultimate service requirement of the joint.

The unit strength of the brazing filler metal may occasionally be higher than that of the base metal. In general, however, the strength of the filler metal is considered lower. The joint strength will vary according to the joint clearance, degree of filler metal-base metal interaction (diffusion and solution), and the presence of defects in the brazement.

There are basically two types of joints used in brazing designs: (1) the lap joint, and (2) the butt joint. These joints are shown in Fig. 11.10.

In the lap joint, the area of overlap may be set so that the joint will be as strong as the weaker member, even when using a low strength filler metal or with the presence of small defects in the joint. An overlap at least three times the thickness of the thinner member will usually yield maximum joint efficiency. Lap joints are generally used because they offer the best joint efficiency and ease of fabrication. They do have the disadvantage of increasing the metal thickness at the joint and of creating a stress concentration at the edges of the lap where there is an abrupt change in cross section.

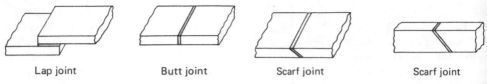

Lap joint   Butt joint   Scarf joint   Scarf joint

Fig. 11.10—Basic lap and butt joints for brazing   *Fig. 11.11—Typical scarf joint designs*

Butt joints are used where the lap joint thickness would be objectionable, and where the strength of the brazed joint will satisfactorily meet the service requirements. The strength of a properly brazed butt joint may be sufficiently high so that failure will occur in the base metal, or it may be below the base metal strength and fail in the braze. The joint strength will depend on the filler metal strength and also on the filler metal-base metal interacton during the brazing cycle. High efficiency will not be obtained with the butt joint when the filler metal in the joint is much weaker than the base metal. To obtain the best efficiency, the brazed joint should be free of defects (no flux inclusions, voids, unbrazed areas, pores, or porosity).

Another means of obtaining high butt joint strength is to utilize minimum joint clearances that are compatible with the base and filler metals involved, as well as with the brazing process to be used. Although the minimum joint clearance produces optimum joint strength, producing such clearances is sometimes considered impractical economically. However, with the current sophisticated metal working techniques, maintaining proper clearances is not ordinarily a major problem. It is important to point out that if high quality and highly reliable brazements are to be manufactured, it is imperative that clearances be controlled.

The diffusion brazing process will also improve the joint strength since this process causes interdiffusion of the filler metal and base metal. This interdiffusion will materially alter the joint by partially or completely eliminating the filler metal as a distinct layer in the joint. The results are increases in joint mechanical properties and joint remelting temperature.

A variation of the butt joint is the scarf joint shown in Fig. 11.11. In this joint the cross-sectional area of the joint is increased without an increase in metal thickness. This joint has several disadvantages which limit its use. The sections are difficult to align and the joint is difficult to prepare, particularly in thin members. Since the joint is at an angle to the axis of tensile loading, the load carrying capacity is similar to the lap joint rather than to the butt joint.

JOINT CLEARANCE

Joint clearance has a potent effect on the mechanical performance of a brazed joint. This applies to all types of loading, such as static, fatigue, and impact, and to all joint designs. Several effects of joint clearance on mechanical performance are (1) the purely mechanical effect of restraint to plastic flow of the filler metal by a higher strength base metal, (2) the possibility of slag entrapment, (3) the possibility of voids, (4) the relationship between joint clearance and capillary force which accounts for filler metal distribution, and (5) the amount of filler metal that must be diffused with the base metal when diffusion brazing.

Clearance is the distance between the faying surfaces of the joint. The clearance between members of similar metals is easily maintained in assemblies where parts are a press or shrink fit. In some parts it is necessary to use spacers, such as wires, shims, or prick punch marks, to assure the proper clearance for optimum flow of brazing alloy into the joint. Clearance between the parts must be considered in terms of conditions at one specific instant, i.e., room temperature or brazing temperature. With similar metals of about equal mass, the room temperature clearance (before brazing) is a satisfactory guide. When brazing dissimilar metals, the one with the higher thermal expansion may tend to increase or decrease the clearance, depending on the relative positions and configurations of the base metals. Therefore, when brazing dissimilar metals (or greatly differing masses of similar metals), consideration must be given to the clearance at brazing temperature, and the room temperature clearance must be designed to achieve the desired clearance at brazing temperature.

The influence of joint thickness on braze shear strength is illustrated in Fig. 11.12, which indicates the change in joint shear strength with joint clearance. Table 11.7 may be used as a guide for clearances at brazing temperature when designing brazed joints for maximum strength.

The joint clearances given in the table are radial clearances for tube type lap joints. The clearances given should be diametral clearance for some applications if there is no provision

in the design to assure alignment and concentricity of the two parts. Excessive joint clearance will result in voids in the joint, particularly when a gas flux is used.

Some specific clearance versus strength data for silver brazed butt joints in steel are

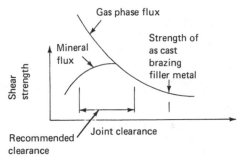

Fig. 11.12—Relationship between joint clearance and shear strength for two fluxing methods

Table 11.7—Recommended joint clearance at brazing temperature

Filler metal AWS classification[c]	mm	in.	Joint clearance[a]
BAlSi group	0.15-0.25	0.006-0.010	For length at lap less than 6.35 mm (1/4 in.)
	0.25-0.61	0.010-0.025	For length at lap greater than 6.35 mm (1/4 in.)
BCuP group	0.03-0.12	0.001-0.005	
BAg group	0.05-0.12	0.002-0.005	Flux brazing (mineral fluxes)
	0.03-0.05	0.001-0.002[b]	Atmosphere brazing (gas phase fluxes)
BAu group	0.05-0.12	0.002-0.005	Flux brazing (mineral fluxes)
	0.00-0.05	0.000-0.002[b]	Atmosphere brazing (gas phase fluxes)
BCu group	0.00-0.05	0.000-0.002[b]	Atmosphere brazing (gas phase fluxes)
BCuZn group	0.05-0.12	0.002-0.005	Flux brazing (mineral fluxes)
BMg group	0.10-0.25	0.004-0.010	Flux brazing (mineral fluxes)
BNi group	0.05-0.12	0.002-0.005	General applications (flux or atmosphere)
	0.00-0.05	0.000-0.002	Free flowing types, atmosphere brazing

a. Clearance on the radius when rings, plugs, or tubular members are involved. On some applications it may be necessary to use the recommended clearance on the diameter to assure not having excessive clearance when all the clearance is on one side. An excessive clearance will produce voids. This is particularly true when brazing is accomplished in a high quality atmosphere (gas phase fluxing).

b. For maximum strength, a press fit of 0.001 mm/mm or in./in. of diameter should be used.

c. See Table 11.3 for an explanation of filler metals.

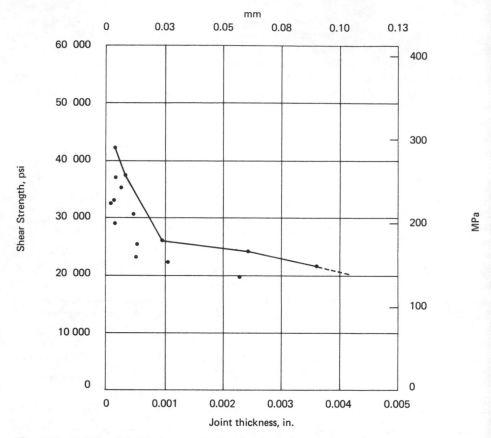

Fig. 11.13—Relationship of shear strength to brazed joint thickness for pure silver joints in 12.7 mm (0.5 in.) diameter steel drill rod

shown in Figs. 11.13 and 11.14.[5] Figure 11.13 shows the optimum shear values obtained with joints in 12.7 mm (0.5 in.) round drill rod using pure silver. The rods were butt brazed by induction heating in a dry 10 percent hydrogen-90 percent nitrogen atmosphere. Testing was accomplished in a specially designed guillotine type fixture. Figure 11.14 shows the relationship of tensile strength to joint thickness for butt brazed joints of the same size. Although this curve shows the normal ascending joint strength with decreasing joint clearances, it is of interest to note that the strength decreased when extremely small clearances were obtained.

Joint strength is related to test specimen design and testing method. Thus, tests must be conducted with the proposed production joint design and brazing procedures to obtain specific design strength data. Variations in the brazing procedures and joint design will alter the effect of joint clearance on the strength properties. For example, a free flowing filler metal brazed in a high quality atmosphere will adequately flow through a joint having a very small clear-

5. The data in Figures 11.13 and 11.14 were obtained with nonstandard test specimens.

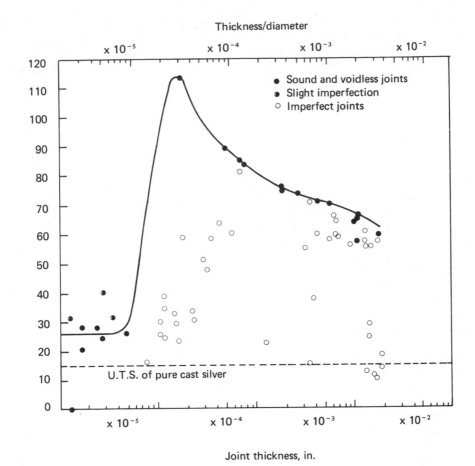

Fig. 11.14—*Relationship of tensile strength to brazed joint thickness of 12.7 mm (0.5 in.) diameter silver brazed butt joints in 4340 steel (1 mm = 0.039 in.; 1 ksi = 6.89 MPa)*

ance. However, when the atmosphere deteriorates, it is often necessary to use larger clearances to obtain adequate flow.

Preplaced filler is brazing alloy placed in the joint, such as foil placed between two plates. In this application, the clearances noted in Table 11.7 generally do not apply. In applications where filler metal is preplaced, the two members of the joint should be preloaded so that the joint clearance decreases during the brazing operation. The decrease causes the filler metal to fill the voids in the normal roughness of the faying surfaces. In some applications, addi-

tional filler metal is added by extending the filler metal shim out beyond the joint.

Effect of Mineral and Gas Fluxes

Flux is usually used in brazing. It may be a mineral type flux, a gas atmosphere type flux, or a combination of both. The type of fluxing will have an important bearing on the joint clearance used to accomplish a given brazement.

A mineral flux must melt at a temperature below the melting range of the brazing filler metal, and it must flow into the joint ahead of the filler metal. The filler metal on melting must force out the flux as the filler metal is drawn into the joint by capillary action. When the joint clearance is too small, the mineral flux may be held in the joint and not be displaced by the molten filler metal. This will produce joint defects. When the clearance is too large, the molten filler metal will flow around pockets of flux, causing excessive flux inclusions.

Gas fluxes are atmospheres that have the same function as the mineral fluxes. Atmospheres also have their effect on the design of clearances for a specific brazement. It is essential to specify the brazing process when the clearance is determined so that they are compatible. The gas fluxes require smaller clearances to obtain optimum strength. In general, when atmosphere furnace brazing a joint in a vertical position, a free flowing filler metal will flow out of a joint having a clearance in excess of 0.08 mm (0.003 in.). The ability of an atmosphere to dissociate the surface oxides of a specific base metal below the brazing temperature will also affect the clearance requirement. When the atmosphere rapidly and thoroughly dissociates the surface oxides of the base metal and filler metal, the filler metal may be drawn out of a 0.05 mm (0.002 in.) clearance joint and onto the adjacent base metal surface. In this case, a smaller clearance can be used.

Effect of Surface Finish

In general, the filler metal is drawn into the joint by capillary action. Thus, if the surface finish of the base metal is too smooth, the filler metal may not distribute itself throughout the entire joint and may leave voids. Very smooth or polished faying surfaces of the joint should be roughened to assure adequate flow of the filler metal throughout the joint, preferably with a clean metallic grit compatible with the base metal. Surfaces to be brazed must not be contaminated with nonmetallic materials (ceramics, nonmetallic blasting materials, minerals from a water rinse bath, etc.), or other detrimental contaminants.

Surfaces that are too rough will cause low joint strength because only the points may braze. A surface roughness of 0.7 to 2μm (30 to 80μin.) RMS is generally acceptable, but tests should be conducted to assure optimum conditions for a specific brazement.

Effect of Base Metal-Filler Metal Interaction

The mutual solubility between base metal and filler metal causes interaction to take place through solution of the base metal with the molten filler metal and diffusion in both the liquid and solid states. This interaction may have some effect on the clearance for a specific brazement. If the interaction is low, the clearances can in general be smaller; however, when the interaction is high, which is usually in the liquid phase, then the clearances required will have to be larger. This also affects the flow through the joint; thus, when the joint is long and interaction is large, the clearance should be increased. For example, when aluminum base metal is brazed with a lower melting alloy of aluminum at a temperature close to the base metal melting point, interaction can be expected.

Aluminum is by no means the only filler metal that exhibits interaction. All filler metals show varying degrees of interaction to different base metals, for example, silver filler metals with copper base metal, copper filler metal with nickel and copper-nickel base metals, gold filler metals with nickel base metals, and nickel filler metals with nickel and cobalt base metals.

Effect of Base Metals

The chemistry of base metals often includes one or more elements in varying quantities whose oxides are not easily dissociated in a specific atmosphere or by a specific mineral flux. Since the metal-metal oxide dissociation of a given base metal by a specific mineral flux or atmosphere is dependent on many factors, it is important to match the proper clearance with the brazing process and flux or atmosphere. For example, when atmosphere furnace brazing a nickel-chromium base metal with no aluminum present, the clearance can be small; but, when a small aluminum addition is incorporated in the base metal, the clearance may have to be increased. With larger aluminum additions, brazing may not take place unless the atmosphere is made more active (lower dew point) or the surface is protected with a barrier nickel coating. The clearances would then require reappraisal to obtain optimum braze joint properties.

The thermal expansion will, of course, have an effect on the joint clearance at brazing temperature when the base metals are dissimilar.

Effect of Brazing Filler Metal

Filler metals have varying degrees of flowability and different viscosities at various brazing temperatures and conditions. At one end of the scale there is a group of free flowing filler metals and at the other end are the sluggish filler metals. The free flowing filler metals generally require a smaller clearance than the sluggish filler metals. Filler metals that have a melting point, such as pure metals (i.e., copper) and eutectic alloys, and self-fluxing filler metals will usually be free flowing, particularly when there is very little interaction with the base metal.

Variations in the quality of fluxes and atmospheres can enhance the free flowing qualities of the filler metals, or they can result in no flow at all. The fluxes may be of poor quality or become oxidized, or atmospheres may be of low relative purity. Thus, the clearances may appear to be improper for a given set of brazing conditions when in fact the flux or atmosphere requires improved control.

Effect of Joint Length and Geometry

The joint length affects the clearance, particularly when there is interaction between the filler metal and the base metal. As the filler metal is drawn into a long joint, it may pick up enough base metal to freeze before it reaches the other end. The more interaction that exists with a specific base metal-filler metal combination under given brazing conditions, the larger the clearance must be as the joint becomes longer. This is only one of a number of reasons why it is important to make the joint length as short as possible, consistent with an optimum joint strength.

Effect of Dissimilar Base Metals

When designing a joint where dissimilar base metals are involved, the joint clearance at the brazing temperature must be calculated from thermal expansion data. Figure 11.15 shows the thermal expansion for some materials. The brazing temperature must be also taken into con-

sideration when designing the joint. Figure 11.16 can be used to find the diametral clearance at brazing temperature between dissimilar metals.

With high differential thermal expansion between two details, the brazing filler metal must be strong enough to resist fracture and the base metal must yield during cooling. Some residual stress will remain in the final brazement as a result of joining at the brazing temperature and subsequently cooling to room temperature. Thermal cycling of such a brazement during its service life will also stress the joint area which may or may not shorten the service life. Whenever possible, the brazement should be designed so that residual stresses do not add to the stress imposed during service.

STRESS DISTRIBUTION

A properly designed and executed high strength brazement will fail in the base metal. In brazements where joints may be lightly loaded, it may be more economical to use simplified joint designs which may break in the brazed joint if overstressed in testing or in service.

In general, the loading of a brazed joint demands the same design considerations as those given any other joint or change in base metal cross section. Brazed joints have a few specific requirements that are usually only apparent under conditions of high dynamic stresses or high static stresses. It is well known that the base metal itself can best withstand high stresses and dynamic loading. A good brazement design will incorporate a joint design that will avoid high stress concentration at the edges of the braze joint and will distribute the stresses uniformly into the base metal. Typical examples of this are shown in the joint design sketches, Figs. 11.17 through 11.20.

A fillet of brazing filler metal is often erroneously incorporated by the designer as a method of eliminating stress concentrations. This is not good brazing design practice because it is seldom possible to have the brazing filler metal consistently form the desired fillet size and contour. When the fillets become too large, shrinkage or piping porosity may be present and will act as a stress concentration. Fillets of brazing filler metal should not be used to replace a base metal fillet to control stress concentration.

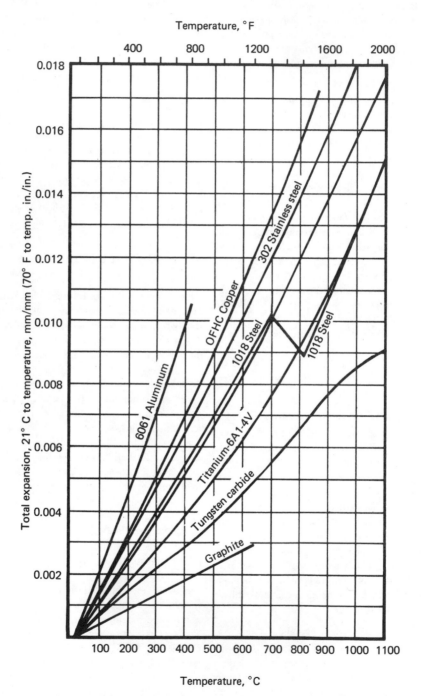

Temperature, °F

Temperature, °C

Fig. 11.15—Thermal expansion curves for some common materials

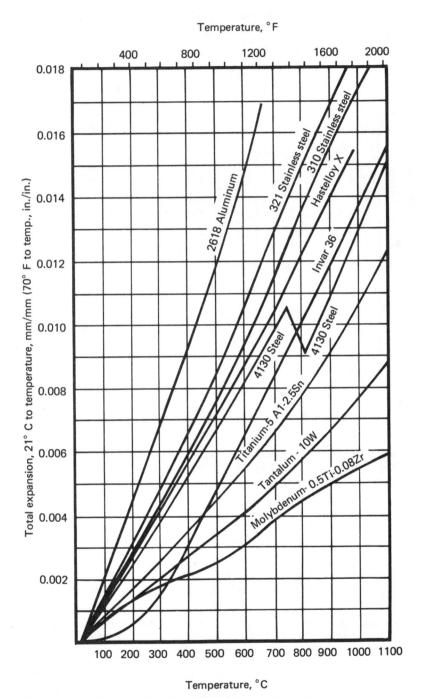

Fig. 11.15 (continued) — Thermal expansion curves for some common materials

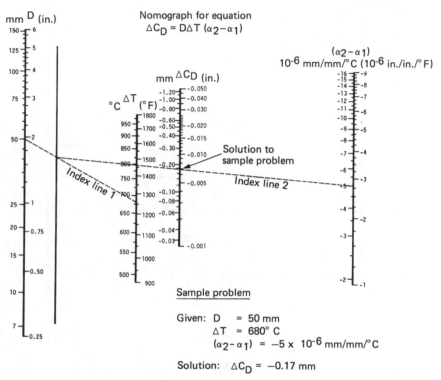

mm D (in.)

Nomograph for equation
$$\Delta C_D = D\Delta T (\alpha_2 - \alpha_1)$$

$(\alpha_2 - \alpha_1)$
10^{-6} mm/mm/°C (10^{-6} in./in./°F)

mm ΔC_D (in.)

°C ΔT (°F)

Solution to sample problem

Index line 1

Index line 2

Sample problem

Given: D   = 50 mm
ΔT = 680° C
$(\alpha_2 - \alpha_1)$ = −5 × 10^{-6} mm/mm/°C

Solution: ΔC_D = −0.17 mm

Notes:

1. This monograph gives change in diameter caused by heating. Clearance to promote brazing filler metal flow must be provided at brazing temperature.

2. D   = nominal diameter of joint, mm (in.)
   ΔC_D = change in clearance, mm (in.)
   ΔT = brazing temperature minus room temperature, °C (°F)
   α_1 = mean coefficient of thermal expansion, male member, mm/mm/°C (in./in./°F)
   α_2 = mean coefficient of thermal expansion, female member, mm/mm/°C (in./in./°F)

3. This nomograph assumes a case where α_1 exceeds α_2 so that scale value for $(\alpha_1 - \alpha_2)$ is negative. Resultant values for ΔC_D are therefore also negative, signifying that the joint gap reduces upon heating. Where $(\alpha_2 - \alpha_1)$ is positive, values of ΔC_D are read as positive, signifying enlargement of the joint gap upon heating.

Fig. 11.16—Nomograph for finding the change in diametral clearance in dissimilar metal joints

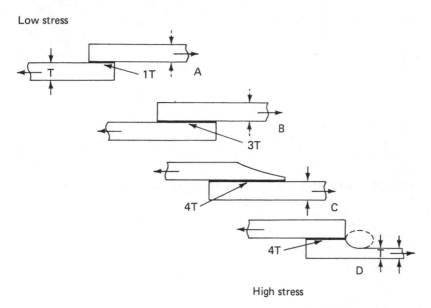

Fig. 11.17—Brazed lap joint designs for use at low and high stresses—flexure of right member in C and D will distribute the load through the base metal

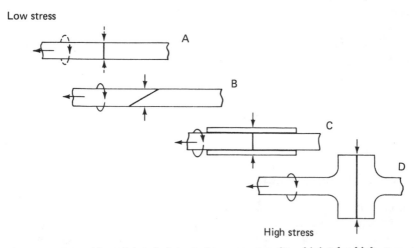

Fig. 11.18—Brazed butt joint designs to increase capacity of joint for high stress and dynamic loading

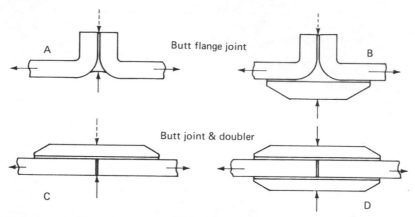

Fig. 11.19 — Butt joint designs for sheet metal brazements — the loading in joint A cannot be symmetrical

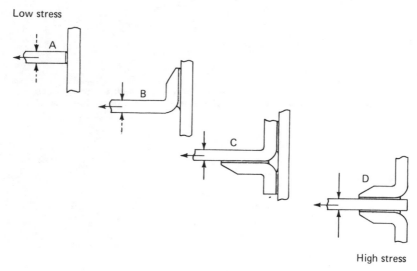

Fig. 11.20 — T-joint designs for sheet metal brazements

BRAZING FILLER METAL PLACEMENT

When designing a brazed joint, the brazing process to be employed and the manner in which the filler metal will be introduced to the joint should be established. In most manually brazed joints, the filler metal is simply fed from the face side of the joint. For furnace brazing and high production brazing operations, the filler metal is preplaced at the joint. Automatic dispensing equipment may be used to do this operation.

Brazing filler metal may be in the form of wire, shims, strip, powder, and paste. Preplaced filler metal is most commonly employed in the form of wire or strips. Figures 11.21 and 11.22 illustrate methods of preplacing the brazing filler metal in wire and sheet forms. When the base metal is grooved to accept preplaced filler

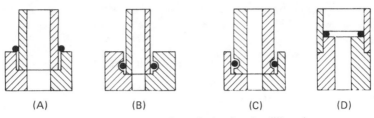

Fig. 11.21—Methods of preplacing brazing filler wire

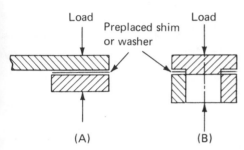

Fig. 11.22—Preplacement of brazing filler shims

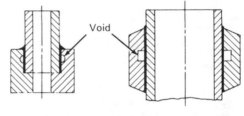

Fig. 11.23—Brazed joints with grooves for pre-placement of filler metal; after the brazing cycle the grooves are void of filler metal

metal, the groove should be cut in the heavier section. When designing for strength, the groove area should be subtracted from the joint area since the brazing filler metal flows out of the groove and into the joint interfaces, as shown in Fig. 11.23.

Powdered filler metal can be applied in any of the locations indicated in Fig. 11.21. It can be applied dry to the joint area and then wet down with binder, or it can be premixed with the binder and applied to the joint. The density of powder is usually only 50 to 70 percent of the solid metal volume, and therefore, the groove volume must be larger for powder.

Where preplaced shims are used, the sections being brazed should be free to move together when the brazing alloy melts. Some type of loading may be necessary to move them together and force excess filler metal and flux out of the joint.

The designer should also consider subsequent inspection needs when designing a brazed joint. When the filler metal is preplaced in internal grooves, the inspector can visually inspect for filler metal flow and complete joint penetra-

tion. When the filler metal cannot be placed in the joint, the filler metal can be placed on one side of the joint and allowed to flow through to the other side. Visual inspection can then be made for flow on the other side of the joint.

ELECTRICAL CONDUCTIVITY

The principal factor to be considered in the design of an electrical joint is its electrical conductivity. The joint, when properly designed, should not add appreciable electrical resistance to the circuit.

Brazing filler metals in general have low electrical conductivity compared to copper. However, a braze joint will not add any appreciable resistance to the circuit if the joint is properly designed.

With butt joints, the brazed joint thickness (length) is very small compared to the length of the conductor. Therefore, the brazed joint will normally contribute very small resistance compared to that of the total electrical path, even though the resistivity of the filler metal is much higher than that of the base metal. A filler metal with low resistivity should be used, provided it will meet all other requirements of the design and fabrication.

From a practical standpoint, there will likely be voids in the brazed joint which will reduce the effective area of the electrical path. For this reason, lap joints are recommended where design will permit. A lap length of at least 1-1/2 times the thickness of the thinner member of the joint is recommended. The joint resistance will be approximately equal to the same length in solid copper.

PRESSURE TIGHTNESS (PRESSURE OR VACUUM)

Joints in brazements required to be pressure tight should be of the lap (shear) type whenever possible. This type of joint provides a large braze area (interface) with less chance of leakage through the joint. Joint clearance is an important factor in obtaining the best quality pressure tight joint. The lower side of the clearance range should be used because a joint having fewer inclusions and leaks can be obtained. Most problems result from excessive clearances, poor flowing filler metal-base metal combinations, and poor brazing technique.

The principal factor other than strength to be considered in the design of pressure or vacuum tight assemblies is proper venting. The heat from the brazing operation expands the air or other gases within the closed assembly rather rapidly. Unless the assembly is well vented (open to the atmosphere), it is likely to be forced apart. Also, the escaping gas may prevent the filler metal from entering and flowing through the joint. If an unvented brazement does braze satisfactorily at brazing temperature, contraction of entrapped gas in the assembly will usually pull the braze fillets to the inside. This may leave air leaks through the joint area.

Dead-end holes may be considered as small pressure containers. They, too, should be vented through small holes leading out to the surface. Similarly, any brazement that is designed with closed segments or areas should be vented. Thus, it is essential to have adequate venting of all en-

closed containers or passages in order to assure good quality joints.

TESTING OF BRAZED JOINTS

The importance of standardizing a method for evaluating the strength of brazed joints cannot be overemphasized. Different designs of test specimens will result in different sets of data. It is readily apparent in Fig. 11.24 that the "apparent joint strength" at a very low overlap distance is high in comparison to the long overlap. Thus, two laboratories that each test a single overlap distance may be testing at opposite ends of the curve and, as a result, come up with different conclusions. It is, therefore, highly desirable to use the entire usable overlap range of the curve to obtain adequate data.

The most sensitive portion of the curve is the low overlap portion where the apparent joint strength is the highest. At this same portion of the curve, the most variation in test results will occur because the joints are very sensitive to all the factors that affect the brazing operation.

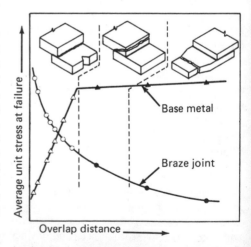

Fig. 11.24—Average unit shear stress in the brazed lap joint and average unit tensile strength in the base metal as functions of overlap distance—(Open symbols represent failures in the filler metal; filled symbols represent failures in the base metal)

The designer, on the other hand, is looking for the load carrying capacity of the joint, and thus is most interested in the right hand portion of the base metal curve. The brazement should be designed, where possible, so that the failure occurs in the base metal but without an excessive overlap.

For further information, refer to the latest edition of AWS C3.2, Standard Method for Evaluating the Strength of Brazed Joints.

BRAZING EQUIPMENT AND TECHNIQUES

Brazing processes are classified by the method used to heat the assembly. Although this is a satisfactory method, some of the processes are now so widely diversified that any one method of classification cannot cover all of the special procedures required. Several methods of heating are used to produce brazed joints. Selection of the method to be used is determined by the type of equipment available, the skill of the operator, the relative cost of labor and materials, and the nature of the part to be brazed. Some of the brazing filler metals and base metals can be brazed by only one of the heating methods.

Any of the methods may be used to braze mild steel when silver base filler metals are used. The use, however, of pure copper as the brazing filler metal requires the use of controlled atmosphere heating. Nickel, palladium, and gold alloy filler metals are most frequently used in an inert, dry hydrogen, or vacuum atmosphere, and a retort type furnace is frequently used. The purchase of equipment of this type, however, for the sole purpose of brazing mild steel might be uneconomical.

A torch is the most common method of heating. A furnace is used for high production work and for critical applications. Induction heating is used where extremely rapid heating is desired. Other heating methods are primarily used for brazed honeycomb fabrication. Some of these brazing media are quartz (infrared) lamps and the electric blanket with ceramic dies. Another heating method is exothermic brazing which utilizes the heat reaction of several chemical compounds.

TORCH BRAZING

Torch brazing is accomplished by the use of a neutral or slightly reducing oxyfuel gas or air-fuel gas flame. The filler metal can be preplaced on the joint and fluxed before heating, or it may be face fed. Flux pastes made with alcohol give better results and should be used in preference to pastes made with water. Many specially prepared fluxes are available. Heat is applied to the joint, first melting the flux, and then continuing until the brazing filler metal melts and flows into the joint. Overheating of the base metal and brazing alloy should be avoided because rapid diffusion and "drop through" of the metal will result. Natural gas is well suited for torch brazing because its relatively low flame temperature reduces the danger of overheating.

Several different types of torches are commonly used for brazing, depending upon the gas mixtures: air-natural gas, air-acetylene, oxy-acetylene, and oxy-hydrogen or other oxyfuel gases.

Air-natural gas torches provide the lowest flame temperature as well as the least heat, depending on the size of the torch. When they are used on heavy parts, there is a tendency to use an oxidizing flame in order to hasten the heating. Acetylene under pressure is used in the air-acetylene torch with air at atmospheric pressure. Suitable torches are available for use with some brazing applications. Both air-natural gas and air-acetylene torches can be used to advantage on small parts and thin sections. When these torches are incorporated in automatic or semiautomatic machines, the lower heating speed may adversely affect production rates.

Torches which employ oxygen with natural gas, or other cylinder gases, such as propane or butane, provide a higher flame temperature. When they are properly applied, excellent results are obtainable with many brazing applications. The torch should be of sufficient size and capacity so that a neutral or slightly reducing flame can be used.

Oxyhydrogen torches are often used for brazing aluminum and other nonferrous alloys. Since the temperature obtained is below that of oxyacetylene, the possibility of overheating the assembly during brazing is reduced. In addition, the use of excess hydrogen provides the joint with additional cleaning and protection during the brazing of some base metals.

A wide range of heat control is obtained with the use of oxyacetylene torches, and high temperatures are possible. These torches offer the most flexible and versatile method of torch heating. Their use is so universal that full information and instructions are readily available from manufacturers. For further information concerning these torches, see Chapter 10.

Specially designed torches having multiple tips or multiple flames can be used to an advantage to increase the rate of heat input. Care must be exercised to avoid local overheating by constantly moving the torch with respect to the work.

Torch heating is limited in use to those brazing filler metals which may be used with flux or are self-fluxing. The filler metals of aluminum-silicon, silver, copper-phosphorus, copper-zinc, and nickel are readily brazed with this type of equipment. With the exception of the copper-phosphorus filler metals they all require fluxes, and for certain applications even the copper-phosphorus filler metals require the use of a flux, as shown in Table 11.5.

The techniques employed for torch brazing differ from those used for oxyfuel gas welding. Operators experienced only in welding techniques may require instruction in brazing techniques. It is good practice, for example, to prevent the inner cone of the flame from coming in contact with the joint except during preheating, since melting of the base metal and dilution with the filler metal may increase its liquidus temperature and make the flow more sluggish. In addition, the flux may be overheated and thus lose its ability to promote capillary flow. Also, some of the low melting constituents of the filler metal may be evaporated.

FURNACE BRAZING

Electric, gas, or oil heated furnaces with automatic temperature control capable of holding the temperature within $\pm 6°C$ ($\pm 10°F$) should be used for furnace brazing. Either fluxes or specially controlled atmospheres to perform the same functions as the fluxes are used for furnace brazing.

Furnace brazing is used extensively when (1) the parts to be brazed can be preassembled or jigged to hold them in the correct position, (2) the brazing filler metal can be placed in contact with the joint, (3) multiple brazed joints are to be formed simultaneously on a completed assembly, (4) many like assemblies are to be joined, and (5) complex parts must be heated uniformly to prevent the distortion that would result from local heating of the joint area.

Parts to be brazed should be assembled with the filler metal and flux if used, located in or around the joints. The assembly is placed in the furnace until the parts reach brazing temperature and brazing takes place. The assembly is then removed. These steps are shown in Fig. 11.2. Many commercial fluxes are available for both general and specific brazing operations. Satisfactory results are obtained if dry powdered flux is sprinkled along the joint. Flux paste is satisfactory in most cases. However, in some cases it retards the flow of brazing alloy. Flux pastes containing water can be dried by heating the assembly at 175 to 200° C (350 to 400° F) for 5 to 15 minutes in drying ovens or circulating air furnaces.

Brazing time will depend somewhat on the thickness of the parts and the amount of fixturing necessary to position them. The brazing time should be restricted to that necessary for the filler metal flow through the joint to avoid excessive interaction between the filler metal and base metal. Normally, one or two minutes at the brazing temperature is sufficient to make the braze. A longer time at the brazing temperature will be beneficial where the filler metal remelt temperature is to be increased and where diffusion will improve joint ductility and strength. Times of 30 to 60 minutes at the brazing temperature are often used to increase the braze remelt temperature.

Furnaces used for brazing are classified as (1) batch type with either air or controlled atmosphere, (2) continuous type with either air or controlled atmosphere, (3) retort type with controlled atmosphere, or (4) vacuum. The temperature of the heating zone must be controlled accurately. Most brazing furnaces have a temperature control of the potentiometer type connected to thermocouples and gas control valves or contractors. The method of heating varies according to the application. Some furnaces are heated by gas or oil, but the majority are electrically resistance heated with silicon-carbide, nickel-chromium, or refractory metal (Mo, Ta, W) heating elements. When a flame is used for heating, it is important that the flame does not impinge directly on the parts.

Air atmosphere furnaces may be used for brazing parts which can be protected with a paste flux. They are not satisfactory when the brazing cycle is beyond the protective capability of available fluxes. The use of flux, however, is economical to reduce the amount of cleaning necessary.

With controlled atmosphere furnaces, a continuous flow of the atmosphere gas is maintained in the work zone to avoid any contamination from outgassing of the metal parts and dissociation of oxides. Since many controlled atmospheres are flammable, and a slightly positive gas pressure must be maintained in the furnace, some gas will escape into the room. Flammable gases should be exhausted or burned to avoid toxicity and explosion hazards. Adequate venting of the working area and efficient protection against explosion are necessary.

Batch type furnaces heat each workload separately. They may be top loading (pit type), side loading, or bottom loading. When a bottom loading furnace is lowered over the work, it is called a bell furnace. Gas or oil fired batch type furnaces without retorts require that flux be used on the

parts for brazing. Electrically heated batch type furnaces are often equipped for controlled atmosphere brazing, since the heating elements can usually be operated in the controlled atmosphere.

Continuous furnaces receive a steady flow of incoming assemblies. The heat source may be gas or oil flames, or electrical heating elements. The parts, either singly or in trays or baskets, are moved through the furnace. Conveyor types, (mesh belts or roller hearth), shaker hearth, pusher, or slot type continuous furnaces are commonly used for high production brazing. Continuous furnaces usually contain a preheat or purging area which the parts enter first. In this area the parts are slowly brought to a temperature below the brazing temperature. The brazing atmosphere gas, if used, flows over and around the parts under a positive pressure. The gas flow removes any entrapped air and starts the reduction of surface oxides. The preheat area is generally separated from the heating zone. Atmosphere gas flows from the brazing zone into the preheat area and the cooling zone.

Retort type furnaces are batch type furnaces in which the assemblies are placed in a sealed retort for brazing. After the air in the retort is purged by controlled atmosphere gas, the retort is placed in the furnace. After the parts have been brazed, the retort is removed from the furnace and cooled. Then the controlled atmosphere is purged from the retort. The retort is opened and the brazed assemblies are removed. An atmosphere is sometimes used in a high temperature furnace to reduce the scaling of the retort.

Vacuum brazing equipment is currently being used to a large extent for brazing stainless steels, super alloys, aluminum alloys, titanium alloys, and metals containing refractory or exotic elements. Vacuum is a relatively economical method of preventing oxidation by removing air from around the assembly. Good surface cleanliness is required for good wetting and the flow of filler metals because no flux is used. Base metals containing chromium and silicon can be vacuum brazed. Otherwise, a very pure, low dew point atmosphere gas would be required. Base metals containing more than a few percent of aluminum, titanium, zirconium, or other elements with particularly stable oxides can generally be brazed only in vacuum. However, a nickel plated barrier is still preferred to obtain optimum quality.

Vacuum brazing furnaces can be classified as three types:

(1) *Hot retort, or single pumped retort furnace.* This consists of a sealed retort usually of fairly thick metal. The work load is placed into the retort, and then the retort is sealed, evacuated, and heated from the outside by a furnace. Most brazing work requires that vacuum pumping be continued throughout the heat cycle to remove the gases that are continuously given off by the workload. The furnaces are either gas fired, or electrically heated. The hot retort type of vacuum furnace is limited in size and maximum operating temperature by the ability of the retort to withstand the collapsing force of atmospheric pressure at brazing temperature. Vacuum brazing furnaces of this type can operate at temperatures as high as 1150° C (2100° F), but most retorts are limited to 870° C (1600° F) or lower.

Leak tightness of hot wall retorts is sometimes a problem, particularly if the materials to be brazed require a pressure of 13.3 Pa (10^{-2} torr) or lower. Argon, nitrogen, or other gas is often introduced into the retort to accelerate cooling.

(2) *Double pumped or double wall hot retort vacuum furnace.* A typical furnace of this type consists of an inner retort containing the work. The inner retort is contained within an outer wall or vacuum chamber. Also within the outer wall are the thermal insulation and electrical heating elements. A moderate pressure, typically 133 to 13.3 Pa (1.0 to 0.1 torr), is maintained within the outer wall, and a much lower pressure, typically 1.3 Pa (10^{-2} torr) or lower, is maintained within the inner retort. Most brazing work requires continuous vacuum pumping of the inner retort throughout the heat cycle to remove the gases that are continuously given off by the workload.

This type of furnace has a particular advantage in that the heating elements and the thermal insulation are not subjected to the high vacuum. Heating elements are typically of nickel-chromium alloy, graphite, stainless steel, silicon carbide, or other commonly used furnace element materials. Thermal insulation material is usually silica or alumina brick, castable, or fiber materials.

(3) *Cold wall vacuum furnace.* A typical cold wall vacuum furnace consists of a single vacuum chamber, with the thermal insulation and electrical heating elements located inside of the chamber. The vacuum chamber is usually water-cooled. The maximum operating temperature of cold wall furnaces is determined by the materials used for the thermal insulation (or heat shield) and the heating elements. The thermal insulation and heating elements are subjected to the high vacuum as well as the operating temperature of the furnace. Heating elements for cold wall furnaces are usually made of molybdenum, tungsten, graphite, tantalum, or other high temperature, low vapor pressure materials. Heat shields are typically made of multiple layers of molybdenum, tantalum, nickel, or stainless steel. Thermal insulation may be high purity alumina brick, graphite, or alumina fibers sheathed in stainless steel. The maximum operating temperature and vacuum obtainable with cold wall vacuum furnaces depends on the heating element material and the thermal insulation or heat shields. Temperatures up to 2200° C (4000° F) and pressures as low as 1.33×10^{-4} Pa (10^{-6} torr) are obtainable.

All three types of furnaces are constructed with side loading (horizontal), bottom loading, or top loading (pit type) configurations. Work zones are usually rectangular for side loading furnaces, and circular for bottom or top loading types.

Vacuum pumps for brazing furnaces may be oil sealed mechanical types for pressures ranging from 13 to 1300 Pa (0.1 to 10 torr). Brazing of base metals containing chromium, silicon, or other rather strong oxide formers usually requires pressures of 1.3 to 0.13 Pa (10^{-2} to 10^{-3} torr) which are best obtained with a high-speed, dry Roots or turbo-mechanical type pump. Vacuum pumps of this type are not capable of exhausting directly to atmosphere and require a roughing vacuum pump. Brazing of base materials containing more than a few percent of aluminum, titanium, zirconium, or other very stable oxide forming elements requires vacuum of 0.13 Pa

$(10^{-3}$ torr) or lower. Vacuum furnaces for this type of brazing usually employ a diffusion pump which typically obtains pressures of 1.3 to 0.0001 Pa $(10^{-2}$ to 10^{-6} torr). The diffusion pump is usually backed by a mechanical vacuum pump or by both a Roots type pump and a mechanical pump.

INDUCTION BRAZING

Induction heating is used extensively on parts that are self-locating or that can be fixtured so that the fixture will not interfere with heating. Induction heating is also used where very rapid heating is required. It is used primarily to produce brazed assemblies economically and also where large numbers of parts can be adapted to special machines.

Parts to be heated act as the short circuited secondary of a transformer where the work coil, which is connected to the power source, is the primary. On both magnetic and nonmagnetic parts, heating is obtained from the resistance of the parts to the currents induced in them by the transformer action. Various ac power frequencies that range from 60 Hz to 450 kHz are used. Heating at 60 Hz is generally slow and tends to penetrate deep into the work. The most common frequencies used are 10 kHz, obtained from a motor-generator set, and 350 to 450 kHz, obtained from a tube type or spark-gap type generator. Most of the heat generated by this method at high frequencies develops near the surface. The interior is heated by thermal conduction from the hot surface. As a rule, the higher the frequency, the shallower the heating.

The equipment is manufactured in sizes from 1 kW to several hundred kilowatts output to be used with either individual or multiple work coils. One generator may be used to energize several individual work stations in sequence with a transfer switch. The assembly may be indexed into position under a work coil or passed through a work coil. Stationary parts may be heated uniformly if they have a uniform wall thickness or symmetrical shape. When the parts are not symmetrical, they should be rotated to obtain uniform heating throughout.

The parts and the coil are sometimes placed in a ceramic chamber (quartz or tempered glass) for brazing in a vacuum, hydrogen, or inert gas; or the work and chamber can be placed inside the coil.

RESISTANCE BRAZING

Resistance brazing is most applicable to joints which have relatively simple configuration. It is difficult to obtain uniform current distribution, and therefore uniform heating, if the area to be brazed is large or discontinuous or is much longer in one dimension. Parts to be resistance brazed should be so designed that pressure may be applied to them without causing distortion at brazing temperature. Wherever possible, the parts should be designed to be self-nesting, to eliminate the need for dimensional features in the fixtures. Parts should also be free to move as the filler metal melts and flows in the joint.

One common source of current for resistance brazing is a stepdown transformer whose secondary circuit can furnish sufficient current at low voltage (2-25 V). The current will range from about 50 A for small, delicate jobs to many thousands of amperes for larger jobs. Commercial equipment is available for resistance brazing.

Electrodes for resistance brazing are made of high resistance electrical conductors, such as carbon or graphite blocks, tungsten or molybdenum rods or inserts, or even steel in some instances. The heat for brazing is mainly generated in the electrodes and flows into the work by conduction. It is generally unsatisfactory to attempt to use the resistance of the workpieces alone as a source of heat.

The pressure applied by a spot welding machine, clamps, pliers, or other means must be sufficient to maintain good electrical contact and to hold the pieces firmly together as the filler metal melts. The pressure must be maintained during the time of current flow and after the current is shut off until the joint solidifies. The time of current flow will vary from about one second for small, delicate work to several minutes for larger work. This time is usually controlled manually by the operator, who determines when the part is brazed by the temperature and extent of filler metal flow.

Brazing filler metal in the form of preplaced wire, shims, washers, rings, powder, or paste is used. In a few instances, face feeding is possible. For copper and copper alloys, the copper-phosphorus filler metals are most satisfactory since they are self-fluxing. The silver base filler metals may be used, but a flux or atmosphere is necessary. A wet flux is usually applied as a very thin mixture just before the assembly is placed in the brazing fixture. Dry flux is a good insulator and will not permit sufficient current to flow.

The parts to be brazed must be clean. The parts, brazing filler metal, and flux are assembled and placed in the fixture and pressure applied. As current flows, the electrodes become heated, frequently to incandescence, and the flux and filler metal melt and flow. The current should be adjusted to obtain uniform and rapid heating in the parts. If too much current is used, the electrodes will become too hot and deteriorate. There is also danger of oxidizing or melting the work. If too little current is employed, the time of brazing will be excessive. A little experimenting with electrode composition, geometry, and voltage will give the best combination of rapid heating with reasonable electrode life. Quenching the parts from an elevated temperature will help flux removal. It should only be done after the assembly has cooled sufficiently to permit the braze to hold the parts together. An exception might be the brazing of insulated conductors; in this case, it is advisable to quench the parts rapidly while they are still in the electrodes to prevent overheating of the adjacent insulation. Water-cooled clamps also prevent damage to the insulation.

DIP BRAZING

Two methods of dip brazing are (1) molten metal bath dip brazing, and (2) molten chemical (flux) bath dip brazing.

Molten Metal Bath Method

This method is usually limited to the brazing of small assemblies, such as wire connections or metal strips. A crucible, usually made of graphite, is heated externally to the required temperature to maintain the brazing filler metal in fluid form. A cover of flux is maintained over the molten filler metal. The size of the molten bath (crucible) and the heating method must be such that the immersion of parts in the bath will not lower the bath temperature below brazing temperature. Parts should be clean and protected with flux prior to their introduction into the bath.

Molten Chemical (Flux) Bath Method

This brazing method requires either a metal or ceramic container for the flux and a method of heating the flux to the brazing temperature. Heat may be applied externally with a torch or internally with an electrical resistance heating unit. Another method involves electrical resistance heating of the flux itself. This method requires the initial melting of the flux by external heating. Suitable controls are provided to maintain the flux within the necessary brazing temperature range. The size of the bath must be such that the introduction of parts for brazing will not lower the flux below brazing temperature.

Parts should be cleaned, assembled, and preferably held in jigs prior to immersion into the bath. Brazing filler metal is preplaced as rings, washers, slugs, paste, or as a cladding on the base metal. Preheat may be necessary to assure dryness of parts and to prevent the freezing of flux on parts which may cause selective melting of flux and brazing filler metal. Preheat temperatures are usually close to the melting temperature of the flux. A certain amount of flux adheres to the assembly after brazing. The flux must be drained off while the parts are hot. Any flux remaining on the parts must subsequently be removed by water or by chemical means.

INFRARED BRAZING

Infrared brazing may be considered as a form of furnace brazing where heat is supplied by light radiation below the visible red rays in the spectrum. The principal heating is done by the invisible radiation from high intensity quartz lamps. Heat sources (lamps) capable of delivering up to 5000 watts of radiant energy are commerically available. The lamps do not necessarily need to follow the contour of the part to be heated, but the heat input varies inversely as the square of the distance from the source. Concentrating reflectors are used to focus the radiation on the parts.

Because the process resembles furnace brazing, the brazing techniques are similar. The major difference is the source of heat for brazing. Cleaning, brazing alloy selection, placement of brazing alloys, fluxing or atmosphere selection, and the other process variables are evaluated and selected in a similar manner to furnace brazing.

Assemblies to be brazed are supported in a position which enables the energy to impinge on the part. In some applications only the assembly itself is enclosed. There are, however, applications whereby the assembly and the lamps are placed in a bell jar or retort that can be evacuated, or in which an inert gas atmosphere can be maintained. The assembly is then heated to a controlled temperature, as indicated by thermocouples.

EXOTHERMIC BRAZING

Exothermic brazing refers to a process by which the heat required to melt and flow a commercial brazing filler metal is generated by a solid state exothermic chemical reaction. An exothermic chemical reaction is any reaction between two or more reactants in which heat is given off due to the free energy of the system. Only the solid state or nearly solid state metal-metal oxide reactions are suitable for use in exothermic brazing units.

All the exothermic mixtures of interest for brazing are capable of propagating from the point of ignition throughout the reaction mass at some specific rate. Local ignition of the exothermic mixture is accomplished by a variety of methods. The easiest and simplest ways are to ignite them with a flame, a resistance heater, or a hot spark. The reaction then propagates like a fuse. Overall, simultaneous ignition of the entire exothermic mixture can be obtained only by heating the mixture evenly to the ambient ignition temperature. Ignition temperature thus signifies the temperature at which the system initiates spontaneous propagation reaction.

Exothermic brazing utilizes simplified tooling and equipment. The process utilizes the reaction heat to bring adjoining metal interfaces to a temperature where preplaced brazing filler metal melts and wets the base metal interface surfaces. The filler metal can be any commercially available brazing alloy having a suitable flow temperature. The only limitations may be the thickness of the base metal and the effect of brazing heat, or any previous heat treatment, on the metal properties.

BRAZING PROCEDURES

PRECLEANING AND SURFACE PREPARATION

Clean, oxide free surfaces are essential to ensure sound brazed joints of uniform quality. Grease, oil, dirt, and oxides prevent the uniform flow and bonding of the brazing filler metal, and they impair fluxing action resulting in voids and inclusions. With the more refractory oxides or the more critical atmosphere brazing applications, precleaning must be more thorough and the cleaned components must be preserved and protected from contamination.

The length of time that cleaning remains effective depends upon the metals involved, the atmospheric conditions, the amount of handling the parts may receive, the manner of storage, and similar factors. Therefore, it is recommended that brazing be done as soon as possible after the parts have been cleaned.

Degreasing is generally done first. The following degreasing methods are commonly used, and their action may be enhanced by mechanical agitation or by applying ultrasonic vibrations to the bath:

(1) Solvent cleaning: petroleum solvents or chlorinated hydrocarbons

(2) Vapor degreasing: stabilized trichloroethylene or stabilized perchloroethylene

(3) Alkaline cleaning: commercial mixtures of silicates, phosphates, carbonates, detergents, soaps, wetting agents and, in some cases, hydroxides

(4) Emulsion cleaning: mixtures of hydrocarbons, fatty acids, wetting agents, and surface activators

(5) Electrolytic cleaning: both anodic and cathodic

Scale and oxide removal can be accomplished mechanically or chemically. Prior degreasing allows intimate contact of the pickling solution with the parts, and vibration aids in descaling with any of the following solutions:

(1) Acid cleaning: phosphate type acid cleaners

(2) Acid pickling: sulfuric, nitric, and hydrochloric acid

(3) Salt bath pickling: electrolytic and nonelectrolytic

The selection of the chemical cleaning agent will depend on the nature of the contaminant, the base metal, the surface condition, and the joint design. For example, copper and silver containing base metals should not be pickled with nitric acid. In all cases, the chemical residue must be removed by thorough rinsing to prevent the formation of other equally undesirable films on the joint surfaces, or subsequent chemical attack of the base metal. Mechanical cleaning removes oxide and scale and also roughens the mating surfaces to enhance capillary flow and wetting of the brazing filler metal. Grinding, filing, machining, and wire brushing can be used. Grit blasting can also be used with clean blasting material that will not leave a deposit on the surfaces that may impair brazing.

FLUXING AND STOP-OFF

When a flux is selected for use, it must be applied as an even coating, completely covering the joint surfaces of the parts. Fluxes are most commonly applied in the form of pastes or liquids to insure adherence to and protection of the metal. Dry powdered flux may be sprinkled on the joint or applied to the heated end of the filler metal rod by simply dipping it into the flux container. The particle size should be small and thoroughly mixed to improve metal coverage and fluxing action. The areas surrounding the joints may be kept free from discoloration and oxidation by applying flux to a wide area on each side of the joint.

The paste and liquid flux should adhere to clean metal surfaces. If the metal surfaces are not clean, the flux will ball up and leave bare spots. The normally thick paste fluxes can be applied by brushing. Less viscous consistencies can be applied by dipping, hand squirting, or automatic dispensing. Required consistency is determined by the types of oxides present, as well as the heating cycle. For example, ferrous oxides generated by a fast heating method are soft and easy to remove, and only limited fluxing action is required. However, when joining copper or stainless steel or when the heating cycle is long, a higher concentration of flux is required. Flux reacts with oxygen, and once it becomes saturated, it loses all its effectiveness. The viscosity of the flux may be reduced without dilution by heating it to 50 to 60° C (120 to 140° F), preferably in a ceramic lined flux or glue pot with a thermostat control. Warm flux breaks down the surface tension and adheres to the metal more readily.

When the danger of flux entrapment must be minimized, a thin coating may be applied to the joint surfaces with a liberal amount of flux placed on the joint edges to prevent the entrance of air. Alternatively, a generous amount of flux is placed on the outside edges of the joint; and upon heating, a sufficient amount of flux will be drawn into the joint by capillary force. This method is only effective where small joints are involved, such as tubing up to 16 mm (5/8 in.) diameter. Type 5 brazing flux is used almost exclusively with torch or gas-air burner brazing by passing the fuel gas through the liquid flux container, thus entraining flux vapors in the fuel gas. Thus, the flux is supplied through the flame. Usually it is used with a small amount of preplaced flux or with a self-fluxing filler metal. Mixtures of powdered filler metal and flux have been successfully used for preplacement of both at the joint.

When control of the filler metal flow to definite areas is essential, it is usually done by controlling the amount of filler metal used and its method of placement. However, in atmosphere brazing or when flux is used to prevent oxidation of surfaces adjacent to the joint, stop-offs are employed to prevent the flow of the filler metal into areas that are not to be brazed. Commercial preparations generally consist of graphite or oxides of aluminum, chromium, titanium, or magnesium, prepared in the form of a water slurry or organic binder mixture. They are painted on the areas where flow of filler metal is not desired.

ASSEMBLY

The parts to be brazed should be assembled immediately after fluxing before the flux has time to dry and flake off. The most economical method of assembly is one in which the parts are self-locating and self-supporting.

When fixtures are essential for the maintenance of proper alignment and dimensional control, their design is affected by the complexity of the parts to be joined, the brazing temperature, and heating method. The mass of a fixture should be kept to a minimum with pinpoint or knife-edge contact to the parts, away from the joint area. These sharp contacts minimize heat loss through conduction to the fixture and subsequent interference with the flow of the brazing filler metal. The fixture must be made of materials with adequate strength at brazing temperature to support the brazement. It must clear the joints, and it must not readily alloy at elevated temperatures with the parts at the points of contact. In torch brazing, the fixture must provide access to the joint for the torch flame as well as the brazing filler metal. In furnace brazing, the fixtures must not be made of materials having objectional vapor pressures. In induction brazing, fixtures are generally made of ceramic materials to avoid introducing extraneous metal in the field of the induction coil. It may be desirable, particularly when ceramics are involved, to have the fixture serve as a heat shield or heat absorber as well.

POSTBRAZE TREATMENT

Parts that are brazed in a suitable atmosphere should be bright and clean, and require no further processing. However, if flux or stop-off material is present, it should be thoroughly removed. If the brazement requires heat treatment, it must be done below the solidus or remelt temperature of the brazing alloy.

Flux Removal

Regardless of the brazing process used, all traces of flux should be removed from the brazement. When sufficient amounts of flux have been used to prevent saturation by oxides and the temperature has been properly controlled, the flux residues may be removed by rinsing with hot water. The inside of assemblies should be flushed with hot water to remove the residual flux. Oxide saturated flux is glass-like and more difficult to remove. If the metal and joint design can withstand quenching, saturated flux can be removed by quenching from an elevated temperature. This treatment cracks off the flux coating. In some cases, it may become necessary to use a warm acid solution, such as 10 percent sulfuric acid solution, or one of the proprietary cleaning compounds which are available commercially for some applications. Nitric acid should not be used on alloys containing copper or silver. The parts should be thoroughly washed after a chemical dip and dried. Fluxes used for brazing some metals, such as aluminum, are not readily soluble in cold water. They are usually removed by rinsing in very hot water, above 82° C (180° F), with a subsequent immersion in nitric acid, hydrofluoric acid, or a combination of the two. A thorough water after-rinse is then nec-essary. Oxidized areas adjacent to the joint may be restored by chemical cleaning or by mechanical methods, such as wire brushing or blast cleaning.

Stop-off Removal

Brazing stop-off materials of the "parting compound" type can be easily removed mechanically by wire brushing, air blowing, or water flushing. The "surface reaction" type can best be removed by a hot nitric acid-hydrofluoric acid pickle, except in assemblies containing copper and silver. Solutions of sodium hydroxide (caustic soda) or ammonium bifluoride can be used in all applications, including copper and silver, because they will not attack the base or filler metals. There are a few stop-off materials that can readily be removed by dipping in a 5 to 10 percent solution of either nitric or hydrochloric acid. These stop-off materials are the easiest to remove completely, particularly when they are internal.

INSPECTION

Inspection of brazements involves many factors. The principles of the brazing process and the brazing operation are of fundamental importance. Some of these are the basic physical and metallurgical properties of brazed joints and base metals, testing methods, and the interpretation of the drawings and specifications. The inspection and frequency of testing methods chosen to evaluate the brazing procedure and the serviceability of the product will be largely dependent on the service requirements of the brazed product. Inspection methods are often specified by the manufacturer, by the ultimate user, or by regulatory codes. Inspection of brazed joints may be conducted on either test specimens or tests of the finished brazed assembly. The tests may be nondestructive or destructive. Each method of testing has a particular capability to disclose one or more of the discontinuities that may be present in the joint. Generally, brazing discontinuities are of three general classes:

(1) Those associated with drawing or dimensional requirements

(2) Those associated with structural discontinuities in the brazed joint

(3) Those associated with the braze metal or the brazed joint

NONDESTRUCTIVE TESTING METHODS

The objectives of nondestructive inspection of brazed joints should be (1) to seek out discontinuities which are evaluated to the requirements of the quality standards or codes, and (2) to obtain clues to the causes of irregularities in the fabricating process. The method selected must be weighed for advantages and limitations.

Visual Inspection

Visual examination is the most widely used of the nondestructive methods of inspection of brazed joints, and it is a convenient preliminary test when other test methods are to be used. To be effective, this type of examination requires that the joint to be inspected be readily examinable. The joint should be free from foreign materials: grease, paint, oil, oxide film, flux, and stop-off. Visual examination should reveal flaws due to damage, misalignment and poor fit-up of parts, dimensional inaccuracies, inadequate flow of brazing alloy, exposed voids in the joint, surface flaws such as cracks or porosity, and heat damage to base metal. Visual inspection will not detect internal flaws, such as flux entrapment in the joint or incomplete filler metal flow between the faying surfaces.

Proof Testing

Proof testing is a method of inspection that subjects the completed joint to loads slightly in excess of those that will be applied during its subsequent service life. These loads can be applied by hydrostatic methods, by tensile loading, by spin testing, or by numerous other methods. Occasionally, it is not possible to assure a serviceable part by any of the other nondestructive methods of inspection, and proof testing then becomes the most satisfactory method.

Leak Testing

Pressure testing is usually used to determine the gas or liquid tightness of a brazed vessel. It may be used as a screening method before using the more sensitive methods to find any gross leaks. A low pressure air or gas test may be done by one of three methods (sometimes used in conjunction with the pneumatic proof test): (1) submerging the pressurized vessel in water and noting any signs of leakage by rising air bubbles; (2) pressurizing the assembly, closing the air or gas inlet source, and then noting any change in internal pressure over a period of time (as corrections for temperature may be necessary); or (3) pressurizing the assembly and checking for leaks by brushing the joint area with a soap solution or a commercially available liquid and noting any bubbles and their source.

A method sometimes used in conjunction with a hydrostatic proof test is to examine the brazed joints visually for indications of the hydrostatic fluid escaping through the joint.

The leak testing of brazed assemblies with freon is an extremely sensitive method. The part under test is pressurized with either pure freon gas or a gas such as nitrogen, containing a tracer, usually Freon 12. Areas are sniffed or probed with a sampling device which is sensitive to the halide ion. The detection of a leak is indicated by a meter or an audible alarm. A leak may be measured quantitatively by this method. Precaution must be taken to avoid contaminating the surrounding air with freon which will decrease the sensitivity of the method.

A less sensitive method is to probe for leaks with a halide torch flame from a butane gas source. The presence of Freon 12 is indicated by a change in the flame color. Precautions must be taken to avoid using this method in the presence of combustible material.

The mass spectrometer method of leak testing is the most sensitive and accurate way of detecting extremely small leaks. A tracer gas such as helium, or hydrogen is used in conjunction with a mass spectrometer in one of three ways: (1) Evacuate the brazed assembly and surround the area to be tested with the tracer gas—the mass spectrometer is coupled to the interior, (2) Pressurize the brazed assembly with the tracer gas and sniff the exterior with a probe connected to the mass spectrometer, or (3) Pressurize the outside of the brazed assembly with a tracer gas in conjunction with the first way. A sensitive sensing device detects the tracer gas and converts it to an electrical signal. The location and size of the leak is determined by the response and the amplitude of the signal, usually displayed on a meter or readout device.

Liquid Penetrant Inspection

This method is used to find cracks, porosity, incomplete flow, and similar surface flaws in a brazed joint. Commercially colored or fluorescent penetrants have the ability to penetrate surface openings by capillary action. After the surface penetrant is removed, any penetrant in a flaw will be absorbed in a white developer that is applied to the surface. Colored penetrant is visible under ordinary light. Fluorescent penetrant flaw indications will glow when observed under an ultraviolet (black) light source. Since penetration of minute openings is involved, interpretation is sometimes difficult because of the irregularities in braze fillets and residues of flux deposits. Some other inspection method must be used to differentiate surface irregularities from joint discontinuities.

Radiographic Inspection

Radiographic inspection is used to detect lack of bond or incomplete flow of filler metal. When used for inspection of brazed joints, the joint should be of uniform thickness and the exposure made straight through the joint. The sensitivity of the method is generally limited to 2 percent of the joint thickness. The X-ray absorption characteristics of certain filler metals, such as gold and silver, are greater than those of most base metals. Therefore, areas in the joint that are void of braze metal show darker than the brazed area on the film or viewing screen.

Ultrasonic Inspection

The ultrasonic testing method uses low energy, high frequency mechanical vibration (sound waves) to detect, locate, or identify discontinuities in brazed joints. There are many methods available, but the principal one uses a single transducer and monitors the wave form of the pulse reflected from the surface of the facing sheet. The applicability of this method depends largely on the design of the joint and the configuration of the adjacent areas of the brazed assembly.

In the application and selection of a particular testing technique, consideration should be given to joint geometry, the metal to be tested, size and frequency of the transducer, the surface condition of the part to be tested, metallurgical grain structure in nonferrous metals, and the thickness of the material.

Thermal Heat Transfer Inspection

This method of inspection will detect lack of bond in brazed assemblies, such as honeycomb and covering skin panel surfaces. With one technique, the surfaces are coated with a developer which is a low melting point powder or liquid. The developer melts and migrates to cool areas upon the application of heat from an infrared lamp source. The bonded areas act like heat sinks, resulting in a thermal gradient to which the developer will react. More sophisticated techniques use thermal sensitive phosphors, liquid crystals, and other temperature-sensitive materials.

Infrared-sensitive electronic devices with some form of readout are available to monitor temperature differences of less than 1°C (1.8°F) that would be characteristic of variations in braze quality.

DESTRUCTIVE TESTING METHODS

Destructive methods of inspection are employed to ascertain whether a brazement design will meet the requirements for the intended service conditions. Destructive methods may be used for partial sampling and a check on the nondestructive methods of inspection by the selection of a production assembly at intervals and testing it to destruction. The methods and frequency are usually established by quality control procedures.

Destructive testing methods may be conducted on either test specimens or finished brazements. The three major types are (1) chemical, (2) metallographic, and (3) mechanical.

Metallographic Inspection

This method requires the removal of sections from the brazed joints and their preparation for macroscopic or microscopic examination. This method is useful for the detection of flaws, such as porosity, poor flow of brazing filler metal, excessive base metal erosion, the diffusion of brazing alloy, improper fit-up of the joint and, in addition, an evaluation of the microstructure of the brazed joint. When defects are found in the brazed joint, it may be an indication that the brazing procedure is out of control or that improper techniques are being used.

Peel Tests

Peel tests are frequently employed for evaluating lap type joints. One member of the brazed specimen is clamped rigidly in a vise, and the free member is peeled away from the joint. This method is used to determine the general quality of the bond and the presence of voids and flux inclusions in the joint. The permissible number, size, and distribution of these discontinuities will depend upon the service conditions intended for the joint and may be defined by specification or code.

Tension and Shear Tests

The tension or shear test method is used to determine quantitatively the strength of the brazed joint or to verify the relative strengths of the joint and base metal parts. This method is widely used for the development of brazing procedures. Random sampling of brazed joints is used for quality control and verification of the brazing procedure.

Torsion Tests

The torsion test is frequently used for inspection to evaluate brazed joints where a stud, screw, or tubular member is brazed to a base member. The base member is clamped rigidly and the stud, screw, or tubular member is rotated to failure in either the base metal or the brazed joint. It is used on a sampling basis to verify the brazing procedure.

SAFE PRACTICES IN BRAZING

Hazards encountered with brazing operations are similar to those associated with welding and cutting processes. Brazing of metals may require temperatures at which some elements vaporize. Personnel and property must be protected against hot materials, gases, fumes, electrical shock, radiation, and chemicals.

The operation and maintenance of brazing equipment or facility should conform to the provisions of American National Standard Z49.1, Safety in Welding and Cutting. This standard provides detailed procedures and instructions for safe practices which will protect personnel from injury or illness, and property and equipment from damage by fire or explosion arising from brazing operations.

It is essential that adequate ventilation be provided so that personnel will not inhale gases and fumes generated while brazing. Some filler metals and base metals contain toxic materials such as cadmium, beryllium, zinc, mercury, or lead, which are vaporized during brazing. Fluxes contain chemical compounds of fluoride, chlorine, and boron, which are harmful if they are inhaled or contact the eyes or skin.

Solvents such as chlorinated hydrocarbons and cleaning compounds, such as acids and alkalies, may be toxic or flammable or cause chemical burn when present in the brazing environment.

Requirements for the purging of furnaces or retorts that will contain a flammable brazing atmosphere are also given in the standard. In addition, to avoid suffocation care must be taken with atmosphere furnaces to insure that the furnace is purged with air before personnel enter it.

BRAZE WELDING

INTRODUCTION

Braze welding is accomplished by the use of a filler metal having a liquidus above 450° C (840° F) and below the solidus of the base metals. Braze welding differs from brazing in that the filler metal is *not* distributed in the joint by capillary attraction. The filler metal is added to the joint as welding rod or as an arc welding electrode.[6] However, the base metals are not melted, only the filler metal. Bonding takes place between the deposited filler metal and the hot unmelted base metals in the same manner as conventional brazing. Joint designs for braze welding are similar to those used for oxyacetylene welding.

Braze welding was originally developed for the repair of cracked or broken cast iron parts. Fusion welding of cast iron requires extensive preheating and slow cooling to minimize the development of cracks and the formation of hard cementite. With braze welding, these conditions are easier to avoid, and fewer expansion and contraction problems are encountered.

To obtain a strong bond between the filler metal and the unmelted base metal, the molten filler metal must wet the hot base metal. The base metal is heated to the required temperature with an oxyfuel gas torch or an electric arc. With torch brazing, flux is generally required to clean both the base metal and previously deposited filler metal and to aid in the precoating (tinning) of the groove faces. Brazing rod is added to produce one or more weld passes, precoating both members as braze welding progresses. Applying flux as required, the joint is finally filled to the desired size.

Although most braze welding is done using an oxyfuel gas welding torch, copper alloy brazing rod, and a suitable flux, the process can be done using a carbon arc, gas tungsten arc, or plasma arc torch without flux. The carbon arc torch is used with galvanized sheet steel. The other two heating methods which employ inert shielding gases are suitable for braze welding with filler metals that have relatively high melting temperatures. Filler metal selection, proper wetting of the base metals, and shielding from air are important considerations for the effective use of the process with any suitable heating torch. Braze welding often has the following advantages over conventional fusion welding processes:

(1) Less heat is required to accomplish bonding, which permits higher joining speeds, lower fuel consumption, and shorter welding time. The process also produces less distortion from thermal expansion and contraction than do most ordinary welding processes.

(2) The deposited filler metal is relatively soft and ductile, providing machinability and low residual stresses.

(3) Welds can be produced with adequate strength for many applications.

(4) The equipment is simple and easy to use.

(5) Metals that are brittle, such as gray cast iron, can be braze welded without extensive preheat.

(6) The process provides a convenient way to join dissimilar metals, for example copper to steel and cast iron; nickel-copper alloys to cast iron and steel.

Disadvantages of braze welding are that

(1) Weld strength is limited to that of the filler metal.

(2) Permissible performance temperatures of the product are lower than those of fusion welds because of the lower melting temperature of the filler metal. With copper alloy filler metal, service is limited to 260° C (500° F) or lower.

(3) The braze welded joint may be subject to galvanic corrosion and differential chemical attack.

(4) The brazing filler metal color may not match that of the base metal.

6. Braze welding of cast iron is sometimes done by the shielded metal arc welding process. See Chapter 2.

EQUIPMENT

Conventional braze welding is done using an oxyfuel gas welding torch and associated equipment described in Chapter 10. In some applications, an oxyfuel preheating torch may be needed. Special applications use carbon arc, gas tungsten arc, or plasma arc welding equipment described in other chapters of the Handbook. Clamping and fixturing equipment might also be needed to hold the parts in place and align the joint.

MATERIALS

Base Metals

Braze welding is generally used for joining cast iron and steel. It can also be used to join copper, and nickel and nickel alloys. Certain other metals can also be braze welded using suitable filler metals that will wet and form a strong metallurgical bond with them. Dissimilar weldments between many of the above metals are possible with braze welding if suitable filler metals are used.

Filler Metals

The commercial torch brazing filler metals are generally brasses containing approximately 60 percent copper and 40 percent zinc. Small additions of tin, iron, manganese, and silicon may be added to improve flow characteristics, decrease volatilization of the zinc, scavenge oxygen, and increase strength and hardness. Nickel is added (10 percent) to whiten the color and increase the weld metal strength. Chemical compositions and properties of four standard copper-zinc welding rods used for braze welding are given in Table 11.8. The minimum joint tensile strength will be approximately 275 to 413 MPa minimum (40 to 60 ksi) depending on the filler metal used. The joint strength will decrease rapidly when the weldment is above 260° C (500° F).

Because a braze weld is a bimetal joint, corrosion must be considered in its application. The completed joint may be subject to galvanic corrosion in certain environments, and the filler metal may be less resistant to certain chemical solutions than the base metal.

Fluxes

Fluxes for braze welding are proprietary compounds developed for braze welding of certain base metals with brass filler metal rods. They are designed for use at higher temperatures, and they remain active for longer times at temperature than similar fluxes used for capillary brazing. The following types of flux are in general use for braze welding of iron and steels:

(1) A basic flux that cleans the base metal and weld beads and assists in the precoating (tinning) of the base metal. It is used for steel and malleable iron.

(2) A flux that performs the same functions as the basic flux and also suppresses the formation of zinc oxide fumes.

(3) A flux that is formulated specifically for braze welding of gray or malleable cast iron. It contains iron oxide or maganese dioxide that combines with the free carbon on the cast iron surface to remove it.

	Table 11.8—Copper-zinc welding rods for braze welding								
AWS classification[a]	Approximate chemical composition, %					Min tensile strength		Liquidus temperature	
	Copper	Zinc	Tin	Iron	Nickel	MPa	ksi	°C	°F
RBCuZn-A	60	39	1			275	40	900	1650
RBCuZn-B	60	37.5	1	1	0.5	344	50	890	1630
RBCuZn-C	60	38	1	1		344	50	890	1630
RBCuZn-D	50	40			10	413	60	935	1715

a. See AWS Specifications A5.7 and A5.8 for additional information.

Flux may be applied by one of the following four methods:

(1) The heated filler rod may be dipped into the flux and transferred to the joint during braze welding.

(2) The flux may be brushed on the joint prior to brazing.

(3) The filler rod may be precoated with flux.

(4) The flux may be introduced through the oxyfuel gas flame.

METALLURGICAL CONSIDERATIONS

The bond between the filler metal and the base metal is the same type with braze welding as with conventional brazing. The clean base metal is heated to a temperature which is sufficient to permit the wetting of its surface by the molten filler metal, producing a metallurgical bond between them. Cleanliness is important to obtain wetting. The presence of dirt, oil, grease, oxide film, carbon, etc. will inhibit wetting.

Wetting is the first action needed to accomplish bonding. Following this, atomic diffusion takes place between the brazing filler metal and the base metal in a narrow zone at the interface. Also, with some base metals, the brazing alloy may slightly penetrate the grain boundaries of the base metal and contribute to bond strength.

Braze welding filler materials are alloys that have adequate as-cast ductility to flow plastically during solidification and subsequent cooling to accommodate shrinkage stresses. They cannot be two-phase alloys, having a low melting grain boundary constituent, that would crack during solidification and cooling.

GENERAL PROCESS APPLICATIONS

The greatest use of braze welding is the repair of broken or defective steel and cast iron parts. Since large components can be repaired in place, significant cost savings are often realized. It is also used to rapidly join thin gage, mild steel sheet and tubing where fusion welding would be difficult. Galvanized steel duct work is braze welded using a carbon arc heat source. The brazing temperature is below the vaporization temperature of the zinc. This minimizes the loss of the protective zinc coating from the steel surfaces and also exposure of the welder to a significant amount of zinc fumes.

The thicknesses of metals that can be braze welded range from thin gage sheet to very thick cast iron sections. Fillet and groove welds are used to make butt, corner, lap, and T-joints.

BRAZE WELDING PROCEDURE

Fixturing

Adequate fixturing is usually required to hold parts in their proper location and alignment for braze welding. In repairing cracks and other defects in cast iron parts, no fixturing may be necessary unless the part is broken or separated.

Joint Preparation

Joint designs are similar to those used for oxyacetylene welding. For thicknesses over 2 mm (3/32 in.) single- or double-V-grooves are prepared with a 90 to 120° included angle to provide large bond areas between the base metal and the filler metal. Square grooves may be used for a thickness of 2 mm (3/32 in.) and less.

The prepared joint faces and adjacent surfaces of the base metal must be cleaned to remove all oxide, dirt, grease, oil, and other foreign material that will inhibit bonding. When braze welding cast iron, the joint faces must also be free of graphite smears caused by prior machining. The graphite smears can be removed by quickly heating the cast iron to a dull red color and then wire brushing it after it cools to black heat. If the casting has been heavily soaked with oil, it should be heated in the range of 320° C (600° F) to 650° C (1200°F) to burn off the oil. The surfaces should be wire brushed to remove any residue.

In production braze welding of cast iron components, the surfaces to be joined are usually cleaned by immersion in an electrolytic molten salt bath. This is the most effective method of removing the free graphite present on machined or broken surfaces.

Preheating

Preheating of the parts to be joined will depend on the type of base metal and the size of the part(s). It may be required to prevent cracking from thermally induced stresses in large cast iron parts. Preheating will aid the braze welding of copper by reducing the amount of heat required from the brazing torch and the time required to complete the joint.

Preheating may be local or general, depending on the size of the part. Preheat temperature should be in the 425 C to 480° C (800-900° F) range for cast iron. Higher temperatures can be used for copper. After the brazing operation is completed on cast iron parts, they should be thermally insulated for slow cooling to room temperature. This will minimize the development of thermally induced stresses in the casting.

Methods of Heating

Braze welding is generally performed with an oxyfuel gas torch, using a neutral or slightly oxidizing flame. A carbon arc and gas tungsten arc can also be used with copper-silicon and copper-aluminum filler metals. No flux is used with electric arc heating.

Technique

The joint to be braze welded must be properly aligned and fixtured in position. Braze welding flux, when required, is applied to preheated filler rod (unless precoated) and also sprinkled on thick joints during heating with an oxyfuel gas torch. The base metal is heated with the flame until the filler metal melts, wets the base metal, and flows onto the joint faces (precoating). The joint faces are precoated with filler metal as braze welding operation progresses along the joint. The joint is filled with one or more passes using operating techniques similar to oxyfuel gas welding. The inner cone of an oxyacetylene flame should not be directed on copper-zinc alloy filler metals nor on iron or steel base metal. The techniques used with electric arc heating torches are similar to those described for oxyfuel gas braze welding except that flux is not generally used.

TYPES OF WELDS

A wide variety of parts can be braze welded with the use of typical weld joint designs. Groove, fillet, and edge welds can be used to join simple and complex assemblies made from sheet and plate, pipe and tubing, rods and bars, castings and forgings. To obtain good joint strength, adequate bond area between the brazing alloy and the base metal is required. Weld groove geometry should provide adequate groove face area so that the joint will not fail along the interfaces.

WELD QUALITY

Braze quality will depend on the care and control exercised during the preparation of the base metal and the braze welding operation. The strength and soundness of the joint will depend on the filler metal used, cleanliness of the base metal, proper flame adjustment and heat input control, and the adequacy of fluxing during the operation to insure complete wetting of the base metal. Using correct procedures, braze welded joints can be produced that are as strong as the adjacent base metal.

SUPPLEMENTARY READING LIST

Aluminum Brazing Handbook. New York: The Aluminum Association, 1971.

Aspden, R.G., and Feduska, W. Fatigue characteristics of single-lap joint of AISI 347 brazed with a Ni-Cr-Si-B-C alloy. *Welding Journal* 37 (3), Mar. 1968, pp. 125-s—128-s.

Brazing Manual, 3rd ed. Miami: American Welding Society, 1976.

Cole, N.C., Gunkel, R.W., and Koger, J.W. Development of corrosion resistant filler metals for brazing molybdenum. *Welding Journal* 52 (10), Oct. 1973, pp. 446-s—473-s.

Gilliland, R.G., and Slaughter, G.M. The development of brazing filler metals for high temperature service. *Welding Journal* 48 (10), Oct. 1969, pp. 463-s—469-s.

Helgesson, C.I. *Ceramic-to-Metal Bonding.* Cambridge, Mass.: Boston Technical Publishers, Inc., 1968.

Kawakatsu, I. Corrosion of BAg brazed joints in stainless steel. *Welding Journal* 52 (6), June 1973, pp. 233-s—239-s.

Patrick, E.P. Vacuum brazing of aluminum. *Welding Journal* 54 (6), Mar. 1975, pp. 159-163.

Pattee, H.E. Joining Ceramics to Metals and Other Materials. Bulletin 178. New York: Welding Research Council, Nov. 1972.

Pattee, H.E. High-Temperature Brazing, Bulletin 187. New York: Welding Research Council, Sept. 1973.

Rugal, V., Lehka, N., and Malik, J.K. Oxidation resistance of brazed joints in stainless steel. *Metal Construction and British Welding Journal,* June 1974, pp. 183-187.

Safety in Welding and Cutting, ANSI Z49.1. Miami: American Welding Society, 1973.

Schultze, W., and Schoer, H. Fluxless brazing of aluminum using protective gas. *Welding Journal* 52 (10), Oct. 1973, pp. 644-651.

Schwartz, M.M. The Fabrication of Dissimilar Metal Joints Containing Reactive and Refractory Metals. Bulletin 210. New York: Welding Research Council, Oct. 1975.

Schwartz, M.M. Brazed Honeycomb Structures. Bulletin 182. New York: Welding Research Council, Apr. 1973.

Schwartz, M.M. *Modern Metal Joining Techniques.* John Wiley & Sons, Sept. 1969.

Slaughter, G.M. Welding and Brazing Techniques for Nuclear Reactor Components. An AEC Monograph, prepared for American Society for Metals. New York: Rowman and Littlefield, Inc., 1964.

Terrill, J.R., *et al,* Understanding the mechanisms of aluminum brazing. *Welding Journal* 50 (12), Dec. 1971, pp. 833-839.

Welding and Brazing, Metals Handbook, Vol. 6, 8th ed. Metals Park, Ohio: American Society for Metals, 1971.

12

Soldering

Prepared by a committee consisting of

W. G. BADER, *Chairman*
   Bell Laboratories

R. E. BEAL
   *Packer Engineering
   Associates*

F. C. DISQUE
   Alpha Metals, Inc.

R. O. JOHNSON
   Hexacon Electric Co.

J. B. LONG
   Tin Research Institute

E. J. MINARCIK
   NL Industries

J. F. SMITH
Lead Industries Association, Inc.

12
Soldering

FUNDAMENTALS OF PROCESS

DEFINITION AND GENERAL DESCRIPTION

Soldering is defined as a group of welding processes which produce coalescence of materials by heating them to a suitable temperature and by using a filler metal (solder) having a liquidus not exceeding 450° C (840° F) and below the solidus of the base metals. The solder is usually distributed between the properly fitted surfaces of the joint by capillary attraction.

The bond between solder and base metal is more than adhesion or mechanical attachment, although these do contribute to the strength. Rather, the essential feature of the soldered joint is that a metallic bond is produced by a metal solvent action. The solder dissolves (not melts) a small amount of the base metal to form a layer of an intermetallic compound. Upon solidification, the joint is held together by the same attraction between adjacent atoms that holds a piece of solid metal together. The ease of wetting is related to the ease with which this solvent action occurs.

SOLDERING PROCESS CONSIDERATIONS

A sound soldered joint is achieved by the selection and use of the proper materials and processes. Each of the following factors contributes to sound soldered joints and should be carefully considered.

Base Metal Selection

Base metals are usually selected for specific property requirements that are needed for the component or part design. These include strength, ductility, electrical conductivity, weight, corrosion resistance, and other properties. When soldering is required, the solderability of the base materials must also be considered in their selection. The selection of flux and the surface preparation will be affected by the solderability of the base materials to be joined.

Solder Selection

The solder is selected to provide good flow, penetration and wettability in the soldering operation, and the desired joint properties in the finished product.

Flux Selection

A flux is intended to enhance the wetting of the base materials by the solder by removing tarnish films from precleaned surfaces and by preventing oxidation during the soldering operation. The selection of the type of flux usually depends on the ease with which a material can be soldered. Rosin fluxes are used with solderable base metals or metals that are precoated with a solderable finish, while inorganic fluxes are often used on metals such as stainless steel. Table 12.1 indicates the relative ease, based on flux requirements, with which a number of alloys and metals can be soldered.

Base metal, alloy, or applied finish	Flux type				Soldering not recommended[a]
	Rosin	Organic	Inorganic	Special flux and/or solder	

Table 12.1—Flux requirements for metals, alloys, and coatings

Base metal, alloy, or applied finish	Rosin	Organic	Inorganic	Special flux and/or solder	Soldering not recommended[a]
Aluminum	–	–	–	X	–
Aluminum-bronze	–	–	–	X	–
Beryllium	–	–	–	–	X
Beryllium-copper	X	X	X	–	–
Brass	X	X	X	–	–
Cadmium	X	X	X	–	–
Cast iron	–	–	–	X	–
Chromium	–	–	–	–	X
Copper	X	X	X	–	–
Copper-chromium	–	–	X	–	–
Copper-nickel	X	X	X	–	–
Copper-silicon	–	–	X	–	–
Gold	X	X	X	–	–
Inconel	–	–	–	X	–
Lead	X	X	X	–	–
Magnesium	–	–	–	–	X
Manganese-bronze (high tensile)	–	–	–	–	X
Monel	–	X	X	–	–
Nickel	–	X	X	–	–
Nickel-iron	–	X	X	–	–
Nichrome	–	–	–	X	–
Palladium	X	X	X	–	–
Platinum	X	X	X	–	–
Rhodium	–	–	X	–	–
Silver	X	X	X	–	–
Stainless steel	–	X	X	–	–
Steel	–	–	X	–	–
Tin	X	X	X	–	–
Tin-bronze	X	X	X	–	–
Tin-lead	X	X	X	–	–
Tin-nickel	–	X	X	–	–
Tin-zinc	X	X	X	–	–
Titanium	–	–	–	–	X
Zinc	–	X	X	–	–
Zinc die castings	–	–	–	–	X

a. With proper procedures, such as precoating, most metals can be soldered.

Joint Design

Joints should be designed to fulfill the requirements of the finished assembly and to permit application of the flux and solder by the soldering process that is utilized. Joints should be designed so that proper clearance is maintained during heating and cooling of the soldering operation. Special fixtures may be necessary or the units can be crimped, clinched, wrapped, or otherwise held together.

Precleaning

All metal surfaces to be soldered should be cleaned before assembly to facilitate wetting of the base metal by the solder. Flux should not be considered as a substitute for precleaning. Precoating may be necessary for base materials that are difficult to solder.

Soldering Process

The soldering process should be selected to provide the proper soldering temperature, heat distribution, and rate of heating and cooling required for the product being assembled. The application of the solder and flux will be dictated by the selection of the soldering process.

Flux Residue Treatment

Generally, flux residues should be removed after soldering.

SOLDERS

TIN-LEAD SOLDERS

Solders of the tin-lead alloy system constitute the largest portion of all solders in use. They are used for joining most metals and have good corrosion resistance to most media. Most cleaning and soldering processes may be used with the tin-lead solders. Fluxes of all types can also be used; the choice depends on the base metal to be joined.

In describing solders, it is customary to give the tin content first. As an example, 40/60 solder is 40 percent tin and 60 percent lead.

The behavior of the various tin-lead alloys can best be illustrated by their constitutional diagram (Fig. 12.1).

The following terms are used to describe this diagram:

(1) *Solidus temperature* is the highest temperature at which a metal or alloy is completely solid, curve ACEDB.

(2) *Liquidus temperature* is the lowest temperature at which a metal or alloy is completely liquid, curve AEB.

(3) *Eutectic alloy* is an alloy that behaves like a pure metal in that it melts at one temperature and not over a range, point E.

(4) *Melting range* is the temperature differential between the solidus ACEDB and the liquidus AEB in which the solder is partially melted.

As shown in the diagram, 100 percent lead melts at 327° C (621° F), point A, and 100 percent tin melts at 232° C (450° F), point B. Solders containing 19.5 percent (point C) to 97.5 percent (point D) tin have the same solidus temperature, which is 183° C (361° F). The diagram shows that the eutectic composition is approximately 63 percent tin and 37 percent lead, point E. This composition becomes completely liquid at 183° C (361° F). Any composition other than a eutectic composition will not become completely liquid until a higher temperature is reached. For example, 50/50 solder has a solidus temperature of 183° C (361° F) and a liquidus temperature of 214° C (417° F), or a melting range of 31° C (56° F), the difference between the solidus and liquidus.

Melting characteristics of the tin-lead solders are shown in Table 12.2.

The 5/95 solder is a relatively high melting temperature solder with a short melting range. Its wetting and flow characteristics are poor compared to solders with higher tin contents.

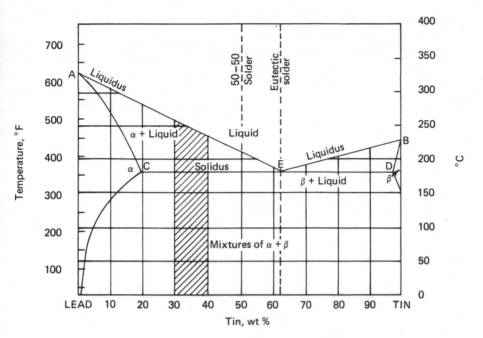

Fig. 12.1 — Constitutional diagram for the tin-lead alloy system

Table 12.2 — Tin-lead solders

ASTM solder classification[a]	Composition, wt. %		Solidus		Liquidus		Melting range	
	Tin	Lead	°C	°F	°C	°F	°C	°F
5	5	95	300	572	314	596	14	24
10	10	90	268	514	301	573	33	59
15	15	85	225	437	290	553	65	116
20	20	80	183	361	280	535	97	174
25	25	75	183	361	267	511	84	150
30	30	70	183	361	255	491	72	130
35	35	65	183	361	247	477	64	116
40	40	60	183	361	235	455	52	94
45	45	55	183	361	228	441	45	80
50	50	50	183	361	217	421	34	60
60	60	40	183	361	190	374	7	13
70	70	30	183	361	192	378	9	17

a. See ASTM Specification B32, *Solder Metal*.

necessitating extra care in surface preparation. This high lead solder has better mechanical properties at 149° C (300° F) than solders containing more tin. The high soldering temperature limits the use of organic base fluxes, such as rosin or those of the intermediate type. This solder is particularly adaptable to torch, dip, induction, or oven soldering. It is used for sealing precoated containers, for coating and joining metals, and for moderately elevated temperature uses.

The 10/90, 15/85, and 20/80 solders have

lower liquidus and solidus temperatures but wider melting ranges than the 5/95 solder. The wetting and flow characteristics are also better. However, to prevent hot tearing, extreme care must be taken to avoid movement of the solder during solidification. Fluxes of all types and all soldering methods are applicable. These solders are used for sealing cellular automobile radiators, for filling joints and dents in automobile bodies, and for the coating and joining of metals.

The 25/75 and 30/70 solders have lower liquidus temperatures than all previously mentioned alloys but have the same solidus temperature as the 20/80 solder. Therefore, their melting ranges are narrower than that of the 20/80 solder. All standard cleaning, fluxing, and soldering techniques can be used with these solders. Torch soldering is widely used.

The 35/65, 40/60, and 50/50 solders have low liquidus temperatures. The solidus temperature is the same as the 20 percent to 30 percent tin solders and the melting ranges are narrower. Solders of this group have the best combination of wetting properties, strength, and economy; and for these reasons they are widely used. They are the general purpose solders being used extensively in sheet metal work and plumbing. They are also used as a rosin-cored wire for radio and television applications.

The 60/40 solder is used wherever temperature requirements are critical, such as in delicate instruments. The composition is close enough to that of the eutectic tin-lead alloy to have an extremely narrow melting range. All methods of cleaning, fluxing, and heating may be used with this solder.

The 70/30 solder is a special purpose solder used where a high tin content is required. All soldering techniques are applicable.

TIN-ANTIMONY SOLDER

The 95 percent tin-5 percent antimony solder has the melting characteristics shown in Table 12.3. It provides a narrow melting range at a temperature higher than the tin-lead eutectic. The solder is used in many plumbing, refrigeration, and air conditioning applications because of its good creep strength.

TIN-ANTIMONY-LEAD SOLDERS

Antimony may be added to a tin-lead solder as a substitute for some of the tin. The addition of antimony up to 6 percent of the tin content increases the mechanical properties of the solder with only slight impairment to the soldering characteristics. All standard methods of cleaning, fluxing, and heating may be used.

TIN-SILVER AND TIN-LEAD-SILVER SOLDERS

The solders containing silver are listed in Table 12.4 with their melting characteristics. The 96 percent tin-4 percent silver solder is free of lead and is often used to join stainless steel for food handling equipment. It has good shear and creep strengths and excellent flow characteristics.

The 62 percent tin-38 percent lead-2 percent silver solder is used when soldering to silver-coated surfaces for electronic applications. The silver addition retards the dissolution of the silver coating during the soldering operation. The addition of silver also increases creep strength.

The high lead solders containing tin and silver provide higher temperature solders for many applications. They exhibit good tensile,

Table 12.3—Tin-antimony solder							
Composition, weight %		Solidus		Liquidus		Melting range	
Tin	Antimony	°C	°F	°C	°F	°C	°F
95	5	232	450	240	464	8	14

Table 12.4—Tin-silver and tin-lead-silver solders

Composition, weight %			Solidus		Liquidus		Melting range	
Tin	Lead	Silver	°C	°F	°C	°F	°C	°F
96	–	4	221	430	221	430	0	0
62	36	2	·180	354	190	372	10	18
5	94.5	0.5	294	561	301	574	7	13
2.5	97	0.5	303	577	310	590	7	13
1	97.5	1.5	309	588	309	588	0	0

Table 12.5—Tin-zinc solders

Composition, weight %		Solidus		Liquidus		Melting range	
Tin	Zinc	°C	°F	°C	°F	°C	°F
91	9	199	390	199	390	0	0
80	20	199	390	269	518	70	128
70	30	199	390	311	592	112	202
60	40	199	390	340	645	141	255
30	70	199	390	375	708	176	318

Table 12.6—Cadmium-silver solder

Composition, weight %		Solidus		Liquidus		Melting range	
Cadmium	Silver	°C	°F	°C	°F	°C	°F
95	5	338	640	393	740	55	100

shear, and creep strengths; they are recommended for cryogenic applications. Because of their high melting range, only inorganic fluxes are recommended for use with these solders.

TIN-ZINC SOLDERS

A large number of tin-zinc solders, some of which are listed in Table 12.5, have come into use for the joining of aluminum. Galvanic corrosion of soldered joints in aluminum is minimized if the metals in the joint are close to each other in the electrochemical series. Alloys containing 70 to 80 percent tin with the balance zinc are recommended for soldering aluminum. The addition of 1 to 2 percent aluminum, or an increase of the zinc content to as high as 40 percent, improves corrosion resistance. However, the liquidus temperature rises correspondingly, and they are therefore more difficult to apply. The 91/9 and 60/40 tin-zinc solders may be used for high temperature applications (above 149° C/300°F), while the 80/20 and the 70/30 tin-zinc solders are generally used to coat parts before soldering.

CADMIUM-SILVER SOLDER

The 95 percent cadmium-5 percent silver solder has the melting characteristics shown in Table 12.6. Its primary use is in applications where service temperatures will be higher than permissible with lower melting solders. At room temperature, butt joints in copper can be made to produce tensile strengths of 170 MPa (25 000 psi). At 219°C (425°F), a tensile strength of 18 MPa (2600 psi) can be obtained. Joining aluminum to itself or to other metals is possible with this solder. Improper use of solders containing cadmium may lead to health hazards. Therefore, care should be taken in their application, particularly with respect to fume inhalation.

CADMIUM-ZINC SOLDERS

These solders are useful for soldering aluminum. Their characteristics are given in Table 12.7. The cadmium-zinc solders develop joints with intermediate strength and corrosion resistance when used with the proper flux. The 40 percent cadmium-60 percent zinc solder has found considerable use in the soldering of aluminum lamp bases. Improper use of this solder may lead to health hazards, particularly with respect to fume inhalation.

ZINC-ALUMINUM SOLDER

This solder, shown in Table 12.8, is specifically for use on aluminum. It develops joints with high strength and good corrosion resistance. The solidus temperature is high, which limits its use

to applications where soldering temperatures in excess of 371°C (700°F) can be tolerated. A major application is in dip soldering the return bends of aluminum air conditioner coils. Ultrasonic solder pots are employed without the use of flux. In manual operations, the heated aluminum surface is rubbed with the solder stick to promote wetting without a flux.

FUSIBLE ALLOYS

Bismuth-containing solders, the so-called fusible alloys, are useful for soldering operations where soldering temperatures below 183°C (361°F) are required. The melting characteristics and compositions of a representative group of fusible alloys are given in Table 12.9.

The low melting temperature solders have applications in cases such as (1) soldering heat

Table 12.7—Cadmium-zinc solders

Composition, weight %		Solidus		Liquidus		Melting range	
Cadmium	Zinc	°C	°F	°C	°F	°C	°F
82.5	17.5	265	509	265	509	0	0
40	60	265	509	335	635	70	126
10	90	265	509	399	750	134	241

Table 12.8—Zinc-aluminum solder

Composition, weight %		Solidus		Liquidus		Melting range	
Zinc	Aluminum	°C	°F	°C	°F	°C	°F
95	5	382	720	382	720	0	0

Table 12.9—Typical fusible alloys

Alloy	Composition, weight %				Solidus		Liquidus		Melting range	
	Lead	Bismuth	Tin	Other	°C	°F	°C	°F	°C	°F
Lipowitz	26.7	50	13.3	10 Cd	70	158	70	158	0	0
Bending (Wood's Metal)	25	50	12.5	12.5 Cd	70	158	74	165	4	7
Eutectic	40	52	–	8 Cd	91	197	91	197	0	0
Eutectic	32	52.5	15.5	– –	95	203	95	203	0	0
Rose's	28	50	22	– –	96	204	107	225	11	25
Matrix	28.5	48	14.5	9 Sb	102	217	227	440	125	223
Mold and pattern	44.5	55.5	–	– –	124	255	124	255	0	0

treated surfaces where higher soldering temperatures would result in the softening of the part, (2) soldering joints where adjacent material is very sensitive to temperature and would deteriorate at higher soldering temperatures, (3) step soldering operations where a low soldering temperature is necessary to avoid destroying a nearby joint that has been made with a higher melting temperature solder, and (4) on temperature-sensing devices, such as fire sprinkler systems, where the device is activated when the fusible alloy melts at relatively low temperature.

Many of these solders, particularly those containing a high percentage of bismuth, are very difficult to use successfully in high-speed soldering operations. Particular attention must be paid to the cleanliness of metal surfaces. Strong, corrosive fluxes must be used to make satisfactory joints on uncoated surfaces of metals, such as copper or steel. If the surface can be plated for soldering with such metals as tin or tin-lead, noncorrosive rosin fluxes may be satisfactory; however, they are not effective below 177°C (350°F).

INDIUM SOLDERS

These solders possess certain properties which make them valuable for some special applications. Their usefulness for any particular application should be checked with the supplier. Melting characteristics and compositions of a representative group of these solders are shown in Table 12.10.

A 50 percent indium-50 percent tin alloy adheres to glass readily and may be used for glass-to-metal and glass-to-glass soldering. The low vapor pressure of this alloy makes it useful for seals in vacuum systems.

Indium solders do not require special techniques during use. All of the soldering methods, fluxes, and techniques used with the tin-lead solders are applicable to indium solders.

SOLDER SPECIFICATIONS

Specifications for solders have been published by the American Society for Testing and Materials (ASTM B32, Solder Metal; ASTM B284, Rosin Flux-Cored Solder; and ASTM

Table 12.10—Typical indium solders

Composition, weight %			Solidus		Liquidus		Melting range	
Tin	Indium	Lead	°C	°F	°C	°F	°C	°F
50	50	–	117	243	125	257	8	14
37.5	25	37.5	138	230	138	230	0	0
–	50	50	180	356	209	408	29	52

Table 12.11—Commercial solder product forms

Pig	Available in 25 and 45 kg (50 and 100 lb) pigs
Ingots	Rectangular or circular in shape, weighing 1.4, 2.3, and 4.5 kg (3, 5, and 10 lb)
Bars	Available in numerous cross sections, weights, and lengths
Paste or cream	Available as a mixture of powdered solder and flux
Foil, sheet, or ribbon	Available in various thicknesses and widths
Segment or drop	Triangular bar or wire cut into any desired number of pieces or lengths
Wire, solid	Diameters of 0.25 to 6.35mm (0.010 to 0.250 in.) on spools
Wire, flux cored	Solder cored with rosin, organic, or inorganic fluxes. Diameters of 0.25 to 6.35 mm (0.010 to 0.250 in.)
Preforms	Unlimited range of sizes and shapes to meet special requirements

B486, Paste Solder) and by the United States Government (Federal Specification QQ-S-571, Solders).

Solders are commercially available in various sizes and shapes which can be grouped into about a dozen classifications. The major groups are listed in Table 12.11. This listing is by no means complete, inasmuch as any desired size, weight, or shape is available on special order.

FLUXES

A soldering flux is a liquid, solid, or gaseous material which, when heated, is capable of promoting or accelerating the wetting of metals by solder. The purpose of a soldering flux is to remove and exclude small amounts of oxides and other surface compounds from the surfaces being soldered. Anything that interferes with the attainment of uniform contact between the surface of the base metal and the molten solder will prevent the formation of a sound joint. An efficient flux removes tarnish films and oxides from the base metal and solder and prevents reoxidation of the surfaces during the soldering process. The flux should be readily displaced by the molten solder.

A functional method of classifying flux is based on the flux's ability to remove metal tarnishes (activity). Fluxes may be classified in three groups: inorganic fluxes (most active), organic fluxes (moderately active), and rosin fluxes (least active).

INORGANIC FLUXES

This class of fluxes consists of inorganic acids and salts which are highly corrosive. These fluxes are used to best advantage where conditions require rapid and highly active fluxing action. They can be applied as solutions, pastes, or dry salts. They function equally well with torch, oven, resistance, or induction soldering methods since they do not char or burn. These fluxes can be formulated to provide stability over the entire soldering temperature range.

The inorganic fluxes have one distinct disadvantage in that the residue remains chemically active after soldering. This residue, if not removed, may cause severe corrosion at the joint.

Adjoining areas may also be attacked by residues from the spray of flux and from flux vapors. Fluxes containing ammonium salts may cause stress corrosion cracking in the soldering of brass.

ORGANIC FLUXES

These fluxes, while less active than the inorganic, are effective at soldering temperatures from 90 to 320°C (200 to 600°F). They consist of organic acids and bases, and often certain of their derivatives such as hydrohalides. They are active at soldering temperatures, but the period of activity is short because of their susceptibility to thermal decomposition. Their tendency to volatilize, char, or burn when heated limits their use with torch or flame heating. When properly used, the residues are relatively inert and can be removed with water. Organic fluxes are particularly useful in applications where controlled quantities of flux can be applied and where sufficient heat can be used to fully decompose or volatilize the corrosive constituents. Caution is necessary to prevent undecomposed flux from spreading to insulating sleeving. Care must also be taken when soldering in closed systems where corrosive fumes may condense on critical parts of the assembly.

ROSIN FLUXES

Nonactivated Rosin

Water-white rosin dissolved in a suitable organic solvent is the closest approach to a noncorrosive flux. Rosin fluxes possess important physical and chemical properties which make them particularly suitable for use in the electrical industry. The active constituent, abietic acid, becomes mildly active at soldering temperatures between 177°C (350°F) and 316°C (600°F). The residue is hard, nonhygroscopic, electrically nonconductive, and not corrosive.

Mildly Activated Rosin

Because of the low activity of rosin, mildly activated rosin fluxes have been developed to increase the fluxing action without significantly altering the noncorrosive nature of the residue. These are the preferred fluxes for military, telephone, and other high reliability electronic products.

Activated Rosin

A third and still more active type of rosin base flux is called activated rosin. These fluxes are widely used in commercial electronics and in high reliability applications where the residue can be completely removed after soldering.

SPECIAL FLUXES

Reaction fluxes are a special group of fluxes that are useful when soldering to aluminum. In practice, the decomposition of the flux provides a metallic film deposited on the aluminum surface in place of the oxide film.

FLUX FORMS

Flux is available as single or multiple cores in wire solder, or in liquid, pastes, and dry powder forms in most of the above categories.

JOINT DESIGN

The selection of a joint design for a specific application will depend largely on the service requirements of the assembly. It may also depend on such factors as the heating method to be used, the fabrication techniques prior to soldering, the number of items to be soldered, and the method of applying the solder.

When service requirements of a joint are severe, it is generally necessary to design so that the strength of the joint is equal to or greater than the load carrying capacity of the weakest member of the assembly. Solders have low strength compared to the metals that are generally soldered; therefore, the soldered joint should be designed to avoid dependence on the strength of the solder. The necessary strength can be provided by shaping the parts to be joined so that they engage or interlock, requiring the solder only to seal and stiffen the assembly.

There are two basic types of joint design used for soldering: the lap joint and the butt joint. Figure 12.2 illustrates a butt joint, a lap joint, a lock seam, and a lap joint applied as a pipe joint.

The lap type of joint should be employed whenever possible, since it offers the best possibility of obtaining joints with maximum strength. It should be used where sealing is required.

An important factor in joint design is the manner in which the solder will be applied to the joint. The designer must consider the number of joints per assembly and the number of assemblies to be manufactured. For limited production using a manual soldering process, the solder may be face-fed into the joint with little or no problems. However, for large production of assemblies containing multiple joints, an automated process such as wave soldering may be advantageous. In this case, the design must provide accessible joints suitable for automated fluxing, soldering, and cleaning.

Clearance between the parts being joined should allow the solder to be drawn into the space between them by capillary action, but not so wide that the solder cannot fill the gap. Joint clearances to 0.075 mm (0.003 in.) are preferred for optimum strength, but variations are allowable in specific instances. For example, when soldering precoated metals, a clearance as low as 0.025 mm (0.001 in.) is possible.

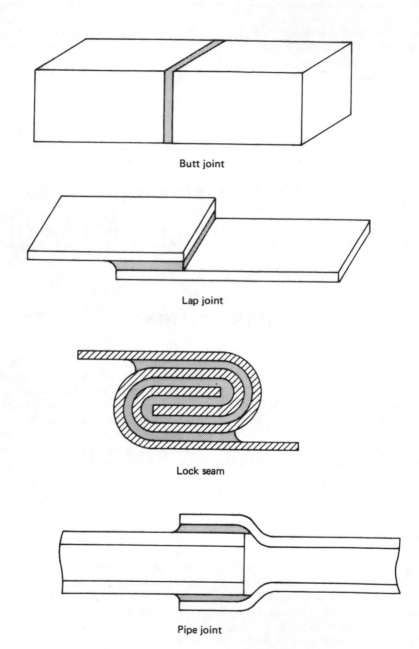

Butt joint

Lap joint

Lock seam

Pipe joint

Fig. 12.2—Typical joint designs used for soldering

PRECLEANING AND SURFACE PREPARATION

An unclean surface will not permit the solder to flow, which makes soldering difficult or impossible and contributes to the formation of a poor joint. Materials such as oil, grease, paint, pencil markings, drawing and cutting lubricants, general atmospheric dirt, oxide, or rust films must be removed before soldering. To insure sound soldered joints, the importance of cleanliness cannot be overemphasized.

DEGREASING

Either solvent or alkaline degreasing is recommended for the cleaning of oily or greasy surfaces. Of the solvent degreasing methods, the vapor condensation of the trichlorethylene type solvents probably leaves the least residual film on the surface. In the absence of vapor degreasing apparatus, immersion in liquid solvents or in detergent solutions is a suitable procedure. Hot alkali detergents are widely used for degreasing. All cleaning solutions must be thoroughly removed before soldering. Residues from hard water rinses may interfere with soldering.

Cleaning methods are often designed for a specific soldering operation, and their suitability for a particular application, therefore, should be investigated thoroughly.

PICKLING

The purpose of pickling or acid cleaning is to remove rust, scale and oxides, or sulfides from the metal so as to provide a clean surface for soldering. The inorganic acids (hydrochloric, sulfuric, phosphoric, nitric, and hydrofluoric), singly or mixed, all fulfill this function, although hydrochloric and sulfuric are the most widely used. The pieces should be washed thoroughly in hot water after pickling and dried as quickly as possible.

MECHANICAL CLEANING

Mechanical cleaning includes the following methods:
(1) Grit or shotblasting
(2) Mechanical sanding or grinding
(3) Filing or hand sanding
(4) Cleaning with steel wool
(5) Wire brushing or scraping

For best results, cleaning should extend beyond the joint area. Shot or steel gritblasting is often effective and preferable to sandblasting because it avoids the embedding of silica particles on the surface, which would interfere with the flow of solder. On the softer metals, such as copper, mechanical cleaning is not recommended.

PRECOATING

The coating of the base metal surfaces with a more solderable metal or alloy prior to the soldering operation is sometimes desirable to facilitate soldering. Coatings of tin, copper, silver, cadmium, iron, nickel, and alloys of tin-lead, tin-zinc, tin-copper, and tin-nickel are used for this purpose. The advantages of precoating are twofold: (1) soldering is more rapid and uniform, and (2) strong acidic fluxes can be avoided during soldering. The precoating of metals which have tenacious oxide films, such as aluminum, aluminum bronzes, highly alloyed steels, and cast iron, is almost mandatory. Precoating of steel, brass, and copper, although not entirely essential, is of great value in some applications.

Precoating of the metal surfaces may be accomplished by a number of different methods. Solder or tin may be applied with a soldering iron or an abrasive wheel, by ultrasonic soldering, by immersion in molten metal, by electrodeposition, or by chemical displacement.

Hot dipping may be accomplished by fluxing and dipping the parts in molten tin or solder. Small parts are often placed in wire baskets, cleaned, fluxed, dipped in the molten metal, and centrifuged to remove excess metal. Coating by hot dipping is applicable to carbon steel, alloy steel, cast iron, copper, and certain copper alloys. Prolonged immersion in molten tin or solder should be avoided to prevent excessive formation of intermetallic compounds at the coating-base metal interface.

Precoating by electrodeposition may be done in stationary tanks, in conveyorized plating units, or in barrels. This method is applicable to all steels, copper alloys, and nickel alloys. The coating metals are not limited to tin and solder. Copper, cadmium, silver, precious metals, nickel, iron, and alloy platings such as tin-copper, tin-zinc, and tin-nickel are also in common use.

Certain combinations of electrodeposited metals (duplex coatings), where one metal is plated over another, are becoming more popular as an aid to soldering. A coating of 0.005 mm (0.0002 in.) of copper plus 0.0075 mm (0.0003

in.) of tin is particularly useful for brass. The solderability of aluminum is assisted by a coating of 0.013 mm (0.0005 in.) of nickel followed by 0.008 mm (0.0003 in.) of tin or by a combination of zincate (zinc), 0.005 mm (0.0002 in.) copper, and tin. An iron plating followed by tin plating is extremely useful over a cast iron surface.

Immersion coatings or chemical displacement coatings of tin, silver, or nickel may be applied to some of the common base metals. These coatings are usually very thin and generally have a poor shelf life.

The shelf life of a coating is defined as the ability of the coating to withstand storage conditions without impairment of solderability. Hot tinned and flow brightened electrotin coatings have an excellent shelf life; inadequate thicknesses of electrotinned or immersion tinned coatings have a limited shelf life. Coating thicknesses of 0.003 mm (0.0001 in.) to 0.008 mm (0.0003 in.) of tin or solder are recommended to assure maximum solderability after prolonged storage.

SOLDERING METHODS AND EQUIPMENT

The proper application of heat is of paramount importance in any soldering operation. The heat should be applied in such a manner that the solder melts while the surface is heated to permit the molten solder to wet and flow over the surface. A number of tools are available as heat sources.

SOLDERING IRONS

The traditional soldering tool is the soldering iron with a copper tip which may be heated electrically or by oil, coke, or gas burners. To lengthen the usable life of a copper tip, a coating of solder-wettable metal, such as iron with or without additional coatings, is applied to the surface of the copper. The rate of dissolution of the iron coating in molten solder is substantially

less than the rate for copper. The iron coating also shows less wear, oxidation, and pitting than uncoated copper.

The selection of soldering irons can be simplified by classifying them in four groups: (1) soldering irons for servicemen; (2) transformer type, low voltage pencil irons; (3) special quick heating and plier type irons; and (4) heavy duty industrial irons.

Regardless of the heating method, the tip performs the following functions:

(1) Stores and conducts heat from the heat source to the parts being soldered
(2) Stores molten solder
(3) Conveys molten solder
(4) Withdraws surplus molten solder

Table 12.12 provides a guide to the selection of the proper soldering iron. The performance of

Table 12.12—Selection of soldering irons

Work to be done	Tip diameter range		Power range, watts
	mm	in.	
Miniature printed circuits, thin substrates, temperature-sensitive components	1–3	1/32–1/8	10–20
Intermittent light assembly work, printed circuits, instruments, jewelry	3–5	1/8–3/16	20–35
Repetitive assembly work, telephone and appliances, art glass	5–6	3/16–1/4	40–60
High speed production soldering, light tinware, general duty, medium electrical, light plumbing	6–10	1/4–3/8	70–150
Medium tinware, light roofing, shipboard repair, heavy electrical, heavy plumbing	16–38	5/8–1-1/2	170–350
Heavy tinware, roofing, radiators, armatures, transformer cans	41–53	1-5/8–2	350–1250

electrical industrial irons cannot be measured solely by the wattage rating of the heating element. The materials used and the design of the iron affect the heat reserve and temperature recovery of the copper tip.

The angle at which the copper tip is applied to the work is important in delivering the maximum heat to the work. The flat side of the tip should be applied to the work to obtain the maximum area of contact. Flux cored solders should not be melted on the soldering tip because this destroys the effectiveness of the flux. The cored solder should be touched to the soldering tip to initiate good heat transfer, and then the solder should be melted on the work parts to complete the solder joint.

TORCH SOLDERING

The selection of a gas torch for soldering is controlled by the size, mass, and configuration of the assembly to be soldered. Flame temperature is controlled by the nature of the gas or gases used. Fuel gas when burned with oxygen will provide higher flame temperatures than when burned with air. The highest flame temperatures are attained with acetylene and lower temperatures with propane, butane, natural gas, and manufactured (city) gas, roughly in the order given. The flame of the fuel gas burned with

oxygen will be sharply defined; with air, the flame will be bushy and flared.

Multiple flame tips, or burners, of shapes suitable to the work are frequently used. They may be designed to operate on oxygen and fuel gas, compressed air and fuel gas, or bunsen type torches.

In adjusting tips or torches, care should be taken to avoid adjustment which results in a "sooty" flame; the carbon deposited on the work will prevent the flow of solder.

DIP SOLDERING

This soldering method utilizes a molten bath of solder to supply both the heat and solder necessary to produce the joints as shown in Fig. 12.3(A). When conducted properly, this method is useful and economical in that an entire unit comprising any number of joints can be soldered in one operation after proper cleaning and fluxing. Fixtures are required to contain the unit and maintain proper joint clearances during solidification of the solder. The soldering pot should be large enough so that at a given rate of production the units being dipped will not appreciably lower the temperature of the solder bath. Pots of adequate size allow the use of lower operating temperatures and still supply sufficient soldering heat.

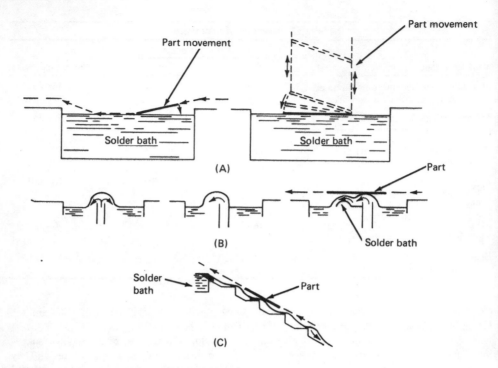

Fig. 12.3—Several soldering techniques used for large production runs: (A) dipping onto a static bath; (B) wave soldering systems, (C) cascade soldering system

WAVE SOLDERING

In wave soldering, as shown in Fig. 12.3(B), the solder is pumped out of a narrow slot to produce a wave or series of waves. The conveyor can be at a small angle to the horizontal to assist drainage of the solder, and double waves or special wave forms may also be used for this purpose. In cascade soldering, as illustrated in Fig. 12.3(C), the solder flows down a trough by gravity and is returned by pump to the upper reservoir. These wave solder systems are excellent in that a virtually oxide-free solder surface is presented to the part, and dislodgement of flux and vapors is also promoted by the flow of the solder.

Integrated wave soldering systems for printed circuit assemblies provide units that can apply the flux, dry and preheat the board, solder components, and clean the completed assembly. Some of these systems have special features where the flux is applied by passing through a wave, by spraying, by rolling, or by dipping. Several systems employ oil mixed with the solder to aid in the elimination of icicles and bridging between conductor paths. Another system features dual waves flowing in the opposite direction to the board travel.

OVEN SOLDERING

There are many applications, especially in high production soldering, where this method produces consistent and satisfactory soldering, but this method is not widely used.

Oven heating should be considered under the following circumstances:

(1) When entire assemblies can be brought to the soldering temperature without damage to any of the components

(2) When production is sufficiently great

to allow expenditure for jigs and fixtures to hold the parts during the soldering

(3) When the assembly is complicated in nature, making other heating methods impractical

Proper clamping fixtures are very important during oven soldering. Movement of the joint during solidification of the solder may result in a poor joint.

Another important consideration is the flux to be used. Rosin and organic fluxes are subject to decomposition when maintained at elevated temperatures for an extended period of time. When a rosin or organic type flux is used, the part must be brought rapidly to the liquidus temperature of the solder. It is sometimes beneficial to dip the parts in a hot flux solution before placing them in the oven. When using rosin base flux, it is generally necessary to use a solder with a tin content of 50 percent or more.

The use of a reducing atmosphere in the oven does not allow joints to be made without flux, because the temperatures at which these atmospheres become reducing are far above the liquidus temperature of the solders. The use of inert atmospheres will prevent further oxidation of the parts, but flux must be used to remove the oxide that is already present.

It is often advantageous to accelerate the cooling of the parts on their removal from the oven. An air blast has been found satisfactory for this.

The ovens should be equipped with adequate temperature controls since the flow of solder is optimum at approximately 45° to 50° C (113° to 122° F) above its liquidus temperature. The optimum heating condition exists when the heating capacity of the oven is sufficient to heat the parts rapidly to the liquidus temperature of the solder.

RESISTANCE SOLDERING

In resistance soldering, the work is placed either between a ground and a movable electrode or between two movable electrodes to complete an electrical circuit. Heat is applied to the joint both by the electrical resistance of the metal being soldered and by conduction from the electrode, which is usually carbon.

Production assemblies may utilize multiple electrodes, rolling electrodes, or special electrodes, depending on which will be advantageous with regard to soldering speed, localized heating, and power consumption.

Resistance soldering electrode tips cannot be tinned, and the solder must be fed into the joint or supplied by preforms or solder coatings on the parts.

INDUCTION SOLDERING

The material that is to be induction soldered must be an electrical conductor. The rate of heating is dependent upon the induced current flow, while the distribution of heat obtained with induction heating is a function of the induced current frequency. The higher frequencies concentrate the heat at the surface. Three types of equipment are available for induction heating: the vacuum tube oscillator, the resonant spark gap system, and the motor generator unit.

Induction soldering is generally applicable for operations with the following requirements:

(1) Large scale production

(2) Application of heat to a localized area

(3) Minimum oxidation of surface adjacent to the joint

(4) Good appearance and consistently high joint quality

(5) Simple joint design which lends itself to mechanization

The induction technique requires that the parts being joined have clean surfaces and that joint clearances be maintained accurately. High grade solders are generally required to obtain rapid spreading and good capillary flow. Preforms often afford the best means of supplying the correct amount of solder and flux to the joint.

When induction soldering dissimilar metals, particularly joints composed of both magnetic and nonmagnetic components, attention must be given to the design of the induction coil in order to bring both parts to approximately the same temperature.

OTHER SOLDERING METHODS

Infrared Soldering

Optical soldering systems are available

which are based on focusing infrared light (radiant energy) on the joint by means of a lens. Lamps ranging from 45 to 1500 watts can be used for different application requirements. The devices can be programmed through a silicon-controlled power supply with an internal timer.

Hot Gas Soldering

The principle is to use a fine jet of inert gas, heated to above the liquidus of the solder. The gas acts as a heat transfer medium and as a blanket to reduce access of air at the joint.

Ultrasonic Soldering

This soldering method has limited use, but units are available for dip soldering pots and hand soldering operations. A transducer produces high frequency vibrations which break up tenacious oxide films on base metals such as aluminum, thereby exposing the base metal to the wetting action of the liquid solder, generally without the use of flux. Ultrasonic units are useful in soldering the return bends to the sockets of aluminum air conditioner coils. Ultrasonic soldering is also used to apply solderable coatings on difficult-to-solder metals.

Spray Gun Soldering

This method of heating is generally used when the contour of the part to be soldered is such that it is difficult to follow with a wiping or drop method or where the part is placed in the assembly in such a way that the solder cannot be applied after the parts are assembled.

Gas fired or electrically heated guns are available and each is designed to spray molten or semimolten solder on the work from a continuously fed solid solder wire.

Two types of guns are used to spray solder. The first uses propane with oxygen or natural gas with air to heat and spray a continuously fed solid solder wire of approximately 3.2 mm (1/8 in.) diameter. About 90 percent of the solder wire is melted by the flame of the gun. The solder contacts the workpiece in a semiliquid form. The workpiece then supplies the balance of the heat required to melt and flow the solder. Adjustments can be made within the spray gun to control the solder spray.

Condensation Soldering

This method utilizes the latent heat of vaporization of a condensing saturated liquid to provide the heat required for soldering. A reservoir of saturated vapor over a boiling liquid provides a constant controlled temperature with rapid heat transfer that is useful for large assemblies as well as temperature sensitive parts.

FLUX RESIDUE TREATMENT

After the joint is soldered, flux residues which may corrode the base metal or otherwise prove harmful to the effectiveness of the joint must be removed. The removal of flux residues is especially important where joints must be subjected to humid environments.

The inorganic type flux residues containing · inorganic salts and acids should be removed completely. Residues from the organic type fluxes that are composed of the very mild organic acids, such as stearic, oleic, and ordinary tallow, or the highly corrosive combinations of urea plus various organic hydrochlorides should also be removed.

Generally, rosin flux residues may remain on the joint unless appearance is the prime factor or if the joint area is to be painted or subsequently coated.

The activated rosin fluxes have a rosin base with small additions of complex organic compounds. These can generally be treated in the same manner as the organic fluxes for structural soldering, but they should be removed for critical electronic applications.

Where flux residue removal procedures are not practical and the nature of the soldered assembly is such that flux corrosion would either interfere with its operation or substantially shorten its life, the use of zinc chloride or other corrosive fluxes must be prohibited. This does not preclude the use of this type of flux in precoating operations where flux residues can be removed before the parts are assembled.

Zinc chloride base fluxes leave a fused residue that will absorb water from the atmosphere. Removal is best accomplished by thorough washing in hot water containing 2 percent of concentrated hydrochloric acid per gallon of water, followed by a hot water rinse. The acid-ified water removes the white crust of zinc oxychloride which is insoluble in water. Complete removal may also require washing in hot water which contains some washing soda (sodium carbonate), followed by a clear water rinse. Occasionally some mechanical scrubbing may also be required.

Acidified rinse water, when used on copper articles, may rise in copper salt content and cause unsightly darkening of soldered joints. When this occurs, the acidified rinse may be regenerated with a small amount of potassium ferrocyanide which precipitates the copper salts from the solution.

The residues from reaction type fluxes used on aluminum are usually removed with a rinse in warm water. If this does not remove all traces of residue, the joint may be scrubbed with a brush and then immersed in 2 percent sulfuric acid, followed by immersion in 1 percent nitric acid. A final warm water rinse is then required.

The residues from the organic fluxes are usually quite soluble in hot water. Double rinsing in warm water is always advisable.

Oily or greasy flux paste residues are generally removed with an organic solvent.

Soldering pastes are usually emulsions of petroleum jelly and a water solution of zinc-ammonium chloride. Because of the corrosive nature of the acid salts contained in the flux, residues must be removed to prevent corrosion of the soldered joints.

If rosin residues must be removed, alcohol or chlorinated hydrocarbons may be used. Certain rosin activators are soluble in water but not in organic solvents. These flux residues require removal by organic solvents, followed by a water rinse.

SUPPLEMENTARY READING LIST

Edwards, R.B. Joint tolerances in capillary copper piping joints. *Welding Journal* 51 (6), June 1972, pp. 321-s—324-s.

Manko, H.H. Soldering fluxes—past and present. *Welding Journal* 52 (3), Mar. 1973, pp. 163-166.

Manko, H.H. *Solders and Soldering*, New York: McGraw-Hill, 1964.

Papers on Soldering. ASTM Special Publication No. 319. Philadelphia: American Society for Testing and Materials, 1962.

Coombs, C.F., Jr., editor. *Printed Circuits Handbook*. New York: McGraw-Hill, 1967.

Soldering Manual. Miami: American Welding . Society, 1978.

Symposium on Solder. ASTM Special Publication No. 189. Philadelphia: American Society for Testing and Materials, 1956.

13

Arc and Oxygen Cutting

Prepared by a committee consisting of

F.H. SASSE, *Chairman*
   *Linde Division, Union
   Carbide Corp.*

R.L. FROHLICH,
   Westinghouse Electric Corp.

R.D. GREEN
   Mapp Products

J.W. HOPKINS
   Sparrows Point Shipyard

W.H. KEARNS
   American Welding Society

J.B. LEWIS
   *Linde Division, Union
   Carbide Corp.*

C.R. McGOWAN
   U.S. Steel Corp.

13

Arc and Oxygen Cutting

INTRODUCTION

A cutting process is one which brings about the severing or removal of metals. This can be done by mechanical means (machining), by melting (arc), and by chemical reaction (oxidation). Although all of these means are used in the welding industry, the latter two are closely allied with welding processes. Arc cutting (AC) is a group of cutting processes which melts the metals to be cut with the heat of an arc between an electrode and base metal. Arc cutting methods typically include air carbon arc, gas metal arc, gas tungsten arc, plasma arc, and shielded metal arc. The two methods of industrial importance are air carbon arc cutting (AAC) and plasma arc cutting (PAC). These two methods are covered in the chapter.

Oxygen cutting (OC) is a group of cutting processes used to sever or remove metals by a high temperature exothermic reaction of oxygen with the base metal. With oxidation resistant metals, the reaction is aided by the use of a chemical flux or metal powder. Typical oxygen cutting processes are oxygen arc, oxyfuel gas, oxygen lance, chemical flux, and metal powder. Oxyfuel gas cutting (OFC) and its chemical flux (FOC) and metal powder (POC) modifications are covered in the chapter. The other two processes are not widely used in the metal fabricating industry. Therefore, they are not discussed here.

OXYFUEL GAS CUTTING

FUNDAMENTALS OF THE PROCESS

Definition and General Description

Oxyfuel gas cutting (OFC) is a group of cutting processes where the severing or removing of metal is accomplished by the chemical reaction of pure oxygen with the metal at ele-

vated temperatures. The necessary temperature is maintained by a fuel gas-oxygen flame. In the case of oxidation resistant metals, the reaction is aided by the addition of chemical fluxes or metal powders to the cutting oxygen stream.

The process has been given various names such as burning, flame cutting, and flame ma-

460

chining. The actual cutting operation is performed by the oxygen stream. The oxygen-fuel gas flame is the mechanism used to raise the base metal to an acceptable preheat temperature range and to maintain the cutting operation.

The OFC torch is a versatile tool that can be generally taken to the work site. It is used to cut steel up to 2 m (7 ft) thick. Because the cutting oxygen jet has a 360° "cutting edge," it provides a rapid means of cutting both straight edges and curved shapes to required dimensions without expensive handling equipment. Cutting direction can be continuously changed during operation.

Principles of Operation

The oxyfuel gas cutting process employs a torch with a tip (nozzle). The functions of the torch are to produce preheat flames by mixing the gas and the oxygen in the correct proportions and to supply a uniformly concentrated stream of high purity oxygen to the reaction zone. The oxygen reacts with the hot metal and also blows the molten reaction products from the joint. Features of cutting torches are shown in Figs. 13.1, 13.2, and 13.3. The cutting

torch body mixes the fuel and pure oxygen for the preheating flame and provides oxygen for the cutting reaction. The torch cutting tip contains a number of preheating flame ports and a cutting oxygen passage.

The preheat flame is used to heat the metal to a temperature where the metal will react with the cutting oxygen. The pure oxygen rapidly reacts with most of the metal in a narrow section.

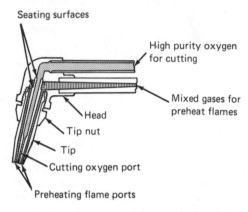

Fig. 13.1—*Typical cutting torch head and tip*

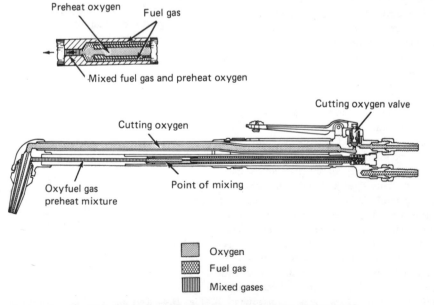

Fig. 13.2—*Typical positive pressure oxyfuel gas cutting torch*

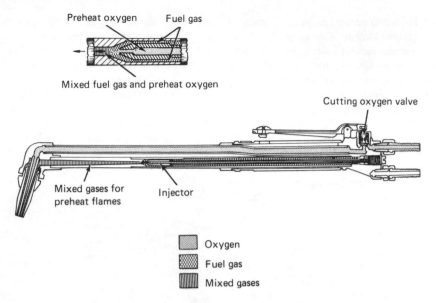

Preheat oxygen   Fuel gas

Mixed fuel gas and preheat oxygen

Cutting oxygen valve

Mixed gases for
preheat flames   Injector

☐ Oxygen
☒ Fuel gas
▨ Mixed gases

Fig. 13.3—Typical injector type oxyfuel gas cutting torch

Metal oxides and molten metal are expelled from the cut by the kinetic energy of the oxygen stream. Movement of the torch across the work-piece at a proper rate produces a continuous cutting action. The torch may be moved manually or by a mechanized carriage.

The accuracy of the manual operation depends largely on the skill of the operator. Mechanized operation generally improves both the accuracy of the cut and the finish of the cut surfaces.

Kerf. When a piece is cut by an OC or AC process, a narrow width of metal is progressively removed. The width of the cut is called the kerf, as shown in Fig. 13.4. Control of the kerf is important in cutting operations where dimensional accuracy of the part and square-ness of the cut edges are significant factors in quality control. With the OFC process, kerf width is a function of the size of oxygen port, type of tip used, speed of cutting, and flow rates of cutting oxygen and preheating gases. As material thickness increases, oxygen flow rates must usually be increased. Cutting tips with larger cutting oxygen ports are required to handle the higher flow rates. Consequently,

the width of the kerf increases as the material thickness being cut increases.

Drag. When the speed of the cutting torch is adjusted so that the oxygen stream enters the top of the kerf and exits from the bottom of the kerf along the axis of the tip, the cut will have zero drag. It is called a "drop cut." If the speed of cutting is increased, or if the oxygen flow is

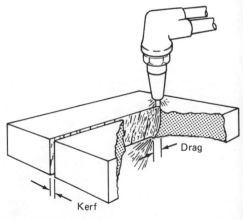

Drag

Kerf

Fig. 13.4—Kerf and drag in oxyfuel gas cutting

decreased, the available oxygen decreases in the lower regions of the cut. With less oxygen available, the oxidation reaction rate decreases, and also the oxygen jet has less energy to carry the reaction products out of the kerf. Then, the most distant part of the cutting stream lags behind the portion nearest to the torch tip. The length of this lag, measured along the line of cut, is referred to as the *drag*. This is shown in Fig. 13.4.

Drag may also be expressed as a percentage of the cut thickness. A ten percent drag means that the far side of the cut lags the near side of the cut by a distance equal to ten percent of the material thickness. If for any reason the piece is not severed, the cut is referred to as a "non-drop cut."

An increase in cutting speed with no increase in oxygen flow usually results in a larger drag. This may cause a decrease in cut quality. There is also a strong possibility of loss of cut at excessive speeds. Reverse drag may be obtained when the cutting oxygen flow is too high or the travel speed is too low. Under these conditions poor quality cuts usually result. Cutting stream lag caused by incorrect torch alignment is not considered to be drag.

Cutting speeds below those recommended for best quality cuts usually result in irregularities in the kerf. The oxygen stream inconsistently oxidizes and washes away additional material from each side of the cut. Excessive preheat flame results in undesirable melting and widening of the kerf at the top. Kerf width is especially important to shape cutting. Compensation must be made in the layout of the work, or the design of the template, for kerf width. Generally, on materials up to 51 mm (2 in.) thick, kerf width can be maintained within ± 0.4 mm (1/64 in.).

Chemistry of Oxygen Cutting

The process of oxygen cutting is based on the ability of high purity oxygen to combine rapidly with iron when it is heated to its ignition temperature, above 870° C (1600° F). The iron is rapidly oxidized by the high purity oxygen and heat is liberated by the reactions.

The balanced chemical equations for these reactions are

(1) $Fe + O \rightarrow FeO$ + heat (267 kJ), first reaction

(2) $3Fe + 2O_2 \rightarrow Fe_3O_4$ + heat (1120 kJ), second reaction

(3) $2Fe + 1.5O_2 \rightarrow Fe_2O_3$ + heat (825 kJ), third reaction.[1]

The tremendous heat release of the second reaction predominates that of the first reaction, which is supplementary in most cutting applications. The third reaction occurs to some extent in heavier cutting applications. Stoichiometrically, 0.29 m³ (10.4 ft³) of oxygen will oxidize 1 kg (2.2 lb) of iron to Fe_3O_4.

In actual operations, the consumption of cutting oxygen per unit mass of iron varies with the thickness of the metal. Oxygen consumption per unit mass is higher than the ideal stoichiometric reaction for thicknesses less than approximately 40 mm (1-1/2 in.), and it is lower for greater thicknesses. The lowest oxygen consumption occurs in a thickness range of 100 to 125 mm (4 to 5 in.). For thick sections, the oxygen consumption is lower than the ideal stoichiometric reaction because only part of the iron is completely oxidized to Fe_3O_4. Some unoxidized or partly oxidized iron is removed by the kinetic energy of the rapidly moving oxygen stream.

Chemical analysis has shown that, in some instances, over 30 percent of the slag is metal. The heat generated by the rapid oxidation of iron melts some of the iron adjacent to the reaction surface. This molten iron is swept away with the iron oxide by the motion of the oxygen stream. The concurrent oxidizing reaction heats the layer of iron at the active cutting front.

The heat generated by the iron-oxygen reaction at the focal point of the cutting reaction (the hot spot) must be sufficient to continuously preheat the material to the ignition temperature. Allowing for the loss of heat by radiation and conduction, there is ample heat to sustain the

---

1. "Thermodynamic Properties of 65 Elements, Their Oxides, Halides, Carbides, and Nitrides," *Bulletin 65*, U.S. Bureau of Mines, 1965.

reaction. In actual practice, the top surface of the material is frequently covered by mill scale or rust. They must be melted away by the preheating flames to expose a clean metal surface to the oxygen stream. Preheating flames help to sustain the cutting reaction by providing heat to the surface. They also shield the oxygen stream from turbulent interaction with air.

The alloying elements normally found in carbon steels are oxidized or dissolved in the slag without markedly interfering with the cutting process. When alloying elements are present in steel in appreciable amounts, their effect on the cutting process must be considered. Steels containing minor additions of oxidation-resistant elements, such as nickel and chromium, can be oxygen cut. However, when oxidation resistant elements are present in large quantities, modifications to the cutting technique are required to sustain the cutting action. This is true for stainless steels.

OXYGEN

Oxygen used for cutting operations should have a purity of 99.5 percent or higher. Lower purity reduces the efficiency of the cutting operation. A one percent decrease in oxygen purity to 98.5 percent will result in a decrease in cutting speed of approximately 15 percent, and an increase of about 25 percent in the cutting oxygen consumption. The quality of the cut will be impaired, and the amount and tenacity of the adhering slag will increase. With oxygen purities below 95 percent, the familiar cutting action disappears, and it becomes a melt and wash action that is usually unacceptable.

PREHEATING FUELS

The functions of the preheat flames in the cutting operation are

(1) To raise the temperature of the steel to the ignition point

(2) To add heat energy to the work to maintain the cutting reaction

(3) To provide a protective shield between the cutting oxygen stream and the atmosphere

(4) To dislodge from the upper surface

of the steel any rust, scale, paint, or other foreign substance that would stop or retard the normal forward progress of the cutting action

A preheat intensity that raises the steel to the ignition temperature rapidly will usually be adequate to maintain cutting action at high travel speeds. However, the quality of the cut will not be the best. High quality cutting can be carried out at considerably lower preheat intensities than those normally required for rapid preheating. Dual range gas controls can be used so that a high intensity preheat is employed for the starting operation. Then, the preheat flames are reduced to lower intensity during the cutting operation to save fuel and oxygen.

A number of commercially available fuel gases are used with oxygen to provide the preheating flames. Some have proprietary compositions. Fuel gases are generally used because of availability and cost. Gasoline is also used with a specially designed torch in which it vaporizes before burning. Properties of some commonly used fuel gases are listed in Table 13.1. To understand the significance of the information in this table, it is necessary to understand some of the terms and concepts involved in the burning of fuel gases. These terms and concepts are discussed in Chapter 10. Also, combustion intensity or specific flame output for various fuel gases is covered in that chapter. This property is an important consideration in fuel gas selection.

Fuel Selection

Some general factors for consideration when selecting a preheat fuel are

(1) Times required for preheating when starting cuts on square edges and rounded corners, and also when piercing holes for cut starts

(2) Effect on cutting speeds for straight line, shape, and bevel cutting

(3) Effect of the above factors on work output

(4) Cost and availability of the fuel in cylinder, bulk, and pipeline volumes

(5) Cost of the preheat oxygen required to burn the fuel gas efficiently

(6) Ability to use the fuel efficiently for

Table 13.1—Properties of common fuel gases

	Acetylene	Propane	Propylene	Methyl-acetylene-propadiene (MPS)	Natural gas
Chemical formula	C_2H_2	C_3H_8	C_3H_6	C_3H_4 (Methylacetylene, propadiene)	CH_4 (Methane)
Neutral flame temperature					
°F	5600	4580	5200	5200	4600
°C	3100	2520	2870	2870	2540
Primary flame heat emission					
btu/ft³	507	255	433	517	11
MJ/m³	19	10	16	20	0.4
Secondary flame heat emission					
btu/ft³	963	2243	1938	1889	989
MJ/m³	36	94	72	70	37
Total heat value (after vaporization)					
btu/ft³	1470	2498	2371	2406	1000
MJ/m³	55	104	88	90	37
Total heat value (after vaporization)					
btu/lb	21 500	21 800	21 100	21 000	23 900
kJ/kg	50 000	51 000	49 000	49 000	56 000
Total oxygen required (neutral flame)					
vol. O₂/vol. fuel	2.5	5.0	4.5	4.0	2.0
Oxygen supplied through torch (neutral flame)					
vol. O₂/vol. fuel	1.1	3.5	2.6	2.5	1.5
ft³oxygen/lb fuel (60°F)	16.0	30.3	23.0	22.1	35.4
m³oxygen/kg (15.6°C)	1.0	1.9	1.4	1.4	2.2
Maximum allowable regulator pressure					
psi	15	Cylinder	Cylinder	Cylinder	Line
kPa	103				
Explosive limits in air, percent	2.5-80	2.3-9.5	2.0-10	3.4-10.8	5.3-14
Volume-to-weight ratio					
ft³/lb (60°F)	14.6	8.66	8.9	8.85	23.6
m³/kg (15.6°C)	0.91	0.54	0.55	0.55	1.4
Specific gravity of gas (60°F, 15.6°C) Air=1	0.906	1.52	1.48	1.48	0.62

other operations, such as welding, heating, and brazing, if required

(7) Ease of handling of fuel containers, when mobility of operation is required

(8) Safety in transporting and handling the fuel gas containers

For best performance and safety, the torches and tips should be designed for the particular fuel selected.

Acetylene

Acetylene is widely used as a fuel gas for oxygen cutting and also for welding. Its chief advantages are availability, high flame temperature, and widespread familiarity of users with its flame characteristics. Combustion with oxygen produces a hot, short flame with a bright inner cone at each preheat port. The hottest point is at the end of this inner cone. Combustion is completed in the long outer flame.

The sharp distinction between the two flames helps to adjust the oxygen-to-acetylene ratio for the desired flame characteristic.

Depending on this ratio, the flame may be adjusted reducing (carburizing), neutral, or oxidizing, as shown in Fig. 13.5. The neutral flame, obtained with a ratio of approximately one part oxygen to one part acetylene, is used for manual cutting. As the oxygen flow is de-

creased, a bright streamer begins to appear. This indicates a reducing flame, which is sometimes used to rough-cut cast iron.

When excess oxygen is supplied, the inner flame cone shortens and becomes more intense. Flame temperature increases to a maximum at an oxygen-to-acetylene ratio of about 1.5 to 1. An oxidizing flame is used for short preheating times and for cutting very thick sections.

The high flame temperature and heat transfer characteristics of the oxyacetylene flame are particularly important for bevel cutting. They are also an advantage for operations in which the preheat time is an appreciable fraction of the total time for cutting, such as short cuts.

Acetylene in the free state should not be used at pressures higher than 103 kPa (15 psi) gage, or 207 kPa (30 psi) absolute pressure. At higher pressures, it may decompose with explosive force when exposed to heat or shock. Chapter 10 contains additional information on acetylene, its production and storage, and on the oxyacetylene flame.

Methylacetylene-Propadiene Stabilized (MPS)

This is a liquified, stabilized acetylene-like fuel that can be stored and handled similarly to liquid propane. MPS is a mixture of

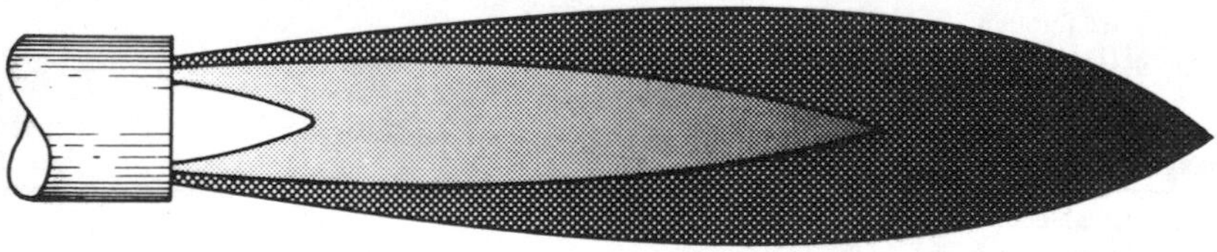

	Neutral flame	Oxidizing flame	Reducing flame
	White blue cone	White cone	Intense white cone
	Nearly colorless	Orange to purplish	White or colorless
	Bluish to orange		Orange to bluish

Fig. 13.5—Oxyacetylene flame

several hydrocarbons, including propadiene (allene), propane, butane, butadiene, and methylacetylene. Methylacetylene, like acetylene, is an unstable, high energy, triple-bond compound. The other compounds in MPS dilute the methylacetylene sufficiently to make the mixture safe for handling. The mixture burns hotter than either propane or natural gas. It also affords a high release of energy in the primary flame cone, another characteristic similar to· acetylene. The outer flame gives relatively high heat release, like propane and propylene. The overall heat distribution in the flame is the most even of any of the gases.

A neutral flame is achieved at a ratio of 2.5 parts of oxygen to 1 part MPS. Its maximum flame temperature is reached at a ratio of 3.5 parts of oxygen to 1 part of MPS. These ratios are used for the same applications as the acetylene flame.

Although MPS gas is very similar in characteristics to acetylene, it requires about twice the volume of oxygen per volume of fuel for a neutral preheat flame. Thus, oxygen cost will be higher when MPS gas is used in place of acetylene for a specific job. To be competitive, the cost of MPS gas must be lower than acetylene for the job.

MPS gas does have an advantage over acetylene for underwater cutting in deep water. Because acetylene outlet pressure is limited to 207 kPa (30 psi) absolute, it cannot be used below approximately 9 m (30 ft) of water. On the other hand, MPS can be used at greater depths, as can hydrogen. For a particular underwater application, MPS, acetylene, and hydrogen should be evaluated for preheat fuel.

Natural Gas

The composition of natural gas varies depending on its source. Its main component is methane (CH_4). When methane burns with oxygen, the chemical reaction is

$$CH_4 + 2 O_2 \rightarrow CO_2 + 2 H_2O$$

One volume of methane requires two volumes of oxygen for complete combustion. The flame temperature with natural gas is lower than with acetylene. It is also more diffused and less intense. The characteristics of the flame for

carburizing, neutral, or oxidizing conditions are not as distinct as the oxyacetylene flame.

Because of the lower flame temperature and the resulting lower heating efficiency, significantly greater quantities of natural gas and oxygen are required to produce heating rates equivalent to those of oxyacetylene fuel. Generally, more time is required to preheat with natural gas than with acetylene. To compete with acetylene, the cost and availability of natural gas and oxygen, the higher gas consumptions, and the longer preheat times must be considered.

The torch and tip designs for natural gas are different than those for acetylene. The delivery pressure of natural gas is generally low and the combustion ratios are different (see Table 13.1).

Propane

Propane is used regularly for oxygen cutting in a number of plants because of its availability and its much higher total heat value (MJ/m^3) than natural gas (see Table 13.1). For proper combustion during cutting, propane requires 4 to 4-1/2 times its volume of preheat oxygen. This requirement is offset somewhat by its high heat value. It is stored in liquid form and it is easily transported to the work site.

Propylene

Propylene, under many different brand names, is used as fuel gas for oxygen cutting. This gas competes with MPS gas for almost any job that uses fuel gas. It is similar to propane in many respects, but it has a higher flame temperature (see Table 13.1). One volume of propylene requires about 2.6 volumes of oxygen for a neutral flame and approximately 3.6 volumes for maximum flame temperature. Cutting tips are similar to those used for MPS.

Gasoline

Gasoline is used as a fuel for OFC with a specially designed cutting torch and associated tips. The flame is highly oxidizing and therefore suitable for cutting applications only. The portability of gasoline and the high temperature flame permit cutting up to and including 360 mm (14 in.) of steel. Although gasoline is fed to the special cutting torch as a liquid, it vaporizes in the cutting tip. About four seconds are required for the tip to heat sufficiently to vaporize the gasoline. The gasoline is stored in a pressurized container. Because of head pressure of the liquid, the torch is not generally used over approximately 7 m (23 ft) above the gasoline tank.

ADVANTAGES AND DISADVANTAGES

Oxyfuel gas cutting has a number of advantages and disadvantages compared to other metal cutting operations, such as sawing, milling, and arc cutting.

Advantages

Several advantages of OFC are as follows:

(1) Steels can generally be cut faster by OFC than by mechanical chip removal processes.

(2) Section shapes and thicknesses that are difficult to produce by mechanical means can be severed economically by OFC.

(3) Basic manual OFC equipment costs are low compared to machine tools.

(4) Manual OFC equipment is very portable and can be used in the field.

(5) Cutting direction can be changed rapidly on a small radius during operation.

(6) Large plates can be cut rapidly in place by moving the OFC torch rather than the plate.

(6) Two or more pieces can be cut from stock simultaneously using multiple-torch flame cutting machines or stack cutting.

(8) OFC is an economical method of plate edge preparation for bevel and groove weld joint designs.

Disadvantages

There are a number of disadvantages with oxyfuel gas cutting of metals. Several important ones are as follows:

(1) Dimensional tolerances are significantly poorer than machine tool capabilities.

(2) The process is essentially limited commercially to cutting steels and cast iron, although other readily oxidized metals, such as titanium, can be cut.

(3) The preheat flames and expelled red hot slag present fire and burn hazards to plant and personnel.

(4) Fuel combustion and oxidation of the metal require proper fume control and adequate ventilation.

(5) Hardenable steels may require preheat, or postheat, or both, to control their metallurgical structures and mechanical properties adjacent to the cut edges.

(6) Special process modifications are needed for OFC of high alloy steels and cast irons.

EQUIPMENT

There are two basic types of OFC equipment: manual and machine. The manual equipment is used primarily for scrap cutting, cutting risers off castings, and other operations that do not require a high degree of accuracy or a high quality cut surface. Machine cutting equipment is utilized for accurate, high quality work, and for large volume cutting, such as in steel fabricating shops. Both types of equipment operate on the same principle.

Manual Equipment

A setup for basic manual OFC requires the following:

(1) One or more cutting torches suitable for the preheat fuel gas to be used and the range of material thicknesses to be cut

(2) The required torch cutting tips to cut a range of material thicknesses .

(3) Oxygen and fuel gas hoses

(4) Oxygen and fuel gas pressure regulators

(5) Sources of oxygen and fuel gases to be used

(6) Flame strikers, eye protection, flame and heat resistant gloves and clothing

(7) Equipment operating instructions from the manufacturer

Torches. The functions of an OFC torch are as follows:

(1) To control the flow of cutting oxygen

(2) To control the flow and mixture of fuel gas and preheat oxygen

(3) To discharge the gases through the cutting tip at the proper velocities and volumetric flow rates for preheating and cutting

These functions are partially controlled by the operator, by the pressures of incoming gases, and by the design of the torch and cutting tips.

For manual cutting, a torch that can be readily manipulated by the operator is preferred. Manual oxygen cutting torches are available in various sizes. Torch and tip selection generally depend on the thickness range of the steel to be cut. Tips used in manual cutting equipment are of varied design, depending on the type of work to be done. For example, for cutting rusty or scaly steel, a tip furnishing a great amount of preheat should be selected.

There are two types of OFC torches: (1) the tip mixing type, in which the fuel and oxygen for the preheating flames are mixed in the tip; and (2) the premixing type, in which the mixing takes place within the torch. If both the oxygen and fuel gas are under appreciable pressure, the premixing torch is designated as a positive pressure type. When the fuel gas is supplied at a low pressure (such as natural gas), and it is drawn into the torch with the aid of an injector, the torch is designated a low pressure or injector torch. The two types of torches are shown in Figs. 13.2 and 13.3, respectively.

Superficially, the oxygen-gasoline cutting torch looks very much like any other cutting torch. The upper tube carries the cutting oxygen. The lower tube carries the preheat oxygen. Inside the lower tube, there is a smaller tube that carries the gasoline. Inside the gasoline tube is a long control rod that emerges at the torch butt in the form of an adjusting knob. The other end of the rod is cone shaped and fits into a mating female cone in the torch head. This is the gasoline valve. Turning the knob withdraws the male cone from the torch head and allows gasoline to flow. The gasoline flows between the control rod and the surrounding tube in a spiral path created by a helical coil which fits between the control rod and the tube. This circular flow permits the flow of gasoline to be unaffected by gravity or torch movement.

Manual Cutting Tips. Cutting tips are precision machined copper-alloy parts of various designs and sizes. They are held in the cutting torch by a tip nut. All oxygen cutting tips have preheat flame ports, usually arranged in a circle around a central cutting oxygen orifice. The preheat flame ports and the cutting oxygen orifice are sized for the thickness range of metal that the tip is designed to cut. Cutting tips are designated as standard or high speed. Standard tips have a straight bore oxygen port, and they are usually used with oxygen pressures from 205 to 415 kPa (30 to 60 psi). High speed tips differ from standard tips in that the exit end of the oxygen orifice flares out or diverges. The divergence allows the use of higher oxygen pressures, typically 415 to 690 kPa (60 to 100 psi), while maintaining a uniform oxygen jet at supersonic velocities. High speed tips are ordinarily used for machine cutting only. They usually permit cutting at speeds approximately 20 percent greater than speeds with standard tips. Both types of tips are shown in Fig. 13.6.

Cutting oxygen orifice size and design are not usually affected by the type of fuel used. However, flame port design does depend on the fuel. Various fuel gases require different volumes of oxygen and fuel, and they burn at different velocities. Therefore, the flame port size and number are designed to provide both a stable flame and adequate preheat for the application with the fuel gas being used. Acetylene tips are usually one piece with drilled or swaged flame ports. They are flat on the flame end. Tips for use with other fuel gases are either one piece, similar to acetylene tips, or two pieces with milled splines on the inner member, as

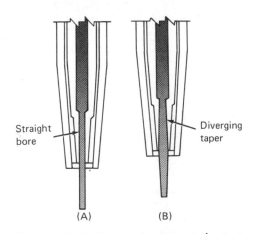

Straight bore

Diverging taper

(A)   (B)

Fig. 13.6 — Oxyfuel gas cutting tips: (a) standard; (b) high speed

illustrated in Fig. 13.7. Tips for use with MPS and propylene gases are usually flat on the flame end. Tips used for natural gas and propane usually have a cupped or recessed flame end.

An oxygasoline cutting tip is a two piece assembly with an inner core and an outer shell. Part of the core piece is a short length of tubing. The inside of the shell contains a short length of slightly smaller diameter tubing which, upon assembly, fits inside the tube of the core. This creates a baffled path for the fuel mixture, and it forces the mixture to flow near the very end of the tip where it is heated. The liquid fuel mixture is vaporized, directed down to the base of the core, and then redirected out of the tip through the flutes of the core.

Cutting tips, although considered consumable items, are precision tools. The tip is considered to have the greatest influence on cutting performance. Proper maintenance of tips can greatly extend their useful life and provide months of additional high quality performance.

The accumulation of slag in and around the preheat and cutting oxygen passages disturbs the preheat flame and oxygen stream characteristics. This can result in an obvious reduction in performance and quality of cut. When this happens, the tip should be taken out of service and restored to a good working condition.

Gas Pressure Regulators. The ability to make a successful material cut not only requires the proper choice of cutting torch and tip for the fuel gas selected but also a means of precisely regulating the proper gas pressures and volumes. Regulators are pressure control devices used to reduce high source pressures to required working pressures by manually adjusted pressure valves. They vary in design, performance, and convenience features. Gas pressure regulators are designed for use with specific types of gases and for definite pressure ranges.

Gas pressure regulators used for OFC are generally similar in design to those used for oxyfuel gas welding (OFW), which are discussed in Chapter 10. Regulators for all other fuel gases are similar in design to acetylene regulators. For OFC, regulators with higher capacities and delivery pressure ranges than those used for OFW may be required for multitorch operations and heavy cutting.

Hoses. Oxygen and fuel gas hoses used for OFC are the same as those used for OFW. They are discussed in Chapter 10.

Other Equipment. Tinted goggles are available for eye protection in a number of different shades. Tip cleaners, wrenches, strikers, fittings, etc., are used.

Mechanzed Equipment

Mechanized OFC will require additional

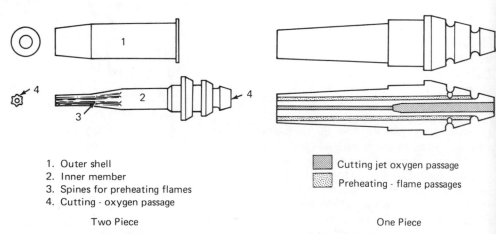

1. Outer shell
2. Inner member
3. Spines for preheating flames
4. Cutting - oxygen passage

Two Piece

Cutting jet oxygen passage
Preheating - flame passages

One Piece

Fig. 13.7—Oxyfuel gas cutting tips, one and two piece designs

facilities depending on the application. Some of these are

(1) A machine to move one or more torches in the required cutting pattern

(2) Torch mounting and adjusting arrangements on the machine

(3) A cutting table to support the work

(4) Means for loading and unloading the cutting table

(5) Automatic preheat ignition devices for multiple torch machines

Mechanized OFC equipment can vary in complexity from simple hand guided machines to very sophisticated numerically controlled units. The mechanized equipment is analogous to the manual equipment in principle but differs in design to accommodate higher fuel pressures, faster cutting speeds, and means for starting the cut. Many machines are designed for special purposes, such as those for making vertical cuts, edge preparation for welding, and pipe cutting and beveling. Many variations of mechanized cutting systems are commercially available. A simple arrangement is shown in Fig. 13.8.

Machine Torches. A typical machine cutting torch consists of a body, similar to a manual torch but with heavier construction and a cut-

ting tip. The heavy torch body encases the oxygen and fuel gas tubes which carry the gases to the end where the cutting tip is secured by a tip nut. The body of the torch may have a rack for indexing the tip to a desired position from the work surface (see Fig. 13.8). A machine torch may have two or three gas inlets. Three inlet torches have separate connections for fuel gas, preheat oxygen, and cutting oxygen. Two inlet torches have a fuel line connection and one oxygen connection with two valves. Three inlet torches permit separate regulation of preheat and cutting oxygen. They are generally recommended when remote control operation is desired.

Machine Cutting Tips. Machine cutting tips are designed to operate at higher oxygen and fuel pressures than those normally used for manual cutting. The two-piece divergent tip is one type used for operation at high cutting speeds (see Fig. 13.6B). Divergent cutting tips are based on the principles of gas flow through a venturi. High velocities are reached as the gas emerges from the venturi nozzle. Divergent cutting tips are precision machined to minimize any distortion of the gases when they exit from the nozzle. They are used for the majority of ma-

Fig. 13.8—A simple mechanized cutting arrangement for beveling a plate edge

chine cutting applications because of their superior cutting characteristics for materials up to 150 mm (6 in.) thick. They are not recommended for cutting materials over 250 mm (10 in.) thick.

Regulators. When natural gas or propane is used as a preheat fuel in machine cutting, fuel and oxygen can be conserved by using combination high-low pressure regulating systems. Because these fuels burn at lower heat transfer intensities than acetylene, high flow rates of fuel and preheat oxygen are required to heat the metal to ignition temperature in a reasonable time. Once the cut is started, less heat is needed to maintain cutting action with an appropriate savings in gas costs.

High-low pressure regulating systems permit the starting gas flow rates to be reduced to a predetermined level when cutting oxygen flow is initiated. This operation may be done manually or automatically depending on the regulator and control system design.

Cutting Machines. Oxyfuel gas cutting machines are either portable or stationary. Portable machines are usually moved to the work. Stationary machines are fixed in location and the work is moved to the machine.

Portable machines are primarily used for straight line cutting, although they can be adapted for circular and shape cutting. Portable machines usually consist of a motor driven carriage with an adjustable mounting for the cutting torch. In most cases, the machine travels on a track, which performs the function of guiding the torch. The carriage speed is adjustable over a wide range. The degree of cutting precision depends upon both the accuracy of the track, or guide, and the fit between the track and the driving wheels of the carriage. Portable machines are of various weights and sizes, depending on the type of work to be done. The smallest machines weigh only a few pounds. They are limited to carrying light duty torches for cutting thin materials. Large portable cutting machines are heavy and rugged. They can carry one or more heavy duty torches and the necessary auxiliary equipment for cutting thick sections.

Generally the operator must follow the carriage to make adjustments, as required, to produce good quality cuts. He adjusts torch height to maintain the preheat flames at correct position from the work surface. The operator ignites the torch, positions it at the starting point, and initiates the cutting oxygen flow and carriage travel. At the completion of the cut, he shuts off the cutting torch and carriage.

Stationary machines are designed to remain in a single location. The raw material is moved to the machine, and the cut shapes are transported away. The work station is composed of the machine, a system to supply the oxygen and preheat fuel to the machine, and a material handling system.

The torch support carriage runs on tracks. The structure either spans the work with a gantry type bridge across the tracks or it is cantilevered off to one side of the tracks. These types of equipment are shown in Figs. 13.9 and 13.10 respectively. They are usually classified according to the width of plate that can be cut (transverse motion). The length that can be cut is the travel distance on the tracks. The maximum cutting length is limited by physical limitations of both operator control and gas and electric power supply lines. An operator station with consolidated controls for gas flow, torch movement, and machine travel is generally part of the machine.

Stationary machines are classified into two main types, straight line cutting machines and shape cutting machines. Straight line cutting machines are used for slitting plate into strip, for plate edge preparation, or for both simultaneously. Their construction is considerably heavier and more rugged than that of small portable machines. They are capable of cutting to close tolerances. The machine usually spans the plate to be cut with one or more torches mounted on the carriage. Torch spacing is adjustable. Some machines have lightweight bridges that carry cross-cutting carriages. They can cut four sides of a plate at the same time.

Shape cutting machines are of two basic designs, a pantograph and a cross-carriage mechanism. In both types of machines, the basic element is a floating bar with one or more torches located on one end and a tracing or driving device on the other. The differences

Fig. 13.9 — Typical gantry type stationary oxyfuel gas cutting machine

Fig. 13.10 — Cantilever type stationary oxyfuel gas cutting machine

in design are the portions of the machine that carry the floating bar.

In the pantograph type machine, the bar is held by two arms that fold open and shut. The entire pantograph assembly is mounted on a carriage riding on rails. In the cross carriage mechanism, the carriage also rides on rails. The bar holding the torches can easily move at right angles to the carriage on which it rides.

Both designs can cut regular or irregular shapes of nearly any complexity and size. Shape cutting machines can mount a number of torches, depending on the size of the machine. In multiple torch operations, many identical shapes can be cut simultaneously. The number depends on the part size, plate size, and the number of available torches (see Fig. 13.10).

Tracing devices are used to guide shape cutting machines. The usual type of shape cutting machine employs a bar with torches on one end and a tracing device on the other end. On some machines this bar is driven at cutting speeds by a small wheel riding on a tracing table. A suitable tracing device is combined with the driving wheel to steer the torch bar around a template or drawing of the shape to be cut. Both pantograph and rectilinear machines use the same tracing and driving principles. A small wheel or magnetic tracer is employed to drive the machine. A small wheel may be manually guided around the outline of a drawing; two drive rollers may follow a shaped vertical strip; or a magnetic tracer may adhere to and follow the outline of a steel template.

An electronic tracing device may also be used. A photoelectric cell follows the outline of a drawing and guides a driving wheel. Traction for driving the machine comes solely from the driving wheel.

A rectilinear or coordinate drive type machine often utilizes a sine-cosine potentiometer to coordinate separate drive motors for longitudinal and transverse motion of the torch. The carriage and the cross arm, each with its own driving motor, are driven in the proper directions, but the speed of movement of the tracing arm remains at the constant preselected

value. This type of construction permits the design and manufacture of an oxygen cutting machine with sufficient rigidity to carry all modern control equipment.

It is possible to feed information to the electric drive motors of the carriage and cross arm from any suitable control. Adaptations of these machines use pilot machines. The pilot machine has a photoelectric cell tracer that can follow reduced scale drawings. Numerical control machines use calculated profile programs placed on either punched or magnetic tapes. These tapes, in turn, control the shape cutting by appropriate signals to the cutting machine drive motors.

Protective Clothing and Equipment

Appropriate protective clothing and equipment for any cutting operation will vary with the size, nature, and location of the work to be performed. Some or all of the following may be required:

(1) Tinted goggles or face shields with filter lens. The recommended filter lenses for various cutting operations are

(a) Light cutting, up to 25 mm (1 in.)—shade 3 or 4

(b) Medium cutting, 25 to 150 mm (1 to 6 in.)—shade 4 or 5

(c) Heavy cutting, over 150 mm (6 in.)—shade 5 or 6

(2) Flame resistant gloves

(3) Safety glasses

(4) Flame resistant jackets, coats, hoods, aprons, etc.

(a) Woolen clothing is preferable to cotton or synthetic materials

(b) Sleeves, collars, and pockets should be kept buttoned

(c) Cuffs should be eliminated

(5) Hard hats

(6) Leggings and spats

(7) Safety shoes

MATERIALS CUT

For most steel cutting, standard oxygen cutting equipment is satisfactory. For high alloy and stainless steel cutting, it may be necessary to use a special OFC process, such as flux injection or powder cutting, or one of the arc cutting processes. The cutting process and type of operation (manual or mechanized) selected depend on the material that is being cut, production requirements, and the ultimate use of the product.

Carbon and Low Alloy Steels

Carbon steels are readily cut by the OFC process. Low carbon steels are cut without difficulty using standard procedures. Typical data for cutting low carbon steel, using commonly available fuel gases, are shown in Tables 13.2 and 13.3. Operating data for cutting with gasoline as preheat fuel with the special torch are given in Table 13.4. The gas flow rates and cutting speeds are to be considered only as guides for determining more precise settings for a particular job. When a new material is being cut, a few trial cuts should be made to obtain the most efficient operating conditions.

High-pressure divergent tips will provide slightly faster cutting speeds with narrower kerf widths, and they will produce equivalent high quality work. Extreme care must be exercised to preserve the original cutting orifice contour, or the tip efficiency will be drastically reduced.

It should be noted that the tables end at 200 to 350 mm (8 to 14 in.) which is the maximum thickness normally encountered for shape cutting in production shops. The division has been made arbitrarily. The cutting of steel plate over approximately 300 mm (12 in.) thick is considered heavy cutting. The characteristics of heavy cutting are discussed later.

Effects of Alloying Elements. Alloying elements have two possible effects on the oxygen cutting of steel. They may make the steel more difficult to cut, or they may give rise to hardened or heat-checked cut surfaces, or both. The effects of alloying elements are roughly evaluated in Table 13.5.

A large quantity of heat energy is liberated in the kerf when steel is cut with an oxygen jet. Much of this energy is transferred to the sides of the kerf, where it raises the temperature of the steel adjacent to the kerf above its critical temperature. Since the torch is moving forward, the source of heat quickly moves on. The mass of cold metal near the kerf acts as a quenching medium, rapidly cooling the hot steel. This quenching action may harden the cut surfaces of high carbon and alloy steels.

The depth of the heat affected zone depends on the carbon and alloy content, on the thickness of the base metal, and the cutting speed employed. Hardening of the heat-affected zones of carbon steels of 0.25 percent carbon maximum is not critical in the thicknesses usually cut. Higher carbon steels and some alloy steels are hardened to a degree that the thickness may become critical.

Typical depths of the heat-affected zones in oxygen cut steel are tabulated in Table 13.6. For most applications of oxygen cutting, the affected metal need not be removed. However, if it is removed, it should be done by mechanical means.

Preheating and Postheating. The material being cut may be preheated to provide desired mechanical and metallurgical characteristics or to improve the cutting operation.

Preheating the work can accomplish several useful purposes.

(1) It can increase the efficiency of the cutting operation by permitting higher travel speed. Higher travel speed will reduce the total amount of oxygen and fuel gas required to make the cut.

(2) It will reduce the temperature gradient in the steel during the cutting operation. This, in turn, will reduce or give more favorable distribution to the thermally induced stresses and prevent the formation of quenching or cooling cracks. Distortion will also be reduced.

(3) It may prevent hardening the cut surface by reducing the cooling rate.

(4) It will decrease migration of carbon toward the cut face by lowering the temperature gradient in the metal adjacent to the cut.

The temperatures used for preheating generally range from 90° to 700° C (200 to 1300° F), depending upon the part size and the type of steel to be cut. The majority of carbon and alloy steels can be cut with the steel heated to the 200° to 315° C (400° to 600° F) temperature range. The higher the preheat temperature, the more rapid is the reaction of the oxygen with the iron. This permits higher cutting speeds.

It is essential that the preheat temperature

Table 13.2 — Data for manual and machine cutting of clean low carbon steel without preheat

Thickness of steel mm	Diameter of cutting orifice, mm	Cutting speed mm/s	Gas flow, L/min			
			Cutting oxygen	Acetylene	Natural gas	Propane
			SI units			
3.2	0.51-1.02	6.8 -13.5	7.1- 21.2	1.4- 4.3	4.3-11.8	1.4- 4.7
6.4	0.76-1.52	6.8 -11.0	14.2- 26.0	1.4- 4.3	4.3-11.8	2.4- 5.7
9.5	0.76-1.52	6.4 -10.1	18.9- 33.0	2.8- 5.7	4.7-11.8	2.4- 7.1
13	1.02-1.52	5.1 - 9.7	26.0- 40.0	2.8- 5.7	7.1-14.2	2.4- 7.1
19	1.14-1.52	5.1 - 8.9	47.2- 70.9	3.3- 6.6	7.1-14.2	2.8- 8.5
25	1.14-1.52	3.8 - 7.6	51.9- 75.5	3.3- 6.6	8.5-16.5	2.8- 8.5
38	1.52-2.03	2.5 - 5.9	51.9- 82.6	3.8- 7.6	8.5-16.5	3.8- 9.4
51	1.52-2.03	2.5 - 5.5	61.4- 89.6	3.8- 7.6	9.4-18.9	3.8- 9.4
76	1.65-2.16	1.7 - 4.7	89.6-142	4.3- 9.4	9.4-18.9	4.3-10.4
102	2.03-2.29	1.7 - 4.2	113 -170	4.3- 9.4	9.4-18.9	4.3-11.3
127	2.03-2.41	1.7 - 3.4	127 -170	4.7-11.6	11.8-23.6	4.7-11.8
152	2.41-2.67	1.3 - 3.0	123 -236	4.7-11.6	11.8-23.6	4.7-14.2
203	2.41-2.79	1.3 - 2.1	217 -293	7.1-14.2	14.2-26.0	7.1-15.1
254	2.41-2.79	0.85- 1.7	274 -331	7.1-16.5	16.5-33.0	7.1-16.5
305	2.79-3.30	0.85- 1.7	340 -401	9.4-18.9	21.2-44.9	9.4-21.2

Thickness of steel in.	Diameter of cutting orifice, in.	Cutting speed in./min	Gas flow, ft³/h			
			Cutting oxygen	Acetylene	Natural gas	Propane
			U.S. customary units			
1/8	0.020-0.040	16-32	15-45	3-9	9-25	3-10
1/4	0.030-0.060	16-26	30-55	3-9	9-25	5-12
3/8	0.030-0.060	15-24	40-70	6-12	10-25	5-15
1/2	0.040-0.060	12-23	55-85	6-12	15-30	5-15
3/4	0.045-0.060	12-21	100-150	7-14	15-30	6-18
1	0.045-0.060	9-18	110-160	7-14	18-35	6-18
1-1/2	0.060-0.080	6-14	110-175	8-16	18-35	8-20
2	0.060-0.080	6-13	130-190	8-16	20-40	8-20
3	0.065-0.085	4-11	190-300	9-20	20-40	9-22
4	0.080-0.090	4-10	240-360	9-20	20-40	9-24
5	0.080-0.095	4-8	270-360	10-24	25-50	10-25
6	0.095-0.105	3-7	260-500	10-24	25-50	10-30
8	0.095-0.110	3-5	460-620	15-30	30-55	15-32
10	0.095-0.110	2-4	580-700	15-35	35-70	15-35
12	0.110-0.130	2-4	720-850	20-40	45-95	20-45

Notes:

1. Preheat oxygen consumptions: Preheat oxygen for acetylene = 1.1 to 1.25 × acetylene flow ft³/h; preheat oxygen for natural gas = 1.5 to 2.5 × natural gas flow ft³/h; preheat oxygen for propane = 3.5 to 5 × propane flow ft³/h

2. Operating notes: Higher gas flows and lower speeds are generally associated with manual cutting, whereas lower gas flows and higher speeds apply to machine cutting. When cutting heavily scaled or rusted plate, use high gas flows and low speeds. Maximum indicated speeds apply to straight line cutting; for intricate shape cutting and best quality, lower speeds will be required.

be fairly uniform through the section in the areas to be cut. If the metal near the surfaces is at a lower temperature than the interior metal, the oxidation reaction will proceed faster in the interior. Large pockets may form in the interior and either produce unsatisfactory cut surfaces or cause slag entrapment that may interrupt the cutting action. If the material is preheated in a furnace, cutting should be started as soon as possible after the material is removed from the furnace to minimize surface cooling.

If furnace capacity is not available for preheating the entire piece, local preheating in the vicinity of the cut will be of some benefit. For light cutting, preheating may be accomplished by passing the cutting torch preheating flames slowly over the line of the cut until the desired preheat temperature is reached. Another method which may give better results is to preheat with a multiflame heating torch mounted ahead of the cutting torch.

To reduce thermally induced internal stresses in the cut parts, they may be annealed, normalized, or stress relieved. Using a proper postheat treatment, most of metallurgical changes caused by the cutting heat can be eliminated. If a furnace of the required size is not available for postheat treatment, the cut surface may be reheated to the proper temperature by the use of multiple flame heating torches.

Cast Iron

The high carbon content of cast iron resists the ordinary OFC techniques used for low carbon steels. Cast irons contain some of the carbon in the form of graphite flakes or nodules, and some in the form of iron carbide (Fe_3C). Both of these constituents hinder the oxidation of the iron. High quality production cuts typical with steels cannot be obtained with cast iron. Most cutting is done to remove risers, gates, or defects; to repair or alter castings; or for scrap.

Cast iron can be manually cut by using an oscillating motion of the cutting torch, as shown in Fig. 13.11. The degree of motion depends on the section thickness. Torch oscillation helps the oxygen jet to blow out the slag and molten metal in the kerf. The kerf is normally wide and rough.

A larger cutting tip and a higher gas flow

than those used for steel are required for cutting the same thickness of cast iron. A reducing preheat flame is used with the streamer extending to the far side of the cast iron section. The excess fuel gas helps to maintain preheat in the kerf as it burns.

Cast iron is also sometimes cut by using the special techniques for cutting oxidation resistant steels. These are waster plate, metal powder cutting (POC), and chemical flux cutting (FOC) which are described later in this chapter.

Oxidation Resistant Steels

The absence of alloying materials in pure iron permits the oxidation reaction to proceed rapidly. As the quantity and number of alloying elements in iron increase, the oxidation rate decreases from that of pure iron. Cutting becomes more difficult. Oxidation of the iron in an alloy steel liberates a considerable amount of heat. The iron oxides produced have melting points near the melting point of iron. However, the oxides of many of the alloying elements in steels, such as aluminum and chromium, have melting points much higher than those of iron oxides. These high melting oxides, which are refractory in nature, may shield the material in the kerf so that fresh iron is not continuously exposed to the cutting oxygen stream. Thus, the speed of cutting decreases as the amount of refractory oxide forming elements in the iron increases. For ferrous metals with high alloy content, such as stainless steel, variations of OFC must be used.

There are several methods for oxygen cutting of oxidation resistant steels which are also applicable to cast irons. The important ones are

(1) Torch oscillation
(2) Waster plate
(3) Wire feed
(4) Powder cutting
(5) Flux cutting

When the above methods are used for cutting oxidation resistant metals, the quality of the cut surface is somewhat impaired. Scale may adhere to the cut faces. Carbon or iron pickup, or both, usually occurs on the cut surfaces of stainless steels and nickel alloy steels. This may affect the corrosion resistance and magnetic properties of the metal. If the corrosion resistance or mag-

Table 13.3—Typical data for cutting low carbon steel using MPS gas [a]

SI Units

| | | | Standard pressure cutting tips | | | | | | | | High pressure cutting tips | | | | | | |
| | | | Oxygen | | | | MPS Gas | | | | Oxygen | | | | MPS Gas | | |
Plate thk mm	Tip size no.	Cutting speed mm/s	Cutting[b] kPa	Flow L/min	Preheat[c] kPa	Flow L/min	kPa	Flow L/min	Kerf width mm	Cutting speed mm/s	Cutting[b] kPa	Flow L/min	Preheat[c] kPa	Flow L/min	kPa	Flow L/min	Kerf width mm
6	68	9-12	240-310	15-20	35-70	2-10	15-40	2-4	0.9	10-13	410-480	26-31	35-70	4-12	15-70	1-5	1.3
10	65	9-12	240-310	20-26	35-70	2-12	15-55	2-5	1.0	10-13	480-550	28-38	35-70	5-12	15-70	2-5	1.4
13	60	8-11	275-345	26-31	35-70	5-12	15-55	2-5	1.1	9-13	550-620	35-45	35-70	6-12	15-70	2-5	1.5
19	56	7-9	275-345	29-33	35-70	5-12	15-70	2-5	1.5	8-11	550-620	55-61	35-70	6-12	15-70	2-5	1.6
25	56	6-8	275-345	29-33	35-70	7-12	15-70	3-5	1.5	8-10	550-620	55-61	35-70	6-12	15-70	2-5	1.6
32	54	6-8	275-410	50-60	70-140	10-17	15-70	4-7	2.0	7-9	480-550	73-80	70-140	10-18	15-70	4-7	2.0
38	54	5-7	275-410	50-60	70-140	10-17	15-70	4-7	2.0	6-8	550-620	80-85	70-140	10-18	15-70	4-7	2.0
51	52	4-6	275-410	70-100	70-140	10-17	15-70	4-7	2.3	6-8	550-620	100-120	70-140	10-18	15-70	4-7	2.3
64	52	4-6	275-410	70-100	70-205	10-24	40-70	4-10	2.3	5-7	550-620	100-120	70-140	10-18	25-70	4-7	2.3
76	49	3-5	275-480	100-120	70-205	10-24	40-70	4-10	2.5	4-6	550-620	145-170	70-140	10-18	40-70	4-7	2.5
102	44	3-4	275-480	150-170	70-205	10-24	40-70	5-10	3.6	4-6	550-620	200-245	70-140	14-18	40-70	5-7	3.0
152	44	2-3	275-480	150-170	70-205	10-24	40-100	5-10	3.6	3-5	550-620	200-245	70-140	14-18	70-100	5-7	3.0
203	38	2-3	275-550	190-240	205-345	24-47	70-100	10-19	4.4	3-4	550-620	280-340	100-200	14-24	70-100	7-10	3.8
254	31	1-2	275-620	260-350	205-345	24-47	70-100	10-19	4.8								
305	28	1-2	275-690	350-450	205-345	35-70	70-100	15-29	5.5								
355	19	1-2	275-830	470-590	205-345	35-70	70-100	15-29	6.4								

a. Methylacetylene-propadiene stabilized

b. Cutting oxygen pressures at the torch. All recommendations are for straight line cutting with three hose torch perpendicular to plate.

c. Preheat pressures measured at regulator based on 7.5m maximum length of 4.8mm I.D. hose. Preheat oxygen 70-200 kPa for equal pressure torch, 200-620 kPa for injector torch.

Table 13.3 (continued) – Typical data for cutting low carbon steel using MPS gas[a]

U.S. customary units

Plate thk in.	Tip size no.	Standard pressure cutting tips								High pressure cutting tips							
		Cutting speed in./min	Oxygen				MPS Gas		Kerf width in.	Cutting speed in./min	Oxygen				MPS Gas		Kerf width in.
			Cutting[b]		Preheat[c]						Cutting[b]		Preheat[c]				
			psi	Flow ft³/h	psi	Flow ft³/h	psi	Flow ft³/h			psi	Flow ft³/h	psi	Flow ft³/h	psi	Flow ft³/h	
¼	68	22-28	35-45	30-40	5-10	5-20	2-6	2-8	0.035	24-31	60-70	55-65	5-10	8-25	2-10	3-10	0.05
⅜	65	21-27	35-45	40-55	5-10	5-25	2-8	2-10	0.04	23-30	70-80	60-80	5-10	10-25	2-10	4-10	0.055
½	60	20-26	40-50	55-65	5-10	10-25	2-8	4-10	0.045	22-29	80-90	75-95	5-10	12-25	2-10	5-10	0.06
¾	56	16-21	40-50	60-75	5-10	10-25	2-10	4-10	0.06	20-26	80-90	115-120	5-10	12-25	2-10	5-10	0.065
1	56	14-19	40-50	60-75	5-10	15-25	2-10	6-10	0.06	18-24	80-90	115-130	5-10	12-25	2-10	5-10	0.065
1¼	54	13-18	40-60	105-120	10-20	20-38	2-10	8-15	0.08	16-22	70-80	155-170	10-20	20-38	2-10	8-15	0.08
1½	54	12-16	40-60	105-120	10-20	20-38	2-10	8-15	0.08	15-20	80-90	170-180	10-20	20-38	2-10	8-15	0.08
2	52	10-14	40-60	145-190	10-20	20-38	2-10	8-20	0.09	14-19	80-90	215-255	10-20	20-38	2-10	8-15	0.09
2½	52	9-13	40-60	150-200	10-30	20-50	6-10	8-20	0.09	12-17	80-90	215-255	10-20	20-38	4-10	8-15	0.09
3	49	8-11	40-70	200-250	10-30	20-50	6-10	10-20	0.10	10-14	80-90	310-365	10-20	20-38	6-10	8-15	0.10
4	44	7-10	40-70	300-360	10-30	20-50	6-10	10-20	0.14	9-13	80-90	420-510	10-20	30-38	6-10	10-15	0.12
6	44	5-8	40-70	300-360	10-30	20-50	6-15	20-40	0.14	7-11	80-90	420-510	10-20	30-38	10-15	10-15	0.12
8	38	4-6	40-80	415-515	30-50	50-100	10-15	20-40	0.17	6-9	80-90	590-720	15-30	30-50	10-15	15-20	0.15
10	31	3-5	40-90	550-750	30-50	50-100	10-15	20-40	0.19								
12	28	3-5	40-100	750-950	30-50	75-150	10-15	30-60	0.22								
14	19	2-4	40-120	1000-1250	30-50	75-150	10-15	30-60	0.25								

a. Methylacetylene-propadiene stabilized

b. Cutting oxygen pressures at the torch. All recommendations are for straight line cutting with three hose torch perpendicular to plate.

c. Preheat pressures measured at regulator based on 25 ft maximum length of 5/16 in. I.D. hose. Preheat oxygen 10-30 psi for equal pressure torch, 30-90 psi for injector torch.

Table 13.4—Operating data for oxygen cutting of low carbon steel using gasoline fuel[a]

Metal thickness		Tip no.[b]	Pressure			
			Gasoline		Oxygen[c]	
mm	in.		kPa	psi	kPa	psi
3.2-6.4	1/8-1/4	0-A	69	10	83-117	12-17
6.4- 25	1/4-1	1-A	69	10	117-172	17-25
25- 51	1-2	2-A	69	10	172-241	25-35
51-102	2-4	3-A	69	10	241-276	35-40
102-152	4-6	4-A	83	12	276-345	40-50
152-203	6-8	5-A	97	14	345-414	50-60
203-254	8-10	6-A	110	16	483-552	70-80
254-305	10-12	7-A	124	18	552-621	80-90
305-356	12-14	8-A	138	20	689-827	100-120

a. This chart offers suggestions for the cutting range of each tip, and suggested gasoline and oxygen pressures.
b. The best combination of tip and pressure depends on operator technique, type and size of metal, desired cutting speed, length of hose, elevation of gasoline tank.
c. Oxygen pressure must always be higher than gasoline pressure.

netic properties of the material are important, approximately 3 mm (1/8 in.) of metal should be machined from the cut edges.

Torch Oscillation. This technique is the one described previously for cast iron cutting. Stainless steels up to 200 mm (8 in.) thick can be severed with a standard cutting torch and oscillation. The entire thickness of the starting edge must be preheated to a bright red color before the cut is started. This technique can also be combined with some of the other cutting methods listed.

Waster Plate. One method of cutting oxidation resistant steels is to clamp a low carbon steel "waster" plate on the upper surface of the material to be cut. The cut is started in the low carbon steel material. The heat liberated by the oxidation of the low carbon steel provides additional heat at the cutting face to sustain the oxidation reaction. The iron oxide from the low carbon steel helps to wash away the refractory oxides from stainless steel. The thickness of the waster plate must be in proportion to the thickness of the material being cut. Several undesirable features of this method are the cost of the waster plate material, the additional setup time, the slow cutting speeds, and the rough quality of the cut.

Wire Feed. With the appropriate equipment, a small diameter low carbon steel wire is fed continuously into the torch preheat flames, ahead of the cut. The end of the wire should melt rapidly into the surface of the alloy steel plate. The effect of the wire addition on the cutting action is the same as that of the waster plate. The deposition rate of the low carbon steel wire must be adequate to maintain the oxygen cutting action. It should be determined by trial cuts. The thickness of the alloy plate and cutting speed are also factors that must be considered in the process. A motor-driven wire feeder and wire guide, mounted on the cutting torch, are needed as accessory equipment.

Metal Powder Cutting. The metal powder cutting process (POC) is a technique for supplementing an OFC torch with a stream of iron rich powdered material. The powdered material accelerates and propagates the oxidation reaction and also the melting and spalling action of hard-to-cut materials. The powder is directed into the kerf through either the cutting tip or single or multiple jets external to the tip. When the first method is used, gas-conveyed powder is introduced into the cutting oxygen prior to its discharge from the tip. When the powder is introduced externally, the gas conveying the powder imparts sufficient velocity to the powder particles to carry them through the preheat envelope into the cutting oxygen stream. The short time in the preheat envelope is sufficient to produce the desired reaction in the cutting zone.

Table 13.5—Effect of alloying elements on resistance of steel to oxygen cutting

Element	Effect of element on oxygen cutting
Carbon	Steels up to 0.25% carbon can be cut without difficulty. Higher carbon steels should be preheated to prevent hardening and cracking. Graphite and cementite (Fe_3C) are detrimental but cast irons containing 4% carbon can be cut by special techniques.
Manganese	Steels of about 14% manganese and 1.5% carbon are difficult to cut and should be preheated for best results.
Silicon	Silicon, in amounts usually present, has no effect. Transformer irons containing as much as 4% silicon are being cut. Silicon steel containing large amounts of carbon and manganese must be carefully preheated and postannealed to avoid air hardening and possible surface fissures.
Chromium	Steels up to 5% chromium are cut without much difficulty when the surface is clean. Higher chromium steels, such as 10% chromium steels, require special technique (see the section Oxidation Resistant Steels) and the cuts are rough when the usual oxyacetylene cutting process is used. In general, carburizing preheat flames are desirable when cutting this type of steel. The flux injection and iron powder cutting processes enable cuts to be readily made in the usual straight chromium irons and steels as well as in stainless steel.
Nickel	Steels containing up to 3% nickel may be cut by the normal oxygen cutting processes; up to about 7% nickel content, cuts are very satisfactory. Cuts of excellent quality may be made in the usual engineering alloys of the stainless steels (18-8 to about 35-15 as the upper limit) by the flux injection or iron powder cutting processes.
Molybdenum	This element affects cutting about the same as chromium. Aircraft quality chrome-molybdenum steel offers no difficulties. High molybdenum-tungsten steels, however, may be cut only by special techniques.
Tungsten	The usual alloys with up to 14% may be cut very readily, but cutting is difficult with a higher percentage of tungsten. The limit seems to be about 20% tungsten.
Copper	In amounts up to about 2%, copper has no effect.
Aluminum	Unless present in large amounts (on the order of 10%) the effect of aluminum is not appreciable.
Phosphorus	This element has no effect in amounts usually tolerated in steel.
Sulfur	Small amounts, such as are present in steels, have no effect. With higher percentages of sulfur, the rate of cutting is reduced and sulfur dioxide fumes are noticeable.
Vanadium	In the amounts usually found in steels, this alloy may improve rather than interfere with cutting.

Table 13.6—Approximate depths of heat-affected zones in oxygen cut steels[a]

Thickness		Depth			
		Low Carbon Steels		High Carbon Steels	
mm	in.	mm	in.	mm	in.
Under 13	Under 1/2	Under 0.8	Under 1/32	0.8	1/32
13	1/2	0.8	1/32	0.8 to 1.6	1/32 to 1/16
152	6	3.2	1/8	3.2 to 6.4	1/8 to 1/4

a. The depth of the fully hardened zone is considerably less than the depth of the heat-affected zone.

Movement when cutting thin cast iron

Movement when cutting heavy cast iron

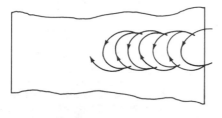

General direction

**Fig. 13.11—Typical cutting torch manip-
ulation for cutting cast iron**

Some of the powders react chemically with the refractory oxides produced in the kerf and increase their fluidity. The resultant molten slags are washed out of the reaction zone by the oxygen jet. Fresh metal surfaces are continuously exposed to the oxygen jet. Iron powder and mixtures of metallic powders, such as iron and aluminum, are used.

Cutting of oxidation resistant steels by the powder method can be done at approximately the same speeds as oxygen cutting of carbon steel of equivalent thickness. The cutting oxygen flow must be slightly higher with the powder

process. Table 13.7 contains powder cutting data for AISI 302 stainless steel.

Powder Cutting Equipment. Dispensers of powder for the POC process are of two general types. One type of dispenser is a vibratory device in which the quantity of powder dispensed from the hopper is governed by a vibrator. Desired amounts of powder can be obtained by adjusting the amplitude of vibration. The vibratory type dispenser is generally used where uniform and accurate powder flow is required. A typical application is precision cutting of materials, such as stainless steel. It is also used when high-quality, sharp top edges are required on cuts in carbon steels.

The other type of dispenser is a completely pneumatic device. In the bottom of a low pressure vessel there is an ejector or fluidizing unit. The powder-conveying gas is brought into the dispenser in a manner that fluidizes the powder. The powder flows uniformly into an ejector unit. Here it is picked up by a gas stream that serves as the transporting medium to the torch.

In addition to the fuel and oxygen hoses, another hose is used to convey the powder to the torch. A special manual powder cutting torch mixes the oxygen and fuel gas and then discharges this mixture through a multiplicity of orifices in the cutting tip. The powder valve is an integral part of the torch. The cutting oxygen lever on the torch also opens the powder valve in proper sequence. The powder carried by the conveying gas is brought through a separate tube into a sealed chamber forward of the preheat gas chamber in the torch head. The powder then enters a separate

Table 13.7—Oxyfuel gas powder cutting data for AISI 302 austenitic stainless steel

Metal thickness		Cutting oxygen Orifice diameter		Cutting oxygen pressure		Cutting speed		Gas consumption				Powder flow	
								Oxygen		Acetylene			
mm	in.	mm	in.	kPa	psi	mm/s	in./min	L/min	ft³/h	L/min	ft³/h	g/min	oz./min
13	1/2	1.02	0.040	344	50	5.9	14	59	125	7	15.	113	4
25	1	1.52	0.060	344	50	5.1	12	106	225	11	23	113	4
51	2	1.52	0.060	344	50	4.2	10	142	300	11	23	113	4
76	3	2.03	0.080	344	50	3.8	9	260	550	15	32	142	5
102	4	2.54	0.100	344	50	3.4	8	319	675	18	38	170	6
127	5	3.05	0.120	414	60	3.0	7	378	800	21	45	198	7
152	6	3.56	0.140	414	60	2.5	6	425	900	30	63	227	8
203	8	3.56	0.140	483	70	1.7	4	472	1000	30	63	227	8
254	10	4.06	0.160	517	75	1.5	3.5	520	1100	35	75	227	8
305	12	4.06	0.160	517	75	1.3	3	566	1200	35	75	227	8

group of passages in a two-piece cutting tip. From there, it discharges at the mouth of the tip in a conical pattern. The powder emerges with sufficient velocity to pass through the burning preheat gas and surrounds the central cutting oxygen stream.

Flux Cutting. This process is primarily intended for cutting stainless steels. The flux is designed to react with oxides of alloying elements, such as chromium and nickel, to produce compounds with melting points near those of iron oxides. A special apparatus is required to introduce the flux into the kerf. With a flux addition, stainless steels can be cut at a uniform linear speed without torch oscillation. Cutting speeds approaching those for equivalent thicknesses of carbon steel can be attained. The tip sizes will be larger, and the cutting oxygen flow will be somewhat greater than for carbon steels.

Flux Cutting Equipment. With the flux process, a flux feed unit is required. The cutting oxygen passes through the feed unit, and it transports the flux to the torch. The flux is held in a dispenser designed to operate at normal cutting oxygen pressures. The flux is transported through a hose from the dispenser to a conventional three-hose cutting torch. A mixture of oxygen and flux flows from the oxygen orifice of the torch tip. Special operating procedures are used to prevent buildup of flux in the cutting oxygen hose and the cutting torch.

GENERAL PROCESS APPLICATIONS

Manual OFC is widely used for rough severing of steel and cast iron shapes. Portability permits taking the equipment to the job site. Structural shapes, pipe, rod, and similar materials can be cut to length for construction or cut up in scrap and salvage operations. In a steel mill or foundry, extraneous projections, such as caps, gates, and risers, are quickly severed from billets and castings. Mechanical fastenings, such as bolts, rivets, and pins, are rapidly severed for disassembly using OFC. Holes can rapidly be made in steel components by piercing and cutting.

Machine OFC is used in many industries and steel warehouses to cut steel plate to size, to cut various shapes from plate, and to prepare plate

edges for welding. Many machine parts such as gears, clevises, frames, and tools are made by the processes.

Machines capable of cutting to tolerances of 0.8 to 1.6 mm (1/32 to 1/16 in.) are used to produce parts that can be assembled into final product form without intermediate machining. They are also used for rapid material removal prior to machining to close tolerances.

Oxyfuel gas cutting is used to cut a wide range of steel thicknesses from approximately 3 to 1525 mm (1/8 to 60 in.). Thicknesses over approximately 500 mm (20 in.) are not generally cut except in steel mill operations where the pieces are cut while still at high temperatures.

OPERATING PROCEDURES

In the operation of OFC equipment, the recommendations of the equipment manufacturer in assembling and using the equipment should be followed. This will prevent damage to the equipment and also insure its proper utilization. Because pure oxygen and a combustible fuel are used for OFC operations, there are certain dangers associated with the process. Safe operating procedures must be followed to avoid injury to personnel and damage to equipment.

Safe Practices

Safe practices for the installation and operation of oxyfuel gas systems for welding and cutting are given in American National Standard Z49.1, latest edition. These practices should be followed by the oxygen cutter in setting up and operating the equipment.

Regulators and Hoses. Only regulators and hoses designed for the particular gas should be used. Oxygen equipment has righthand threads, and the hose is green. Fuel gas equipment has lefthand threads, and the hose is red. All connections must be tight and leak free. The oxygen regulator must be clean. The pressure adjustment should be backed off completely before opening the cylinder or manifold valve.

Flashback. A flashback is the burning of the flame in or beyond the torch mixing chamber. It is a serious condition, and corrective action must be taken to extinguish it. The torch oxygen valve should be turned off immediately and then the fuel gas valve. One cause of flashback is failure to purge the hose lines before lighting the torch; another cause is the overheating of the torch tip.

Operating the Torch

The recommended lighting procedure for a torch depends on the type of equipment and fuel being used. In all cases, only a sparklighter or other recommended lighting devices should be used. Shaded or tinted eye protection should be worn.

Both equal and positive pressure gas torches may be lighted in either of the following ways:

(1) Open the fuel gas valve slightly, and light the torch with a sparklighter. Adjust the fuel gas until a stable flame is maintained at the end of the tip. Open the oxygen valve slowly and increase the flow until the desired flame is attained. The intensity of the flame may be adjusted by slightly increasing or decreasing the volumes of both gases.

(2) Open the fuel gas valve slightly to make sure fuel gas is flowing from the tip. Then open the oxygen valve slightly. Light the gases with a sparklighter. Adjust the oxygen and fuel gas in successive steps to obtain the desired flame.

Low or universal pressure gas torches use an injector or venturi mixing method with relatively low fuel pressures. To light this type of torch, open the fuel gas valve about 3/4 turn, and then open the oxygen valve slightly. Light the torch with a sparklighter, and adjust the preheat oxygen valve to obtain the desired flame.

Special procedure must be followed to light an oxygen-gasoline torch. The manufacturer's operating instructions must be followed to avoid injury.

A torch is extinguished by rapidly closing the torch valves in the sequence recommended by the manufacturer. After extinguishing the torch, close both the oxygen and the fuel valves on the cylinders or pipelines. Open the torch fuel gas valve to bleed off the fuel gas from the regulator, hose, and torch. Back out the regulator adjusting screw, and then close the torch fuel gas valve. Repeat this procedure for the oxygen system.

Flame Adjustment

Flame adjustment is a critical factor in attaining satisfactory torch operation. The amount of heat produced by the flame depends on the intensity and type of flame used. Three types of flames can be set by properly adjusting the torch valves (see Fig. 13.5).

A reducing flame with acetylene or MPS is indicated by trailing feathers on the primary flame cone or by long yellow-orange streamers in the secondary flame envelope. Propane and propylene base fuels and natural gas have a long rounded primary flame cone. The reducing flame is often used for the best finish and for stack cutting of thin material.

A neutral flame with acetylene or MPS is indicated by a sharply defined, dark primary flame cone and a pale blue secondary flame envelope. Propane and propylene base fuels and natural gas have a short and sharply defined cone. This flame is obtained by adding oxygen to a reducing flame. It is the most frequently used flame for cutting.

An oxidizing flame for acetylene or MPS has a light color primary cone and a smaller secondary flame shroud. It also generally burns with a harsh whistling sound. With propane and propylene base fuels and natural gas, the primary flames cones are longer, less sharply defined, and have a lighter color. This flame is obtained by adding some oxygen to the neutral flame. This type of flame is frequently used for fast low quality cutting, and selectively in piercing and quality beveling.

Cutting Procedures

Manual Cutting. Several methods can be used to start a cut on an edge. The most common method is to place the preheat flames halfway over the edge, holding the end of the flame cones 1.5 to 3 mm (1/16 to 1/8 in.) above the surface of the material to be cut. The tip axis should be aligned with the plate edge. When the top corner reaches a reddish yellow color, the cutting oxygen valve is opened and the cutting process starts. Torch movement is started after the cutting action reaches the far side of the edge.

Another starting method is to hold the torch halfway over the edge, with the cutting oxygen turned on, but not touching the edge of the material. When the metal reaches a reddish yellow color, the torch is moved onto the material and cutting starts. This method wastes oxygen, and starting is more difficult than with the first method. It should only be used for cutting thin material where preheat times are very short.

A third method is to put the tip entirely over the material to be cut. The preheat flame is moved back and forth a short distance along the line of cut until it reaches ignition temperature. Then, with the tip at the starting point, cutting is initiated. This method has the advantage of producing sharper corners at the beginning of the cut.

Once the cut has been started, the torch is moved along the line of cut with a smooth, steady motion. The operator should maintain as constant a tip-to-work distance as possible. The torch should be moved at a speed that produces a light ripping sound and a smooth spark stream.

For plate thicknesses of 13 mm (1/2 in.) or more, the cutting tip should be held perpendicular to the plate. For thin plate, the tip can be tilted in the direction of the cut. Tilting increases the cutting speed and helps prevent slag from freezing across the kerf. When cutting material in a vertical position, start on the lower edge of the material and cut upward.

It is often necessary to start a cut at some point other than on the edge of a piece of metal. This technique is known as piercing. Piercing usually requires a somewhat larger preheat flame than the one used for an edge start. In addition, the flame should be adjusted slightly oxidizing to increase the heat energy.

The spot to be pierced should be located in the scrap area. The material surface can be brought to starting temperature very rapidly if the ends of the primary flame cones are allowed to touch the plate surface. A few vertical movements of the tip will indicate the proper distance. When a few sparkles are observed, the material is usually at ignition temperature. A slight rotary motion of the tip will insure that the center of the spot is also hot.

The cutting oxygen valve is opened and the torch tip lifted slowly away from the plate. Torch motion is slowly initiated along the cut line. Cutting action should progress through the plate as the torch moves forward. If the cutting oxygen is turned on too quickly and the torch is not lifted, slag may be blown into the tip and plug the gas ports.

Machine Cutting. Operating conditions for machine oxygen cutting will vary depending on the fuel gas and the style of cutting torch being used. Tip size designations, tip design, and operating pressures differ among manufacturers. Operating data can be obtained from the torch manufacturers.

Start up and shutdown procedures for machine OFC are essentially the same as those previously given for hand torch operation. However, proper adjustment of operating conditions is more critical if high speed, high quality cuts are to be obtained. The manufacturer's or supplier's cutting chart should be used to select the proper tip size for the material thickness to be cut. In addition to the tip size, initial fuel and oxygen pressure settings and travel speeds should be selected from the chart. Frequently, gas flow rates, drill size of the oxygen orifice, preheat cone lengths, and kerf width are also listed. Opeating conditions should then be adjusted to give the desired cut quality.

Proper tip size and cutting oxygen pressure are very important in making a quality machine cut. If the proper tip size is not used, maximum cutting speed and the best quality of cut will not be achieved. The cutting oxygen pressure setting is an essential condition; large deviations from the recommended setting will greatly affect performance. For this reason, some manufacturers specify the pressure setting at the regulator when operating with a given length of hose. When longer or shorter hoses are used, an adjustment in pressure should be made. An alternative is to measure oxygen pressure at the torch inlet. Cutting oxygen pressure settings are then adjusted at the torch rather than at the regulator.

Other adjustments, such as the preheat fuel and oxygen pressure settings and the travel speed, are also important. Once the regulators have been adjusted, the torch valves are used to throttle gas flows to give the desired preheat flame. If sufficient flow rates are not obtained, pressure settings at the regulator can be increased to compensate. Cleanliness of the nozzle, type of base metal, purity of cutting oxygen, and other factors have a direct effect on performance.

Manufacturers differ in their recommended travel speeds. Some give a range of speeds for specific thicknesses while others list a single speed. In either case, the settings are intended only as a guide. In determining the proper speed for an application, begin the cut at a slower speed than that recommended. Gradually increase the speed until cut quality falls below the required level. Then reduce the speed until the cut quality is restored, and continue to operate at that speed.

Heavy Cutting

Heavy cutting is considered the cutting of steel over approximately 300 mm (12 in.) thick. The basic reactions that permit oxygen cutting of heavy steel are the same as those for the cutting of thinner sections. Thicknesses ranging from 300 to 1525 mm (12 to 60 in.) may be cut using heavy duty torches. Preheat and cutting oxygen flow increase, and cutting speed decreases as thickness increases.

For heavy cutting, the most important factor is oxygen flow. Tip size and operating pressure must provide the necessary cutting oxygen flow required for the thickness being cut. Oxygen cutting pressures in the range of 152 to 386 kPa (22 to 56 psi), measured at the cutting torch, have been found adequate for the heaviest cutting using the proper tip size and equipment. The oxygen flow at the torch entry is of paramount importance in comparing results of different cutting opera-tions. By predicting performance on the basis of oxygen flow rate rather than pressure, heavy cutting data can be plotted as a continuous curve.

In terms of flow, it is possible to arrive at an approximate demand constant that will be useful as a guide in selecting equipment suitable for a given job. These demand constants may vary, but in terms of thickness they usually fall within the approximate range of from 89 to 139 L of oxygen per mm of thickness (80 to 125 ft^3 per in.) Table 13.8 gives the range of operating conditions that cover normal heavy cutting operations.

Heavy cutting covers a wide variety of operations such as ingot cropping, scrap cutting, and riser cutting. The data of Table 13.8 may not be entirely suitable for all heavy cutting operations, although the values given have been used success-fully. They may be used as a guide in selecting the correct equipment and operating conditions. The actual values for most efficient operation of a specific cutting application are always best found by trial cuts.

When heavy cutting is performed with the torch in a horizontal position, the cutting oxygen pressure may have to be increased to aid in re-moval of the slag produced from the kerf.

Recommendations as to the speed of travel are not included in Table 13.8, but speeds from 0.85 to 2.5 mm/s (2 to 6 in./min) are used in the range of thicknesses covered. A speed of 1.3 mm/s (3 in./min) is possible for thicknesses up to at least 910 mm (36/in.). The correct speed is obtained by observing the operating conditions

Table 13.8—Data for heavy oxyfuel gas cutting of low carbon steel

Material thickness		Cutting oxygen					
		Orifice diameter		Flow rate		Pressure at torch	
mm	in.	mm	in.	L/min	ft^3/h	kPa	psi
305	12	3.74- 5.61	0.147-0.221	472- 708	1000-1500	386-228	56-33
406	16	4.32- 7.36	0.170-0.290	614- 944	1300-2000	372-172	54-25
508	20	4.93- 8.44	0.194-0.332	803-1180	1700-2500	359-152	52-22
610	24	5.61- 8.44	0.221-0.332	944-1416	2000-3000	331-200	48-29
711	28	6.35- 9.53	0.250-0.375	1087-1652	2300-3500	283-179	41-26
813	32	6.35- 9.53	0.250-0.375	1274-1888	2700-4000	352-207	51-30
914	36	7.37-10.72	0.290-0.422	1416-2120	3000-4500	276-179	40-26
1016	40	7.37-10.72	0.290-0.422	1605-2360	3400-5000	317-207	46-30
1118	44	7.37-11.90	0.290-0.468	1792-2600	3800-5500	352-179	51-26
1219	48	8.44-11.90	0.332-0.468	1888-2830	4000-6000	276-193	40-28

carefully and making suitable adjustments while actual cutting is in progress.

Because heavy pieces usually have a scale-covered surface, techniques of starting the cut differ from those usually used with clean thin material. The start of the reaction is made more slowly on the rougher edges. Figure 13.12 indicates correct and incorrect starting procedures. Figure 13.12(A) shows the desirable starting position with the preheat flames on the top corner and extending down the face of the material. The cutting reaction starts at the top corner. It proceeds down the face of the material to the bottom as the torch moves forward. Figures 13.12(B), (C), (D), (E), and (F) show problems occurring from incorrect procedures.

When the cut proceeds properly with correct oxygen flow and forward speed, the reaction will proceed to the end of the cut without leaving a skipped corner. Figure 13.13 illustrates various correct and incorrect terminating conditions and also proper drag conditions. Conditions producing a drop cut are depicted in Fig. 13.13(A).

In general, the conditions required for successful heavy cutting on a production basis are

(1) An adequate gas supply sufficient to complete the cut. This is necessary because a lost cut on heavy materials is extremely difficult, if not impossible, to restart.

(2) Equipment of sufficient size structurally to maintain rigidity and to carry the equipment needed, and of sufficient capacity to handle the range of speeds and gas flows required.

(3) Skilled personnel that are trained in proper heavy cutting techniques.

Stack Cutting

If data on machine OFC speeds and gas requirements are plotted against the material thickness, the requirements are not directly proportional to material thickness, t. Gas consumption per unit of thickness, Δt, decreases as the thickness, t, increases. Consequently, cutting costs per Δt may decrease as t increases when Δt is below a specific value, depending on the material being cut. Stacking of material for cutting can be more economical than cutting individual pieces, particularly when the material thickness is under

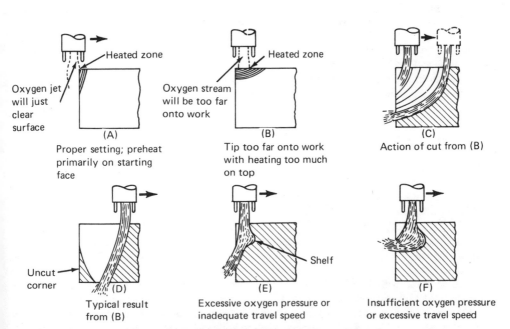

Fig. 13.12—Typical conditions for heavy cutting

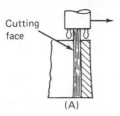

Cutting face

(A)

No drag permits stream to break through face uniformly at all points. Typical of balanced conditions

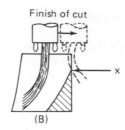

Finish of cut

(B)

Drag causes action to carry through at x and to pass beyond material, leaving uncut corner. Typical of insufficient oxygen or excessive speed

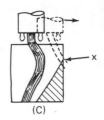

(C)

Forward drag causes stream to break through at x and become deflected, leaving uncut corner. Typical of high cutting oxygen pressure or too little speed

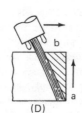

(D)

If cutting face is such that breakthrough at bottom, a, lies ahead of entry point, b, and at no point does face extend beyond a, action will sever from a, upward

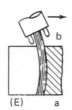

(E)

Angular tip disposition similar to (D) showing limit of effectiveness. Similar to (A)

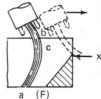

(F)

If conditions are such that a and b are in line or otherwise disposed, but c lies ahead of a, stream will break at x, leaving uncut corner, similar to (C)

Fig. 13.13—Terminating conditions for heavy cutting—A, B, and C with torch vertical; D, E, and F with torch angled in direction of cutting

6 mm (1/4 in.). Stack cutting is limited to sheet and plate up to 13 mm (1/2 in.) thick because of the difficulty in clamping heavier material in a tight stack. A stack cutting operation is shown in Fig. 13.14.

Stack cutting is also a means of cutting sheet material that is too thin for ordinary OFC methods. Sheet thicknesses of 0.9 mm (20 gage) and over are the most practical. Stack cutting is used in place of shearing or stamping, particularly where volume does not justify expensive dies. The flame cut sheet edges are square with no burrs.

Successful stack cutting requires clean, flat sheet or plate. Dirt, mill scale, rust, and paint may interrupt the cut and ruin the stack of material. The stack must be securely clamped, particularly at the cut location, with the edges aligned at the cut starting point. Piercing of stacks with the OFC torch to start a cut is impractical. Holes must be drilled through the stacks to start an interior cut.

The total thickness of the stack is determined by the cutting tolerance requirement and the thickness of the top piece. With a cutting tolerance of 0.8 mm (1/32 in.), stack height should not exceed 50 mm (2 in.); with a 1.6 mm (1/16 in.) tolerance, the thickness may be up to 100 mm (4 in.). The maximum practical limit of thickness is about 150 mm (6 in.).

When stack cutting material less than 5 mm (3/16 in.) thick, a waster plate 6 mm (1/4 in.) thick is used on top. It insures better starting, a sharper edge on the top production piece, and no buckling of the top sheet.

Starting the cut must be done with extreme care so that it will extend through the stack. One method of starting is to align the sheet edges exactly in a vertical line. A vertical strip along the aligned face is preheated with a hand torch to ignition temperature. The machine torch is quickly positioned at the starting point and cutting init-

Fig. 13.14—Typical stack cutting operation with the plates clamped by vertical welds

iated. Another procedure is to position each sheet so that its edge projects slightly over the edge below. This is advantageous for sheared sheet stacked with the burr down. Cutting is initiated on the top plate (waster plate) and progresses from one sheet to the other through the stack. A third method is to run a vertical weld bead down the stack to form a continuous strip of metal. The cut is started through the weld bead and progresses into the stack.

Even when extreme care has been exercised, there is always the possibility of an interruption of cutting with possible loss of the entire stack. The application of flux cutting and powder cutting processes greatly minimizes this hazard. These methods assist in propagating the oxidation reaction through the cut. Appreciable air gaps that otherwise might inhibit cutting can be tolerated between plates. The use of divergent tips with high velocity cutting jets appears to aid this transfer action also. High alloy steels, including stainless steels, can be stack cut by using one of these methods.

Regardless of procedure employed, the economy of a stack cutting operation must be carefully compared with the total costs involved, including such items as material preparations, stack makeup, clamping devices, and increased skill and care requirements.

Plate Edge Preparation

Bevel, V-, J-, and U-groove joint designs are used for welding steel components together. The preparation of the edges to be welded together can be done by oxygen cutting or gouging. Single and double bevels are produced by using standard cutting tips and torches, usually mechanized, for straight line beveling. Oxygen gouging is done by using specially designed cutting tips on mechanized cutting torches to produce J- and U-groove joints.

Plate Beveling. The beveling of plate edges before welding is necessary in many applications to insure proper dimensions and fit, and also to allow proper welding techniques. Beveling may be done by using a single torch or multiple

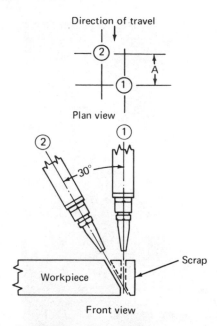

Fig. 13.15 — Cutting a single bevel edge preparation with a root face

The best quality of cut face is not always obtained at the highest possible cutting speed. The cut face finish can usually be improved by operating at lower speeds. When speed is reduced to obtain improved surface finish, the preheat flames should be decreased to prevent excessive melt-down of the top edge of the faces.

Figures 13.15, 13.16, and 13.17 illustrate the torch positions to cut the three basic beveled edges. In each case, torch position spacings A and B are governed by plate thickness, tip size, and speed of cutting. The cutting torches are positioned at spacings that are practical without interrupting the cutting action of any of the three cutting oxygen streams. When the lengths of A or B, or both, are too great, the cutting action of the trailing torch does not span the kerf of the leading torch. This causes the oxygen stream to be deflected into the kerf of the leading torch, and it gouges the cut face. This produces a rough surface and usually a light adherence of dross to the underside of the prepared edge.

torches operating simultaneously. Although single beveling can be done manually, beveling is best done by machine for accurate control of the cutting variables. When cutting bevels with two or three torches, plate riding devices should be used to insure constant tip position above the plate (see Fig. 13.8).

In single torch beveling, the amount and type of torch preheat is a dominant factor. With bevel angles of less than 15°, the loss of preheat efficiency is small. When the bevel angle is above 15°, the heat transfer from the preheat flames to the plate decreases rapidly as the bevel angle increases. Considerably greater preheat input is required, particularly for thicknesses up to 25 mm (1 in.). Best results are obtained by positioning the tip very close to the work and using high oxygen to fuel ratios.

An auxiliary torch (with only preheat flames burning) mounted perpendicular to the work or an auxiliary adapter, which divides the preheat and applies a portion of it at right angles to the work, may be used to obtain faster beveling speeds. Either method actually consumes less total preheat gases than a single angled tip.

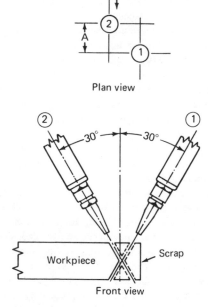

Fig. 13.16 — Cutting a double bevel edge preparation with no root face

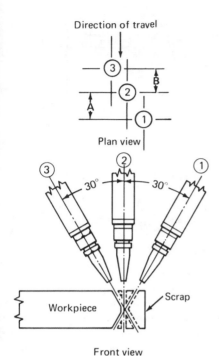

**Fig. 13.17—Cutting a double bevel edge
preparation with a root face**

The positioning of the torches in a lateral direction for multibevel cutting is usually accomplished by trial and error. However, this can be costly and result in lengthy reworking or possible scrap. The use of a simple machined template, which is typical of the desired edge geometry, is quite useful for torch alignment. A kerf-centering device is attached to each cutting tip, as shown in Fig. 13.18. The torches are then properly angled and adjusted to the edge template. The multiple torch cutting head is now ready to duplicate the template profile.

To obtain close dimensional tolerance when preparing plate edges, precise torch-conveying equipment is necessary. For reproducibility, accuracy, and maximum efficiency, large gantry and rail type cutting machines are used. Such apparatus may be classified in the same category as a machine tool. A plate is placed on a flat cutting table between the rails of a three-gantry type cutting machine, as shown in Fig. 13.19. The machine can prepare all four edges of the plate without repositioning it. It can also cut the plate into smaller segments at the same time.

Grooving Plate Edges. Gouging is used to prepare the edges of heavy plate for welding. The objective of this edge preparation is to obtain a contour similar to a "J." When two pieces of plate prepared in this manner are butted together, the resulting groove is in the shape of a "U." J-grooving is usually done with a straight gouging tip. The torch is mounted on a tractor type cutting machine. The process utilizes an oxyfuel gas flame applied in a single pass to plates 38 to 100 mm (1-1/2 to 4 in.) thick and to plates of greater thickness in a dual pass combination. With single pass J-grooving, the desired J-groove contour can be produced repeatedly. The root face width and the groove radius can be maintained to specified dimensions. A suitable installation is required to successfully reproduce single pass J-grooves to reasonable tolerances. Proper operating techniques and conditions are also required.

Variables that must be controlled include travel speed, cutting oxygen and preheat gas flows, and gouging tip angle and distance settings. The basic components of a J-grooving installation are a special J-grooving tip, adequate gas and oxygen supplies, a tip positioning device, a plate follower, and a motor-driven carriage to

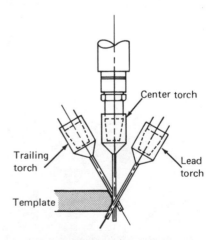

**Fig. 13.18—Kerf centering and bevel
angle setting method**

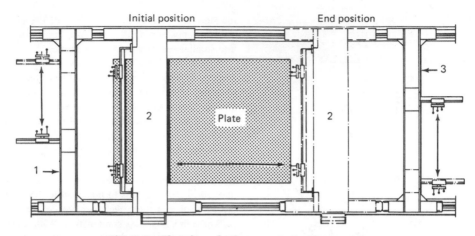

Fig. 13.19—Plan view of a three-gantry cutting machine

act as a motivating force. A typical installation is shown in Fig. 13.20.

The four variables encountered when positioning a J-grooving tip are indicated on Fig. 13.21. The variables are the impinging angle, θ on Fig. 13.21(A); the lateral angle, ψ on Fig. 13.21(B); the slot angle, ϕ on Fig. 13.21(C); and the slot distance, α on Fig. 13.21(B).

Operating conditions and variables for 38

to 100 mm (1-1/2 to 4 in.) thick plates, shown in Table 13.9, will produce J-grooves with a 4.8 to 6.4 mm (3/16 to 1/4 in.) root face; 9.5 to 19 mm (3/8 to 3/4 in.) groove radius; and a 5° to 7-1/2° bevel angle. The larger dimensions are for the thicker plates. At the start of the J-grooving pass, there is a short distance before the groove actually assumes its final depth and contour. Therefore, it is necessary to use a starting tab lightly tack

Fig. 13.20—Single pass oxyfuel gas J-grooving installation for horizontal plate

A—Impinging angle

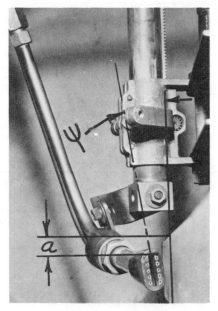

B—Lateral angle and slot distance

C—Slot angle

Fig. 13.21—Variables encountered in positioning J-grooving tips

Table 13.9—Operating conditions for single-pass oxy-fuel gas J-grooving

Plate thickness		Cutting oxygen flow		J-grooving speed		Impinging angle Θ, deg	Lateral angle Φ, deg	Slot angle angle Ψ, deg	Slot distance α	
mm	in.	L/min	ft³/h	mm/s	in./min				mm	in.
38	1-1/2	780	1650	19	.44	15	4	2	11	7/16
51	2	685	1450	10	24	18	4	4	13	1/2
64	2-1/2	2030	4300	17	40	18	4	4	16	5/8
76	3	2360	5000	11	27	15	2	4	13	1/2
89	3-1/2	2360	5000	11	27	15	2	4	13	1/2
102	4	3300	7000	11	27	18	2	8	25	1

Table 13.10—Gas consumptions and weld areas for single-pass oxyfuel gas J-grooving

SI units

Plate thickness, mm	Groove cross section area,[a] mm²	Total gas consumption, liters per meter of J-groove length					
		Oxygen	Acetylene	Oxygen	Propane	Oxygen	Natural gas
38	240	73	3.3	76	2.0	73	2.5
51	420	124	11.7	130	6.1	123	9.3
64	580	205	11.7	212	6.1	210	9.3
76	870	358	14.5	376	10.5	361	19.8
89	1060	358	14.5	376	10.5	361	19.8
102	1420	497	14.5	514	10.5	508	19.8

U.S. customary units

Plate thickness, in.	Groove cross section area,[a] in.²	Total gas consumption, ft³ per linear ft of J-groove					
		Oxygen	Acetylene	Oxygen	Propane	Oxygen	Natural gas
1-1/2	0.375	7.9	0.35	8.2	0.22	7.9	0.27
2	0.650	13.4	1.26	14.0	0.66	13.2	1.00
2-1/2	0.900	22.1	1.26	22.9	0.66	22.6	1.00
3	1.35	38.6	1.56	40.5	1.13	38.9	2.13
3-1/2	1.65	38.6	1.56	40.5	1.13	38.9	2.13
4	2.20	53.5	1.56	55.3	1.13	54.7	2.13

a. Weld area equals two times groove area

welded to the edge of the plate. It must be long enough to ensure that a full-depth groove is being made before the plate itself is reached. Typical lengths vary between 200 and 300 mm (8 and 12 in.); the shorter tabs are for the thinner plates, and the longer tabs are for the thicker plates. In J-grooving, it is essential that the original plate edges be perpendicular to the plate faces since the bevel angle is adjusted relative to the plate edge.

Using the data presented in Table 13.10, it is possible to estimate J-grooving costs and to predict the subsequent welding requirements. As shown in Table 13.10, the U-groove area that is formed when two J-grooves are placed in welding position is twice the J-groove area. The U-groove area is approximately the same as the weld metal area.

Single pass J-grooves are limited to plates having a maximum 100 mm (4 in.) thickness.

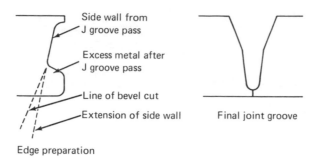

Side wall from
J groove pass

Excess metal after
J groove pass

Line of bevel cut

Extension of side wall

Edge preparation

Final joint groove

Fig. 13.22—Two pass technique of J-groove preparation

Thicker plates can be J-grooved by utilizing a two-pass technique in which a standard J-grooving pass is followed by a cutting operation to remove excess metal along the lower edge of the plate, as shown in Fig. 13.22.

Underwater cutting

Underwater cutting is used for salvage work and for cutting below the water line on piers, dry docks, and ships. The two methods most widely used are oxyfuel gas cutting and oxygen arc cutting.

The technique for underwater cutting with OFC is not materially different from that used in cutting steel in open air. An underwater OFC torch embodies the same features as a standard OFC torch with the additional feature of supplying its own ambient atmosphere. In the underwater cutting torch, fuel and oxygen are mixed together and burned to produce the preheat flame, and a cutting jet is provided to supply the oxygen to cut the steel. In addition, the torch provides an air bubble around the cutting tip. The air bubble is maintained by a flow of compressed air around the tip, as shown in Fig. 13.23. The air shield stabilizes the preheat flame and at the same time displaces the water from the cutting area. Oxyfuel gas cutting underwater without the air shield around the preheat flame is not practical.

The underwater cutting torch has connections for three hoses to supply compressed air, fuel gas, and oxygen. A combination shield and spacer device is attached at the cutting end of the torch. The adjustable shield controls the formation of the air bubble. The shield is adjusted so

that the preheat flame is positioned at the correct distance from the work. This feature is essential for underwater work with poor lighting by operators wearing cumbersome diving suits. Slots in the shield allow the burned gases to escape. A short torch is used to reduce the reaction force produced by the compressed air and cutting oxygen pushing against the surrounding water.

As the depth of cutting increases, the gas pressures must be increased to overcome both the added water pressure and the frictional losses in the longer hoses. Approximately 3.5 kPa (1/2 psi) for each 300 mm (12 in.) of depth must be added to the basic gas pressure requirements used in air for the thickness being cut. Hydrogen, MPS, and natural gas are the best all-purpose preheat gases because they can be used at depths that a diver can descend to and perform satisfactorily. Although most other gases normally associated with flame cutting can be used, they are usually restricted to shallower depths. Acetylene can be used to depths up to approximately 6 meters (20 ft) without exceeding the maximum safe operating pressure of 100 kPa gage (15 psi).

The oxyfuel gas cutting torch experiences no great difficulty underwater in progressively severing steel plate in thicknesses from 13 mm (1/2 in.), up to approximately 152 mm (6 in.). Under 13 mm (1/2 in.), the constant quenching effect of the surrounding water lowers the efficiency of preheating. This requires much larger preheating flames and preheat gas flows.

Cutting oxygen orifice size is considerably larger for underwater cutting than that used in air. A special apparatus for lighting the preheat flames

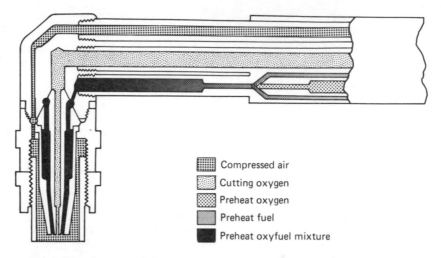

Compressed air
Cutting oxygen
Preheat oxygen
Preheat fuel
Preheat oxyfuel mixture

Fig. 13.23—Basic design of an underwater oxyfuel gas cutting torch

under water is also needed. The recommendations of the manufacturer should be followed for setting up and operating underwater OFC equipment.

QUALITY OF CUTTING

Acceptable quality of OFC depends on the job requirements. Salvage operations and severing members for scrap do not require high quality cutting. The process is used to rapidly complete the operations with little regard to the quality of the cut surfaces. When the cut materials are used in fabrications with no other processing of the cut surfaces, the quality of the surfaces may be significant.

Cutting quality may include such things as

(1) Proper angle of the cut surface with adjacent surfaces

(2) Flatness of the surface

(3) Sharpness of the preheat edge

(4) Dimensional tolerances of the cut shape

(5) Adherence of tenacious slag

(6) Cut surface defects, such as cracks and pockets

Close control of these items is generally confined to machine OFC. Good control of torch position, initiation of the cut, travel speed, and template stability are required for high quality cutting. Also, consistent maintenance and cleanliness of the equipment is needed.

With the proper equipment in good condition, a well trained operator, and reasonably clean and well-supported work, shapes can be cut to tolerances of 0.8 to 1.6 mm (1/32 to 1/16 in.) from material not more than 51 mm (2 in.) thick. Correct cutting tip, preheat flame adjustment, cutting oxygen pressure and flow, and travel speed must be used.

Regardless of operating conditions, drag lines are inherent to oxygen cutting. They are the lines that appear on the cut surface, shown in Fig. 13.24, resulting from the way that the iron oxidizes in the kerf. Light drag lines on the cut surface are not considered detrimental. The amount of drag is important. If it is too great, the corner at the end of the cut may not be completely severed and the part will not drop.

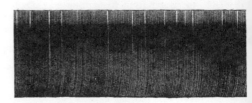

Fig. 13.24—Typical high quality production cut surface

Cut surface quality is dependent on many variables, the most significant being

(1) Type of steel

(2) Thickness of the material

(3) Quality of steel (freedom from segregations, inclusions, etc.)

(4) Condition of the steel surface

(5) Intensity of the preheat flames and the preheat oxyfuel gas ratio

(6) Size and shape of the cutting oxygen orifice

(7) Purity of the cutting oxygen

(8) Cutting oxygen flow rate

(9) Cleanliness and flatness of the exit end of the nozzle

(10) Cutting speed

For any given cut, the variables listed should be evaluated so that the required quality of cut may be obtained with the minimum aggregate cost in oxygen, fuel gas, labor, and overhead. Figures 13.25 and 13.26 show typical edge conditions resulting from variation in the cutting procedure for material of uniform type and thickness.

Dimensional tolerance and surface roughness must be considered together when judging the quality of a cut because they are somewhat dependent on each other. Most specifications include dimensional tolerances. These include straightness of edge, squareness of edge, and permissible variation in plate width. All of these are primarily a function of the cutting equipment and its mechanical operation. When the torch is

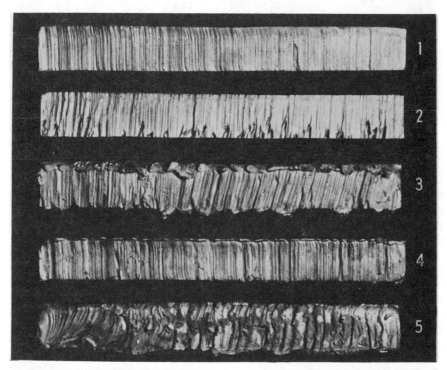

Fig. 13.25—Typical edge conditions resulting from oxyfuel gas cutting operations: (1) good cut in 25 mm (1 in.) plate—the edge is square, and the drag lines are essentially vertical and not too pronounced; (2) preheat flames were too small for this cut, and the cutting speed was too slow, causing bad gouging at the bottom; (3) preheating flames were too long, with the result that the top surface melted over, the cut edge is irregular, and there is an excessive amount of adhering slag; (4) oxygen pressure was too low, with the result that the top edge melted over because of the slow cutting speed; (5) oxygen pressure was too high and the nozzle size too small, with the result that control of the cut was lost

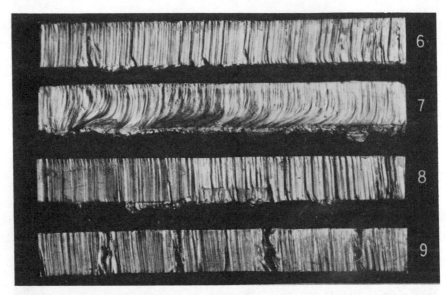

Fig. 13.26—Typical edge conditions resulting from oxyfuel gas cutting operations: (6) cutting speed was too slow, with the result that the irregularities of the drag lines are emphasized; (7) cutting speed was too fast, with the result that there is a pronounced break in the dragline, and the cut edge is irregular; (8) torch travel was unsteady, with the result that the cut edge is wavy and irregular; (9) cut was lost and not carefully restarted, causing bad gouges at the restarting point

held rigidly and advanced at a constant speed, as in machine OFC, dimensional tolerances can be maintained within reasonable limits. The degree of longitudinal precision of a machine cut depends primarily on such factors as the condition of the equipment, trueness of guide rails, clearances in the operating mechanism, and the uniformity of speed control of the drive unit. In addition to equipment, dimensional accuracy is dependent on the control of thermal expansion of the material being cut. Lack of dimensional tolerance may result from buckling of the material (thin plate or sheet), warpage resulting from the heat being applied to one edge, or shifting of the material while it is being cut.

The OFC operation should be planned carefully to minimize the effect of the variables on dimensional accuracy. For instance, when trimming opposite edges of a plate, warpage will be minimized if both cuts are made simultaneously in the same direction. Distortion can often be controlled, when cutting irregular shapes from plates, by inserting wedges in the kerf following the cutting torch to limit movement of the metal from thermal expansion and contraction. In cutting openings in the middle of a plate, distortion may be limited by making a series of unconnected cuts. The section remains attached to the plate in a number of places until cutting is almost completed. The connecting locations are finally cut through. The intermittent cutting will reduce cut quality somewhat. Thin material is often stack-cut to eliminate warping and buckling. Another technique is to cut the thin plate while it is partially submerged in water to remove the heat.

ARC CUTTING

FUNDAMENTALS OF ARC CUTTING

Definitions and General Descriptions

Arc cutting is a group of cutting processes where the severing or removing of metals is effected by melting them with the heat of an arc generated between an electrode and the base metal.

Arc cutting, with or without the use of a compressed gas, is based on the conversion of electrical energy into heat within the confines of an electric arc maintained between the material being cut and an electrode. This localized conversion, with a practical degree of control, is utilized for arc cutting by a number of methods.

Plasma Arc Cutting. Plasma arc cutting employs an extremely high-temperature, high-velocity constricted arc between an electrode contained within the torch and the piece to be cut. The arc is concentrated by a nozzle onto a small area of the workpiece. The metal is continuously melted by the intense heat of the arc, and then it is removed by the jet-like gas stream issuing from the torch nozzle. Modern plasma arc cutting equipment provides a degree of arc constriction and control which can produce a smooth cut surface with minimum bevel. Usually, a nontransferred arc feature is utilized to facilitate arc starting.

Air Carbon Arc Cutting. Air carbon arc cutting is a method for cutting or removing metal by melting it with an electric arc and then blowing away the molten metal with a high velocity jet of compressed air. The air jet is external to the consumable carbon-graphite electrode. It strikes the molten metal immediately behind the arc. Air carbon arc cutting and metal removal differ from plasma arc cutting in that they employ an open (unconstricted) arc, which is independent of the gas jet.

Oxygen Arc Cutting.[2] Oxygen arc cutting is a method of cutting, piercing, and gouging

_____

2. These methods are not widely used commercially, and they cannot always be obtained as a complete equipment package. They are used for special applications only. The articles in the Supplementary Reading List will provide more information.

metals by the use of an electric arc and a stream of oxygen. A consumable electrode is used that consists of a ferrous metal tube covered with a nonconductive mineral covering. Oxygen is conveyed through the bore of the tube to the area heated by the arc, producing an oxidation reaction in a manner similar to oxyfuel gas cutting. The method is not well suited for cutting nonferrous metals, although electrodes are available with special coverings for such applications. With special coverings, the process is then similar to oxyfuel gas flux cutting, with the molten metal flow assisted by the electrode covering. Oxygen arc cutting is primarily used in underwater applications.

Gas Metal Arc Cutting.[2] Gas metal arc cutting obtains heat from an electric arc formed between a continuously fed electrode wire and the plate, usually in the presence of inert gas shielding. Arcing occurs between the forward side of the wire and the advancing kerf edge. A force within the kerf ejects the molten metal from the kerf in all cutting positions. The force is developed from a pressure gradient produced in the shielding gas by both the pumping action of the arc and the metal vapor from the electrode.

Gas Tungsten Arc Cutting.[2] Gas tungsten arc cutting is similar to GTA welding in that the same basic power supply and electrode holder are used. For cutting, current and shielding gas flow are increased. The moderately high velocity jet of gas (usually an argon-hydrogen mixture) emitted from the shielding gas cup blows away the molten metal to form a kerf. Currents ranging from 200 to 600 A are used to cut stainless steel and aluminum up to 13 mm (1/2 in.) thick.

Shielded Metal Arc Cutting.[2] Shielded metal arc cutting uses standard covered electrodes and electrode holders without compressed gas. The molten material is removed by the force of gravity. High current density is used to melt through the metal.

Carbon Arc Cutting.[2] Carbon arc cutting is the oldest of the arc cutting methods. It employs the heat of an arc either between a carbon (graphite) electrode and the workpiece. Molten metal is removed by arc forces and gravity. The process is used primarily in small shops where more efficient equipment is not available.

PLASMA ARC CUTTING

Introduction

The basic plasma arc cutting torch is similar in design to a plasma arc welding torch. For welding, a plasma gas jet of low velocity is used to melt base and filler metals together in the joint.[3] For the cutting of metals, a high velocity plasma gas jet is used to melt the metal and blow it away to form a kerf. The basic arrangement and terminology for a plasma arc torch are shown in Fig. 13.27.

All plasma arc torches constrict the arc by passing it through an orifice as it travels away from the electrode toward the workpiece. As the orifice gas passes through the arc, it is heated rapidly to high temperature, expands, and accelerates as it passes through the constricting orifice. The intensity and velocity of the arc plasma gas are determined by such variables as the type of orifice gas and its entrance pressure, constricting orifice shape and diameter, and the plasma energy density on the work.

Principles of Operation

The basic plasma arc cutting circuitry is shown in Fig. 13.28. The process operates on direct current, straight polarity (dcsp), electrode negative, with a constricted transferred arc. In the transferred arc mode, an arc is struck between the electrode in the torch and the workpiece. The arc is initiated by a pilot arc between the electrode and the constricting nozzle. The nozzle is connected to ground (positive) through a current limiting resistor and a pilot arc relay contact. The pilot arc is initiated by a high frequency generator connected to the electrode and nozzle. The welding power supply then maintains this low current arc inside the torch. Ionized orifice gas from the pilot arc is blown through the constricting nozzle orifice. This forms a low resistance

path to ignite the main arc between the electrode and the workpiece. When the main arc ignites, the pilot arc relay may be opened automatically to avoid unnecessary heating of the constricting nozzle.

Plasma arc cutting (PAC) was originally developed for severing nonferrous metals using inert gases. Modifications of the process and equipment to permit the use of oxygen in the orifice gas permit efficient cutting of steel.

Because the plasma constricting nozzle is exposed to the high plasma flame temperatures (estimated at 10 000 to 14 000° C), the nozzle must be made of water-cooled copper. In addition, the torch should be designed to produce a boundary layer of gas between the plasma and the nozzle.

Several process variations are used to improve the PAC quality for particular applications. They are generally applicable to materials in the 3 to 38 mm (1/8 to 1-1/2 in.) thickness range. Auxiliary shielding, in the form of gas or water, is used to improve cutting quality.

Dual Flow Plasma Cutting. Dual flow plasma cutting provides a secondary gas blanket around the arc plasma, as shown in Fig. 13.29. The usual orifice gas is nitrogen. The shielding gas is selected for the material to be cut. For mild steel, it may be carbon dioxide (CO_2) or air; for stainless steels, CO_2; and an argon-hydrogen mixture for aluminum. For mild steel, cutting speeds are slightly faster than with conventional PAC, but the cut quality is not satisfactory for many applications.

Water Shield Plasma Cutting. This technique is similar to dual flow plasma cutting. Water is used in place of the auxiliary shielding gas. Cut appearance and nozzle life are improved by the use of water in place of gas for auxiliary shielding. Cut squareness and cutting speed are not significantly improved over conventional PAC.

Water Injection Plasma Cutting. This modification of the PAC process uses a symmetrical impinging water jet near the constricting nozzle orifice to further constrict the plasma flame. The arrangement is shown in Fig. 13.30. The water jet also shields the plasma from turbulent mixing with the surrounding atmosphere.

---

3. Plasma arc welding is covered in Chapter 9.

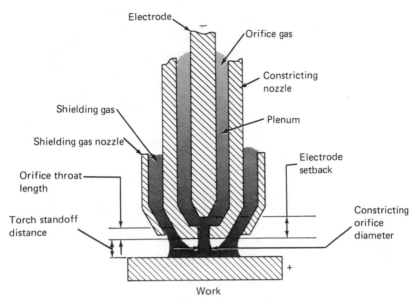

Fig. 13.27—Plasma arc torch terminology

The end of the nozzle can be made of ceramic which helps to prevent double arcing.[4]

The water constricted plasma produces a narrow, sharply defined cut at speeds above those of conventional PAC. Because most of the water leaves the nozzle as a liquid spray, it cools the kerf edge, producing a sharp corner. The kerf is clean with little or no dross. When the orifice gas and water are injected tangentially, the plasma gas swirls as it emerges from the nozzle and water jet. This can produce a high quality perpendicular face on one side of the kerf. The other side of the kerf is beveled. In shape cutting applications, the direction of travel must be selected to produce a perpendicular cut on the part and the bevel cut on the scrap.

Equipment and Consumables

Plasma arc cutting requires a torch, a control unit, a power supply, one or more cutting gases, and a supply of clean cooling water. Equip-

ment is available for both manual and mechanized PAC.

Cutting Torch. A cutting torch consists essentially of an electrode holder which centers the electrode tip with respect to the orifice in the constricting nozzle. The electrode and nozzle are water cooled to prolong their lives. Plasma gas is injected into the torch around the electrode and

---

4. Double arcing results when the arc jumps from the electrode to the nozzle and then to the work. This usually destroys the nozzle.

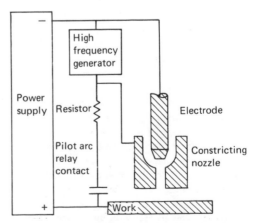

Fig. 13.28—Basic plasma arc cutting circuitry

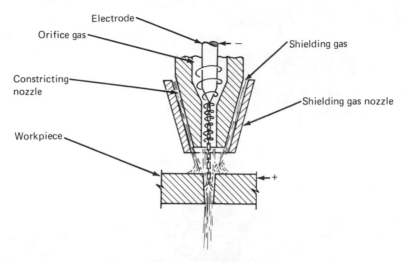

Fig. 13.29 — Dual flow plasma arc cutting

exits through the nozzle orifice. Nozzles with various orifice diameters are available for each type of torch. Orifice diameter depends on the cutting current; larger diameters are required at higher currents. Nozzle design depends on the type of PAC and the metal being cut.

Both single and multiple port nozzles may be used for PAC. Multiple port nozzles have the auxiliary gas ports arranged in a circle around the main orifice. All of the arc plasma passes through the main orifice with a high gas flow rate per unit area. These nozzles produce better quality cuts than single port nozzles at equivalent travel speeds. However, cut quality decreases with increasing travel speed.

Torch designs for introducing shielding gas or water around the plasma flame are available. PAC torches are similar in appearance to gas tungsten arc welding electrode holders, both manual and machine types. A typical mechanized PAC torch is shown in Fig. 13.31.

Mechanized PAC torches are mounted on shape cutting machines similar to mechanized oxyfuel gas shape cutting equipment. Cutting may be controlled by photoelectric tracing, numerical control, or computer.

Controls. Control consoles for PAC may contain solenoid valves to turn gases and cooling water on and off. They usually have flow-meters for the various types of cutting gases used and a water flow switch to stop the operation if cooling water flow falls below a safe limit. Controls for high-power automatic PAC may also contain programing features for upslope and downslope of current and orifice gas flow.

Power Sources. Power sources for PAC are specially designed units with open-circuit voltages in the range of 120 to 400 V. A power source

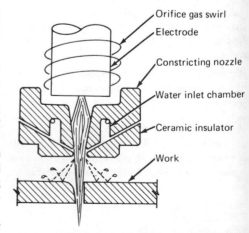

Fig. 13.30 — Water injection plasma arc cutting arrangement

Fig. 13.31—Typical plasma arc cutting torch for mechanized cutting operations

is selected on the basis of the design of PAC torch to be used, the type and thickness of the work metal, and the cutting speed range. Their volt-ampere output characteristic must be the typical drooping type (Fig. 1.16, page 23).

Heavy cutting requires high open-circuit voltage (400 V) for capability of piercing material as thick as 50 mm (2 in.). Low current, manual cutting equipment uses lower open-circuit voltages (120-200 V). Some power sources have the necessary connections to change the open-circuit voltage as required for specific applications.

The output current requirements range from about 70 to 1000 A depending on the material, its thickness, and cutting speed. The unit may also contain the pilot arc and high frequency power source circuitry. Additional information on power sources is contained in Chapter 1.

Gas Selection. Cutting gas selection depends on the material being cut and the cut surface quality requirements. Most nonferrous metals are cut by using nitrogen, nitrogen-hydrogen mixtures, or argon-hydrogen mixtures. Titanium and zirconium are cut with pure argon because of their susceptibility to embrittlement by reactive gases.

Carbon steels are cut by using compressed air (80 percent N_2-20 percent O_2) or nitrogen for plasma gas. Nitrogen is used with the water injection method of PAC. Some systems use nitrogen for the plasma forming gas with oxygen injected into the plasma downstream of the electrode. This arrangement prolongs the life of the electrode by not exposing it to oxygen.

For some nonferrous cutting with the dual flow system, nitrogen is used for the plasma gas

with carbon dioxide (CO_2) for shielding. For better quality cuts, argon-hydrogen plasma gas and nitrogen shielding are used.

Applications

Plasma arc cutting can be used to cut any metal. Most applications are for carbon steel, aluminum, and stainless steel. It can be used for stack cutting, plate beveling, shape cutting, and piercing.

In stack cutting, the plates should be clamped together as closely as possible. However, PAC can usually tolerate wider gaps between carbon steel plates than OFC. When high PAC speeds are used, there is less distortion of the top plate. Several plates of 2 to 6 mm (1/16 to 1/4 in.) thickness can be economically stack cut.

Plate and pipe edge beveling is done by using techniques similar to those for OFC. One to three PAC torches are used depending on the joint preparation required.

For shape cutting, PAC torches are used on shape cutting machines similar to those used for OFC. Generally, plasma arc shape cutting machines can operate at higher travel speeds than OFC machines. Because of the fumes and heat produced by the cutting action, water tables are sometimes used with PAC shape cutting machines. The water just touches the bottom of the plate where it traps the fumes, slag, or dross as they emerge from the bottom of the kerf. It also helps reduce noise.

Plasma arc cutting of carbon steel plate can be done faster than with OFC processes in thicknesses below 75 mm (3 in.) if the appropriate equipment is used. For thicknesses under 25 mm

(1 in.), PAC speeds can be up to five times that for OFC. Over 37 mm (1-1/2 in.) thickness, the selection of PAC or OFC will depend on other factors such as equipment costs, load factor, and applications for cutting thinner plates and nonferrous metals.

The economic advantages of PAC, as compared to OFC, are more likely to be apparent where long, continuous cuts are made on a large number of pieces. This type of cutting might be used in shipbuilding, tank fabrication, bridge construction, and steel service centers. The comparative economy of short cut lengths which require frequent starts will depend on the number of torches that can be used practically in multiple piece production.

Some limits on the number of PAC torches that can operate simultaneously are the power demand on the plant utility lines, the limited visibility imposed on the operator by lens shade requirements for high current PAC, and the potential damage to the equipment and material from delayed detection of a torch malfunction. Therefore, for simultaneously cutting several small shapes from a carbon or stainless steel plate, the reliability of a PAC installation must be compared to a similar OFC installation.

Plasma arc cutting may be modified so it can be used underwater. The torch must be sealed to prevent water from reaching uninsulated torch parts and electrical connections. Electrical circuitry modifications may be required to improve operation of the high frequency generator and the pilot arc. A constant flow of filtered air should be maintained through the torch while it is underwater, but not cutting, to keep water out of the torch.[5]

Process Variables

The normal process variables for cutting a particular material are
(1) Torch design
(2) Process variation (dual gas flow, water injection)

5. For information on PAC of stainless steel underwater, see C.H. Wodke, et al, "Development of underwater plasma arc cutting," *Welding Journal* 55 (1), Jan 1976, pp. 15-24.

(3) Constricting nozzle design and orifice size
(4) Orifice gases and flow rates
(5) Torch standoff distance
(6) Travel speed
(7) Current
(8) Power

Generally, proper operating conditions for cutting various materials and thicknesses are established by the manufacturers for their particular torch designs and associated equipment. Many of the process variables are interrelated. Therefore the manufacturer's recommendations should be used as a guide for setting up and operating the equipment. Typical conditions for PAC of aluminum, stainless steel, and carbon steel are given in Tables 13.11, 13.12, and 13.13 respectively. These data should be considered as general information. Cut quality or cutting speed may be improved by varying the operating conditions. Generally, cut quality is better at lower speeds. Torch standoff distance ranges between 6 and 16 mm (1/4 and 5/8 in.).

Manual PAC may be done with low currents (in the range of 70 to 100 A) at low cutting speeds. Good quality cuts can be made in most nonferrous metals up to 25 mm (1 in.) thick and carbon steel up to 13 mm (1/2 in.) thick.

Operating Procedures

In automatic shape cutting operations, the operator has only to press the start button after selecting the proper process conditions. The cutting control then regulates the sequence of operations. One important variable not discussed in the previous section is direction of travel. Most plasma cutting torches impose a swirl on the orifice gas flow pattern by injecting the gas through tangential holes or slots (see Figs. 13.29 and 13.30). One feature of the gas swirl is a more efficient transfer of arc energy to one side of the kerf. With a clockwise swirl, for example, the right side of the cut (facing in the direction of travel) will be reasonably square but the left side will be beveled. Therefore, the travel direction must be selected to place the scrap material on the left, as illustrated in Fig. 13.32. Reverse-swirl torch components can be used if a square left side is desired—for example when cutting

Table 13.11—Typical conditions for plasma arc cutting of aluminum alloys

Thickness		Speed		Orifice Diam [a]		Current (dcsp), A	Power kW
mm	in.	mm/s	in./min	mm	in.		
6	1/4	127	300	3.2	1/8	300	60
13	1/2	86	200	3.2	1/8	250	50
25	1	38	90	4.0	5/32	400	80
51	2	9	20	4.0	5/32	400	80
76	3	6	15	4.8	3/16	450	90
102	4	5	12	4.8	3/16	450	90
152	6	3	8	6.4	1/4	750	170

a. Plasma gas flow rates vary with orifice diameter and gas used from about 47 L/min. (100 ft³/h) for a 3.2 mm (1/8 in.) orifice to about 120 L/min. (250 ft³/h) for a 6.4 mm (1/4 in.) orifice. The gases used are nitrogen and argon with hydrogen additions from 0 to 35%. The equipment manufacturer should be consulted for each application.

Table 13.12—Typical conditions for plasma arc cutting of stainless steels

Thickness		Speed		Orifice Diam [a]		Current (dcsp), A	Power kW
mm	in.	mm/s	in./min	mm	in.		
6	1/4	86	200	3.2	1/8	300	60
13	1/2	42	100	3.2	1/8	300	60
25	1	21	50	4.0	5/32	400	80
51	2	9	20	4.8	3/16	500	100
76	3	7	16	4.8	3/16	500	100
102	4	3	8	4.8	3/16	500	100

a. Plasma gas flow rates vary with orifice diameter and gas used from about 47 L/min. (100 ft³/h) for a 3.2 mm (1/8 in.) orifice to about 94 L/min. (200 ft³/h) for a 4.8 mm (3/16 in.) orifice. The gases used are nitrogen and argon with hydrogen additions from 0 to 35%. The equipment manufacturer should be consulted for each application.

Table 13.13—Typical conditions for plasma arc cutting of carbon steel

Thickness		Speed		Orifice Diam. [a]		Current (dcsp), A	Power kW
mm	in.	mm/s	in./min	mm	in.		
6	1/4	86	200	3.2	1/8	275	55
13	1/2	42	100	3.2	1/8	275	55
25	1	21	50	4.0	5/32	425	85
51	2	11	25	4.8	3/16	550	110

a. Plasma gas flow rates vary with orifice diameter, from about 94 L/min. (200 ft³/h) for a 3.2 mm (1/8 in.) orifice to about 140 L/min. (300 ft³/h) for a 4.8 mm (3/16 in.) dual flow orifice. The gases used are usually compressed air, nitrogen with up to 10% hydrogen additions, or nitrogen with oxygen added downstream from the electrode (dual flow). The equipment manufacturer should be consulted for each application.

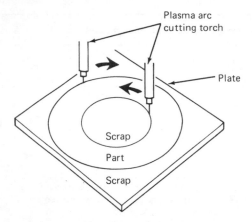

Fig. 13.32—Relationship of torch travel direction to the part with clockwise swirl of the orifice gas

opposite edges with two torches moving in the same direction.

For manual operation, the operator selects the orifice gas flow rate and current according to the recommended procedures. The torch is held at the cut start location on the workpiece, and the arc is initiated by pressing the torch switch. The torch is then manually guided across the workpiece at the desired cut location. The power and gas are automatically turned off when the torch switch is released. The control may reinitiate the pilot arc to allow immediate restarting for repetitive operation.

The constricting nozzle of the torch must not be allowed to contact the work. This would cause nozzle damage if the arc goes from the electrode to the nozzle body and then to the work (double arcing), rather than from the electrode to the work through the nozzle orifice. Cutting torch designs usually minimize double arcing by insulating or recessing the nozzle.

Quality

The factors that determine cut quality are surface smoothness, kerf width, parallelism of the cut faces, dross formation on the bottom edge of cut faces, squareness of the cut, and sharpness of the top edge. These factors are controlled by the type of material being cut, the equipment design and setup, and the operating variables.

Generally, quality cuts are obtained with moderate power and low cutting speeds. However, low speed cutting may offset the otherwise economical features of PAC. Therefore, required cut quality should be defined before applying the process.

Plasma cuts in plates up to approximately 75 mm (3 in.) thick may have a surface smoothness very similar to that produced by oxyfuel gas cutting. Surface oxidation is almost nonexistent with modern mechanized equipment that uses water injection or water shielding. On thicker plates, low travel speeds produce a rougher surface and discoloration. On very thick stainless steel, 125 to 180 mm (5 to 7 in.) in thickness, the plasma arc process has little advantage over oxyfuel gas powder cutting.

Kerf widths of plasma arc cuts are 1-1/2 to 2 times the width of oxyfuel gas cuts in plates up to 50 mm (2 in.) thick. For example, a typical kerf width in 25 mm (1 in.) stainless steel is approximately 5 mm (3/16 in.). Kerf width increases with plate thickness. A plasma cut in 180 mm (7 in.) stainless steel made at approximately 3 mm/s (4 in./min) has a kerf width of 28 mm (1-1/8 in.).

The plasma jet tends to remove more metal from the upper part of the kerf than from the lower part. This results in beveled cuts wider at the top than at the bottom. A typical included angle of a cut in 25 mm (1 in.) steel is four to six degrees. This bevel occurs one one side of the cut when orifice gas swirl is used. The bevel angle on both sides of the cut tends to increase with cutting speed.

Dross is the material that melts during cutting and adheres to the bottom edge of the cut face. With present mechanized equipment, dross-free cuts can be produced on aluminum and stainless steel up to approximately 75 mm (3 in.) thickness and on carbon steel up to approximately 40 mm (1-1/2 in.) thickness. With carbon steel, selection of speed and current are more critical. Dross is usually present on thick materials.

Top edge rounding will result when excessive power is used to cut a given plate thickness or when the torch standoff distance is too large. It may also occur in high speed cutting of materials less than 6 mm (1/4 in.) thick.

Metallurgical Effects

In most applications of PAC, the material at the cut surface is heated to the melting temperature range and ejected by the force of the plasma jet. This produces a heat-affected zone along the cut surface, as with fusion welding operations. The heat not only alters the structure of the metal in this zone but also introduces internal tensile stresses from the rapid expansion, upsetting, and contraction of the metal at the cut surface.

The depth to which the arc heat will penetrate the workpiece is inversely proportional to cutting speed. The heat-affected zone on the cut face of a 25 mm (1 in.) thick stainless steel plate severed at 21 mm/s (50 in./min) is 0.08 to 0.13 mm (0.003 to 0.005 in.) deep. This measurement was determined from microscopic examination of the grain structure at the cut edge of a plate.

Because of the high cutting speed and quenching effect of the base plate, the cut face passes through the critical 650° C (1200° F) temperature very rapidly. Thus, there is virtually no chance for chromium carbide to precipitate along the grain boundaries and reduce corrosion resistance. Measurements of the magnetic properties of Type 304 stainless steel made on base metal and on plasma arc cut samples indicate that magnetic permeability is unaffected by arc cutting.

Metallographic examination of cuts in aluminum plates indicates that the heat-affected zones in aluminum are deeper than those in stainless steel plate of the same thickness. This results from the higher thermal conductivity of aluminum. Microhardness surveys indicate that the heat effect penetrates about 5 mm (3/16 in.) into a 25 mm (1 in.) thick plate. Age hardenable aluminum alloys of the 2000 and 7000 series are crack sensitive at the cut surface. Cracking appears to result when grain boundry eutectic film melts and separates under stress. Machining to remove the cracks may be necessary on edges that will not be welded.

Hardening will occur in the heat-affected zone of a plasma arc cut in high-carbon steel if the cooling rate is very high. The degree of hardening can be reduced by preheating the workpiece to reduce the cooling rate of the cut face.

Various metallurgical effects may occur when long, narrow, or tapered parts, or outside corners are cut. The heat generated during a preceding cut may reach and adversely affect the quality of a following cut.

Safety

Operators and persons in the vicinity of the cutting operation must be protected from arc glare, spatter, fumes, and noise. Safety procedures are given in AWS publication A6.3, Recommended Safe Practices for Plasma Arc Cutting. Two common accessories are available for mechanized plasma cutting machines to aid in fume and noise control. One is the water table. It is simply a cutting table filled with water up to the bottom surface of the plates. The high-speed gases emerging from the plasma jet produce turbulence in the water. Almost all of the fume particles are trapped in a manner similar to the operation of a scrubber in an air pollution control system.

The second device is a water muffler to reduce noise. The muffler is a nozzle attached to the torch body that produces a curtain of water around the front of the torch. It is always used in conjuction with a water table. Water from the table is pumped through the nozzle. The combination of the water curtain at the top of the plate and the water contacting the plate bottom encloses the arc in a sound-deadening shield. The noise output is reduced by roughly 20 dBA. This equipment should not be confused with water injection or water shielding PAC variations.

AIR CARBON ARC CUTTING

Principles of Operation

In air carbon arc cutting (AAC) a high current arc is established between a carbon-graphite electrode and the metal workpiece to be melted. A compressed air jet is continuously directed at the point of melting to eject the molten metal away, as illustrated in Fig. 13.33. Metal removal is continuous as the carbon arc is advanced in the cut. The process is used for severing and gouging. Gouging is sometimes used for weld groove preparation and for the removal of a weld root or a defective weld zone.

Both dc and ac power are used with the appropriate electrodes. The electrode tip is heated to a high temperature by the arc current, but it

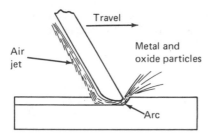

Fig. 13.33—Cutting action in the air carbon arc process

does not melt. The electrode is consumed during cutting, as carbon is lost by oxidation or sublimation of the tip.

Because metal removal is primarily by melting rather than oxidation, the process can be used to cut the commonly used industrial metals. However, it would not be suitable for cutting metals that react rapidly with oxygen or nitrogen, or both, at elevated temperatures. The cut surfaces would be severely embrittled.

Equipment and Consumables

Air carbon arc cutting requires an electrode holder, cutting electrodes, a power source, and an air supply. The process may be done manually or mechanically.

Electrode Holders. Electrode holders may be designed for manual, semiautomatic, or automatic operation. Manual holders for gouging are similar to conventional heavy duty shielded metal arc welding holders, as shown in Fig. 13.34.

Gouging or cutting electrode holders are equipped with air passages and orifices to direct the air stream in line with the electrode. A valve is provided for turning the air on and off.

Semiautomatic electrode holders for gouging are designed for mounting on a machine carriage. The operator manually feeds the electrode as it is consumed during cutting.

For automatic operation, a voltage controlled unit is used which maintains constant arc length by voltage signals through a solid-state electronic control. Voltage controlled units can produce consistent grooves with depth tolerances of ± 0.64 mm (0.025 in.). Automatic holders are mounted on machine carriages as shown in Fig. 13.35. They can also be mounted in fixed positions with the work pieces moving under them.

Electrodes. Carbon electrodes are made from mixtures of carbon and graphite. There are three basic types normally used for AAC: dc copper coated, dc plain, and ac copper coated. Round electrodes are the most frequently used shape. Flat, half round, and special are also available to produce specially designed groove geometries.

DC copper coated electrodes are most widely used because of their comparatively long electrode life, stable arc characteristics, and groove uniformity. These electrodes are made from mixtures of graphite and special additives, and a suitable binder. The mixtures are baked at appropriate temperatures to produce dense, homogeneous, graphite electrodes of low electrical resistance. The electrodes are then plated with a controlled thickness of high-purity copper. Elec-

Fig. 13.34—Manual air carbon arc electrode holder with the electrode

Fig. 13.35—Typical automatic air carbon arc cutting equipment

trodes are produced in diameters from 4.0 to 19.1 mm (5/32 to 3/4 in.). Jointed electrodes are available for continuous operation without stub loss. They are furnished with a female socket and a matching male tenon. Their diameters range from 7.9 to 19.1 mm (5/16 to 3/4 in.).

DC plain electrodes, of limited use, are generally restricted to diameters of less than 9.5 mm (3/8 in.). During cutting, these plain electrodes erode more rapidly than the copper coated electrodes. They are manufactured like the copper coated electrodes except for the copper plating. Plain electrodes are available in sizes ranging from 3.2 to 25.4 mm (1/8 to 1 in.) diameter.

AC copper coated electrodes have certain rare-earth metals incorporated in them to provide arc stabilization with alternating current. These electrodes, plated with a controlled thickness of copper, are available in 4.8 mm (3/16 in.), 6.4 mm (1/4 in.), 9.5 mm (3/8 in.), and 12.7 mm (1/2 in.) diameters.

Power Sources. Most standard welding power sources can be used for the air carbon arc cutting process. The open-circuit voltage should

be sufficiently higher than the required arc voltage to allow for the voltage drop in the circuit. The arc voltages used in air carbon arc gouging and cutting range from 35 to 55 V. An open-circuit voltage of at least 60 V is adequate. The actual arc voltage in air carbon arc gouging and cutting is governed to a large extent by the size of the electrode and the application. Recommended power sources are given in Table 13.14. Additional information on power sources is given in Chapter 1.

Electrical leads in the cutting circuit should be standard welding cables recommended for arc welding. Cable size is determined by the maximum cutting current that will be used.[6]

Air Supply. Compressed air supplied under pressure ranging from 560 to 700 kPa (80 to 100 psi) is normally required for air carbon arc gouging. Light duty electrode holders allow for gouging with as little as 280 kPa (40 psi) at 8.5 liter/min (3 ft³/min). Compressed nitrogen or inert gas

6. See Table 2.1, page 52 for recommended cable sizes.

Table 13.14 — Power sources for air carbon arc cutting and gouging

Type of current	Type of power source	Remarks
dc	Constant current motor-generator, rectifier, or resistor grid unit	Recommended for all electrode sizes
dc	Constant potential motor-generator or rectifier	Recommended for 6.4 mm (1/4 in.) and larger diameter electrodes only. May cause carbon deposit with small electrodes. Not suitable for automatic torches with voltage control.
ac	Constant current transformer	Recommended for ac electrodes only
ac or dc	Constant current rectifier	DC supplied from three phase transformer-rectifier supplies is satisfactory, but dc from single phase sources gives unsatisfactory arc characteristics. AC output from ac/dc units is satisfactory provided ac electrodes are used.

may be used where compressed air is not available.

The air stream must be of sufficient volume and velocity to properly remove the melted slag from the kerf. The orifices in the air carbon arc torches are designed to provide an adequate air stream for gouging. However, poor quality gouging may result if the air pressure falls below the minimum specified by the torch manufacturer or if the volume of air is restricted by hoses or fittings that are too small.

While gouges or cuts made with insufficient air may not always look particularly bad, they may be loaded with slag and carbon deposits. For this reason, it is important that the air pressure be at or above the minimum specified for the type of torch being used. The inside diameter of all hoses and fittings must be large enough to allow a sufficient volume of air to reach the electrode holder.

Hoses and fittings with an inside diameter of 6.4 mm (1/4 in.) are sufficient for small light duty holders. A minimum 9.5 mm (3/8 in.) inside diameter is recommended when gouging with holders having a capacity for 9.5 mm (3/8 in.) or larger electrodes. Automatic gouging holders should be equipped with hoses and fittings with a minimum inside diameter of 12.7 mm (1/2 in.).

Applications

The carbon arc cutting process can be used to sever and gouge carbon, low alloy, and stainless steels; cast iron; and aluminum, magnesium, copper, and nickel alloys. Gouging may be used to prepare plate and pipe edges for welding. Two edges may be butted together and a U-groove gouged along the joint, as shown in Fig. 13.36. The root of a weld may be gouged out to sound metal before completing the weld on the second side. Similarly, defective weld metal may be gouged out for repair. Another application is the removal of old surfacing material before a part is resurfaced.

Process Variables

Air carbon arc cutting electrodes are designed to operate with ac or dc, or both, depending on the material being cut. Table 13.15 gives the recommended electrodes and types of current for cutting several alloys. They may not be suitable for all alloys of a particular metal.

The operating current ranges for commonly used AAC electrodes are shown in Table 13.16. The actual current used for a given electrode size will depend on the operating conditions such as

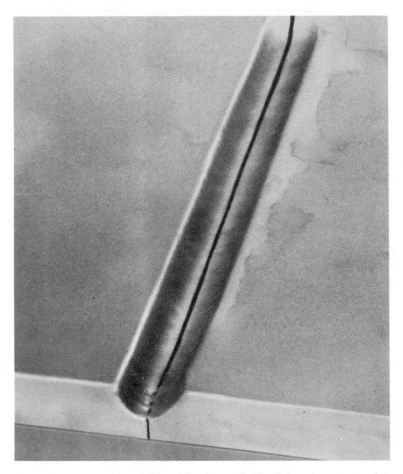

Fig. 13.36—Plates prepared for welding by mechanized air carbon arc gouging

the material being cut, type of cut, cutting speed, cutting position, and cut quality requirements. The recommendations of the manufacturer should be followed for the operation and maintenance of the equipment and consumables.

Operating Procedures

Manual Gouging. The graphite electrode should be gripped so that a maximum of 150 mm (6 in.) extends from the electrode holder to the work. For aluminum alloys, this extension should be 100 mm (4 in.). The air jet should be turned on before gouging and the electrode holder should be held so that the electrode slopes back from the direction of travel with the air blast be-

hind the electrode. Under proper operating conditions, the air stream should sweep beneath the electrode tip and blow away the molten metal. The arc length must provide sufficient clearance for this action. The arc may be struck by touching the electrode lightly to the work. The electrode should not be drawn back once the arc is struck. The gouging technique is different from that of arc welding because metal is removed instead of deposited. Proper arc length must be maintained by moving the torch in the direction of the cut fast enough to keep up with metal removal. The steadiness of progression determines the smoothness of the resulting cut surface.

In air carbon arc gouging it is important that

Table 13.15 — Electrode and current recommendations for air carbon arc cutting of several alloys

Alloy	Electrode type	Current type	Remarks
Carbon, low alloy, and stainless steels	dc	dcrp	
	ac	ac	Only 50% as efficient as dcrp
Cast irons	ac	dcsp	At middle of electrode current range
	ac	ac	
	dc	dcrp	At maximum current only
Copper alloys:			
copper 60% or less	dc	dcrp	At maximum current
copper over 60%	ac	ac	
Nickel alloys	ac	ac	
	ac	dcsp	
Magnesium alloys	dc	dcrp	Before welding, suface must be cleaned.
Aluminum alloys	dc	dcrp	Electrode extension should not exceed 100 mm (4 in.). Before welding, surface must be cleaned.

Table 13.16 — Suggested current ranges for the commonly used AAC electrode types and sizes

Electrode diam		Dc electrode with dcrp, A		Ac electrode with ac, A		Ac electrode with dcsp, A	
mm	in.	min	max	min	max	min	max
4.0	5/32	90	150	—	—	—	—
4.8	3/16	150	200	150	200	150	180
6.4	1/4	200	400	200	300	200	250
7.9	5/16	250	450	—	—	—	—
9.5	3/8	350	600	300	500	300	400
12.7	1/2	600	1000	400	600	400	500
15.9	5/8	800	1200	—	—	—	—
19.1	3/4	1200	1600	—	—	—	—
25.4	1	1800	2200	—	—	—	—

the air stream be directed at the arc in a manner that effectively blows the molten metal out of the kerf. This action properly takes place when the air stream is directed from behind the electrode. The air stream will then blow away all molten metal from the groove as it sweeps under the tip of the electrode and passes through the arc (see Fig. 13.33).

If the electrode is held in the holder in a manner which places the air orifices above and in front of the electrode, the air stream will not properly clean the groove. In the vertical position, gouging should be done in a downward direction.

This permits gravity to assist in removing the molten metal. Vertical gouging may be done in the opposite direction, but it is more difficult. Gouging in the horizontal position may be done either to the right or to the left, but the air jet must be behind the electrode.

For gouging in the overhead position, the electrode should be nearly parallel to the electrode holder. The electrode holder should be held at an angle that will prevent molten metal from dripping on the cutting operator's glove.

The depth and contour of the groove are controlled by the electrode angle and travel speed.

Grooves up to 25 mm (1 in.) deep may be made. A small push angle and slow travel speed will produce a narrow, deep groove. A large push angle and fast speed will produce a wide, shallow groove. The width of the groove is determined by the size of the electrode used, and it is usually about 3 mm (1/8 in.) greater than the electrode diameter, if the electrode is not weaved. A wider groove can be made with a small electrode by oscillating it in a circular or weaving motion. A steady rest is recommended for gouging to ensure a smoothly gouged surface. It is particularly advantageous in overhead work.

Proper travel speed depends on the size of the electrode, type of metal, amperage, and air pressure. Proper speed, which produces a smooth hissing sound, will usually result in a good cut.

Mechanized Gouging. Table 13.17 gives the typical operating conditions for mechanized gouging. The equipment manufacturer should be consulted for specific applications.

Severing. In general, the technique for this operation is similar to manual gouging. The electrode is held at a steeper push angle, and it is directed at a point that will permit the tip of the electrode to pierce the metal being severed.

For cutting thick nonferrous metals, the electrode should be held in the vertical plane with a push angle of 45° and with the air jet above it. With the electrode in this position, the metal may then be severed by moving the arc up and down through the metal with a sawing motion.

Cut Quality

In manual operations, the smoothness and uniformity of the surface depend largely on operator skill under proper AAC operating conditions. To produce a smooth, uniform cut the operator must feed and advance the electrode at a nearly uniform rate. Automatic equipment traveling at constant speed can produce groove depths within a tolerance of ± 0.7 mm (0.03 in.).

Carbon deposits may occur at the beginning or at intervals along the groove. They may be caused by improper air jet flow, by excessive travel speed, or the use of a constant voltage power source for an 8 mm (5/16 in.) diameter electrode or smaller.

A rough groove surface may result if the travel speed is too slow with intermittent arc action. Metal or slag may adhere to the edges of the groove when ejection is inadequate. Proper air pressure, volume, and direction are required to properly eject the molten metal or slag from the groove as cutting progresses.

Compared to oxygen cutting, there is less energy input to the metal. Therefore, distortion of the workpiece should be lower with AAC.

Metallurgical Effects

To avoid difficulties with carburized metal, users of the air carbon arc cutting process must be aware of the metallurgical events that occur during gouging and cutting. When the carbon electrode is positive (reverse polarity), the current flow carries ionized carbon atoms from the electrode to the base metal. The free carbon particles are rapidly absorbed by the melted base metal until a high carbon content is reached. Since this absorption cannot be avoided, it is important that all molten (carburized) metal be removed from the kerf, preferably by the air blast.

When the air carbon arc cutting process is used under improper conditions, the carburized molten metal left behind in the kerf or groove can usually be recognized by its dull gray-black color. This contrasts with the bright blue color of a properly made groove. Inadequate air flow may leave small pools of carburized metal in the bottom of the groove. Irregular electrode travel, particularly in a manual operation, will produce ripples in the groove wall that tend to trap carburized metal. Finally, an improper electrode angle may cause small beads of carburized metal to remain along the edge of the groove.

The effect of carburized metal that remains in the kerf or groove through a subsequent welding operation depends on many factors including the amount of carburized metal present, the welding process to be employed, the kind of base metal, and the weld quality required. Although it seems likely that filler metal deposited during welding would assimilate small pools or beads of carburized metal on the kerf in steel base metal, experience shows that traces of high carbon metal (containing approximately one percent of carbon) may remain along the weld bond line. These imperfections become more significant stress con-

Table 13.17—Typical operating conditions for mechanized gouging [a, b, c]

U-Groove Width (mm)	U-Groove Width (in.)	Depth (mm)	Depth (in.)	Electrode diam (mm)	Electrode diam (in.)	Current (A)	Voltage (V)	Electrode feed rate (mm/s)	Electrode feed rate (in./min)	Travel speed (mm/s)	Travel speed (in./min)
6.4	1/4	1.6	1/16	4.8	3/16	200	43	2.6	6.2	35	82
7.1	9/32	3.2	1/8	4.8	3/16	200	40	2.9	6.7	16	38
8	5/16	4.8	3/16	4.8	3/16	200	42	2.9	6.7	12	27
		6.4[e]	1/4[e]								
8	5/16	2.4	3/32	6.4	1/4	270	40	1.7	4.0	23	54
		3.2	1/8			300	42	1.7	4.0	22	51
		4.8	3/16			300	40	2.9	6.7	20	38
		6.4	1/4			320	42	2.6	6.2	13	30
		9.5	3/8			320	46	1.5	3.6	7	15
9.5	3/8	3.2	1/8	7.9	5/16	320	40	1.3	3.0	28	66
		4.8	3/16			400	46	1.8	4.3	20	46
		6.4	1/4			420	42	1.6	3.8	13	31
		12.7	1/2			540	42	2.4	5.6	12	27
11	7/16	3.2	1/8	9.5	3/8	560	42	1.8	4.2	35	82
		3.2	1/8					1.4	3.3	28	65
		4.8	3/16					1.1	2.6	18	41
		6.4	1/4					1.3	3.0	13	30
		12.7	1/2					1.4	3.2	6	15
		17.5	11/16					1.5	3.5	5	12
14.3	9/16	3.2	1/8	12.7	1/2	1200	45	1.3	3.0	14	34
		6.4	1/4							9	22
		9.5	3/8							9	21
		12.7	1/2							8	19
		15.9	5/8							7	15
		19	3/4							5	13
20.5	13/16	3.2	1/8	15.9	5/8	1300	42	1.1	2.5	19	45
		6.4	1/4							13	30
		9.5	3/8							8	20
		12.7	1/2							6	15
		15.9	5/8							5	13
		19.1	3/4							5	11
		25.4	1							4	10

a. All values are for dc, copper-coated electrode using dcrp (electrode positive).
b. Air pressures throughout are 550-690 kPa (80-100 psi); 690 kPa (100 psi) is recommended for 12.7 mm (1/2 in.) and 15.9 mm (5/8 in.) electrodes.
c. A push angle of 45° is used for these settings. When jointed electrodes are used, a push angle of 55° is often preferred.
d. Combination of settings and multiple passes may be used for grooves deeper than 19 mm (3/4 in.).
e. To make 6.4 mm (1/4 in.) deep groove, use two 3.2 mm (1/8 in.) deep passes.

centrations with increasing demands for weld strength and toughness performance.

There is no evidence that the copper from copper coated electrodes is transferred to the kerf surface in the base metal.

Although carburized metal on the kerf surface can be removed by grinding, it is much more efficient to conduct the air carbon arc gouging and cutting operations properly within prescribed conditions and thus completely avoid the retention of this undesirable metal.

The machinability of low carbon and nonhardenable steels is not affected by the air carbon arc cutting process. With cast iron and high carbon steels, however, this process will generally cause sufficient hardening to make the cut surface nonmachinable. Nevertheless, because the heat-affected zone is usually very shallow, it is possible for some cutting tools to get under the hardened zone and remove the hard surface layer.

Safety

A description of general safety procedures can be found in the ANSI Z49.1, Safety in Welding and Cutting. In particular, fire is a potential hazard if combustible material is present in the work area. The operator and others in the area must be protected from arc glare and spatter.

Fumes are a potential health hazard. When the process is used in an enclosed or semi-enclosed area, exhaust ventilation should be provided and the operator should be equipped with a respirator. Noise from the operation may exceed safe levels in some circumstances. When necessary, ear protection should be provided for the operator.

SUPPLEMENTARY READING LIST

Oxyfuel Gas Cutting

Anthes, C.C. Recent developments in oxyfuel gas cutting. *Welding Journal* 39 (10), Oct. 1960, pp. 1022-1027.

Canonico, D.A. Depth of heat-affected zone in thick pressure vessel steel plate due to flame cutting (technical note). *Welding Journal* 47 (9), Sept. 1968, pp. 410-s—419-s.

Couch, M.F., and Silknitter, D.H. Economic evaluation of fuel gases for oxyfuel gas cutting in steel fabrication. *Welding Journal* 46 (10), Oct. 1967, pp. 825-832.

Fay, R.H. Heat transfer from fuel gas flames. *Welding Journal* 46 (8), Aug. 1967, pp. 380-s—383-s.

Hembree, J.D., *et al*. New fuel gas—stabilized methylacetylene-propadiene. *Welding Journal* 42 (5), May 1963, pp. 395-404.

Kandel, C. Underwater cutting and welding. *Welding Journal* 25 (3), Mar. 1946, pp. 209-212.

Khuong-Huu, D., White, S.S., and Adams, C.M. Jr. Combustion of liquid hydrocarbon fuels for oxygen cutting. *Welding Journal* 37 (3), Mar. 1958, pp. 101-s—106-s.

Milton, C.B. Single-pass J-grooving in heavy plate with an oxyacetylene flame. *Welding Journal* 40 (4), Apr. 1961, pp. 331-338.

Slottman, G.V., and Roper, E.H. *Oxygen cutting*. New York: McGraw-Hill, 1951.

Spies, G.R., Jr. Specific problems associated with oxygen cutting of bevels. *Welding Journal* 39 (6), June 1960, pp. 584-591.

Worthington, J.C. Analytical study of natural gas-oxygen cutting, theory and application. *Welding Journal* 39 (3), Mar. 1960, pp. 229-235.

Metal Powder Cutting

Babcock, R.S. Powder washing for metal removal. *Welding Journal* 30 (12), Dec. 1951, pp. 1092-1097.

Babcock, R.S. Recent developments in powder processes. *Welding Journal* 32 (10), Oct. 1953, pp. 986-991.

Powell, C.W. Powder cutting and scarfing of stainless steels. *Welding Journal* 29 (2), Feb. 1950, pp. 308-310.

Chemical Flux Cutting

Bellew, G.E. Use of flux injection for cutting stainless steel. *Welding Journal* 27 (3), Mar. 1948, pp. 181-187.

Plasma Arc Cutting

Alban, J.F. Revival of a lost art: plasma arc gouging of aluminum. *Welding Journal* 55 (11), Nov. 1976, pp. 954-959.

Couch, R.W., Jr., and Dean, D.C., Jr. High quality water arc cutting. *Welding Journal* 50 (4), April 1971, pp. 233-237.

Hebble, C.M., Jr. Cutting with low current broadens application of plasma process. *Welding Journal* 52 (9), Sept. 1973, pp. 587-589.

O'Brien, R.L., Wickham, R.J., and Keane, W.P. Advances in plasma arc cutting. *Welding Journal* 43 (12), Dec. 1964, 1015-1021.

O'Brien, R.L. Arc Plasmas for Joining Cutting, and Surfacing. Bulletin No. 131. New York: Welding Research Council, July 1968.

Shamblin, J.E., and Armstead, B.H. Plasma arc cutting. *Welding Journal* 43 (10), Oct. 1964, pp. 470-s—472-s.

Skinner, G.M., and Wickham, R.J. High quality plasma arc cutting and piercing. *Welding Journal* 46 (8), Aug. 1967, pp. 657-664.

Spies, G.R., Jr. Comparison of plasma and oxyfuel gas cutting. *Welding Journal* 44 (10), Oct. 1965, pp. 815-828.

Wodtke, C.H., Plunkett, W.A., and Frizzell, D.R. Development of underwater plasma arc cutting. *Welding Journal* 55 (1), Jan. 1976, pp. 15-24.

Air Carbon Arc Cutting

Christensen, L.J. Air carbon arc gouging. *Welding Journal* 52 (12), Dec. 1973, pp. 782-791.

Panter, D. Air carbon arc gouging. *Welding Journal* 56 (5), May 1977, pp. 32-37.

Stepath, M.D., Coughlin, W.J., and Nelson, H.B. Recent metal removal developments with compressed air carbon arc process. *Welding Journal* 38 (8), Aug. 1959, pp. 755-759.

Oxygen Arc Cutting

Campbell, H.C. The theory of oxyarc cutting. *Welding Journal* 26 (10), Oct. 1947, pp. 889-903.

Hughey, H.G. Stainless steel cutting. *Welding Journal* 26 (5), May 1947, pp. 393-400.

Gas Metal Arc Cutting

Babcock, R.S. Inert gas metal arc cutting. *Welding Journal* 34 (4), April 1955, pp. 309-315.

Gas and Electric Arc Cutting

Metals Handbook, Forming, Vol. 4, 8th ed. Metals Park, Ohio: American Society for Metals, 1969, pp. 278-300, 301-304.

Gas Tungsten Arc Cutting

Conner, G.A. Tungsten arc cutting of stainless steel. *Welding Journal* 39 (3), Mar. 1960, pp. 215-222.

Wait, J.D., and Resh, S.H. Tungsten arc cutting of stainless steel shapes in steel warehousing operations. *Welding Journal* 38 (6), June 1959, pp. 576-581.

Shielded Metal Arc Cutting

Thielsch, H., and Quass, J. Shielded metal arc cutting and grooving. *Welding Journal* 33 (5), May 1954, pp. 438-446.

14

Surfacing

Prepared by a committee consisting of

R.D. KERR, *Chairman,*
  Babcock and Wilcox Company

D.L. GRAVER
  International Nickel Company

N.S. WAMACK
  Combustion Engineering, Inc.

517

14
Surfacing

FUNDAMENTALS OF THE PROCESS

DEFINITIONS AND GENERAL DESCRIPTION

Surfacing is the deposition of filler metal on the surface of a base metal. Its purpose is to provide the properties or dimensions necessary to meet a given service requirement. There are several types of surfacing. They may be categorized as cladding, hardfacing, buildup, and buttering.

The desired properties, in the same order, are corrosion resistance, wear resistance, dimensional control, and metallurgical needs.

Cladding

Cladding is a relatively thick layer of filler metal applied to a carbon or low alloy steel base metal for the purpose of providing a corrosion-resistant surface when that surface is to be exposed to a corrosive environment. As a rule, the strength of the cladding is not included in the design of the component.

Hardfacing

Hardfacing is a form of surfacing that is applied for the purpose of reducing wear or abrasion, impact, erosion, galling, or cavitation. As with cladding, the strength of hardfacing is not included in the design of a component.

Buildup

This term, as it is normally used, connotes the addition of weld metal to a base metal surface,

the edge of a joint, or a previously deposited weld metal for the restoration of the component to the required dimensions. In this case, the strength of the weld metal is a necessary consideration in the component design.

Buttering

This term also connotes the addition of one or more layers of weld metal to the face of the joint or surface to be welded. It differs from buildup in that its use is for metallurgical reasons, not dimensional control. An example is high nickel alloy weld metal deposited on a carbon or low alloy steel base metal, later to be welded to a high alloy steel base metal. The buttered member can be heat-treated after buttering, or it can be left in the as-welded condition, if that is permitted for the completed joint. The strength of buttering must be taken into consideration in the design of the joint.

PRINCIPLES OF OPERATION

Cladding

Cladding is usually done by the shielded metal arc or submerged arc welding processes. The gas shielded welding processes are also used, but only to a limited extent. Filler metals are available as covered electrodes, coiled electrode wire, and strip electrodes. Many of these are discussed in AWS Specifications A5.4, 5.6, 5.7, 5.9, 5.11, 5.14, and 5.22. Information concerning the properties of filler metals not included in the AWS specifications must be obtained from the supplier of the material.

Cladding must resist several types of corrosion. First, it must resist general corrosion (uniform thinning) in the corrosive environment. The general corrosion rate of a material is expressed in millimeters or mils per year. The most common rating system is: *excellent,* 0.13 mm/year (5 mils/year); *good,* 0.13 to 0.51 mm/year (5-20 mils/year); *acceptable,* 0.51 to 1.27 mm/year (20-50 mils/year); and *unacceptable,* greater than 1.27 mm/year (50 mils/year).

Cladding must also resist localized corrosion, such as pitting, crevice corrosion, intergranular corrosion, and stress corrosion cracking. The corrosion resistance of the cladding is, in many cases, the limiting factor in the life of a component. For this reason, corrosion resistance is the primary consideration in selecting an alloy, a welding process, and the welding procedure for cladding.

If maximum corrosion resistance is to be obtained, it is important that the surface of the base metal, which otherwise would be exposed to the corrodent, be completely covered with cladding and the cladding itself be sound. It is also important that the surface of the cladding be smooth and the composition be rich enough in the necessary alloying elements to resist corrosion. In those cases where the cladding will be used with the surface in the as-deposited condition (i.e., without machining to make the surface smooth), any corrosion testing intended to demonstrate the serviceability of the cladding should be done on a similar surface.

Cladding is usually of stainless steel or one of the nickel base alloys, although there is some use of certain copper base alloys. Some very specialized cladding is done with silver, using the gas tungsten arc welding process, and with lead, using the oxyfuel gas process. Special techniques are employed, particularly with lead, in order to avoid the toxicity of the metal. Initially, the type of filler metal to be used for cladding in a given application is usually determined on the basis of previous experience with a wrought material of similar composition. It is important, therefore, that actual corrosion tests be conducted to verify

the serviceability of the weld cladding. This can be especially important in severely corrosive applications since the structure of wrought base metal is quite different from that of weld metal, as are the heat treatments involved in producing the two. Some standard tests which may be used to determine the effects of heat treatment on corrosion resistance are ASTM A-262, for stainless steel, and ASTM G-28, for nickel-rich alloys which contain chromium.

Laboratory corrosion testing of weldments has recently been reviewed by Henthrone[1] and by Brautigam.[2] Each of these articles contains extensive lists of references.

Hardfacing

Hardfacing usually is deposited by manual, semiautomatic, or automatic arc or gas welding processes. Suitable surfacing metals are available in the form of bare welding rods, covered electrodes, coiled wire, paste, and powders. Specifications for surfacing metals are difficult to develop since evaluation tests have not yet been well standardized. Thirty-one classifications of rods and electrodes from six well-known families of alloys are included in AWS Specifications A5.13 and A5.21. In addition, tubular tungsten carbide surfacing rods and electrodes are included in AWS Specification A5.21. Because the identifying characteristics of certain metals have yet to be defined, several other useful grades have not been included in these specifications. The important properties of hardfacing are

(1) Hardness
    (a) Gross hardness (macrohardness)
    (b) Microhardness (the hardness of individual constituents in a heterogeneous structure)
    (c) Hot hardness (resistance to the weakening effect of service at elevated temperature)

---

1. Henthorne, Corrosion testing of weldments. *Corrosion* 32(2), Feb. 1976, pp. 39-46.

2. Brautigam, F. C. Selective corrosion of weld metal in high nickel alloys and stainless steels. *Corrosion* 31 (3), March 1975, pp. 101-103.

(d) Creep resistance (resistance to plastic deformation when loaded at elevated temperatures)

(2) Abrasion resistance

    (a) Under low stress (wear)

    (b) Under high stress (grinding)

(3) Impact resistance

    (a) Resistance to plastic deformation under repeated impact load (related to yield strength)

    (b) Resistance to cracking under impact loading (related to ductility but including work-hardening considerations)

(4) Heat resistance

    (a) Resistance to tempering

    (b) Retention of strength when hot (including hot hardness)

    (c) Creep resistance (time factor added to hot strength)

    (d) Oxidation or resistance to corrosion by hot gases

    (e) Resistance to thermal fatigue

(5) Corrosion resistance

(6) Frictional properties and galling tendencies

    (a) Friction coefficients

    (b) Galling tendency

    (c) Surface films

    (d) Lubricity

    (e) Plasticity

The result of any surfacing operation is intended to be an increase in the service life of the component, but because conditions that cause wear are often varied and complex, it is difficult to establish a simple correlation between service life and surfacing properties. The best approach seems to be a careful analysis of the service conditions encountered, followed by a logical application of the pertinent physical, mechanical, and wear test data. The usual alternative, an arbitrary ranking of alloys by reputation or casual service tests, is not acceptable from an engineering point of view. The engineering principles which should form the basis for the selection of hardfacing materials are, however, only partially understood. Certain hardfacing materials which have been found acceptable for particular applications are addressed later in this chapter.

The following advantages of hardfacing are well recognized:

(1) Additional resistance to wear or corrosion exactly where it is needed

(2) Easy use of very hard, wear-resistant alloys

(3) Ready application in the field

(4) Economical use of expensive alloys

(5) A hard surface layer to resist wear, supported by a tough substrate to carry the load

There are several important points to be considered in the use of hardfacing, aside from the characteristics of the surfacing material and the base metals. Some of these points are the geometry of the part to be surfaced; the cost of the material and the labor; the available knowledge of techniques which might be necessary to prevent cracks in the surfacing or to minimize distortion from the thermal stresses of welding; and the definition of the quality level required of the deposit.

The quality required may vary widely, depending upon the particular application. Engine valve facings, for example, may require perfection in the soundness of the surfacing applied to the valves. (Figure 14.1 shows typical surfaces applied to engine valves.) To meet high quality standards, careful control is necessary in the manufacture of the surfacing rods, and a considerable amount of training may be necessary before a welder can do acceptable work. At the other extreme, the surfacing of certain areas on earth-handling equipment may be done by less experienced welders using filler metals with less exacting requirements. Deposits which contain some cracks and blowholes may be acceptable, and increased service life may be obtained despite them.

Hardfacing for Abrasion Resistance

Abrasion resistance still is the most important application of hardfacing. In any specific instance, filler metals for such applications should be selected only after a careful analysis of the conditions causing the wear. This analysis is dominated by the factors of abrasion and impact, but heat, sometimes corrosion, and even fatigue may also be involved. There usually is a rough correlation between hardness and abrasion resistance, yet neither hardness nor alloy content, sometimes erroneously used as a criterion, are a reliable index of abrasion resistance. The best approach in selecting the surfacing material is to identify certain important factors, then compare the various surfacing alloys on the basis of their response to these factors under carefully controlled conditions. The factors to be identified include the kind of abrasive to be encountered and its character (hardness, sharpness, and toughness); the amount of impact to be encountered; the kind of support provided to the deposit; the levels of stress involved; the nature of that stress (tensile, compression, or shear); the operating temperature; and any other significant environment conditions.

The data used for evaluation, whether from service performance or laboratory testing, should be examined critically. The tests themselves should give reproducible results that correlate with service performance.

Abrasion is of three recognizably different types (1) low-stress scratching abrasion or erosion, (2) high-stress grinding abrasion, and (3) gouging abrasion. These abrasion types may be exemplified by (1) sand sliding down a chute or a plowshare working in sandy soil, (2) action in a ball mill where abrasive ore fragments are crushed between metal faces, and (3) the gouging and grooving actions of sharp rocks on dipper teeth and gyratory crusher faces. They may also be exemplified in machine shop operations as (1) polishing with loose abrasive on a soft cloth, (2) lapping, and (3) metal removal by a cutting tool or a grinding wheel.

Hardness is usually an asset in resisting erosion. Tungsten carbide, high chromium irons, and martensitic cast irons are excellent for resisting this kind of wear.

There are examples of grinding abrasion in which relatively soft, but tough, alloys outwear harder materials. Impact may also be present. To cope with the various combinations of wear and impact, a group of alloys ranging from tough austenitic manganese steel to the martensitic irons (including air-hardening steels) currently are used.

Gouging abrasion also may place a premium on toughness, at the expense of extreme hardness.

The combination of very high stresses and shock loading is likely to be involved in such cases, and toughness to resist the heavy impact loads may dominate the choice. Massive parts must be tough enough to resist brittle fracture when their surfaces are protected with brittle but more abrasion-resistant alloys. The design must be such that the surfacing is adequately supported and, preferably, in compression rather than in tension or shear. Under these conditions, hardfacing can provide impressive economies. Conditions of this sort are sometimes so severe that great toughness is imperative, even in the surfacing material. In these cases, austenitic manganese steel is the preferred base metal and surfacing is a matter of replacing the worn areas with a suitable manganese steel filler metal. Sometimes, this repair is supplemented with bars, plates, and special shapes of the same alloy.

Hardfacing for Impact

Impact may be classified as light (where kinetic energy is absorbed elastically), medium (where the energy is absorbed both elastically and plastically), and heavy (where the surface of even the strongest materials must either deform or fracture). Impact classification permits a logical selection of surfacing alloys, martensitic

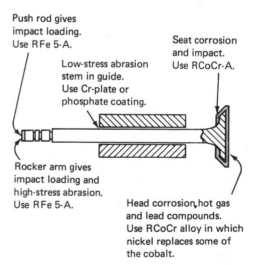

Push rod gives impact loading. Use RFe 5-A.

Low-stress abrasion stem in guide. Use Cr-plate or phosphate coating.

Seat corrosion and impact. Use RCoCr-A.

Rocker arm gives impact loading and high-stress abrasion. Use RFe 5-A.

Head corrosion, hot gas and lead compounds. Use RCoCr alloy in which nickel replaces some of the cobalt.

Fig. 14.1—Typical surfaces that may be applied to engine valves, subject to corrosion, impact, and abrasion

cast irons, martensitic steels, and austenitic steels, respectively.

The toughness of certain austenitic cast irons is intermediate between that of the martensitic cast irons and that of the martensitic steels. Typical of these are compositions containing about 15 percent chromium. The austenitic irons lack the high compressive strength of the martensitic irons and, thus, are less suitable for light impact, although they are more resistant to thermally induced cracking and moderate impact.

Buildup

Surfacing is sometimes applied for reasons other than the increase of wear or corrosion resistance of the part. Such applications fall into either of two categories, buildup or buttering. Buildup is intended to alter or restore the dimensions or shape of the object. The composition and properties of the weld metal, in such a case, usually are similar, in general, to those of the base metal to be built up; a typical example is given in Fig. 14.2.

Buttering

The primary purpose of buttering is to satisfy some metallurgical consideration. It is used especially when joining dissimilar metals and also for joining carbon steel to a low alloy steel when stress relief of the completed weld is to be avoided. Such a weld, however, is usually not considered a dissimilar metal weld.

Buttering is deposited on the surface of one or both members of a joint prior to depositing the weld metal that actually joins the two members. However, it may be used on only one side of a dissimilar metal joint to reduce or eliminate the pickup in the final weld of unwanted elements that otherwise would be encountered with that base metal. Buttering may also be used to provide a transition between materials with substantially different coefficients of expansion but which must endure cycling temperatures in service. Similarly, buttering may be used to provide a barrier layer which will slow the migration of undesirable elements from the base metal to the weld metal during postweld heat treatment or in service at elevated temperatures. An example of this is a buffer layer of ingot iron laid between a low alloy steel base metal and a stainless steel cladding for corrosion resistance.

Buttering may also be helpful in solving other problems associated with postweld heat treatment. Welds sometimes must be made between a material which requires a high-temperature, postweld heat treatment (PWHT) and a material which will be harmed by that heat treatment. The problem is solved by buttering the material which requires the PWHT with a material that can withstand the heat treatment and can be welded to the other material without a PWHT. The buttered component is heat-treated prior to making the final weld. A common example of this process is a stainless steel-clad, low alloy steel nozzle buttered with a nickel-chromium-iron alloy before welding to stainless steel piping, using a nickel-chromium-iron alloy filler metal. Applications of this sort are encountered in the nuclear power field.

Almost any of the welding processes may be used for buttering. The requirement that must be met is the production of a sound weld deposit having the necessary chemical composition and mechanical properties. Usually, the process which meets this requirement, gives the highest deposition rate, and produces an acceptable weld profile, is chosen. High deposition rates usually go hand-in-hand with high dilution rates; and, for this reason, special attention will be required to obtain the desired chemical composition for the first bead and even for the first layer or two of the deposit.

Fig. 14.2—Automatic submerged arc resurfacing of worn steel mill table rollers in place

SPECIAL CONSIDERATIONS

In most cases, special considerations are necessary for surfacing that are not required for welding a joint. Included in these considerations are the following:

(1) The chemical composition and mechanical properties of the surfacing material usually are quite different from those of the base metal on which it is deposited.

(2) A relatively large area of base metal usually is covered in surfacing.

(3) The smallest possible amount of weld metal is desired for surfacing. For this reason, there is frequently an extraordinarily large gradient in alloy and carbon content and mechanical properties across the fusion line between base metal and the weld metal.

DILUTION CONTROL

Probably the single biggest difference between welding a joint and depositing surfacing is that dilution frequently is such a point of concern in surfacing. Figure 14.3 defines dilution and illustrates that the percentage of dilution equals the amount of base metal melted (B) divided by the sum of the filler metal added and base metal melted (A+B), the quotient of which is multiplied by 100.

From a metallurgical point of view, the composition and properties of surfacing are strongly influenced by the dilution obtained. Because of this influence, the amount of dilution which will be obtained with each welding process must be considered to select properly the combination of filler material and welding process required for any particular application. A good example of this is the surfacing of low alloy steel with stainless steel SMAW electrodes. This process normally gives 15 to 50 percent dilution after the first two or three beads in a layer. Thus, if an E308 (19 % Cr-9% Ni) electrode were used on low alloy or carbon steel, a first layer deposit containing about 12 percent Cr and 6 percent Ni would be obtained. This deposit would have poor mechanical properties and corrosion resistance. On the other hand, if an E309 electrode (25%

Cr-12% Ni) were used under the same welding conditions, a deposit containing about 16 percent Cr and 8 percent Ni would be obtained. This deposit would have better corrosion resistance and, probably, better mechanical properties (ductility); dilution nearer the 15 percent level would still further improve both.

It must be remembered that stainless steel deposits which are fully austenitic are susceptible to microfissuring or hot cracking and those which contain martensite are likely to be hard and brittle. A sound, ductile, corrosion-resistant overlay usually is low in carbon and contains 3 to 15 percent of ferrite, while surfacing for wear resistance usually is higher in carbon and is largely martensitic. The Schaeffler or DeLong diagrams for stainless steel weld metal can be used to predict the structure which will be obtained with the various stainless steel filler metals on different steel base metals at any level of dilution.[3]

Deposition of the surfacing material on different base metals can also cause phase changes just as easily as a change in the filler metal or dilution rate. Hardsurfacing materials which give good results on carbon steel may give unacceptable results on 13 percent manganese steel, due to the effect of manganese in stabilizing austenite. The solution, in this case, would be to use a welding procedure which would give lower dilution on the 13 percent manganese steel and, thereby, allow hard, wear-resistant martensite to form.

Filler metals, such as the nickel base alloys and cupronickels, also are used to surface steel. They are, essentially, single-phase alloys. The soundness and mechanical properties of their deposits depend on the chemical composition, largely controlled by dilution, of those deposits. Specific information on such deposits usually can be obtained from the manufacturer of the filler metal or may be approximated by studying

3. See Chapter 65, *Welding Handbook*, Section 4, 6th ed. for a discussion of the use of stainless steel filler metals with other steels. Section 4 is scheduled for revision in the 7th edition.

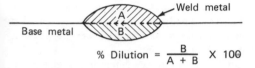

% Dilution = $\frac{B}{A+B}$ X 100

Fig. 14.3—Dilution of weld metal. The bead, A, and penetration, B, are a single homogeneous alloy, but are shaded differently to illustrate the calculation of dilution

the properties of base metals whose compositions are similar to that predicted for a surfacing layer.

Most surfacing is done using one of the consumable electrode arc welding processes. Because of the importance of dilution, it is necessary that the effect of each consumable electrode arc welding variable be known. Many of the welding variables that affect dilution and, therefore, require close control in surfacing are not usually so controlled when arc welding a joint.

Amperage

Increasing the amperage (current density) increases dilution. The arc becomes stiffer and hotter, penetrating more deeply and melting more base metal.

Polarity

Direct current straight polarity gives less penetration and, hence, lower dilution than dc reverse polarity. Alternating current gives dilution which is intermediate between dcsp and dcrp.

Electrode Size

The smaller the electrode, the lower the amperage, as a rule, and, therefore, the lower the dilution. For a given amperage, however, the larger the electrode (the lower the current density), the lower the dilution.

Electrode Extension

A long electrode extension decreases dilution (for consumable electrode processes) by increasing the melting rate of the electrode (I^2R heating) and diffusing the energy of the arc as it impinges on the base metal. Conversely, a short electrode extension increases dilution, within limits.

Bead Spacing or Pitch

Tight bead spacing (more overlap) reduces dilution because some of the previously deposited bead is melted, rather than so much of the base metal. Wider bead spacing (less overlap) increases dilution.

Electrode Oscillation

Greater width of electrode oscillation reduces dilution. The stringer bead gives maximum dilution. The frequency of oscillation also affects dilution. As a rule, the higher the frequency of oscillation, the lower the dilution. There are three basic oscillation patterns, as shown in Fig. 14.4.

Pendulum oscillation is characterized by a slight hesitation at both sides of the bead where it produces slightly greater penetration and somewhat higher dilution. The arc length is continually changing with pendulum oscillation, resulting in varying arc characteristics.

Straight-line oscillation gives approximately the same results as pendulum oscillation but provides the advantage of unchanging arc characteristics since the arc length is maintained constant. Straight-line, constant-evelocity oscillation produces the lowest level of dilution and provides for movement on a horizontal path so that the arc is maintained constant. The opti-

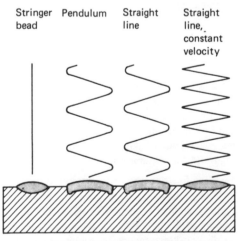

Fig. 14.4—Basic surfacing oscillation techniques and bead configurations

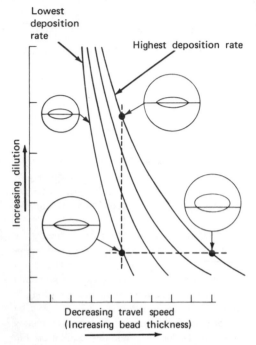

Lowest deposition rate

Highest deposition rate

Increasing dilution

Decreasing travel speed
(Increasing bead thickness)

Fig. 14.5—Effect of travel speed on dilution (other conditions unchanged)

mum movement is programmed to have no end dwell so that the deeper penetration resulting from hesitation at either side is eliminated. Sometimes, slightly more penetration is desired on the tie-in side of the bead. Some oscillation equipment has provisions for controlling dwell time on one or both sides of the oscillation.

Travel Speed

A decrease in travel speed decreases the amount of base metal melted and increases the amount of filler metal added, thus decreasing dilution. This reduction in dilution is brought about by the change in bead shape and thickness and by the fact that the arc force is expended on the weld puddle rather than the base metal. Figure 14.5 shows this effect.

Welding Position and Work Inclination

The position of welding in which the surfacing is applied has an important bearing on the amount of dilution obtained. Depending on the welding position or work inclination, gravity will cause the puddle to run ahead of, remain under, or run behind the arc. The more the puddle stays ahead of or under the arc, the less the penetration into the base metal and the lower the dilution; thus, the puddle acts as a cushion, absorbing some of the arc energy before it can impinge on the base metal. This absorption of arc energy flattens and spreads the puddle and, hence, the weld bead. If the puddle is too far ahead of the arc or too thick, there will be insufficient melting of the surface of the base metal and bonding will not take place.

The order of decreasing dilution for some welding and work positions is
(1) Vertical-up
(2) Horizontal
(3) Uphill
(4) Flat
(5) Downhill

Uphill or downhill welding can be achieved by inclining the part to be surfaced or by placing the arc off-center for rotating cylindrical parts, as in Fig. 14.6.

Arc Shielding

The shielding medium, gas or flux, has a significant effect on dilution. It influences the fluidity and surface tension of the weld pool. These, in turn, determine the extent to which the weld metal will "wet" the base metal and feather in well along the edges of the bead, forming a nicely shaped weld bead. The shielding medium also has a significant effect on the type of welding current which can be used. The list below ranks, in general, the different shielding media in order of decreasing dilution:
(1) Helium (highest)
(2) Carbon dioxide
(3) Argon
(4) Granular fluxes without alloy addition
(5) Granular fluxes with alloy addition (lowest)

Auxiliary Filler Metal

The addition of filler metal, other than the electrode, to the weld pool during surfacing can greatly reduce dilution. The extra metal, added separately as powder, wire, strip, or with the flux, reduces dilution by both increasing the total amount of filler metal and reducing the amount of base metal that is melted. This process is accomplished by using some of the arc energy to melt auxiliary filler metal instead of base metal. The greater the amount of the filler metal, the lesser the dilution. The various methods of adding these different forms of filler metal are discussed with each specific welding process.

CONTAMINATION

Some alloying elements, even in small amounts, can influence the serviceability of a surfaced component. These elements, taken as a group, can cause cracking, reduce corrosion resistance, or reduce strength, ductility, and toughness. Carbon, for example, can reduce the corrosion resistance of austenitic stainless steel cladding. Small amounts of carbon added to the stainless steel cladding through dilution with moderate or high carbon base metals, or by other means, makes the cladding metal susceptible to intergranular corrosion. Also, the diffusion of carbon from ferritic base metals into austenitic stainless steel cladding can create hard and brittle

carbides in the cladding and in components which must be heat treated after cladding or must operate at high temperatures for long times. Such carbides can lead to mechanical failure of the component during service.

Lead, phosphorus, and tin must be controlled in nickel base alloys, as copper must be controlled to prevent cracking in ferritic materials. Low-dilution surfacing techniques can help in this regard; however, at times, it may be necessary to resort to a buffer layer of a different material to produce a serviceable component. Section 4, 6th edition of the *Welding Handbook* contains more information on the effect of minor alloying elements in various base metals and weld metals.

HEAT TREATMENT

Some components may require heat treatment after surfacing. In such cases, the specific heat treatment that is applied must take into account the needs of the surfacing metal as well as

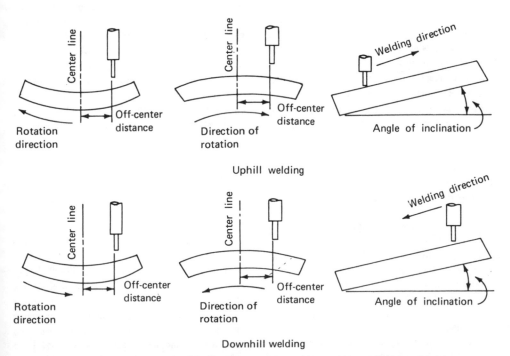

Fig. 14.6—Uphill and downhill welding of flat plate and rotating cylindrical parts

those of the base metal and, also, the requirements of any construction code that might be applicable.

Austenitic stainless steel-clad pressure vessels built to the ASME code, for example, may require a stress relieving heat treatment after cladding. This may be considered necessary for the base metal, though it can be damaging to the corrosion resistance and even the mechanical properties of the cladding. In such cases, it may be necessary to apply special controls to the metals used for cladding and the techniques used in depositing those metals. Similar considerations may even be necessary for the base metal itself. Damage to corrosion resistance of austenitic stainless steels centers on sensitization (carbide precipitation), whereas damage to the mechanical properties of the cladding centers on the formation of the sigma phase. Ferrite, a certain amount of which is desirable in austenitic stainless steel cladding, tends to reduce the speed with which damaging carbides form, but it increases the speed with which sigma phase forms. In very severe cases, sigma phase can reduce corrosion resistance, but its primary effect is to reduce the ductility and impact resistance of the cladding.

Heat treatment can also affect the base metal, as well as the cladding. Underclad cracking in low alloy steel is an example. Cladding of these steels with austenitic stainless steels, using high heat input methods, produces a highly stressed, large grain structure in certain areas of the heat-affected zone. Upon stress relief, tensile stresses produce grain boundary separation in this area. Special surfacing techniques which provide low heat input or reduce the grain size in this area may be required to prevent these grain boundary separations.

Carbon migration is another problem that may be encountered with postweld heat treatment or during longtime, high-temperature service. It occurs when a difference exists between the alloy content of the cladding and the base metal on which the cladding is deposited. The carbide-forming elements are of chief concern. In stainless steel, these elements include chromium, molybdenum, and the iron itself. The higher solubility of carbon in the austenitic cladding also has a bearing on the problem. Among the alloying elements, chromium is of primary importance. When the component is exposed, for a sufficient period of time, to temperatures above some certain level, carbon migrates from the metal with the lower chromium content to the metal with the higher chromium content (or from the material with the lower carbon solubility limit to the one with the higher limit). If the difference in alloy content is great enough, carbon will even migrate from the lower carbon member to the higher carbon member. Migration of the carbon will produce a decarburized band near the fusion line in the member having the lower chromium content. The width of the band and the extent of migration will depend on the difference in alloy (chromium) content and the time and temperature of exposure.

Insofar as the effect of carbon migration is concerned, it is logical to expect some reduction of strength in the decarburized band. This reduction and the stresses resulting from differences in the expansion rate of the two metals can produce failure when the component is cycled a sufficient number of times over a wide range of temperatures. To avoid such a problem in applications of this type, nickel base filler metals have been used instead of stainless steel for surfacing. The expansion rate of nickel-chromium-iron weld metal is reasonably well matched to that of low alloy steel and does not promote carbon migration as does stainless steel. When stainless steel is desired, however, a buffer layer of an extremely low carbon ferritic steel has been used successfully in surfacing applications.

SURFACE APPEARANCE

Another area requiring special attention during some surfacing operations is that of surface smoothness. Surfacing usually produces the finished surface of a component, and subsequent machining is not always possible or even economically feasible. For this reason, the as-deposited weld metal frequently must be smooth enough to meet the requirements of the intended service and to permit any nondestructive testing that must be carried out during fabrication. Thus, the filler metals and welding techniques may have to be selected specifically to provide the required surface smoothness.

THERMAL STRESSES

The success of a surfacing application sometimes depends upon the magnitude of the internal stresses and whether or not the external stresses are shear, tensile, or compressive. Residual stresses from the welding operation may add to or oppose any stresses encountered in service and, thus, accentuate or diminish any tendency of the surfacing to crack. Hardfacing is often not stress-relieved, and the residual stresses which result from thermal expansion and contraction during welding may be formidable. Whether or not these stresses cause warping or cracking depends largely on the strength and ductility of the surfacing metal and the base metal.

Among the surfacing alloys, the austenitic stainless steels are the toughest. They have high ductility and moderate strength. In this respect, they contrast somewhat with the martensitic steels which may be low in ductility but are high in strength. The high carbon irons are relatively strong in compression, but they are brittle and, hence, weak in tension. Surfacing with these high carbon irons is likely to crack on base metals that are not extraordinarily hard but are thick enough to resist distortion.

FURTHER CONSIDERATIONS

The metallurgical characteristics of each type of metal and the welding considerations involved with each process are discussed in the other volumes and chapters of the *Welding Handbook*. The information found there and in the supplementary reading list at the end of this chapter will help in the selection of surfacing materials, processes, and procedures. In the end, however, special welding tests which simulate actual conditions may be required to make sure that the materials, processes, and procedures are acceptable for the intended application.

SURFACING METHODS

Surfacing can be applied with a variety of joining processes, including welding, brazing, and thermal spraying.[4] For this reason, filler metals are provided in several different forms. Most of the processes involve fusion welding, where the filler metal and surface of the base metal melt, join, and then resolidify to provide a continuous metallurgical bond. Chemical reactions between the molten metal and the atmosphere may be prevented by a flux which covers the molten pool, by gas shielding, or by including protective elements (e.g., deoxidizers) within the filler metal. Table 14.1 summarizes most of the surfacing methods and materials.

The oxyfuel gas process involves slower heating and cooling rates for the base metal, tends to facilitate greater precision of placement, involves little or no dilution of the overlay by the base metal, and can be done with portable and relatively inexpensive equipment. Arc welding, on the other hand, is faster, is less expensive overall, and may require less operator skill. It is more commonly used but imposes sharper thermal gradients, and the surfacing may be more susceptible to cracking. For small jobs and field use, the choice is usually between the oxyfuel gas process and one of the arc welding processes. The selection of one or the other is based on the factors referred to above.

For factory use, especially where surfacing is extensive, automatic methods are likely to be more economical than manual methods, even though large capital outlays for equipment are required.

FILLER METAL FORMS

The surfacing filler metals are usually in the form of rods, filled tubes, strip, granules, or powders. Fluxes may be in the central core portion of tubes, on the surface of covered electrodes, introduced as a granular blanket, or mixed in an agglomeration with powdered filler metal.

4. Thermal spraying is the subject of Chapter 29, *Welding Handbook*, Sec. 2, 6th ed. This chapter will be revised in Volume 3 of the 7th edition.

For gas welding, rods and powders are commonly used. Solid rods may be drawn wires (for ductile alloys) or castings. Tubular rods are composites. The usual sheath material is mild steel formed by special machines from strip, filled with alloying elements, and, perhaps, supplemented with fluxing ingredients. An important composite type has granular tungsten carbide as the filling. All types of rods should be bare and clean for gas welding.

The tubular rod design can be produced in the form of continuous coils. Tubular wires, solid drawn wires, or drawn strips serve as the electrodes for semiautomatic and automatic arc surfacing. They are usually provided with a flux blanket or a gas shield at the arc, but some are adapted for bare arc welding. The tube design is quite versatile, and it provides the only way that brittle filler metals can be furnished for automatic arc welding. Where both forms are available for the same filler metal, the tubular product in some cases shows superior usability. A number of the stainless steels are available in tubular form, sometimes with their composition adjusted to compensate for dilution from the base metal.

The powder form is also quite versatile. Practically any metal or alloy can be made available at relatively low cost. The character of the powder is important; each process requires a specific powder size range and shape for optimum results. The powder alloys are usually arc melted after precise formulation and then atomized or cast and granulated. The alloy powders may be simple mechanical mixtures as well as prefused alloys. Composition control of mechanical mixtures, usually by a metering device adjustable by the welder, is considered inherently less precise than that of arc furnace melting under metallurgical control.

SURFACING BY ARC WELDING

A number of arc welding processes are used for surfacing. While some are suitable for small, precise overlays, as in oxyacetylene welding, many offer high deposition rates which make

Table 14.1—Surfacing methods and materials

| Process | Mode of application | Hardfacing | | Cladding and buildup | | |
		Filler metal form	Applicable filler metals	Filler metal form	Applicable filler metals	Remarks
		Welding processes				
Oxyfuel gas	Manual and automatic	Bare cast or tube rod	Co, Ni, and Fe base; tungsten-carbide composites	—	—	—
	Manual	Powder	Co, Ni and Fe base; tungsten-carbide composites	—	—	—
Bare metal arc	Manual and semiautomatic	Bare solid or tube wire	Austenitic manganese steel	—	—	Limited use
Shielded metal arc	Manual	Flux covered cast rod, wire, or tube rod	Co, Ni, and Fe base; tungsten-carbide composites	Flux covered wire	Stainless steel, Ni, Cu, and Fe base	—
Flux cored electrode (self-shielded)	Semiautomatic	Flux cored wire	Fe base	Flux cored wire	Stainless steel, Fe base	—
	Automatic	Flux cored wire	Fe base	Flux cored wire	Stainless steel, Fe base	—
Gas tungsten arc (GTA)	Manual	Bare cast or tube rod	Co, Ni, and Fe base; tungsten-carbide composites	Bare solid wire or rod	Stainless steel, Ni, Cu, and Fe base	—
	Automatic	Bare tube wire; extra long bare cast rod; tungsten-carbide powder with bare solid wire	Co, Ni, and Fe base; tungsten-carbide composites	Bare solid wire	Stainless steel, Ni, Cu and Fe base	—

Table 14.1 (Cont.)—Surfacing methods and materials

Process	Mode of application	Hardfacing		Cladding and buildup		
		Filler metal form	Applicable filler metals	Filler metal form	Applicable filler metals	Remarks
			Welding processes			
Gas metal arc (GMA)	Semiautomatic, automatic			Bare solid wire	Stainless steel, Ni, Cu, and Fe base	—
Submerged arc						
Single wire electrode	Semiautomatic	Bare solid or tube wire	Fe base	—	—	Limited use
	Automatic	Bare tube or solid wire	Fe base	Bare solid or tube wire	Stainless steel, Ni, Cu, and Fe base	Limited use with Cu base alloys
Multiple electrode	Automatic	Bare tube or solid wire	Fe base	Solid or tube wire	Stainless steel, Ni, Cu, and Fe base	Limited use with Cu base alloys
Series arc	Automatic	Solid wire	Fe base	Solid wire	Stainless steel, and Ni base	Used primarily for cladding large vessels
Strip electrode and auxiliary strip	Automatic	—	—	Bare strip	Stainless steel, and Ni base	Used primarily for cladding large vessels
Auxiliary powder	Automatic	Bare solid or tube wire with metal powder	Fe & Co base	Bare solid wire with metal powder	S.S. and Ni base	—

Table 14.1 (Cont.)—Surfacing methods and materials

Process	Mode of application	Hardfacing Filler metal form	Hardfacing Applicable filler metals	Cladding and buildup Filler metal form	Cladding and buildup Applicable filler metals	Remarks
Welding processes						
Plasma arc (transferred arc)						
Powder	Automatic	Powder with or without tungsten-carbide granules	Fe, Co, Ni base; tungsten-carbide composites	Powder	Stainless steel, Ni, and Cu base	Used primarily for production hardfacing
Hot wire	Automatic	Bare solid or tube wire	Fe base	Bare solid or tube wire	Stainless steel, Ni, Cu and Fe base	Used primarily for cladding large vessels and related components
Electroslag	Automatic	—	—	Plate or wire	Stainless steel, Ni, and Fe base	Use on heavy sections only
Coating processes						
Flame spray (metallizing)	Semiautomatic and automatic	Powder, rod	Fe, Co, Ni base; tungsten-carbide composites	Bare solid wire	Stainless steel, Ni, Cu base, Al, Zr, etc.	Some hardfacing alloys may be fused after spraying to produce a fused bond
Detonation gun plating	Automatic (proprietary)	Powder	Tungsten-carbide with selected matrices; selected oxides	—	—	—
Plasma spraying (metallizing)	Semiautomatic and automatic	Powder	Fe, Co, Ni base; tungsten-carbide composites, ceramics	—	—	—
Brazing		Wire, sheet, powder	Fe, Co, Ni, Cu Base	Wire, sheet, powder	Ag, Au, Ni, and alloys	Limited use

very extensive surfacing practical. Large parts can be surfaced economically with relatively low heat buildup and minimum distortion. Arc welding is the only process recommended for application on manganese steel and is the preferred process for other work where excessive heat buildup must be avoided.

Manual, semiautomatic, and automatic arc welding methods are all used extensively for surfacing. Several of the arc welding processes are adaptable to more than one mode of application; the variety of useful methods available is shown in Table 14.1.

For many applications, the appropriate welding method will be obvious. For others, two or more options may be reasonable; for these, careful weighing of the cost factors involved can lead to maximum economy.

The high temperature of the arc makes melting of a base metal layer likely. Admixture with the molten filler metal may produce an undesirable composition at the base metal-weld metal interface.

Surfacing by Semiautomatic Arc Welding

Semiautomatic welding often has an economic advantage over manual methods because it can continue for a considerable time without interruption, so the percentage of arc time and deposition rates can be high. It is particularly advantageous for extensive overlays or when large numbers of parts are involved. The equipment outlay is lower than for other automatic methods, because only an electrically controlled wire feed device is needed besides the power source.

A bare electrode is used. Therefore, the arc and the weld pool are in contact with air unless a shielding gas is supplied. To prevent the problems associated with oxidation, some electrodes contain components that provide deoxidation. Much surfacing is done with tubular "open-arc" electrodes, thereby gaining the advantages of good visibility of the weld pool, lower heat input to the base, less penetration and dilution, and less distortion than with submerged arc methods. However, while the open arc is an aid to deposit control, it is very intense, and protection for the welder's skin and eyes is important.

Some alloys require that the melt be protected by a flux cover or a gaseous shield to prevent adverse effects (porosity, embrittlement, etc.) from oxygen, nitrogen, or by hydrogen (from moisture) pickup. These can be combined with the semiautomatic process, but then auxiliary equipment for handling gas or flux or sometimes water-cooling of the nozzle are needed.

While the welder controls the arc travel and metal placement, the wire feed motor is controlled by electrical feedback from the arc voltage drop. The arc voltage can be set and has some effect on welding behavior.

With most tubular electrodes, reverse polarity, direct current is recommended or preferred. Current ranges usually overlap but do not extend so high as those for submerged arc welding with its larger wire sizes. Most surfacing is done downhand, as shown in Fig. 14.7.

The deposition rates for a given electrode size are greatly affected by welding current, voltage, and electrode extension. Increased current, other factors being equal, increases penetration and dilution as well as deposition rate. As a practical matter, because of the useful current ranges and effect of current density, the smaller diameter welding electrodes deposit metal faster than larger diameters. Voltage, or wire speed, should be adjusted to obtain a stable arc with minimum spatter; excessive voltages increase loss of carbon and other elements by vaporization. Travel should provide the deposit thickness desired, avoiding excessive molten pool time.

A great advantage of the semiautomatic arc welding processes is that the equipment may be transported to the work, rather than having to carry the work to a shop, as is usually the case with fully automatic equipment. Because of this feature, semiautomatic arc welding is greatly favored for maintenance, rebuilding, and other surfacing applications.

Surfacing by Automatic Arc Welding

This process uses machine control of the arc position and motor feed of the filler metal (usually wire drawn continuously from coils or drums) with electrical feedback to control the arc length. The demands on the operator are only for the setup and the correct operation of the controls. The method lends itself best to flat or symmetrical curved surfaces and large scale or repetitive work. Compared to the simple, semi-automatic wire feed device, the equipment is heavy, not usually portable, and considerably more expensive. The significant savings are in manpower. Specialized units are available (e.g., machines for crusher roll rebuilding, track link surfacing, and tractor roller reclamation). Sometimes, a special work positioner is needed in addition to the automatic machine.

Automation is possible with all types of gas shielded arc welding as well as the submerged

arc process, the most popular method used for surfacing. Surfacing differs from other applications of these processes chiefly in its focus on relatively thin, broad area deposits. For this reason, multiple arc variations of the processes and the other special adaptations to surfacing should receive consideration in planning for large scale overlays. With these variations, it is possible to attain deposition rates up to 45 kg (100 lb) of filler metal per hour of arc time.

Except for some use of coiled strip and solid electrodes for corrosion-resistant alloy surfacing and the feeding of granular alloy mixtures ahead of the arc, most submerged arc surfacing employs tubular electrodes.

Fig. 14.7 — Semiautomatic surfacing of a row of pulverizing hammers using a high-chromium iron filler metal in the form of a composite electrode in coils

Manual Arc Surfacing With Covered Electrodes

Low equipment cost, great versatility, and general convenience make manual shielded metal arc welding very popular. The welding machine, which is essentially a power conversion device, is usually the main item of equipment needed. It may be a motor-generator, a transformer, a transformer-rectifier combination, or a fuel operated engine combined with a generator. The arc power may be either direct or alternating current. The filler metal is in the form of covered electrodes. (Bare electrode arc welding is a rarity today, though it is feasible with austenitic manganese steel electrodes). Welding can be done in almost any position or location and is practicable for a variety of work, ranging from very small to quite large. For some applications, it is the only feasible method; and, for many others (especially where continuous methods do not offer significant benefits), it is the economical choice (see Chapter 2).

The operation is under the observation and control of the welder, who can easily cover irregular areas and often correct for adverse conditions. It is also helpful if the welder exercises judgment in other matters, such as holding the arc power down to minimize cracking; keeping a short arc and avoiding excessive puddling to minimize the loss of expensive alloying elements in the filler metal; minimizing dilution with base metal; and restricting moisture pickup, especially with low hydrogen electrodes. This process is used extensively for hardfacing, buttering, buildup, and cladding.

Surfacing of carbon and low alloy steels, high alloy steels, and many nonferrous metals may be done with the shielded metal arc process. Base metal thicknesses may range from below 6 mm (1/4 in.) to 450 mm (18 in.) or more. The surfacing metals employed include low and high alloy steels, the stainless steels, nickel base alloys, cobalt base alloys, and copper base alloys.

The welding conditions for surfacing are not fundamentally different from those used in welding a joint. The arc and weld pool are shielded by the slag or the gases, or both, produced by the electrode. The type of covering on the electrode has considerable effect on the characteristics of the weld metal. Surfacing can be done in all positions on work ranging in size from very small to quite large.

Table 14.2 shows how the various shielded metal arc process variables affect the three most important surfacing characteristics: dilution, deposition rate, and deposit thickness. After reviewing Table 14.2 and Chapter 2 of this volume, the influence of some of the variables on dilution, deposition rate, and deposit thickness are self-explanatory, because the variables for welding basically apply in the same manner to surfacing.

The table indicates only general trends and does not cover questions of weldability or weld soundness. These factors may make it unwise to change only the indicated variable; this in turn may mean that the desired change in dilution, deposition rate, or deposit thickness may not be achieved. For example, a given welding procedure with a small electrode diameter may produce high dilution. The table indicates that a change to a larger size electrode will decrease dilution. This is true, however, only if the amperage, travel speed, position, etc., also remain constant. In many cases, a larger amperage valve must be used with the larger electrode size to obtain acceptable weld quality. In this case, the dilution may remain constant or even increase with the change to the larger electrode size.

The process usually achieves a deposition rate from 0.5 to 2 kg (1 to 4 lb) per hour at dilution levels from 30 to 50 percent.

Gas Tungsten Arc Surfacing

This process employs a nonconsumable tungsten electrode shielded by argon or helium. A wrought, tubular, or cast welding rod is fed into the arc area where heat for melting the filler metal is supplied by an arc between the tungsten electrode and the work. Fluxes are not used with this process. Though slower than other arc welding methods, it provides overlays of excellent quality (see Chapter 3).

A common use is the semiautomatic application of cobalt base alloys, which are available only in welding rod form and not as continuous wire. Weld quality is comparable to that obtained from oxyacetylene welding, but the carbon content of the deposit will tend to be lower, since no carburizing flame is involved. Dilution is somewhat greater but much less than is ordinarily obtained with gas metal arc welding. This method requires very little preheating and is faster than oxyacetylene surfacing. It results in less heat buildup, minimizes distortion, affects the base metal to a lesser degree, and avoids the adverse effect of the oxyacetylene flame on some base metals.

In one variation of surfacing with gas tungsten arc welding, hardfacing filler metal in the form of granules rather than wire is fed into the arc area. As the base metal is melted by the tungsten arc, the granules (usually tungsten carbide) are embedded in the fused metal matrix. This method is frequently used for surfacing drill pipe joints. The carbide particles are essentially undissolved and well-placed at the pipe surface.

The various types of metals which are surfaced with the gas tungsten arc process include almost every major ferrous and nonferrous alloy. The base metal thicknesses usually range from 6 mm (1/4 in.) to 100 mm (4 in.), but much thicker sections can be surfaced by this process. The surfacing materials used with this process include the high alloy steels, chromium stainless steels, nickel and nickel base alloys, copper and copper base alloys, and cobalt and cobalt base alloys. Most materials are used in wire or rod form, but

Table 14.2—The effect of shielded metal arc variables on the three most important characteristics of surfacing

Variable	Change of Variable[a]	Influence of change on		
		Dilution	Deposition Rate	Deposit Thickness
Polarity	AC	Intermediate	Intermediate	Intermediate
	DCRP	High	Low	Thin
	DCSP	Low	High	Thick
Electrode type	10, 11	1 (Highest)	2	2
E XX___	12, 14, 15, 16, 18	2	2	2
from AWS	24	3	1 (Highest)	1 (Thickest)
Spec. A5.1	13	4 (Lowest)	3 (Lowest)	3 (Thinnest)
or A5.5				
Amperage	High	High	High	Thick
	Low	Low	Low	Thin
Technique	Stringer	High	No effect	Thick
	Weave	Low	No effect	Thin
Bead spacing	Narrow	Low	No effect	Thick
	Wide	High	No effect	Thin
Electrode diameter	Small	High	High	Thick
	Large	Low	Low	Thin
Arc length	Long	Low	No effect	Thin
	Short	High	No effect	Thick
Travel speed	Fast	High	No effect	Thin
	Slow	Low	No effect	Thick
Position	Flat	4	No effect	4
	Uphill	3	No effect	3
	Downhill	4	No effect	4
	Horizontal	2-4	No effect	1 (Thickest)
	Vertical-up[b]	1 (Highest)	No effect	5 (Thinnest)
	Vertical-up[c]	5 (Lowest)	No effect	2

a. This table assumes that only one variable at a time is changed. However, for acceptable surfacing conditions, a change in one variable may require a change in one or more other variables.

b. The arc directed on work (forehand welding).

c. The arc directed on surfacing buildup (backhand welding).

some are deposited using powdered filler material. Powdered filler material is usually used in hardfacing applications (tungsten carbide), while wire and rod filler metal forms are used for all surfacing applications. The size of filler material should be selected to give the desired quality deposits. Wires from 0.76 mm (0.5 in.) to 4.76 mm (3/16 in.) are usually used with this process.

Argon, helium, and mixtures of these gases are used for gas tungsten arc surfacing. While most surfacing is done with argon, generally giving lower dilution, other gases and gas mixtures should be considered. These gases can greatly affect bead shape and surface quality. Chapter 3 should be consulted for further information on shielding gases and their effect on the process.

The equipment used for gas tungsten arc surfacing is the same as that employed for welding. Both manual and automatic equipment are available. The automatic equipment often has attachments that oscillate the arc either by mechanical or by magnetic means.

The gas tungsten arc surfacing procedures are essentially the same as those used for welding, with due consideration for dilution control and deposit thickness. The gas tungsten arc process is basically a low deposition rate process that produces a high quality deposit with minimum dilution and low part distortion. A modification of the process, in which electrical resistance heats the filler wire before it enters the weld pool, is used to increase filler deposition rate and, at the same time, maintain minimum dilution. Both stringer and weave bead techniques are used for gas tungsten arc surfacing. The weave technique is used where minimum dilution is wanted.

Surfacing is possible in almost all positions with the gas tungsten arc process. The position of the component to be surfaced must be considered when selecting the welding conditions because the position will greatly affect weld dilution. A summary of the effect of welding variables on gas tungsten arc surfacing is given in Table 14.3. This process is used for hardfacing, buttering, and cladding.

Gas Metal Arc Surfacing

This process is a type of automatic surfacing and is basically like that described in Chapter 4.

The arc and the weld pool are shielded by a flow of inert gas from an external source. Alloying elements used to produce the desired deposit analysis are contained in the electrode, which may be either solid or tubular. Shielding gases include helium, argon, and carbon dioxide which may be used singly, in various mixtures, or with small amounts of oxygen. The type of shielding has some effect on deposit composition and other characteristics.

The cost of gases and equipment for gas shielding adds to the expense of surfacing. Open arc welding is entirely satisfactory for many hardfacing overlays, largely because fluxing and deoxidizing materials are contained in the core of the tubular electrode. Where gas shielding is added, its effects should be clearly warranted.

The shielded gases used in this process minimize oxidation and alloy loss. Their use contributes to superior weld quality and makes feasible the deposition of some metals, such as the aluminum bronzes, that cannot be applied by other semiautomatic or automatic methods. Some of the carbon and alloy steel filler metals, particularly if available only as solid electrodes without the benefit of internal fluxing ingredients, require gas shielding for good quality overlays. In some applications, shielding with carbon dioxide has been found to modify the arc and deposit characteristics in such a way as to make its use desirable, even though not required, for weld perfection. Variations of this process that have distinct characteristics include auxiliary wire feed, pulsed arc, and short circuiting transfer.

The auxiliary wire modification feeds extra filler metal into the weld puddle; the arc energy absorbed by melting it reduces penetration and dilution while the deposition rate is increased about 1-1/2 times. By choosing an auxiliary filler metal differing in composition from that of the electrode, deposit modifications can be made. This method is likely to be favored over the high deposition submerged arc process when suitable granular fluxes are not available for the alloy being deposited. A notable application of the gas metal arc auxiliary wire process is artillery shell banding with gilding metal where dilution by base metal must be held below 3 percent.

Another modification features short circuiting metal transfer. Using a quite small solid electrode, metal transfer occurs only while the arc is extinguished, at a rate from about 20 to 200 times per second. Deposition rates are slightly higher than with covered electrodes, while dilution and workpiece distortion are minimized. This modification is usually employed with semi-automatic equipment.

The pulsed arc technique is a suitable method for out-of-position surfacing and also for metals that are very fluid in the molten state. This tech- nique provides dilution characteristics almost as good as those obtained by the globular transfer mode; arc stability and the deposition rate are as good as those of the spray transfer mode. The high current pulse transfers the molten metal to the workpiece and the low background current allows the molten metal just transferred to cool rather quickly. Deposition rates are lower than with continuous high current operation.

The different types of base metals on which the gas metal arc process can be used for surfacing are numerous. Carbon steels, low alloy steels,

Table 14.3—The effect of gas tungsten arc variables on the three most important characteristics of surfacing

Variable	Change of variable[a]	Influence of change on		
		Dilution	Deposition Rate	Deposit Thickness
Current type	AC	Average	Average	Average
	DC	Lower or higher	Lower or higher	Lower or higher
Polarity	DCSP	High	High	Thick
	DCRP	Low	Low	Thin
Shielding gas	Argon	Lowest	Lowest	Thinnest
	Helium	Highest	Highest	Highest
Amperage	High	High	High	Thick
	Low	Low	Low	Thin
Technique	Stringer	High	No effect	Thick
	Weave	Low	No effect	Thin
Bead spacing (pitch)	Narrow	Low	No effect	Thick
	Wide	High	No effect	Thin
Electrode extension	Short	No effect	No effect	No effect
	Long	No effect	No effect	No effect
Filler wire diameter	Small	High	Low	Thin
	Large	Low	High	Thick
Voltage	High	Low	No effect	Thin
	Low	High	No effect	Thick
Travel speed	Fast	High	No effect	Thin
	Slow	Low	No effect	Thick
Position	Flat	4	No effect	4
	Uphill	3	No effect	3
	Downhill	4	No effect	4
	Horizontal	2-4	No effect	1 (Thickest)
	Vertical-up[b]	1 (Highest)	No effect	5 (Thinnest)
	Vertical-up[c]	5 (Lowest)	No effect	2
Auxiliary wire(s)		Low	High	Thicker

a. This table assumes that only one variable at a time is changed. The table indicates only general trends and does not cover questions of weldability or weld soundness. These factors may make it unwise to change only the indicated variable so that the desired change in dilution, deposition rate, or deposit thickness may not be achieved.

b. Arc directed on work (forehand welding).

c. Arc directed on surfacing buildup (backhand welding).

high alloy steels, and many nonferrous metals can be surfaced with this process. The thickness of the base metals may range normally from about 6 mm (1/4 in.) to 457 mm (18 in.). The base metal tensile strengths in most surfacing applications range up to 620 MPa (90 000 psi). The surfacing electrodes are too numerous to list except by broad alloy classification. The common classifications for the wires are the high alloy steels, the chromium-stainless steel alloys, the nickel and nickel base alloys, the copper and copper base alloys, the titanium and titanium base alloys, and the cobalt and cobalt base alloys. This surfacing process is basically a high dilution process when operating in the spray transfer mode and can be expected to deposit 5.4 to 6.75 kg/h (12 to 15 lb/h) at dilution levels of about 30 to 50 percent.

The equipment used for gas metal arc surfacing is the same as that discussed in Chapter 4; except, in numerous surfacing applications, equipment is needed to oscillate (or weave) the electrode. The oscillating equipment may be very simple (providing only one basic pattern), or it may be very sophisticated (providing numerous patterns). Oscillators may be mechanical or electronic.

The equipment may be either automatic or semiautomatic. The power source output is usually dc, but pulsed dc current may be used for applications where better control of electrode melting rate and bead shape is required. A typical semiautomatic gas metal arc surfacing operation is illustrated in Fig. 14.7. Figure 14.8 shows an automatic gas metal arc surfacing operation.

Table 14.4 shows how the various gas metal arc parameters affect dilution, deposition rate, and deposit thickness. The table assumes that only one of the listed variables is changed at a time. For example, when the electrode diameter is changed, no compensating changes are made in amperage, voltage, travel speed, wire extension, etc. In actual practice, this assumption may or may not be true, depending on the reason for making the change.

After reviewing Table 14.4 and Chapter 4 of this volume, the influence of some of the variables on dilution, deposition rate, and deposit thickness become self-explanatory, because the characteristics for fillet and groove welding,

detailed in Chapter 4, basically apply to surfacing. Some of the variables listed in the table are peculiar to surfacing.

Technique. A stringer or weave bead technique greatly influences dilution. If both a stringer bead and a weave bead are deposited on base material and the dilution is measured, the weave bead would have lower dilution (all other variables, except travel speed, being equal). The reason for this is that weaving allows more liquid metal between the arc and the base material, acting as a cushion and absorbing arc energy that would otherwise cause deeper penetration into the base metal.

Electrode Extension. The length of electrode extension (stickout) beyond the contact tip is very important in surfacing. Depending on the other welding variables, the base material, the surfacing metal, and the desired deposit results, the electrode extension will vary from a normal length of approximately eight times the electrode diameter to approximately 51 mm (2 in.). The long extension at least accomplishes three things: (1) increased deposition rate at any given amperage due to I^2R heating of the electrode; (2) softening of the arc energy impinging on the base material, resulting in less penetration and dilution; and (3) evaporation of contaminants from the electrode. Surfacing procedures should be developed with known extensions. A worn contact tip can inadvertently lengthen the electrode extension. An unsuspecting welder or welding operator may be unaware of this change in electrode extension and mistakenly increase his amperage.

Electrode Diameter. For a given amperage, the electrode diameter is critical to the dilution control and deposition rate. At 300 A, for example, a 1.6 mm (1/16 in.) electrode will produce less dilution and a lower deposition rate than a 0.9 mm (0.04 in.) electrode. The effects of electrode diameter on dilution control are caused, in part, by the mode of arc transfer and current density in the electrode (amperes per unit of cross-sectional area). At a given amperage, the 1.6 mm (1/16 in.) electrode may operate with globular transfer; however, at the identical amperage, the smaller electrode may operate with spray transfer. If two electrode diameters

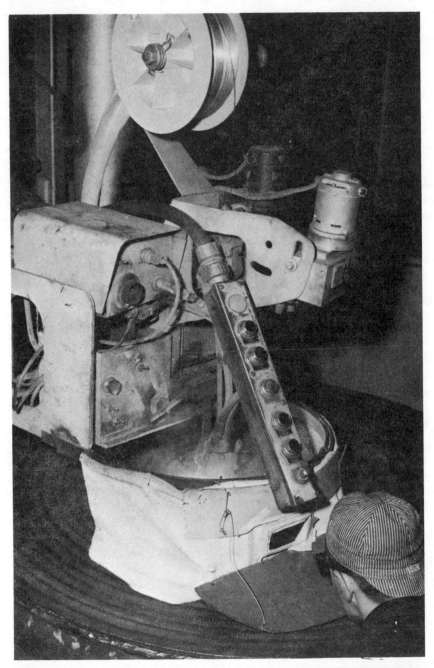

Fig. 14.8—Heat exchanger tube sheet of mild steel being surfaced with aluminum bronze by means of automatic gas metal arc welding

operating at a given amperage have the same mode of metal transfer, the small diameter electrode will have a greater dilution and a higher deposition rate because of the higher current density.

Flux Cored Arc Surfacing

The flux cored arc process is a modification of the gas metal arc process. Therefore, the base materials on which this process can be used for surfacing are essentially identical to those used for gas metal arc surfacing. Generally, flux cored electrodes are available in the same broad alloy classifications, although, in some cases, not as many are available as gas metal arc electrodes. For some alloys, flux cored electrodes are the only ones available because the alloys are not readily workable into wire form. Alloy powders can easily be added to the core with an appropriate sheath alloy. Flux cored electrodes may or may not require auxiliary shielding gas, depending upon the specific alloys. For example, self-

Table 14.4—The effect of gas metal arc variables on the three most important characteristics of surfacing

Variable	Change of variable[a]	Influence of change on		
		Dilution	Deposition rate	Deposit thickness
Polarity	DCRP	High	Low	Thin
	DCSP	Low	High	Thick
Shielding gas	Argon	Lowest	Lowest	Thinnest
	Helium	Highest	Highest	Thickest
	CO_2	Intermediate	Intermediate	Intermediate
Arc transfer	Spray	1 (Highest)	1 (Highest)	1 (Thickest)
	Globular	3	3	3
	Short circuit	4 (Lowest)	4 (Lowest)	4 (Thinnest)
	Pulsed	2	2	2
Amperage	High	High	High	Thick
	Low	Low	Low	Thin
Technique	Stringer	High	No effect	Thick
	Weave	Low	No effect	Thin
Bead spacing	Narrow	Low	No effect	Thick
	Wide	High	No effect	Thin
Electrode extension	Short	High	Low	Thin
	Long	Low	High	Thick
Electrode diameter	Small	High	High	Thick
	Large	Low	Low	Thin
Voltage	High	Low	No effect	Thin
	Low	High	No effect	Thick
Travel speed	Fast	High	No effect	Thin
	Slow	Low	No effect	Thick
Position	Flat	3	No effect	4
	Uphill	2	No effect	3
	Downhill	4	No effect	4
	Horizontal	2-4	No effect	1 (Thickest)
	Vertical-up[b]	1 (Highest)	No effect	5 (Thinnest)
	Vertical-up[c]	5 (Lowest)	No effect	2
Auxiliary wire(s)		Low	High	Thicker

a. This table assumes that only one variable at a time is changed. The table indicates only general trends and does not cover questions of weldability or weld soundness.
b. Arc directed on work (forehand welding).
c. Arc directed on surfacing buildup (backhand welding).

shielded stainless steel electrodes are available. The cost of cases and their handling equipment add to the expense of surfacing. Where gas shielding is required, its effect should be clearly warranted.

The equipment used for surfacing by the flux cored arc process is the same as that used for gas metal arc surfacing except for the electrode feed rolls. Chapter 5 describes the process and equipment in more detail.

In general, all the process variables for flux cored arc surfacing are the same as those for gas metal arc surfacing and need not be repeated, except where a significant difference occurs. When an auxiliary shielding gas is employed, it is usually CO_2 or a mixture of CO_2 and another gas, such as argon. With CO_2 shielding, it is only possible to transfer metal across the arc in the globular transfer mode, not the spray transfer mode. In general, this method produces more dilution and higher deposition rates than does gas metal arc surfacing. The process has two disadvantages over gas metal arc: (1) it produces a slag, similar to shielded metal arc surfacing, that has to be removed before subsequent beads can be deposited; and (2) the cored electrode cannot be fed around as small a radius as an equivalent solid electrode. One advantage of flux cored electrodes is that the manufacturer can tailor, or customize, the electrode chemical composition more easily than can be done with solid electrodes. Table 14.4 also applies for flux cored arc surfacing.

Submerged Arc Surfacing

For this method, the electrode is a continuous wire, either solid or tubular, or a solid strip, and the arc zone is protected by a blanket of granular flux. The electrode compositions are usually formulated to work with a specific flux grade. Because of its many advantages, the submerged arc, single electrode process is the most widely used automated technique for surfacing. It utilizes high welding currents and has the advantage of very high deposition rates. A wide range of electrodes give good results, and they can usually be used with either direct or alternating current. Deposits, characteristically, are of high quality and, with many filler metals, are practically flawless. There is no spatter that might require cleaning, and the absence of the ultraviolet radiation of an exposed arc is advantageous in a shop. The process may be readily adapted to many applications that can utilize larger electrodes than are used in the semiautomatic method. Chapter 6 covers the process and equipment in more detail.

The high quality of the deposit is advantageous for overlays requiring strength, toughness, or abrasion resistance. However, it should be realized that this method, with its relatively deep penetration and protective flux cover, puts more heat into the workpiece than other arc welding processes. With the dilution that takes place and the gain or loss of different elements from the molten flux, the full properties of the surfacing filler metal are ordinarily not attained until two or more layers are deposited.

Though quality, in the sense of freedom from porosity and slag pockets, is assured, the high thermal gradients may cause cracking. Crack avoidance may require preheating and postheating, usually with special equipment. With some surfacing deposits, such as the high carbon chromium irons, a fine pattern of surface cracking is normal on large parts and on those parts subjected to high restraint. This behavior is characteristic of brittle hardfacing alloys and is not always undesirable.

Those methods that feed extra filler metal to the arc zone make more efficient use of the heat of the arc, while increasing deposition rates and decreasing dilution from the base metal. The extra filler metal may be one or more wires or strips, the latter being used primarily for stainless steels and nickel base alloys. Figure 14.9 shows additional filler metal being added in the form of strip.

A variation of the process feeds powder onto the surface of the work ahead of the arc, then the blanketing flux is added. The arc heat melts the granular filler metal mix as well as the electrode and the base metal; all three materials combine to make the deposited metal. The powder filler metal is formulated for this admixture, based on a fixed set of welding conditions. Arc power, bead size, travel speed, electrode feed speed, and alloy powder metering must all be correct and carefully controlled if the intended composition is to be obtained. Weaving of the electrode during deposition reduces dilution and melting of the base metal. Expected advantages of the process are less expensive filler metal, high deposition rates, low penetration, and little dilution from the base metal.

Materials. The different types of base materials for which the submerged arc process is normally used for surfacing are carbon and low alloy steels, stainless steel, cast iron, and nickel and nickel alloys. The thickness range of the base metal can vary from approximately 13 mm (1/2 in.) to 457 mm (18 in.). The materials used for surfacing are (1) consumable electrodes (bare and composite), (2) filler metal (bare, composite, and metal powder), and (3) flux. The electrode and filler metal may be any alloy that can be produced in the required form, but the most common are the high alloy steels, the austenitic stainless steels, the nickel base alloys, the copper base alloys, and the cobalt base alloys. The fluxes

used for surfacing are essentially the same as those used for welding, which are described in Chapter 6.

Equipment. The equipment used for surfacing is the same as that used for welding. However, an oscillator and filler metal feeding equipment may be needed if the techniques that require them are used. The submerged arc process is basically an automatic process for surfacing. Figure 14.2 shows a typical submerged arc surfacing operation done in place with automatic equipment.

Process Variations. Single electrode stringer beads can reasonably be expected to deposit about 6.3 kg/h (14 lb/h) at approximately 15 to 50 percent dilution, using 350 A with special techniques. The oscillating technique can typically deposit 11.75 kg/h (25 lb/h) with a 25 to 40 percent dilution. Beads as wide as 89 mm (3.5

Fig. 14.9—Surfacing of a low alloy steel component with corrosion-resistant alloy strips; one is the electrode and the other is auxiliary filler metal

in.) may be made with oscillation. A variation of the oscillating method employs a small diameter electrode operating on dcsp power to take advantage of resistance heating and a higher melting rate. This method will deposit about 6.75 kg/h (15 lb/h), with a 20 percent dilution, and the deposits will be about 25 mm (1 in.) wide.

Two or more electrodes (arcs), operating parallel and feeding into the same weld puddle with or without oscillation, are sometimes used for high deposition rate surfacing. The deposition rate can be increased considerably with additional electrodes, while maintaining the same dilution expected with a single electrode.

Two-electrode surfacing (one arc, series connection) will deposit about 11.75 to 22.5 kg/h (25 to 50 lb/h) with 10 to 20 percent dilution. Deposition rates on the high side of this range are obtained by feeding a cold filler wire into the arc. Beads approximately 32 to 38 mm (1-1/4 to 1-1/2 in.) wide can be produced with series electrode operation. Constant current ac power is preferred to achieve a uniform penetration pattern.

Using a strip rather than a round electrode, the submerged arc welding process is capable of depositing a relatively thin, flat surfacing bead at deposition rates of 27 to 45 kg/h (60 to 100 lb/h) or more. Dilution is typically about 20 percent. The technique also works well when an auxiliary strip is fed into the arc to provide additional filler metal. With this refinement, the deposition rate is increased and the dilution can be held to between 10 and 15 percent. Strip dimensions usually are 50 mm (2 in.) or 200 mm (4 in.) by 1 mm (0.04 in.) or 1.5 mm (0.06 in.) for the electrode and 41 mm (1-5/8 in.) by 1.25 mm to 1.5 mm (0.05 to 0.06 in.) for the cold strip.

The large cross section of the strip electrode permits currents as high as 1500 A to be used. Normal conditions call for approximately 1200 A, 32 V, and a travel speed of 6.5 mm/s (15 in./min) which will produce a cladding thickness of 4.8 mm (3/16 in.). Thickness can be varied between 4 and 9 mm (5/32 and 3/8 in.) by adjusting the travel speed, the electrode feeding rate, and the dimensions of the cold strip. Savings are derived not only from higher deposition rates, but also from lower filler metal consumption. Flux consumption is reduced by about 66 percent compared to surfacing with conventional elec-trodes. The low penetration of the process reduces dilution and permits a thinner weld cladding to be deposited.

Direct current, straight or reversed polarity, is used although ac can be used. In any case, a constant potential power source is the most suitable.

Another version of submerged arc surfacing provides additional filler metal in the form of powder. The granular metal powder is metered onto the work ahead of the flux covering. The arc, which is usually oscillated, melts all the granular metal and the electrode to produce the deposit. Arc power, bead size, travel speed, electrode feed speed, and alloy powder metering must all be correct and carefully controlled if the intended composition is to be obtained. This method is very useful to produce alloy compositions that cannot be economically fabricated into wire or strip form. The amount of granular metal applied bears a fixed relationship to the electrode used and, normally, ranges from 1.5 to 3 times the weight of the electrode. Most frequently, the electrode is mild steel and the granular metals supply the other elements to produce a specific alloy.

The granular metal process features deposition rates up to four times the rate possible with an electrode alone, with no increase in welding current. Penetration can be controlled to give about 15 percent dilution, since the arc does not impinge directly on the base metal. Most of the available power is used to melt the filler metal and electrode. Deposition rates in excess of 45 kg/h (100 lb/h) may be obtained. Advantages of the process are less expensive filler metal, high deposition rates, low penetration, and minimum dilution with the base metal. One significant limitation is the need for very precise process control.

Effect of Process Variables. Table 14.5 shows the effect of changing the welding variables on dilution, deposition rate, and deposit thickness. The table assumes that only one variable at a time is changed. The effects of changing some variables for submerged arc surfacing are the same as those with gas metal arc surfacing.

The submerged arc fluxes used for surfacing are important to the success of the operation. In addition to the variables listed in Table 14.5, the fluxes used also affect dilution, deposition rate, and deposit thickness. The flux controls the molten weld puddle fluidity. Fluidity and some of the variables in Table 14.5 control bead shape. One of the major effects of fluidity is the control of bead edge-wetting characteristics due to the molten weld puddle surface tension. Surface tension determines whether or not the liquid meniscus wets, like water, or rolls over, like mercury. A flux that is suitable for single electrode surfacing may not be acceptable for multiple electrode or strip electrode surfacing applications because of the higher amperages used and the difference in fluidity of the molten weld metal. Basically, the flux, the electrode(s), and the process variation (single electrode, multiple electrode, strip, etc.) must be selected and controlled for a given application.

Table 14.5—The effect of submerged arc variables on the three most important characteristics of surfacing

Variable	Change of variable[a]	Effect of the change on		
		Dilution	Deposition rate	Deposit thickness
Power supply and connection	AC	Intermediate	Intermediate	Intermediate
	DCRP	Highest	Lowest	Thinnest
	DCSP	Lowest	Highest	Thickest
Amperage	High	High	High	Thick
	Low	Low	Low	Thin
Technique	Stringer	High	No effect	Thick
	Weave	Low	No effect	Thin
Bead spacing	Narrow	Low	No effect	Thick
	Wide	High	No effect	Thin
Electrode extension	Short	High	Low	Thin
	Long	Low	High	Thick
Electrode diameter	Small	High	High	Thick
	Large	Low	Low	Thin
Voltage	High	Low	No effect	Thin
	Low	High	No effect	Thick
Travel speed	Fast	High	No effect	Thin
	Slow	Low	No effect	Thick
Position	Flat	Intermediate	No effect	Intermediate
	Uphill	Highest	No effect	Thickest
	Downhill	Lowest	No effect	Thinnest
Process variations	Single electrode	2	5 (Lowest)	5 (Thinnest)
	Single electrode and filler wire	3	5	4
	Single electrode and hot filler wire	4	4	4
	2 Wire series	3	4	4
	2 Wire series and cold wire	4	3	3
	Multiple wire	2	2	2
	Strip electrode	1 (Highest)	2	3
	Hot and cold strip	5 (Lowest)	1 (Highest)	1 (Thickest)
	Powder addition	4	3	3

a. This table assumes that only one variable at a time is changed. The table indicates only general trends and does not cover questions of weldability or weld soundness. These factors may make it unwise to change only the indicated variable so that the desired change in dilution, deposition rate, or deposit thickness may not be achieved.

Plasma Arc Surfacing

The plasma arc process is used for hardfacing and cladding operations. Plasma arc powder surfacing (the filler material is in powder form) is used for hardfacing applications while plasma arc hot wire surfacing is used for cladding operations.

Powder Surfacing. Plasma arc powder surfacing uses ultrahigh temperatures of 5500 to 22 000°C (10 000 to 40 000°F) to deposit hardfacing materials. Both transferred arcs (struck between the electrode and the workpiece) and nontransferred arcs are used for surfacing. While filler metal is supplied in powder form, the process differs from thermal spraying in that deposits are homogeneous and fused to the base metal.

Properly fused plasma-deposited layers are similar in metallurgical structure to gas tungsten arc welded overlays. A number of cobalt, nickel, and iron base alloys are available as homogeneous powders. Tungsten carbide particles may be added, under different conditions, to the alloy powders being applied, or directly to the weld pool.

Plasma arc powder surfacing is performed downhand when surfacing round or curved items. The arc is offset slightly from the vertical center line, as with other processes. Layers can be applied in a considerable range of thicknesses, with deposit thickness, dilution, and deposition rate being interrelated. These variables are, to a considerable extent, dependent upon the characteristics of the alloy being applied. While heat input into the base metal is low compared to other welding processes, some distortion can be expected and must be considered. As with other surfacing methods, procedures to prevent cracking are generally determined by the properties of the alloy and the size of the part. Particular advantages of plasma arc powder surfacing are (1) the ability to deposit a wide range of engineering materials, including refractories; (2) suitability for surfacing lower melting point base materials; (3) controlled application of very thin as well as relatively thick layers; and (4) close control over surface finish that helps to minimize grinding and machining. Normal deposition rates of about 5.5 kg/h (12 lb/h) are reported.

Equipment costs for the plasma process are relatively high and considerable exacting technology is involved. The process is fully mechanized and particularly suited to high production hardfacing of new parts. Applications include hardfacing of flow control valve parts, tool joints, extruder screws, and lawn mower components.

Hot wire surfacing. The plasma hot wire surfacing process is capable of the application of cladding at high deposition rates and very low dilution levels. The process consists of two basic systems combined for form and fuse a surfacing deposit to the base metal. One system heats the filler metal to almost melting temperature and deposits it on the surface of the base metal. The plasma torch (second system) melts the surfaces of the base metal and the filler metal to completely melt and fuse the molten metal to the base plate.

The main advantage gained by combining the two systems is that molten metal is deposited on the surface of the base metal with minimal dilution. The amount of heat added by the plasma torch can be held to the temperature required to fuse the filler material to the base metal. Deposit dilution is, therefore, easily controlled and heat distortion of the work is held to a minimum. The process is used for overlaying pressure vessels and related components with stainless steel, nickel base alloys, and various bronzes. Equipment cost is high but, for many applications, considerably lower surfacing costs can result. Chapter 9 contains a complete description of the plasma arc process and equipment.

SURFACING BY OXYACETYLENE WELDING

Manual Surfacing With Welding Rods

This method, described in Chapter 10, has great usefulness for smooth, precise, and extremely high quality surfacing deposits. The freedom from base metal dilution can be very important where filler and base metals differ considerably, as with cobalt base filler metals applied to steels. Here, the addition of iron is known to be detrimental to the overlay properties.

The oxyacetylene welding operation lends itself to very close scrutiny and control by the welder. Tiny areas can be surfaced. Grooves and other recesses may be accurately filled, and very thin layers may be smoothly applied. Rods containing tungsten carbide may be used with minimum solution of the wear-resistant particles and with controlled particle distribution. The preheating and slow cooling tend to minimize cracking, even with extremely wear-resistant but brittle overlays.

The base metal should be clean because dirt or oxide may interfere with the uniform wetting action desired. A flux is seldom necessary with most alloys. The flame adjustment should be made as recommended by the manufacturer for the specific filler metal, to insure proper melt behavior and intended deposit properties. Most surfacing filler metals are applied by a reducing flame that prevents loss of carbon. This may actually add some carbon to the deposit. The effect of the flame on the composition of the deposit is anticipated in the formulation of the welding rod during manufacture. A degree of welding skill is needed for high quality oxyacetylene deposits because improper flame adjustment or manipulation and excessive oxide can cause defects. The welding rod used, often a casting, must be of good quality.

In a typical application, a low melting point high carbon filler metal, such as a high chromium iron or a Cr-Co-W alloy, is deposited on a low or medium carbon steel that has a high melting point. Using a reducing flame, the base metal is preheated uniformly over a small area. The reducing flame carburizes the surface, lowers its melting point, and, finally, melts a film on the surface. This result is signalled by a "sweat" or glistening appearance. The tip of the surfacing rod, which should be preheating on the fringe of the flame, is now moved into the hot center of the flame and melted. It should wet the surface as it melts and spread smoothly over the heated area. As the flame is moved to the edge of the surfaced area, the welding rod tip is moved before it, progressively melting the rod as needed. This method of application is known as the forehand method. It minimizes the dilution of the base metal that occurs when the backhand method is used. Many welders use a weaving motion of the flame and the rod to maintain the balance between surface heating and rod melting.

Special preparation of the base metal may be helpful. Grooves or recesses can be ground or machined where surfacing is needed and filled with the molten deposit. This not only aids in precise positioning but also tends to protect the edge of the deposit from chipping under impact.

Surfacing deposits that have a wide plastic temperature range, such as the austenitic high chromium irons, can be wiped or "struck off" to shape with a steel straight edge while the weld is still only partially solidified.

Because of preheating requirements and the nature of the flame and the process, oxyacetylene surfacing does not have the capability for high deposition rates offered by most of the arc welding processes. However, for many applications it is clearly the most satisfactory process. Well-established uses include the surfacing of steam valves, automotive diesel engine valves, chain saw bars, plowshares, and other agricultural implements.

Although ductility may be decreased, the pickup of carbon from a reducing flame is generally beneficial for those alloys that rely on carbides for abrasion resistance. Oxyhydrogen welding, now seldom used, has features much like those of the oxyacetylene technique, but it cannot carburize the metal.

Manual Surfacing With Powdered Filler Metal

This variation of the gas welding process utilizes an oxyacetylene torch fitted with a powder hopper and a powder feed control device. The surfacing powder is aspirated into the gas stream and exits with the gas from the torch tip, sometimes resulting in a sprayed deposit that is only mechanically bonded to the surface. However, by using the proper technique, a fusion bond may be obtained by "sweating" the base metal and applying the surfacing alloy. The application and fusion of the surfacing alloy with the base metal occurs in one operation.

The equipment is inexpensive, and excellent results can be obtained by even an inexperienced operator. One advantage of this method is that many alloys not readily obtainable in rod or wire form are available as suitable powders. Smooth, thin, porosity-free deposits can be made in one pass. Deposition rates will vary with torch size. Large torches with tips producing bulbous flames are able to deposit up to 4.5 kg (10 lb) of metal per hour. The thickness is controlled by the rate of powder flow and movement of the torch.

Semiautomatic Surfacing

Where the number and symmetry of parts are such that an orderly sequence of preheating, deposition, and postheating can be arranged, it may be economical and efficient to set up a semiautomatic installation. For this installation, specially designed burners and mechanized part handling are utilized. The operator, using a welding rod (6 ft.) or more long, intermittently feeds the hardfacing filler metal as it is required. Good judgment is needed for deposition control, but manual skill is less important, productivity is higher, and a more uniform product results than with manual gas welding. This method has been well developed in some plants where repetitive production of many small parts is involved. The facing of truck and engine valves is an example. The burner arrangement assures directional solidification. Deposit quality is held at a high level. Cast welding rods, made by butt welding shorter pieces together, are commercially available for the important alloys used for valve facing. A somewhat different application utilizes long, tungsten-carbide filled welding rods for the hardfacing of feed mill hammers; the narrow hammers are fixtured in a series to provide a large flat surface for welding.

Since specialized burners and fixtures are needed for mechanized hardfacing by gas welding, it is advisable to consult equipment manufacturers when planning in this field.

BASE METALS FOR HARDFACING OVERLAYS

Frequently, the base metals for hard-surfaced parts are dictated by structural design or forming considerations. The weldability and suitability of such materials must be considered on an individual basis. However, there are many cases where the overlay properties are of chief concern and there is considerable latitude in base metal selection.

For a simple operation, the best base metal selection is usually unalloyed carbon steel. The practical range extends from AISI-SAE 1020 up to 1095 (0.20 to 0.95 percent carbon), though welding difficulties increase with carbon content. A favorable base metal, combining good weldability and strength after welding, is 1045 steel. If the steel is clean and well-killed, welding difficulties are minimized. The microstructure of 1045 steel usually consists of ferrite and pearlite.

The amount of pearlite can be increased by raising carbon content, and the pearlite itself can be made somewhat harder by a modest addition of alloys. The weldability of the base metal tends to suffer from these modifications, and they should not be imposed unless their strengthening is needed.

If a very tough base is needed, austenitic manganese steel is probably the toughest available; it is surprisingly economical in the form of castings. Its nature should be understood by the welder since embrittlement from careless procedure is possible, but it is weldable and provides a yield strength of about 380 MPa (55 000 psi). (Refer to Chapter 66, *Welding Handbook*, Section 4, for detailed information.[5]) Higher yield strength grades are also available. The contribution of manganese to the fusion zone must be considered because beads of filler metal next to the base metal may be rendered austenitic by manganese pickup.

---

5. Section 4 is scheduled for revision in the 7th edition.

METALLURGY OF SURFACING ALLOYS

The primary purpose of a classification system is to provide logical designations and a rational basis for selection of hardfacing alloys. Well-defined types of alloys are desirable, and the engineering data associated with these types are important. Available hardfacing information only partially meets this need. Composition and microstructure both have an important influence on the performance of hardfacing alloys. Neither, when used alone, provides a satisfactory basis for classification. Together they can provide a useful framework for the selection of wear-resistant alloys.

Nearly all of the generally available surfacing alloys have a base of iron, nickel, cobalt, or copper. Carbon is the most important auxiliary element, and further alloying with chromium, molybdenum, tungsten, manganese, and silicon is common. Vanadium, nitrogen, titanium, and other elements may be included. Carbon in low percentages influences matrix hardness. Above 1 percent, it forms hard carbides that increase brittleness and resistance to low-stress abrasion. Above 1.7 to 2.0 percent carbon, the iron base alloys are cast irons. Many of the very abrasion-resistant alloys are in this group.

Alloying elements form hard carbides. In order of decreasing hardness, they are (1) titanium, (2) tungsten, (3) chromium, (4) molybdenum, and (5) iron. High percentages of tungsten or chromium with 2 to 4 percent carbon form distinctive carbide crystals that are harder than quartz. Tungsten carbide for hardfacing consists of a fused mixture of W_2C and WC, ranging from 3.0 to 4.0 percent carbon. It is the most abrasion-resistant constituent commonly employed for surfacing. Chromium carbide is usually the Cr_7C_3 compound found in high chromium (20 to 30 percent) irons. It is softer and more brittle than tungsten carbide but less expensive and better adapted to fine dispersion. Iron carbide (Fe_3C), or cementite, is present in many hardfacing alloys. It is usually modified by the presence, in moderate percentages, of Cr, Mo,

or W as alloying elements. At higher levels, these elements produce double carbides, such as $(Fe,W)_6C$, $(Fe,Mo)_6C$, and related complex compounds containing chromium, such as those that may be found in high-speed steel.

Carbide-forming elements also contribute to matrix properties. Tungsten, molybdenum, vanadium, and chromium confer high-temperature strength and secondary hardening in the 480 to 650° C (900 to 1200° F) range. Chromium makes a unique contribution to oxidation and scaling resistance. A 25 percent chromium content provides effective protection up to 1090° C (2000° F). The presence of vanadium is undesirable at high temperatures, since its low-melting oxide acts as a flux by removing protective scales.

Desired hot strength may be obtained by the selection of iron, nickel, or cobalt as the alloy base. The cobalt base alloys, protected with 25-30 percent chromium and strengthened with 5-15 percent tungsten, are highest in hot hardness and creep resistance at temperatures above 650° C (1200° F). The abrasion resistance of this group depends on carbon content. Nickel, cobalt, and chromium confer corrosion resistance and, also, promote oxidation resistance. Nickel is approximately one-third as effective as chromium. The nickel-chromium alloys containing boron also have high hot hardness.

The matrix of iron base alloys can exhibit a wide range in properties. Low carbon steels with a ferrite matrix are of little use in surfacing. As the carbon increases towards 0.8 percent, the weld deposit becomes more pearlitic and harder (to approximately 400BHN). Pearlite forms near 480 to 650° C (900 to 1200° F) as the weld deposit cools from the higher temperatures where the matrix is austenitic. Alloy additions in small quantities make pearlite finer, harder, and slower to form. Larger alloy additions prevent pearlite formation entirely, whereupon transformation may occur below 480° C (900° F), forming the harder structures of bainite or martensite. Even larger alloy percentages (12 percent manganese or 18 percent Cr + 8 percent Ni for example) will prevent transformation entirely and confer an austenitic matrix at ordinary temperatures.

The matrix of the nickel or cobalt base alloys is generally of the austenitic type. The hardness of these alloys depends largely on the amount and size of the carbides or other compounds that are dispersed in the matrix. Thus, the properties and the rational selection of surfacing alloys depend on a knowledge of both composition and microstructure.

Martensite is the largest and strongest matrix constituent. It is characteristic of tool steel that has been quenched from red heat. In weld deposits, it occurs upon normal cooling if the alloy content is properly balanced. Martensitic steels for surfacing applications depend on such a balance and on a carbon content from 0.25-1.5 percent. The lower carbon martensites are tougher, while the higher carbon grades are more abrasion-resistant. Martensite has a high yield strength, making it desirable for resisting light and medium impact.

The properties of bainite fall between those of martensite and pearlite. Bainite formation is less likely to occur in surfacing deposits and, thus, is not included in the classification.

Pearlite is moderately hard and tough. Low alloy steels, with from 0.4-0.9 percent carbon, are pearlitic steels when air-cooled from welding heat, whereas they may form martensite if quenched. They are the least expensive of the hardsurfacing alloys.

Whether a weld deposit of alloy steel is martensitic, austenitic, or a mixture of these structures depends on alloy content, carbon level, and weld-cooling rates. Precise compositions associated with sharply defined properties and well-described cooling rates are rare in hardfacing unless the alloy grade has seen considerable development as a tool steel or a structural type. The more vague descriptions of "air hardening," "semiaustenitic," and "work hardening" are common for surfacing steels. This entire group is popular because it includes good combinations of toughness and abrasion resistance at relatively low cost. These alloys serve a large percentage of the hardfacing market.

Austenite is soft, strong, and tough. It has a remarkable capacity to work harden as it is deformed. The air-hardening steels usually contain at least a small percentage of austenite; and, as alloy content rises, the austenitic proportion also increases until the steel may become entirely austenitic at high alloy levels.

Wholly austenitic steels usually depend on high manganese or chromium and nickel content. They have low or moderate yield strengths and, thus, may deform under medium impact, but their great toughness makes them excellent for heavy impact applications. Because of their stability above 760° C (1400° F), the Cr-Ni grades are preferred for elevated temperature service.

Ferrite is soft and weak (e.g., soft, pure iron) and can be avoided in hardsurfacing by using sufficient carbon to produce the harder structures previously listed. Ferritic steels are used as a base for hardsurfacing rather than as overlays.

Carbide as a matrix is possible in the cast iron range. Some martensitic irons (e.g., 3.0C-4.5 Ni-1.5Cr) have a matrix of cementite containing islands of martensite. Such alloys have high compressive strength, good resistance to light impact, and good abrasion resistance.

The high chromium irons may be martensitic or reverse the pattern and exhibit spine like crystals of Cr_7C_3 in a mixture of either martensite or austenite, depending on alloy content and cooling rate. The high carbon (e.g., 2.5 percent) Cr-Co-W alloys have structures similar to the austenitic high chromium irons, but they are complicated by additional hard compounds. Except for the high chromium alloys, the austenitic irons are usually the carbide matrix type. The carbide matrix may be prominent and extensive, or it may diminish to a network, depending on carbon content. It confers resistance to low-stress abrasion, but not to high-stress crushing or grinding conditions. Weld deposits of the austenitic irons are brittle because of the carbide network, but they are less prone to cracking from weld-cooling conditions than the martensitic irons.

SURFACING FILLER METALS

GENERAL

Since surfacing can be defined as the deposition of filler metal on a metal surface to obtain desired properties or dimensions, any filler metal can be used, at times, as a surfacing material. Table 14.6 classifies the commonly used surfacing filler metals by basic types with their important features and typical applications. A number of surfacing filler metals are covered by AWS Specifications A5.13 and A5.21, but other filler metals are not included because they have not achieved adequate industrial standardization (they are either proprietary alloys or they have limited availability).

The issuance of two specifications is explained by a difference in manufacturing. For solid drawn or cast bare rods and electrodes, the grade can be defined by chemical composition of the bare rod. For covered electrodes, chemical composition must be that of the deposited metal, because the covering may have made some contribution. Such compositions are defined in AWS A5.13.

Filler metals can also be formulated on the basis of the melt that results from welding with a tube filled with powder or granules. Such tube products can be either bare or covered and are defined in AWS A5.21. Deposit compositions specified for composite filler metals are substantially the same as similar filler metals in A5.13.

Additionally, the AWS 5.21 Specification defines composite tungsten-carbide welding rods and electrodes as consisting of tungsten-carbide granules in a mild steel tube or sheath. Classification of these products is based on the mesh size of the tungsten-carbide granules contained in "RWC" rods and "EWC" electrodes. Specified mesh size and weight percent of tungsten-carbide granules for rods are given in AWS A5.21. Specifications for electrodes are similar.

DESIRED CHARACTERISTICS

A combination of properties, including hardness, abrasion resistance, corrosion resistance, impact resistance, and heat resistance, must be considered when selecting filler metal for surfacing applications. Hardness requirements may be in the hot or "red hardness" range as well as at normal temperatures. Abrasion resistance is sometimes, but not always, related to hardness and depends upon both the type of wear and the individual constituents present in the surfacing metal. Corrosion resistance depends on the service conditions and the soundness and composition of the weld deposit, with the stipulation that the original filler metal may be altered by dilution with the base metal.

Impact resistance (somewhat different from that indicated by conventional notched-bar tests) depends on yield strength, to resist plastic flow under battering blows, and toughness, to resist spalling and cracking under deformation. Oxidation resistance, which is needed in high-temperature applications, depends chiefly upon chromium content. Metal-to-metal wear applications may involve seizing and galling, which are welding phenomena that are inhibited by high elastic strength and surface films. Soft copper base alloys (leaded bronzes) and those that develop tenacious oxide films (aluminum bronzes) may thus be used for friction surfaces.

Toughness and abrasion resistance tend to be incompatible properties in an alloy (with the notable exception of austenitic manganese steel) and usually force a compromise selection. The harder alloys are prone to cracking, and many overlays of these are fissured. Overlay cracking from the thermal stresses of welding is usually undesirable, but it is sometimes acceptable if vertical cracking relieves stresses and prevents lateral spalling of the deposit.

Table 14.6—Classification of surfacing alloys

Classification by basic types	Important Features	Successful applications
Tungsten carbide deposits		
Granules or inserts	Maximum abrasion resistance	Oil well rock drill bits and tool joints
Coarse granule tube rods	Worn surfaces become rough	A wide range of severely abrasive conditions
Fine granule tube rods	Best performance when gas welded	
High chromium irons	Excellent erosion resistance	
Multiple alloy type	Hot hardness from 410-650° C (800-1200°F) with W & Mo	Abrasion by hot coke
Martensitic type	Can be annealed and rehardened	Erosion by 510° C (1000°F) catalysts in refineries
Austenitic type	Oxidation resistant	Agricultural equipment in sandy soil
Martensitic alloy irons	Excellent abrasion resistance	General abrasive conditions with light impact
Chromium-tungsten type	High compressive strength	Machine parts subject to repetitive metal-to-metal wear and impact
Chromium-molybdenum type	Good for light impact	
Nickel-chromium type		
Austenitic alloy irons	More crack-resistant than martensitic irons	General erosion conditions with light impact
Chromium-molybdenum type		
Nickel-chromium types		
Chromium-cobalt-tungsten alloys	Hot strength and creep resistance	
High carbon (2.5%) type	Brittle and abrasion-resistant	Hot wear and abrasions above 650° C (1200°F)
Medium carbon (1.4%) type	Tough and oxidation-resistant	Exhaust valves of gasoline engines.
Low carbon (1.0% type)		Valve trim of steam turbines.
Nickel base alloys	Good hot hardness and erosion resistance	
Nickel-chromium-boron type		Oil well slush pumps
Nickel-chromium-molybdenum-tungsten type		
Nickel-chromium-molybdenum type	Corrosion resistance	
Nickel-chromium type	Resistant to exhaust gas erosion	Exhaust valves of trucks, buses, and aircraft
Copper base alloys	Oxidation resistant	
	Antiseizing; resistant to frictional wear	Bearing surfaces
Martensitic steels		General abrasive conditions with medium impact
High carbon (0.65-1.7%) type	Fair abrasion resistance	
Medium carbon (0.30-0.65%) type	Good resistance to medium impact	Hot working dies
Low carbon (below 0.30%) type	Tough, economical	
Semiaustenitic steels	Tough, crack resistant	
Pearlitic steels	Crack-resistant and low in cost	General low-cost hardfacing
Low alloy steel	Suitable for buildup of worn areas	Base for surfacing or a buildup to restore dimensions
Simple carbon steel	A good base for hardfacing	
Austenitic steels	Tough; excellent for heavy impact	
13% manganese-1% molybdenum type	Fair abrasion and erosion resistance	General metal-to-metal wear under heavy impact
13% manganese-3% nickel type	Lower yield strength	Railway trackwork
13% manganese-nickel-chromium type	High yield strength for austenitic types	

CLASSIFICATION OF ALLOYS

The magnitude and variety of the surfacing field have resulted in a great variety of products, making selection of the best filler metal difficult. Careful analysis of service conditions—matching them against weld deposit properties, supplemented by established, reliable, and valid service test data—provides the best method of selecting the proper alloy.

Classification can be based on many factors, including hardness, composition, service conditions, and abrasion resistance, in a specific test. For this discussion, the most useful method is a combination of chemical composition and structure of the as-deposited metal, since most surfacing deposits are used in this condition. A basic distinction exists between ferrous and nonferrous base alloys.

An older and unsophisticated classification based on total alloy content (excluding iron) should be avoided. It implies increased merit with increased alloy percentage; there is no need for this oversimplification and it can be very misleading. The alloying elements should be named and percentages given whenever filler metals are to be characterized by composition.

MAJOR FERROUS CLASSIFICATIONS

Martensitic and Pearlitic Steels

Carbon content is usually in the range of 0.10 to 0.50 percent, but may be as high as 1.50 percent. Other elements, selected for their hardenability contribution, are used in moderate amounts to promote martensite formation as the weld deposit cools. Molybdenum and nickel (rarely above 3 percent), and chromium (up to about 15 percent) are the favored alloys; manganese and silicon are usually present, chiefly for deoxidation.

The carbon in these steels is the major element influencing properties. These steels are relatively tough; the lower carbon alloys are tougher and more crack-resistant than the higher carbon types. They are capable of being built up to form thick, crack-free deposits. The deposits have high strength and some ductility. Abrasion resistance is moderate but tends to incease with carbon content and hardness. Lower hardness deposits may be machinable with tools, while grinding is advisable with higher hardnesses. This classification represents the largest volume usage, on a weight basis, of surfacing filler metal.

The moderate price, good welding behavior, and broad range of properties of these steels make them popular for many bulk-surfacing uses, such as the buildup of shafts, rollers, and other machined surfaces; simple moderate abrasion applications involving impact; and buildup of badly worn surfaces before finishing with a more expensive and more highly alloyed deposit.

High-speed steels are basically martensitic steels with tungsten, molybdenum, and vanadium additives to improve hardness up to about 650° C (1200° F). They are similar to the molybdenum-type high-speed tool steels and are used in applications where a high-speed tool steel deposit is desired. The air-hardening deposits need no heat treatment.

Pearlitic steels are similar to the martensitic steels but they contain less alloy additions. Because of this lower alloy content, they form pearlite (a structure softer than martensite) on cooling. Pearlite steels are useful as build-up overlays.

The low alloy steels, which represent the largest volume use of surfacing filler metal, may be either martensitic or pearlitic in weld deposits. Mixed structures are common. As the potent "hardenability" elements are increased (particularly chromium), there is a tendency to raise the amount of retained austenite in the deposit. In such medium alloy steels, the untempered martensite that forms as the weld cools is relatively brittle and prone to cracking. The austenite is softer and tougher. Some grades, which may be termed "semiaustenitic," are formulated to exploit this toughening while utilizing martensite for hardness.

Hardsurfacing deposits are seldom heat-treated. The deposit properties are dependent on composition and weld cooling rate. A notable exception is the buildup or repair of special tool steels, such as dies. Here, the filler metal is usually closely matched to the base metal, and the full tool steel techniques of heat treatment and handling are invoked to achieve a high-integrity product.

Austenitic Steels

The two major types of austenitic steels are those based on high manganese (related to Hadfield manganese steel) and the Cr-Ni-Fe stainless steels. Both are tough, crack-resistant, and capable of crack-free deposition in thick, multiple layer deposits. They are relatively soft as deposited (150 to 230 BHN), but they rapidly work-harden from deformation or impact.

The stainless grades of filler metal, notably types 308, 309, 310, and 312, are used for corrosion-resistant overlays and for joining or build-up purposes.[6] They are rarely employed for wear resistance. As a separation layer between carbon or low alloy steels and manganese or surfacing overlays, they contribute enough alloying elements to minimize or prevent the formation of brittle martensite in the dilution zone. The 309 and 310 grades are good heat-resistant alloys and serve for surface protection against oxidation up to about 1100° C (2000° F). Modified 308 types are used for railway trackwork rebuilding (with additives of 0.3 to 0.6 percent carbon and 4 percent manganese) and for various hot-wear applications (with higher carbon and possibly molybdenum).

The manganese steel types usually depend on 12 to 15 percent manganese to ensure the austenitic structure, 0.5 to 0.9 percent carbon (the base 13 percent manganese steel may contain between 1.0 and 1.4 percent carbon), and nickel (2.75 to 5.0 percent) or molybdenum (0.6 to 1.4 percent) to enhance toughness. The molybdenum type has higher yield strength and flow resistance. These grades are widely used to rebuild railway trackwork, provide metal-to-metal wear resistance coupled with impact, and protect the surface or replace worn areas where abrasion is associated with severe impact. Surfacing of power shovel dippers is a common and typical application. However, these grades are seldom appropriate for hot wear because temperatures above 430° C (800° F) may embrittle the high manganese austenite.

More complex alloy grades have been developed to provide higher yield strength, better resistance to reheating, and less tendency for a vulnerable dilution zone when deposited on carbon steels. The following combinations may be encountered: 15 percent manganese, 15 percent chromium, 1 percent nickel; 4 percent nickel, 4 percent chromium, 14 percent manganese; and related types with up to 2 percent vanadium. The 4Ni-4Cr type has high yield strength and resistance to the effects of welding heat. This complex grade is also considered superior for joining cast and wrought manganese steel where carbon and phosphorus respectively should not exceed 0.9 and 0.035 percent in order to avoid weld metal fissuring.

The welding of carbon steel to the 13 percent manganese type should be done cautiously. A low manganese fusion zone may develop martensite and become brittle unless the lowered manganese is compensated with other equally effective alloying elements. The 309 and 310 stainless grades, as well as the complex manganese types, are capable of this.

Arc welding is almost universally used in this area because the greater heat of gas welding tends to embrittle manganese steel base metal. The manganese steel electrodes are very popular in the mining, mineral processing, and earth-moving industries, primarily for surfacing and rebuilding austenitic manganese steel castings.

Irons

The iron base alloys with high carbon are termed irons because they have the characteristics of cast irons. They have a moderate-to-high alloy content to confer air-hardening properties or to create special hard carbides in the surfacing deposits. Although the carbon range is from 2.0 to 5.5 percent, 3.5 to 5 percent is typical. The irons resist abrasion better than the two previously described alloy types and are preferred up to the point where they lack toughness to withstand the associated impact. They are usually limited to one or two pass overlays, but tension cracks are common, especially if large areas are covered.

The highest alloy group is the high-chromium irons with about 24 to 33 percent chromium. This high chromium, combined with high car-

---

6. Corrosion resisting chromium and chromium-nickel steel filler metals are covered by AWS Specifications A 5.4 and A 5.9.

bon, produces hard Cr_7C_3 type carbides in the structure. Frequently, 4 to 8 percent manganese or 2 to 5 percent nickel is added to promote an austenitic matrix. Also, tungsten, molybdenum, or vanadium may be added to increase hot hardness and abrasion resistance.

Though resistance to low-stress abrasion (for example, a plowshare working in sandy soil) is high for all of this group, the irons with an austenitic matrix are inferior for high-stress grinding abrasion (a ball mill). The high-chromium irons that undergo a martensitic transformation are good for both types of abrasion, especially when heat-treated. High chromium confers oxidation resistance, and in some applications, these alloys resist hot wear quite well, as in rolling mill or piercing mill guides. However, they are inferior to the Cr-Co-W type alloys in hot hardness above 600° C (1100° F).

Lower alloy irons, typically with 15 percent chromium and molybdenum or nickel, have an austenitic matrix and are very popular for their general abrasion resistance (they have more crack resistance than the martensitic irons). They have excellent resistance to low-stress abrasion in proportion to the content of hard carbide. Since low-stress abrasion usually dominates situations where impact is low or absent, these lower alloys are applicable.

Martensitic irons comprise a group whose alloy content is low enough to permit at least partial transformation to hard martensite as the deposit cools to room temperature. However, alloying must be sufficient to prevent pearlite transformation; balance must be maintained for the intended purpose. Chromium, nickel, molybdenum, and tungsten are the customary control elements. The usual matrix is a complex carbide containing islands of martensite with some retained austenite.

The high-chromium martensite irons have an aggregate austenite-martensite matrix. The presence of considerable martensite confers excellent resistance to high-stress abrasion (recognizable by the way imposed stress fractures the abrasive), very high compressive strength, and, consequently, high resistance to light impact. At the same time, resistance to low-stress abrasion is outstanding and metal-to-metal wear resistance may be high. With some grades, high hardness is good up through 650°C (1200°F). Disadvantages are the sensitivity to composition variables and the martensite cracking tendency. Though compressive strength is high, tensile strength is low. Most cracking is from tension resulting from either thermal stresses or the deformation of a soft base under the harder overlay. Proper support of the martensitic irons is important.

The structure and properties of the hard facing deposit can be modified by welding methods and cooling rates. Gas welding with the recommended reducing flame, which has a feather two to three times as long as the inner cone, tends to add carbon to the deposit, increasing brittleness and hardness. Arc welding, however, tends to burn out carbon, increases dilution with the base metal, and may produce tougher but less wear-resistant deposits unless the electrode composition is formulated to compensate for this.

A cross section through a chromium-molybdenum martensitic iron gas weld deposit shows that advantage is taken of carbon diffusion in surfacing by carburizing the steel surface with the torch until a thin layer of iron reaches the eutectic composition and melts at 1090-1150° C (2000-2100° F). The appearance of this layer, which glistens when it melts, is termed "sweating" and signals to the welder that the overlay can be readily flowed and bonded onto the heated surface.

MAJOR NONFERROUS CLASSIFICATIONS

Tungsten-Carbide Composites

This surfacing filler metal is supplied in the form of mild steel tubes filled with crushed and sized granules of cast tungsten carbide, usually in the proportions of 60 percent carbide and 40 percent tube by weight. The carbide, a mixture of WC and W_2C, is very hard, surprisingly tough, and very abrasion-resistant. Deposits containing large, undissolved amounts of carbide have more resistance to all types of abrasion than any other welded overlay. Various grades supply different granule sizes in the filler, usually designated by the screen mesh size of the particles. Differential wear of the deposits will make them rough in proportion to the size of the particles. A grade with 8/12 granules (finer than 8 mesh and coarser than 12 mesh) might be used for nonskid surfacing of horseshoes; the much finer 40/120 grade is better for plowshares. The 20/30 and 30/40 sizes are popular for general use.

As the heat of welding melts the steel tube, the molten metal dissolves some of the tungsten carbide to form a matrix of high-tungsten steel. The main function of this abrasion-resistant matrix is to anchor and support the granules undissolved by oxyacetylene welding. However, arc welding, especially with fine-granule electrodes, may dissolve so much of the carbide that abrasion resistance is impaired, though hot hardness is increased. Hot wear applications are unusual even though hardness is high up to approximately 540°C (1000°F). Use above 650°C (1200°F) is limited by softening of the matrix and oxidation of the carbide.

Oxyacetylene welding is preferred for critical applications such as oil well drilling bits. The process can add carbon to the matrix, supplementing the tungsten and carbon contributed by moderate solution of the granules. Such weld deposits have higher abrasion resistance than any other type of hardfacing.

Welding operations can have a pronounced effect on weld deposit properties, since the matrix composition depends on the amount of carbide granules dissolved during welding. Arc welding tends to dissolve the tungsten carbide more readily and, in an extreme case with very fine granules, may dissolve them all (Fig. 14.10 C). Such a matrix, though hard, is inferior to the composite that contains a volume of granules anchored in a hard, strong matrix (Figs. 14.10 A and B). Fusion with the base metal and consequent weld dilution are also associated with arc welding. Satisfactory arc welds can be made, but the welder should understand what occurs during fusion and minimize puddling time if optimum results are to be obtained. Because of its lower cost, arc welding is generally used for the hardfacing of earth handling and mining equipment.

Cobalt Base Surfacing Metals·

These alloys usually contain 26 to 33 percent chromium, 3 to 14 percent tungsten, and 0.7 to 3.0 percent carbon. Three grades are available, with hardness, abrasion resistance, and crack sensitivity increasing as the carbon and tungsten contents increase.

These alloys have high oxidation, corrosion, and heat resistance. The one percent carbon grade is outstanding for exhaust valves in internal combustion engines. High hardness and creep resistance are retained at temperatures above 540° C (1000° F), and the alloys are useful for some types of service at temperatures as high as 980° C (1800° F). The cobalt base, combined with the chromium, provides good corrosion resistance in many applications; resistance to metal-to-metal wear is also very good.

The cobalt base alloys are not subject to hardening transformations like steel and have negligible response to heat treatment. Occasionally, a stress-relief treatment may be advisable to minimize cracking.

Oxyacetylene welding may increase the carbon content of the deposit, while arc welding tends to reduce carbon and, at the same time, dilute the deposit with elements from the base metal. These changes will be reflected in properties of the deposited metal.

For oxyacetylene welding, a 3× feather-to-cone reducing flame is recommended. Preheating the cleaned surface with a neutral flame up to at least 430° C (800° F) is advisable for heavy sections. For shielded metal arc welding, dc reverse polarity is used with a short arc length. For a 6.4 mm (1/4 in.) diameter electrode, a current of approximately 190 to 200 A is recommended. All deposits should be cooled slowly to prevent cracking.

Nickel Base Surfacing Metals

This is a broad group with many different categories and varying heat and corrosion resistances. The most common nickel base surfacing alloys are those containing 0.3 to 1.0 percent carbon, 8 to 18 percent chromium, 2.0 to 4.5 percent boron, and 1.2 to 5.5 percent each of silicon and iron. There are three alloys in this group, with hardness and abrasion resistance

(A) Oxyacetylene overlay from a welding rod containing 100% of the granules sized 30/40 mesh. (10X)

(B) Oxyacetylene overlay from a welding rod containing 100% of the granules sized 40/120 mesh. (10X)

(C) Arc weld from an electrode containing 100% of the granules sized 40/120 mesh. (10X)

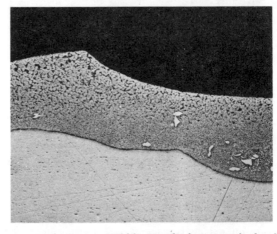

Fig. 14.10 — Variation in distribution and abundance of tungsten-carbide granules in composite hard surfacing overlays

increasing with the carbon, boron, silicon, and iron contents. They may be spray-coated. The general comments made on the cobalt base filler metals also apply to this group, although hot hardness and resistance to high-stress abrasion are lower.

Many nickel base filler metals are also used for surfacing applications. Among these are nickel base alloys with additions of copper, chromium, molybdenum, chromium-molybdenum, and chromium-molybdenum-tungsten.

The important nickel base alloys are
(1) Nickel-chromium alloys, chiefly of the 80 percent nickel-20 percent chromium grade
(2) Nickel-chromium-iron compositions corresponding to well-known heat-resistant alloys
(3) Nickel-chromium-iron-silicon-boron alloys
(4) Nickel-chromium-molybdenum-tungsten alloys
(5) Nickel-iron-molybdenum alloys
(6) Nickel-copper alloys

For some purposes, the nickel base alloys provide the best performance obtainable. For some applications requiring corrosion resistance, the stainless steels are frequently as satisfactory and less expensive.

Where erosion resistance is the primary requirement, the high chromium irons should receive first consideration because of their lower cost. The microstructure of gas welds from both types exhibits carbide spines of the Cr_7C_3 type. The very hard chromium borides that occur in the nickel base alloy do not clearly confer superiority in abrasion resistance.

As protective overlays, these nickel base alloys, when available as continuous electrodes, are best applied by gas shielded arc welding. In this way, carbon pickup is avoided, as well as complications resulting from the use of flux. The automatic deposition lends itself to coating cylindrical vessels that must be protected from corrosion.

Nickel base alloys containing chromium and boron are adapted to flame spray applications in powder form. When sprayed, they are subsequently fused by close application of a hot gas flame to produce a thin, hard overlay. The spray technique permits covering irregular contours more uniformly than is possible with conventional welding. The alloy for this use is available as a loose powder and as a plastic-bonded powder in "wire" form. It is also commonly used as a cast rod (nominally 70 percent Ni, 17 percent Cr, and 3.75 percent B) for oxyacetylene welding.

Copper Base Surfacing Metals

The copper base alloys are employed as overlay surfaces chiefly to resist corrosion, cavitation erosion, and metal-to-metal wear. They are nonmagnetic and practically nonsparking. Where copper alloy bearing metals are normally employed as homogeneous parts, it is sometimes more economical to apply them as overlays on a less expensive base metal, usually iron.

There are numerous, commonly available, copper base alloys used for surfacing; the common ones are classified in AWS Specification A5.13.[7] Most of these alloys are resistant to atmospheric attack, salt and fresh water corrosion, alkaline solutions (except those that are ammoniacal), and many acids (especially those of the reducing type). They have poor resistance to sulfur compounds that produce a corrosive copper sulfide, but they are generally good with other saline solutions. They have poor abrasion resistance under high-stress grinding conditions, and they should not be subjected to other than mild abrasion. They are unsuitable at elevated temperatures above 200-260° C (400-500° F).

For "soft" bearings, where the bearing is expected to wear in preference to the mating face, the phosphor bronzes, the softer aluminum bronzes, and the brasses may be appropriate. They are usually selected to be 50 to 75 percent Brinell hardness numbers softer than the other face. For the reverse situation, where the bearing surface is to be the harder of the two, the hard aluminum bronzes are used. Silicon bronze is not intended for bearing use.

Welding techniques will affect properties. Iron pickup from a steel base metal is a hardener.

7. Additional information on copper and copper alloy filler metals is included in AWS Specifications A5.6 and A5.7.

To limit this effect at the surface, an overlay should be about 6 mm (1/4 in.) thick, preferably in a recess with fillets greater than a 6 mm (1/4 in.) radius, and consist of three or more layers. Gas and gas tungsten arc methods are preferred. Manual shielded metal arc and gas metal arc processes require minimum amperage (to limit pickup), and they also require a fast, wide push-weaving technique for the initial layer. The opposite effect (softening from loss of alloys during deposition) is minimized by inert gas shielding. Temperature control of the base metal can be important and should not be neglected.

The gas metal arc process is recommended for large area repair of copper base alloys; for minor repairs, the gas tungsten arc method with thoriated electrodes is favored. The copper-zinc alloys can be deposited only by gas welding, but gas metal arc overlays of phosphor-bronze on low (red) brass and aluminum-bronze on high (yellow) brasses are feasible.

SELECTION OF SURFACING METALS

In general, the loss of metal from a part will involve one or more of the following: sliding or rolling friction, shock and impact, heat, abrasion, and corrosion. Other factors, such as smoothness of the deposit desired or the ability of the surfaced part to form an efficient tool, also affect the choice. Previous discussion of each classification indicates the relative merits of the various groups. For specific applications, selection should be based on careful analysis of all pertinent factors, including previous service performance if available, from a clearly comparable situation Reliable service validation, like the acceptance of the one percent carbon-chromium-cobalt-tungsten alloy for exhaust valves, is practical for many applications. However, careful analysis is better than casual service tests which are frequently misleading because of uncontrolled variables. Well-controlled and statistically valid tests are expensive and time consuming and are justifiable for only very important applications.

Since most surfacing is a compromise between abrasion resistance and impact resistance as opposing properties, the initial selection is usually the most abrasion-resistant alloy that will give an acceptable deposit and will probably withstand the expected impact. If service causes impact failure, a tougher but less wear-resistant alloy is then substituted. If wear is rapid but toughness is clearly adequate, a step in the other direction can be made.

Alloy cost should not weigh heavily, since welding labor cost is usually more critical. Expensive alloys may more than pay with longer life when they are properly selected.

SUPPLEMENTARY READING LIST

Almquist, G., and Egeman, N. Nickel cladding with strip electrodes. *Welding Journal* 45 (4), Apr. 1966, pp. 275-283.

Arnoldy, R.F., and Kachelmeier, E.J. Development and selection of filler metals for bulk welding. *Welding Journal* 48 (2), Feb. 1969, pp. 109-113.

Bernstein, A., *et al.* Surfacing with stainless steel strip electrodes. *Welding Journal* 54 (9), Sept. 1975, pp. 647-655.

Campbell, H.C., and Johnson, W.C. Granular metal filler metals for arc welding. *Welding Journal* 46 (3), Mar. 1967, pp. 200-206.

Campbell, H.C., and Johnson, W.C. Cladding and overlay welding with strip electrodes. *Welding Journal* 45 (5), May 1966, pp. 399-409.

Garrabrant, E.C., and Zuchowski, R.S. Plasma arc hot wire surfacing, a new high deposition process. *Welding Journal* 48 (5), May 1969, pp. 385-395.

Garriott, F.E. Surfacing with bronze. *Welding Journal* 46 (4), Apr. 1967, pp. 279-287.

Meyer, J.J. Plasma hot wire surfacing. *Welding Journal* 55 (2), Feb. 1976, pp. 97-100.

Recommended Practices for Plasma Arc Welding, AWS C5.1.73. Miami: American Welding Society, 1973.

Van Dyke, L., and Wittstock, G. Submerged arc welding and surfacing. *Welding Journal* 51 (5), May 1972, pp. 317-325.

Vinckier, A.G., and Pense, A.W. A Review of Underclad Cracking in Pressure Vessel Components. Bulletin 197. New York: Welding Research Council, Aug. 1974.

Welding Handbook
Index of Major Subjects

	Seventh Edition Volume	Sixth Edition Section	Chapter

	Seventh Edition Volume	Sixth Edition Section	Chapter
Forge welding		3B	61
Friction welding		3A	50
Gas metal arc welding	2		4
Gas tungsten arc welding	2		3
Gas welding	2		10
High frequency welding		3A	45
Laser beam welding		3B	55
Oxyfuel gas welding	2		10
Percussion welding		2	27
Plasma arc welding	2		9
Projection welding		2	26
Seam welding, resistance		2	26
Shielded metal arc welding	2		2
Spot welding, resistance		2	26
Stud welding	2		8
Submerged arc welding	2		6
Thermit welding		3B	57
Ultrasonic welding		3B	59
Upset welding		2	27
Welding specifications		1	11
Welding standards		1	11
Welding symbols		2	21
Wrought iron, weldability of		4	61

Z

Zinc, weldability of		4	72
Zirconium, weldability of		4	74

INDEX